Relativistic Jets from Active Galactic Nuclei

Edited by
Markus Böttcher,
Daniel E. Harris, and
Henric Krawczynski

Related Titles

Belusevic, R.

Relativity, Astrophysics and Cosmology

2008

ISBN: 978-3-527-40764-4

Shore, S. N.

Astrophysical Hydrodynamics

An Introduction

2007

ISBN: 978-3-527-40669-2

Thompson, A. R., Moran, J. M., Swenson, G. W.

Interferometry and Synthesis in Radio Astronomy

2001

ISBN: 978-0-471-25492-8

Rybicki, G. B., Lightman, A. P.

Radiative Processes in Astrophysics

1979

ISBN: 978-0-471-82759-7

Relativistic Jets from Active Galactic Nuclei

Edited by

Markus Böttcher, Daniel E. Harris, and Henric Krawczynski

WILEY-VCH Verlag GmbH & Co. KGaA

The Editors

Prof. Markus Böttcher
Astrophysical Institute
Department of Physics and Astronomy
339, Clippinger
Athens, OH 45701-2979
USA

Dr. Daniel E. Harris
Harward Smithsonian CfA
High Energy Astrophysics Division
60 Garden Street
Cambridge, MA 02138-1516
USA

Prof. Henric Krawzcynski
Physics Department, CB 1105
Washington University in St. Louis
1 Brookings Drive
St. Louis, MO 4899
USA

Library of Congress Card No.:
applied for

British Library Cataloguing-in-Publication Data:
A catalogue record for this book is available from the British Library.

Bibliographic information published by the Deutsche Nationalbibliothek
The Deutsche Nationalbibliothek lists this publication in the Deutsche Nationalbibliografie; detailed bibliographic data are available on the Internet at http://dnb.d-nb.de.

Typesetting le-tex publishing services GmbH, Leipzig
Printing and Binding Markono Print Media Pte Ltd, Singapore
Cover Design Adam Design, Weinheim

Printed in Singapore
Printed on acid-free paper

Print ISBN 978-3-527-41037-8
ePDF ISBN 978-3-527-64176-5
oBook ISBN 978-3-527-64174-1
ePub ISBN 978-3-527-64175-8
Mobi ISBN 978-3-527-64177-2

Contents

Preface

The idea of writing a text book on extragalactic jets was initially beset by many doubts. Why produce a book on a subject that is only poorly understood at best? Most text books are the result of several years teaching a course on a relatively stable field of endeavor, but our understanding of jets is still evolving rapidly. Furthermore, the subject matter is very broad, ranging from cutting edge simulations, to relativistic MHD, and of course observations at the limits of our best instruments. We asked ourselves, is it worth our time to pursue this project? Yet in the end, we cast aside these doubts and convened a group of experts who each agreed to contribute a chapter. The reader of this book will see that although we do not know with certainty what jets are made of and how they are launched, accelerated and collimated, there is a vast body of observational evidence which has already been confronted with a wide range of theoretical models. Thus, although the theory of jets is beset with uncertainties, it is highly constrained at the same time. This book gives an overview of a mature and rich field in which key observational and theoretical breakthroughs as well as paradigm shifts can happen at any time.

The book is intended to be used by graduate students, teachers, and researchers who are interested in active galactic nuclei (AGN) and their relativistic outflows. The introductory chapters covering aspects of special relativity and radiation processes may be of interest to a broader audience, e.g., graduate students working in the fields of radio astronomy or high-energy astrophysics. Some of the more advanced chapters of this book will also be of interest to researchers in the field, as they give up-to-date reviews of the rapidly evolving subject matter.

We decided to limit the book to relativistic jets from AGN. Although jets from microquasars may share much the same physics as our AGN jets, we felt that we could not treat the subject matter of galactic binaries adequately given the constraints of the project. We also decided not to deal with jets from gamma-ray bursts mainly because we felt that while again the underlying physics may be similar, there are so few observables available that jet models for GRB are difficult to disprove by observation. Stellar jets are another subject we decided to exclude since they are not believed to be relativistic.

We thank the staff at Wiley for their patience and assistance. We are very grateful to E. Mandel (SAO) who kindly wrote the scripts to facilitate construction of the index.

September 2011

M. Böttcher
D. E. Harris
H. Krawczynski

List of Contributors

Miguel A. Aloy
Departamento de Astronomía y Astrofísica
Universidad de Valencia
Edificio de Investigación Dr. Molinder
46100-Burjassot (Valencia), Spain
miguel.q.aloy@uv.es

Matthew G. Baring
Rice University
Department of Physics and Astronomy
MS 108 P.O. Box 1892
Houston, Tx 77251-1892, USA
baring@rice.edu

Markus Böttcher
Astrophysical Institute
Department of Physics and Astronomy
Ohio University
Athens, OH 45701, USA
boettchm@ohio.edu

Alan H. Bridle
National Radio Astonomy Observatory
520 Edgemont Road
Charlottesville, VA 22903, USA
abridle@nrao.edu

Marshall H. Cohen
Caltech
MS 105-24
Pasadena, CA 91125, USA
mhc@astro.caltech.edu

Daniel E. Harris
Harward Smithsonian CfA
High Energy Astrophysics Division
60 Garden Street
Cambridge, MA 02138-1516, USA
harris@xfa.harvard.edu

Rony Keppens
Centre for Plasma Astrophysics
Mathematics Department, K.U. Leuven
Celestijnenlaan 200B
3001 Heverlee, Belgium
rony.keppens@wis.kuleuven.be

Serguei Komissarov
School of Mathematics
University of Leeds
Leeds LS29JT, UK
serguei@maths.leeds.ac.uk

Henric Krawczynski
Physics Department, CB 1105
Washington University in St. Louis
1 Brookings Drive
Saint Louis, MO 63130, USA
krawcz@wuphys.wustl.edu

Zakaria Meliani
Laboratory Universe and Theories
– LUTh
Observatoire de Paris
5 place Jules Janssen
92195 Meudon cedex, France
zakaria.meliani@obspm.fr

Petar Mimica
Department d'Astronomia i Astrofisica
University Valencia
Edifici d'Investugacio Jeroni Munoz
C/ Dr. Moliner, 50
E-46100 Burjassot (Valencia), Spain
petar.mimica@uv.es

Eric Perlman
Department of Physics and Space
Sciences
Florida Institute of Technology
150 West University Boulevard
Melbourne, FL 32901, USA
eperlman@fit.edu

Anita Reimer
Leopold-Franzens-Universität
Innsbruck
Institut für Theoretische Physik und
Institut für Astro- und Teilchenphysik
Technikerstrasse 25
6020 Innsbruck, Austria
anita.reimer@uibk.ac.at

Christopher S. Reynolds
Joint Space Science Institute
University of Maryland College Park
College Park
MD 20742-2421, USA
chris@astro.umd.edu

Rita Sambruna
NASA/GSFC
Code 662, X-ray Astrophysics
Laboratory
Greenbelt, MD, 20771, USA
rita.m.sambruna@nasa.gov

Glossary and Acronyms

AMR	Adaptive Mesh Refinement
AMRVAC	Adaptive Mesh Refinement Versatile Advection Code
ATCA	Australian Telescope Compact Array
BLR	Broad-Line Region (as seen in AGN spectra)
BeppoSAX	A small X-ray satellite launched in 1996
Big blue bump (BBB)	A spectral feature of some AGN thought to arise from the accretion disk
CJF	Caltech–Jodrell Bank Flat Spectrum VLBI Survey
CSO	Compact Symmetric Object (describing a small radio source)
CSS	Compact Steep Spectrum (describing a small radio source)
Chandra	One of NASA's "Great Observatories"; for X-ray imaging and spectroscopy
EBL	Extragalactic Background Light
EVLA	Extended Very Large Array. The upgraded workhorse of radio astronomy in New Mexico
Fermi	A NASA mission for MeV and GeV observations, a.k.a. FGST: Fermi Gamma-Ray Space Telescope
FIDO	Fiducial observer
FSRQ	Flat-Spectrum Radio Quasar
Faraday rotation	The rotation of the angle of the linear polarization during propagation through a magnetized plasma
GEMS	Gravity and Extreme Magnetism SMEX
GMVA	Global Millimeter VLBI Array
GPS	Gigahertz-peaked spectrum. A radio source with peak flux density in the GHz range
GRB	Gamma-ray burst
HALCA	Highly Advanced Laboratory for Communications and Astronomy
HBL	High-frequency (peaked) BL Lac object

H.E.S.S.	High-Energy Spectroscopic System. A GeV/TeV gamma-ray observatory in Namibia (a.k.a. HESS)
HETE II	High-Energy Transient Explorer II
HST	Hubble Space Telescope
HYMORS	Hybrid Morphology (Radio) Source. Usually one side is like FR I, the other FR II
High-Polarization Quasar	Quasar with a high degree of optical polarization (HPQ)
IGM	Intergalactic medium
ISM	Interstellar medium
IXO	International X-ray Observatory
LDQ	Lobe-Dominated Quasar (radio morphology)
LOFAR	Low-Frequency Array for Radio Astronomy
MAGIC	A GeV/TeV gamma-ray observatory in the Canary Islands
MERLIN	Multielement Radio-Linked Interferometer
MOJAVE	Monitoring of Jets of AGN with VLBI Experiments
NLR	Narrow-line region
OSSE	Oriented Scintillation Spectrometer Experiment
OVV	Optically Violent Variable
PARAMESH	Parallel Adaptive Mesh Refinement
Poynting Flux	Energy flux carried by electromagnetic fields, $\boldsymbol{S} = (\frac{1}{\mu_0})\boldsymbol{E} \times \boldsymbol{B}$ (in SI units)
RHD	Relativistic hydrodynamics
RMHD	Relativistic magnetohydrodynamics
RRFID	Radio Reference Frame Image Database
SED	Spectral energy distribution
SKA	Square Kilometer Array
SSA	Synchrotron self-absorption
SSC	Synchrotron self-Compton (emission)
TANAMI	Tracking Active Galactic Nuclei with Austral Milliarcsecond Interferometry
UHECR	Ultra-high-energy cosmic ray
VCS1	VLBA Calibrator Survey No. 1
VERITAS	Very Energetic Radiation Imaging Telescope Array System. A GeV/TeV gamma-ray observatory in Arizona
VIPS	VLBA Imaging and Polarimetry Survey
VLA	Very Large Array. Twenty-seven element radio interferometer in New Mexico
VLBA	Very Long Baseline Array
VLBI	Very Long Baseline Interferometry
VSOP	VLBI Space Observatory Program
ZAMO	Zero Angular Momentum Observer

Part One Introduction

1
Introduction and Historical Perspective

Markus Böttcher, Daniel E. Harris, and Henric Krawczynski

Definition: In this book, the term "relativistic jet" designates highly collimated outflows from supermassive black holes, or in more general terms from a "central engine" in the centers of active galactic nuclei. The jets transport energy and momentum from the central engine to remote locations. In some sources, energy and momentum are dissipated in hot spots hundreds of kiloparsec away from the galactic nucleus and in radio lobes which surround the jets and the hot spot complexes.

1.1
A Brief History of Jets

Jets are collimated outflows associated with supermassive black holes (SMBH) in the nuclei of some types of active galactic nuclei (AGN). The first recorded observation of a jet was in 1918 within the elliptical galaxy M 87 in the Virgo cluster: "A curious straight ray lies in a gap in the nebulosity in p.a. 20°, apparently connected with the nucleus by a thin line of matter. The ray is brightest at its inner end, which is $11''$ from the nucleus" [1]. At that time, the extended feature was a mere curiosity and its nature was not understood. It was not until well after World War II, when technical improvements provided for increasingly better angular resolutions and lower noise receivers, that it was demonstrated that many galaxies exhibited extended radio emission consisting of a nuclear component, jets, hot spot complexes, and radio lobes. Implementation of radio interferometry developed quickly during the same period. In particular, so-called Very Long Baseline Interferometers (VLBI) showed that there were compact high-temperature radio cores in AGN.

According to our current understanding, jets originate in the vicinity of a SMBH (with several million to several billion solar masses) located at the center of the AGN. The jets are powered by these black holes and possibly by their associated accretion disks, and the jets themselves transport energy, momentum, and angular momentum over vast distances [2–4], from the "tiny" black hole of radius $r = 10^{-4} M_{\rm BH}/10^9\,M_\odot$ pc to radio hot spots, hot spot complexes and lobes which may be up to a megaparsec or more away. Thus the study of jets must encompass a

Relativistic Jets from Active Galactic Nuclei, First Edition. Edited by M. Böttcher, D.E. Harris, H. Krawczynski.

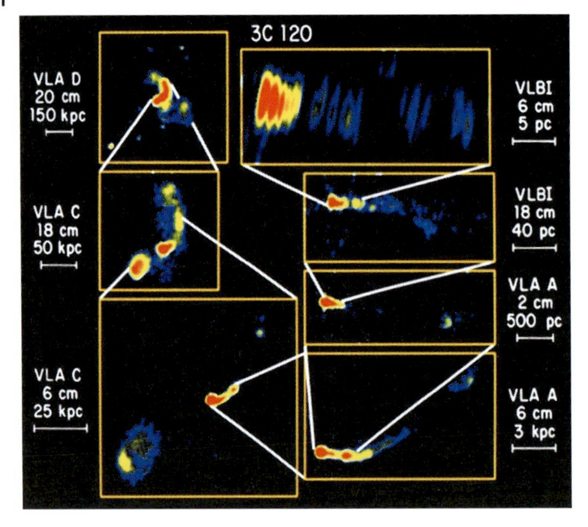

Figure 1.1 This montage of radio observations of 3C 120 demonstrates the range of physical sizes (projected of course), the knotty behavior (common to most jets), and quite pronounced bending (not a universal property of jets). "Knotty" is a somewhat poorly defined description, but essentially means that jets usually have regions of high brightness interspersed with low-brightness segments: often so low that these segments are not detected. The apparent bending of a jet does not mean it bends that much in 3D since if the jet is coming towards us a physically small bend may be sharply amplified in projection. This figure is a "false-color" version with red indicating the highest brightness and blue the lowest. It first appeared in Harris, Mossman, and Walker [5]; reproduced with permission from the National Radio Astronomy Observatory.

range of scales covering a factor of 10^{10}! Some sense of this vast range of scales is provided by Figure 1.1.

It is not yet known what jets are made of and which of the components (atoms, protons, electrons, positrons, Poynting flux) carries the dominant fraction of the jet energy and momentum. In a similar fashion, it is not firmly established how jets form and accelerate and what collimates them over vast distances.

1.1.1
Synchrotron Emission as the Primary Process for Continuum Radio Sources

During the 1950s, a fascinating interplay between theory and observation led to key advances in our understanding of the nature of continuum radio sources. When discrete cosmic radio sources were first observed, the general notion was to interpret them as radio stars since it was already known that the Sun produced radio waves. Thus as early as 1950, Alfvén and Herlofson [6] suggested that the radio emission of the object Cygnus A (which we know call a radio galaxy) could be interpreted as synchrotron radiation from cosmic ray electrons gyrating in a star's magnetic field. A few years later, Shklovsky [7] championed the idea that both the radio emission and the optical light from the diffuse component of the Crab Nebula – a supernova remnant in our Galaxy, were segments of the same synchrotron spectrum. With detections of optical linear polarization from both the Crab Nebula and from the jet in M 87 [8], the community was quickly convinced that synchrotron radiation was the primary emission mechanism for the ever-increasing number of continuum radio sources. Theorists were soon publishing plausible physical parameters for synchrotron models for the radio and optical nonthermal emissions, for example Burbidge [9] for the AGN M 87 in the nearby Virgo galaxy cluster.

1.1.2
Occurrence/Ubiquity of Radio Jets

The discovery of radio emission from extragalactic jets began with the identification by [10] of "a star of about thirteenth magnitude and a faint wisp or jet" near the accurate radio positions of two bright components of 3C 273 derived from a lunar occultation observation [11]. Schmidt's identification of 3C 273B and the measurement of a cosmological redshift of $z = 0.158$, which implied an extragalactic origin, famously marks the start of the quasar "industry".

His identification of 3C 273A with the tip of the "faint wisp or jet" also marks the start of a radio jet industry that grew more slowly. The next step came when the work in [12] showed that a compact extranuclear radio component in 3C 274 coincided with the brightest knot in the optical jet of the object M 87, but it was not until the one-mile and five-kilometer interferometers at Cambridge began systematic high-resolution studies of 3C sources that further radio jets were found in the low-power plumed source 3C 66B [13] and in the higher-power "classical double" 3C 219 [14]. Radio jets were soon recognized in images taken with the Westerbork Synthesis Radio Telescope of other low-power radio galaxies: B0844+319 and 3C 129 [15] and retrospectively in 3C 449 [16] and 3C 83.1 [17].

The first example of an AGN radio jet remaining well collimated and evidently stable over several hundred kiloparsecs was found in the giant radio galaxy NGC 6251 [18]. These early discoveries showed that the jet phenomenon ran the gamut of radio powers and structure types, but it remained unclear whether the sources with detected radio jets were in some sense exceptional. That question was answered when the Very Large Array (VLA) came into operation in the late 1970s. Radio jets were soon found in radio-loud AGN of all known sizes and powers, including Seyfert, classical radio galaxies, and quasars. The VLA was able to show the *ubiquity* of radio jets because its 27-element design provided the sensitivity to detect weak jets in only brief observations, the dynamic range to do so in the presence of bright unresolved emission from the AGN, and the angular resolution to separate jets convincingly from other extended radio emission. This allowed simple but well-defined quantitative criteria to be applied for identifying a linear radio feature as a "jet" [19, 20].

The mass and luminosity range of objects known to generate relativistic jets was also extended to include the galactic "microquasars", for example SS 433 [21] and 1E140.7-2942 [22].

1.1.3
Origin of the Notion that SMBHs Reside in All Galactic Nuclei

It was not long after the realization that radio galaxies (which roughly account for $\approx$ 1% of all galaxies, and favor massive ellipticals) produce jets, which transmit prodigious amounts of power out of the central region, that it was also understood that jets had to be coupled with processes involving a black hole of high mass, that is, more than a million solar masses [23]. The two available sources of energy to power jets and thus hot spots and lobes are potential energy (in the gravitational field of a SMBH) and rotation of the SMBH itself. Furthermore, the role of AGN in influencing galaxy formation and evolution was recognized quite early (e.g., [24]). As other developments established a relationship between the properties of galaxies and the masses of their black holes [25, 26], a great deal of effort was expended on trying to ascertain why some galaxies and quasars are radio-loud, whereas others are radio-quiet ("quiet" is a relative term, which indicates a bimodality in the distribution of the ratio of radio intensity to optical intensity rather than a complete absence of radio emission).

1.1.4
Working Out of Relativistic Effects

Perhaps the most important development after the realization that jets carry power over vast distances and that the emission we see is nonthermal came from Very Long Baseline Interferometry (VLBI), see Section 5.2. After primitive radio interferometry demonstrated the basic source structure of extragalactic radio sources [27], astronomers developed VLBI techniques in which phase information was sacrificed in order to obtain baselines which were too long to be coherently connected. Thus the order of the day was to develop more stable atomic clocks and more reliable and higher-density magnetic tape drives so as to record wider bandwidths. Since almost all radio sources have emission on both sides of the nucleus, but a majority of bright jets are on one side only (Section 5.2.1), the notion of "Doppler boosting" became the established paradigm. As the VLBI data improved, it became apparent that individual emission regions at the parsec scale could be tracked and had apparent proper motions larger than the velocity of light. This effect was quickly explained as a bulk relativistic motion of an emitting plasma moving close to our line-of-sight (Section 5.2.5). Most of the salient relativistic effects pertaining to jets were worked out in Konigl's thesis [28].

1.1.5
Microquasars

"Microquasars" is a term applied to those galactic X-ray binary systems that have been found to eject luminous material, which appears to be moving at a considerable fraction of the speed of light. In one case (SS 433) the material traces out a helical path and emission lines of ionized common elements have been found in the optical and X-ray spectra. A detailed precessing jet model provides an accurate value of the velocity: $0.26c$. In addition to the spectral lines, there is a continuous spectrum of synchrotron emission so VLBI techniques have augmented the optical and X-ray study of line emission with direct imaging of the twin jets.

No line emission has been observed for any other galactic microquasar, nor for any extra galactic jet, large or small. Although many researchers have attempted to demonstrate that microquasar jets and AGN jets are basically the same (physical) phenomenon, there are significant differences. Chief among these is the difference in velocity of ejection. While some microquasars have features believed to be moving at 0.8 or $0.9c$, we always see both sides of the jet whereas for AGN jets, Doppler boosting often enhances the approaching jet and the receding jet is below detection thresholds.

The study of X-ray binaries and their jets has led to a vast amount of published literature and the classification of many aspects of their behavior. In this book, we will not attempt to deal with these jets as such, but refer to them from time to time if relevant. Interested readers can consult [29].

1.2 Jets at Optical, UV, X-Rays and γ-Rays

At the higher frequencies, the synchrotron E^2 half-lives are on the order of several years in typical fields of 0.1–1 mG. Since we have optical polarization from many jets and, at least in the case of knot HST-1 in the M 87 jet where we have strong variability at all wavelengths, we are confident that the high-frequency radiation from some jets comes from synchrotron emission. This translates into the fact that every emission region must also be an acceleration region.

1.2.1 HST Optical/UV Jets

During the latter part of the 1900s, there were several ground-based studies of the brightest optical jets. These efforts were plagued by limited angular resolution (poor seeing) and often resulted in ambiguous results with respect to spectral properties and degree of polarization.

The quasar jet 3C 273 was almost as bright as the M 87 jet, and did not have to contend with the optical light of the encompassing galaxy, and thus was an obvious target. Therefore, once the Hubble Space Telescope (HST) was repaired, detailed morphology, robust polarization data and spectral properties could be obtained. Figure 1.2 shows a comparison of the relative brightness distributions for three different bands.

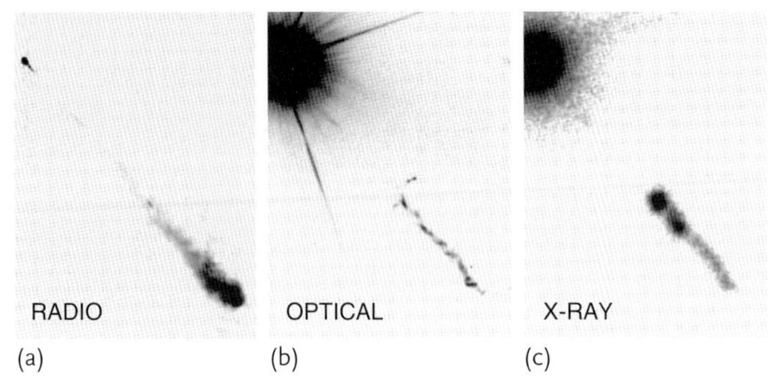

Figure 1.2 The famous jet of the quasar 3C 273. (a) The radio panel is from the MERLIN archives at a frequency of 1.6 GHz; (b) the optical is from the Hubble Space Telescope at 6000 Å; and (c) the X-ray image is from a set of observations made by Chandra. It is filtered for the energy band 0.3–6 keV and has been slightly smoothed. The distance from the quasar (upper left) to the tip of the jet is 22.3″ which corresponds to 60 kpc (in projection) at the distance (750 Mpc) of this quasar. The large apparent diameter of the quasar in the optical and X-ray images is an instrumental effect caused by the low-level wings of the point spread functions. Note how the jet is almost invisible for quite some distance before abruptly appearing with a very bright X-ray knot which is weak in the radio. Although there are some "wiggles", this jet is basically straight, unlike that of 3C 120 (Figure 1.1).

1.2.2 X-Ray Jets

The Einstein Observatory, launched in 1978, was the first device that had sufficient angular resolution ($\approx 5''$) and sensitivity to detect X-ray knots and hot spots associated with jets. *ROSAT* (launched in 1990) had similar capabilities; both satellites had microchannel plate detectors in addition to the proportional counters. So, with some concerted effort, a handful of jets had been detected before the launch of Chandra: the hot spots of Cygnus A and the jets of Cen A, M 87, and 3C 273.

With the advent of Chandra with its much improved angular resolution ($\leq 1''$), many more jets were detected. The early "surveys" targeted the brighter radio jets thought to be close to our line-of-sight, and of course the first detection was achieved by chance: the quasar PKS 0637 was thought to be a bright unresolved source and was chosen as a target for focusing the mirrors (i.e., determining the best position for the detectors; see Section 7.2.1).

X-ray data on knots and hot spots have provided key insights into several facets of jet physics. If the X-rays from a particular feature are dominated by synchrotron emission, Lorentz factors of the radiating electrons, γ of order 10^7–10^8 are required in the nominal fields on the order of hundreds of microgauss (μG). These extremely high-energy electrons have short half-lives and so are more restrictive than the optical data as to how large a source can be in the sense of how far these electrons can travel from their acceleration site. Additionally, the ability to estimate physical parameters for these very high-energy electrons provides insights into loss mechanisms and high-energy cutoffs in the electron distributions.

Another aspect of these X-ray detections came from the hot spots of FR II radio galaxies. It was shown that synchrotron self-Compton (SSC) emission provided a good model for the X-ray intensities observed from the high-brightness radio hot spots of the radio galaxy Cygnus A. Since the photon energy density could reliably be measured from the observed radio spectrum and the (resolved) source size, it was possible to estimate the average magnetic field strength without first assuming equipartition. While this represented an important advance, many more sources with hot spot detections have become available, and of these, only a relative few match the SSC predictions as well as the example of Cygnus A.

The case of X-ray emission from the knots in the jets of FR II radio galaxies and quasars is still a topic of debate. Since there appears to be a high-frequency cutoff in the optical for many knots, it was difficult to fit a simple synchrotron spectrum to the radio-optical-X-ray data. A relatively simple model was advanced [30, 31] to overcome this difficulty: the X-rays were identified as the result of inverse-Compton scattering between the relativistic electrons and the photons of the microwave background. To get enough intensity from this process it is necessary to assume that the bulk Lorentz factor of kiloparsec-scale quasar jets are on the order of 10 or greater in many cases. Further, it turns out that very low-energy electrons (i.e., γ on the order of 100) are those that produce the observed X-rays and therefore, one must rely on a blind extrapolation of the electron spectra from the typical energies of several thousand (estimated from centimeter radio data) down to a few hundred or less. This is usually performed by assuming the observed radio spectrum can be extrapolated down to a MHz or so without any cutoffs or breaks (see Section 7.5.5).

1.2.3
Jets in γ-Rays

While jets in AGN can often be resolved on parsec and larger scales, most radio-loud AGN possess bright cores on subparsec scales, which are spatially unresolved. It is commonly believed that γ-ray emission, present in a special class of radio-loud AGN called blazars, is produced in this unresolved core region. One of the major surprises of the EGRET experiment on board the satellite-borne Compton Gamma-Ray Observatory (1991–2000) was the discovery that high-energy (100 MeV up to a few GeV) γ-ray emission is a common property of blazars, and that power emitted in the γ-ray band typically surpasses the power emitted at longer wavelengths. At even higher energies, ground-based experiments can detect γ-rays in the TeV energy range. Pioneering observations with the Whipple Cherenkov Telescope between 1992 and 1996 showed that some blazars (e.g., Mrk 421) are not only sources of extremely energetic TeV γ-rays, but that the TeV emission shows time variability on extremely short ($\sim$ min) time scales. The fast flux variability constrains (via causality arguments) the size of the emission region. The fact that TeV γ-rays are observed and are not absorbed in $\gamma\gamma \rightarrow e^+e^-$ pair production processes of TeV γ-rays interacting with IR and optical photons leads to the conclusion that the jets have to be highly relativistic with bulk velocities exceeding 99%

of the speed of light (see the discussion in Section 3.2.1.2). Modeling the broad-band radio to γ-ray energy spectra of TeV-bright blazars indicates even higher jet velocities: 99.98% of the speed of light. Intensive observational campaigns involving a large number of ground-based and spaceborne observatories have resulted in rich data sets with detailed information about the time dependence of the broad-band energy distributions. The blazar phenomenon will be discussed in detail in Chapter 8.

1.2.4 Gamma-Ray Bursts

Gamma-ray bursts (GRBs) have been known for almost 40 years as short flashes of γ-rays which dominate the entire γ-ray sky for durations of $\lesssim 1$ s to a few minutes. With the successful operation of the Italian-Dutch BeppoSAX satellite in the late 1990s, it became possible to associate GRBs with afterglows at longer (X-rays, optical) wavelengths. This also allowed for the determination of redshifts of their counterparts, establishing that these sources must be extragalactic. Once the distance to a GRB is known, one can convert the observed γ-ray flux into a luminosity, assuming that the γ-rays are emitted isotropically. This calculation yielded total released energies up to $\sim 10^{54}$ erg, that is, surpassing the typical radiative energy release of a supernova by a factor of ~ 1000. The fast flux variability on time scales $\Delta t \sim\ll 1$ min, suggests a very small emission region of size $R \ll c\Delta t \sim 1.8 \times 10^{12}$ cm. As for blazars, the fast flux variability and enormous brightness of GRBs implies a lower limit on the velocity of the emitting plasma (the only way to avoid high optical depth to pair production). In the case of GRBs, the lower limits typically lie between 99.995 and 99.99995% of the speed of light (i.e., bulk Lorentz factors Γ of 100–1000). This velocity implies that clocks in the comoving frame go between 100 and 1000 times slower than in the observers frame. Relativistic aberration leads to an extremely narrow emission cone with a $\lesssim 1°$ opening angle. This makes GRBs the most extreme examples of relativistic jets known. Common traits of AGN and GRB jets and breakthroughs in the theoretical understanding of the jets from these two types of objects will be discussed in Chapter 10.

1.3 The Role of Simulations

As the physics of jets are governed by classical physics (the special and general theories of relativity and classical electrodynamics) and by well-understood components of quantum mechanics, one might think that it is straightforward to simulate the formation and propagation of jets and their electromagnetic emission. Quite the opposite is the case, and it is only quite recently that numerical simulations are becoming an increasingly useful tool to complement analytical modeling of jet processes. As an example, the interested reader may consult a seminal – and still very instructive – review article on the theory of AGN jets [32], which hardly mentions any simulation results, and in the few cases in which it does, it does so for demonstrative purposes.

A major difficulty of numerical calculations is the wide range of scales (spatial and temporal), which are relevant for the properties of jets. Jets form in the surroundings of supermassive black holes with a characteristic size of the Schwarzschild radius

$$r_{\mathrm{Sch}} = \frac{2G\,M_{\mathrm{BH}}}{c^2} = 3 \times 10^{13} \frac{M_{\mathrm{BH}}}{10^8\,M_\odot}\ \mathrm{cm}\,, \tag{1.1}$$

which is comparable to the diameter of Earth's orbit, and carry energy over distances of several 100 kpc (3.1×10^{23} cm) – often exceeding the size of the host galaxy. Simulating a jet all the way from its base and covering its propagation over its full length is already a formidable task, given that the simulations should be in 3D, as 1D and 2D simulations cannot capture the topology of true plasma flows. Another issue that complicates simulations is that the processes of jet formation, acceleration, and confinement depend on processes on much smaller scales. The viscosity of the accretion disk and the magnetic field in the accretion disk are thought to be generated by the magnetorotational instability on spatial scales $\ll R_S$. The stability of the jet, the shocks in the jet, and the particle acceleration processes in the jet are ultimately governed by processes on the order of a gyroradius r_g of mildly relativistic electrons and/or protons with

$$r_g = \frac{c p_\perp}{e B} \approx 3.3 \times 10^7 \frac{c p_\perp}{10\,\mathrm{GeV}} \left(\frac{B}{1\,\mathrm{G}}\right)^{-1}\ \mathrm{cm}\,, \tag{1.2}$$

where $p_\perp$ is the momentum of the particle perpendicular to the magnetic field, and B is the magnetic field. The magnetic field is thought to be on the order of $B \sim 10^4$ G in the accretion disk of a $10^8\,M_\odot$ black hole accreting with the Eddington rate [33], and $B \sim 1$ G at the base of the jet where gamma-rays are emitted [34].

A full simulation of jets and their emission would start with a dark matter plus magnetohydrodynamics simulation of the interstellar and/or intracluster plasma based on a cosmological simulation, which grows primordial density perturbations into the environments of AGNs. Subsequently, the accretion and jet formation process would be simulated, best with a general relativistic magnetohydrodynamic simulation. Last but not least, one would simulate the jet flow including the formation of shocks, the acceleration of high-energy particles, the various emission processes of these high-energy particles, and the interaction of the jet with the ambient material.

Although no one has carried through such a comprehensive simulation, numerical studies have achieved substantial progress over the last 30 years by focusing on well-defined problems (see Chapters 4, 10, and 11 for detailed descriptions). As examples, we would like to highlight a few results:

- Early hydrodynamical jet simulations improved our understanding of the structure of radio galaxies (e.g., [35] and references therein).

- Magnetohydrodynamical simulations of differentially rotating accretion disks have led to rediscovery of the magnetorotational instability, which produces viscosity in accretion disks (e.g., [36–38]).
- General relativistic magnetohydrodynamical simulations include all ingredients which are deemed to be important for the process of jet formation: general relativity, hydrodynamics, and magnetic fields [39]. Simulations are now able to reproduce how highly relativistic jets may form and accelerate, and are beginning to validate earlier analytic models (e.g., [40–42]).
- Codes have been developed to simulate the acceleration of high-energy particles [43, 44] and their emission [45, 46]. The codes can be used to determine the properties of jets even though they do not contain jet formation, nor the formation of shocks in the jets.
- Recent particle-in-cell simulations are starting to be able to simulate astrophysical shocks and the acceleration of particles in these shocks [47, 48].

Although in many cases, simulations reproduce observed characteristics, there may be fundamental aspects of real jets which are not included in the simulations. Conceptually, one might imagine that it would be possible to compare jets dominated by Poynting flux, protons, and pairs. We could then perform a test of each by attempting to bend the jets with an external force provided by the "wind" of a thermal plasma such as occurs for tailed radio galaxies in clusters of galaxies. As the growth of computing capabilities has followed Moore's law for the last half century (i.e., doubled approximately every two years) and is expected to continue in this way for quite some time to come, we can expect an interesting decade in which jet simulations will make enormous progress and will enable comparisons of simulated and observed characteristics, and of simulated and analytic results.

1.4 Jet Composition

By the term "jet composition" we mean "What is the means of transporting energy over vast distances?" We are fairly confident that we know the essential ingredients of the emitting volumes we call "knots" and "hot spots": a relativistic plasma consisting of at least a magnetic field (average value $\geq 1\,\mu$G), a rather wide (in energy) distribution of relativistic electrons (and/or positrons), and photons of the CMB and of other sources more particular to the local environment. A relatively straightforward argument [49] convinces us that the particles responsible for the electromagnetic radiation observed cannot be the means of transporting energy: electrons with $\gamma \geq 2000$ would lose their energy long before they reached the end of many long jets. Thus knots and hot spots should be viewed as products of the jet: sites where jet energy is transferred to the radiating plasma.

1.4.1 Options

The standard list of possibilities for jet composition is quite short. It basically consists of cold (thermal) or hot (relativistic) protons, cold electrons/positrons, and Poynting flux. Occasionally some other transporter has been suggested such as neutrons or low-frequency EM radiation, but these have not been embraced by the community, primarily because of difficulties associated with the genesis of the jet, and because of problems of how to deflect and bend jets (both of which have been observed in several jets).

1.4.2 Constraints

Most of the published works on jet composition are based on attempts to find evidence for the particle content: either pairs or normal plasma. The general approach is to seek an estimate of the total power requirements of the jet, and then come up with some estimate of how many electrons are in the jet (e.g., [50, 51]). If the energy transport relies on pairs, there have to be more electrons (and positrons) than if most of the energy is carried by protons. Sikora and Madejski [52] argue that pairs outnumber protons, but protons are the chief energy carrier.

Attempts to estimate the total number of electrons are fraught with uncertainty. As well as a blind faith that there are no spectral breaks at low energies, it is usually assumed that a steep power law describes the electron distribution, which is arguably the case for emitting regions, but not necessarily for the transport mechanism. Since a steep power law is assumed, estimates of the total number of electrons are extremely sensitive to the low-energy cutoff of the power law ($\gamma_{\rm min}$, see for example [53]).

Various constraints from synchrotron theory have been invoked, including using the rotation measure to study parsec-scale jets with VLBI data. Wardle *et al.* [54] argued that on the parsec scale, the jets being studied had to consist of e^-, e^+ pairs: a pair plasma does not produce Faraday rotation.

Reynolds [55, 56] and others (e.g., [50, 51]) consider the effects of synchrotron self-absorption (SSA) in order to get an estimate of the magnetic field strength. However, it is notoriously difficult to find convincing values for the peak flux density and peak frequency of an absorbed component.

Sikora *et al.* [57] use various arguments to suggest that while jets are initially dominated by Poynting flux, they quickly become particle-dominated with protons as the primary transporter of energy.

In spite of the cunning arguments employed, we consider the question of jet content to be undecided.

1.5 Some Things (We Think) We Know, and Some (We Know) We Don't

We end this introductory chapter with a brief compilation of knowns and unknowns about jets. Knowns include:

- The jet phenomenon bridges many orders of magnitude in size: jets originate on subparsec scales and can propagate over many 100 kpc.
- Black holes are extreme powerhouses. While stars generate (or liberate) energy through nuclear fusion of lighter elements into heavier elements, mass-accreting black holes convert potential energy of interstellar matter into electromagnetic radiation and into jet energy. The jets carry away a substantial fraction of the accretion energy.
- The jet emissions we observe come from relativistic electrons.
- Hot (e.g., $\gamma > 2000$) electrons/positrons cannot be the agent of energy transport over huge distances.
- The polarized radio and optical jet emissions are synchrotron emission.
- The emitting plasmas of most/all jets are moving relativistically.
- Black holes are not only passive sinks of matter at the centers of galaxies and galaxy clusters. Their radiation and their jets have an impact on their host through heating and stirring of the interstellar and intracluster gas.

However, although theoretical concepts have evolved substantially over the last decades, the most important mechanisms concerning AGN accretion, jet formation, jet acceleration and collimation, and the various emissions from AGN jets are still the subject of a lively debate. Relevant questions include:

- How does black hole accretion work and how are jets launched?
- Which effects regulate the activity of AGNs and turn jets on and off?
- What mechanisms cause the spectacular flares of electromagnetic radiations from blazars?
- What is the composition of the "fluid" responsible for the energy transport at different distances from the supermassive black holes?
- What maintains the collimation of jets?
- What is the emission process for the spatially resolved X-rays from high-power sources such as radio quasars?
- How do jets accelerate particles to TeV energies?
- Are jets the sources of ultra-high-energy cosmic rays?

References

1 Curtis, H.D. (1918) *Pub. Lick Obs.*, **13**, 31.

2 Rees, M.J. (1971) *Nature*, **229**, 312.

3 Blandford, R.D. and Rees, M.J. (1974) *Mon. Not. R. Astron. Soc.*, **169**, 395.

4 Scheuer, P.A.G. (1974) *Mon. Not. R. Astron. Soc.*, **166**, 513.

5 Harris, D.E., Mossman, A.E., and Walker, R.C. (2005) in *X-Ray and Radio Connections* (eds L.O. Sjouwerman and K.K Dyer), published electronically by NRAO, http://www.aoc.nrao.edu/events/xraydio (accessed 7 September 2011), presented 3–6 February 2004 in Santa Fe, New Mexico, USA, (E7.16).

6 Alfvén, H. and Herlofson, N. (1950) *Phys. Rev.*, **78**, 616.

7 Shklovsky, I.S. (1953) *Proc. Acad. Sci. USSR*, **91**, 475.

8 Baade, W. (1956) *Astrophys. J.*, **123**, 550.

9 Burbidge, G.R. (1956) *Astrophys. J.*, **124**, 416.

10 Schmidt, M. (1963) *Nature*, **197**, 1040.

11 Hazard, C., Mackey, M.B., and Shimmins, A.J. (1963) *Nature*, **197**, 1037.

12 Hogg, D.E., MacDonald, G.H., Conway, R.G., and Wade, C.M. (1969) *Astron. J.*, **74**, 1206.

13 Northover, K.J.E. (1973) *Mon. Not. R. Astron. Soc.*, **165**, 369.

14 Turland, B. (1975) *Mon. Not. R. Astron. Soc.*, **172**, 181.

15 van Breugel, W.J.M. and Miley, G.K. (1977) *Nature*, **265**, 315.

16 Högbom, J.A. and Carlsson, I. (1974) *Astron. Astrophys.*, **34**, 341.

17 Miley, G.K., Wellington, K., and van der Laan, H. (1975) *Astron. Astrophys.*, **38**, 381.

18 Waggett, P.C., Warner, P.J., and Baldwin, J.E. (1977) *Mon. Not. R. Astron. Soc.*, **181**, 465.

19 Bridle, A.H., and Perley, R.A. (1984) *Annu. Rev. Astron. Astrophys.*, **22**, 319.

20 Bridle, A.H. (1986) *Can. J. Phys.*, **64**, 353.

21 Hjellming, R.M. and Johnston, K.J. (1981) *Nature*, **290**, 100.

22 Mirabel, I.F., Rodriguez, L.F., Cordier, B., Paul, J., and Lebrun, F. (1992) *Nature*, **358**, 215.

23 Zel'dovich, Y.B. and Novikov, I.D. (1964) *Sov. Phys. Dokl.*, **158**, 811.

24 Burbidge, G.R., Burbidge, E.M., and Sandage, A.R. (1963) *Rev. Mod. Phys.*, **35**, 947.

25 Ferrarese, L. and Merritt, D. (2000) *Astrophys. J.*, **539**, L9.

26 Gebhardt, K. (2000) *Astrophys. J.*, **539**, L13.

27 Jennison, R.C. and Das Gupta, M.K. (1956) *Philos. Mag.*, **1**(1), 65.

28 Konigl, A. (1980) Relativistic effects in extragalactic radio sources. PhD Thesis. California Institute of Technology.

29 Belloni, T. *et al.* (2010) *The Jet Paradigm: From Microquasars to Quasars*, Springer.

30 Tavecchio, F., Maraschi, L., Sambruna, R.M., and Urry, C.M. (2000) *Astrophys. J. Lett.*, **544**, L23.

31 Celotti, A., Ghisellini, G., and Chiaberge, M. (2001) *Mon. Not. R. Astron. Soc.*, **321**, L1–5.

32 Begelman, M.C., Blandford, R.D., and Rees, M.J. (1984) *Rev. Mod. Phys.*, **56**, 255.

33 Ghosh, P. and Abramowicz, M.A. (1997) *Mon. Not. R. Astron. Soc.*, **292**, 887.

34 Krawczynski, H. *et al.* (2001) *Astrophys. J.*, **559**, 187.

35 Williams, A.G. (1991) Numerical simulations of radio source structure, in *Beams and Jets in Astrophysics*, Cambridge Astrophysics Series (ed. P.A. Hughes), Cambridge University Press.

36 Velikhov, E. (1959) *J. Exp. Theor. Phys. (USSR)*, **36**, 1398.

37 Chandrasekhar, C. (1961) *Hydrodynamic and Hydromagnetic Stability*, Oxford University Press, Oxford. and Dover Publications Inc., 1981.

38 Balbus, S.A. and Hawley, J.F. (1991) *Astrophys. J.*, **376**, 214.

39 Spruit, H.C. (2010) Theory of magnetically powered jets, in *The Jet Paradigm*, Lecture Notes in Physics, vol. 794, Springer-Verlag, Berlin, Heidelberg, p. 233, ISBN 978-3-540-76936-1, arXiv:0804.3096.

40 Komissarov, S.S. (2009) *Magnetic acceleration of relativistic jets,* Procs. Steady Jets and Transient Jets, 7–8 April 2010, Bonn, Germany, Memorie della Society Astronomica Italiana, vol. 82, p. 95, 2011, arXiv:1006.2242.

41 Hawley, J.F. (2009) *Astrophys. Space Sci.*, **320**, 107.

42 McKinney, J.C. (2006) *Mon. Not. R. Astron. Soc.*, **368**, 1561.

43 Stecker, F.W., Baring, M.G., and Summerlin, E.J. (2007) *Astrophys. J. Lett.*, **667**, L29.

44 Kirk, J.G., Rieger, F.M., and Mastichiadis, A. (1998) *Astron. Astrophys.*, **333**, 452.

45 Böttcher, M. (2010) *Models for the spectral energy distributions and variability of blazars,* in Procs. Fermi Meets Jansky (eds T. Savolainen, E. Ros, R.W. Porcas, and J.A. Zensus), Bonn, Germany, 21–23 June 2010, MPI for Radio Astronomy, Bonn, p. 41, arXiv:1006.5048.

46 Krawczynski, H., Coppi, P.S., and Aharonian, F. (2002) *Mon. Not. R. Astron. Soc.*, **336**, 721.

47 Sironi, L. and Spitkovsky, A. (2009) *Astrophys. J. Lett.*, **707**, L92.

48 Kato, T.N. and Takabe, H. (2010) *Astrophys. J.*, **721**, 828.

49 Harris, D.E. and Krawczynski, H. (2007) Revista Mexicana de Astronomia y Astrofisica, Serie de Conferencias, 27, Contents of Supplementary CD, p. 188.

50 Celotti, A. and Fabian A.C. (1993) *Mon. Not. R. Astron. Soc.*, **264**, 228.

51 Dunn, R.J.H., Fabian, A.C., and Celotti, A. (2006) *Mon. Not. R. Astron. Soc.*, **372**, 1741.

52 Sikora, M. and Madejski G. (2000) *Astrophys. J.*, **534**, 109.

53 Blundell, K.M., Fabian, A.C., Crawford, C.S., Erlund, M.C., and Celotti, A. (2006) *Astrophys. J.*, **644**, L13.

54 Wardle, J.F.C., Homan, D.C., Ojha, R., and Roberts, D.H. (1998) *Nature*, **395**, 457.

55 Reynolds, C.S., Fabian, A.C., Celotti, A., and Rees, M.J. (1996) *Mon. Not. R. Astron. Soc.*, **283**, 873.

56 Reynolds, C. (1997) thesis, The matter content of the jet in M87, http://www.astro.umd.edu/~chris/publications/thesis/node98.html (accessed 7 September 2011).

57 Sikora, M., Begelman, M.C., Madejski, G.M., and Lasota, J.-P. (2005) *Astrophys. J.*, **625**, 72.

Part Two Theory Basics

2
Special Relativity of Jets

Markus Böttcher

This chapter will introduce the fundamental aspects of special relativity, which are needed for the exposition in the rest of this book. We will assume that the reader is familiar with the concept of inertial frames and the postulates of special relativity. Based on the Lorentz invariance of four-vector scalar products, we will develop a formalism that will allow us to evaluate Lorentz transformations in a rather straightforward way. We will develop tools to study the kinematics of two-particle interactions, as needed for the discussion of radiation mechanisms in Chapter 3, and aspects of Lorentz boosts and relativistic aberration resulting from relativistic motion in astrophysical jets. We will then use those results to develop observational diagnostics to measure relativistic motions in jets, which will be used and discussed in more detail in Chapter 5 on radio observations of astrophysical jets. This chapter is not meant as a complete introduction to special relativity. In particular, we will not cover any aspects of relativistic dynamics as they will not be needed in the further discussions in this book. More complete introductions to all aspects of special (and general) relativity may be found, for example, in [1–6], to name just a few of a large number of excellent monographs on the topic.

2.1
Space-Time, Four-Vectors, and Lorentz Invariance

Special relativity is based on the concept of *four-vectors*, including time as a zeroth component in the form

$$\underline{x} \equiv \begin{pmatrix} ct \\ x \\ y \\ z \end{pmatrix} . \tag{2.1}$$

In the following, we denote four-vectors by underlining the symbols, to distinguish them from the (three-dimensional) spatial components, styled as bold italic sym-

Relativistic Jets from Active Galactic Nuclei, First Edition. Edited by M. Böttcher, D.E. Harris, H. Krawczynski.

bols. In general, we denote any four-vector as

$$\underline{a} \equiv \begin{pmatrix} a_t \\ a_x \\ a_y \\ a_z \end{pmatrix} \equiv \begin{pmatrix} a_t \\ \boldsymbol{a} \end{pmatrix} . \tag{2.2}$$

The appropriate choice of a four-vector scalar product that will remain invariant under relativistic velocity transformations (*Lorentz transformations*), is given by

$$\underline{a} \cdot \underline{b} \equiv a_t b_t - a_x b_x - a_y b_y - a_z b_z = a_t b_t - \boldsymbol{a} \cdot \boldsymbol{b} . \tag{2.3}$$

We will derive almost all the predictions of special relativity by using the postulate that *four-vector scalar products as defined in (2.3) yield the same value in any inertial frame in which they are evaluated.*

In order to exploit the Lorentz invariance of four-vector scalar products, we need to generalize the four-vector concept from space-time vectors $\underline{x} = (ct, \boldsymbol{x})$ to other vector quantities needed for relativistic mechanics. In particular, we need to find appropriate special relativistic four-vector generalizations for the velocity $\boldsymbol{v}$, and the momentum $\boldsymbol{p}$.

Let us start with the generalization of the velocity. Consider a particle moving through space along any trajectory parameterized by $\boldsymbol{x}(t)$. In Newtonian mechanics, the velocity $\boldsymbol{v}$ is then defined as $\boldsymbol{v} = d\boldsymbol{x}(t)/dt$. If we attempt to simply generalize this to the four-vector concept, we might choose $\underline{v} = d\underline{x}(t)/dt = (c, \boldsymbol{v})$. Now, however, we realize that the time component of this construct, c, will not change under transformation to any moving frame, while the three-dimensional velocity, v, will change. Hence, it is obvious that the scalar product $\underline{v} \cdot \underline{v}$ will *not* be invariant under Lorentz transformation. Therefore, our simplistic choice of $\underline{v}$ is not an appropriate four-vector generalization of the velocity. The problem lies in the fact that the time t used to parameterize the particle's trajectory is dependent on the arbitrary choice of a specific reference frame. We can overcome this problem by parameterizing the trajectory in the frame of the particle itself. The time elapsed in the particle's own rest frame is called the *proper time*, τ. Due to relativistic time dilation, it is related to the time in any other inertial frame by

$$d\tau = \frac{dt}{\gamma} , \tag{2.4}$$

where γ is the Lorentz factor, and we define β as the ratio of the particle's velocity to the speed of light:

$$\beta \equiv \frac{v}{c} ,$$
$$\gamma \equiv \frac{1}{\sqrt{1-\beta^2}} . \tag{2.5}$$

For any physical body or particle, $v < c$ and hence $\beta < 1$, and $\gamma \geq 1$. The Lorentz factor γ depends on the relative velocity between the arbitrary inertial frame and the particle's rest frame. A more appropriate representation of the particle trajectory is therefore $\underline{x}(\tau)$, and the four-vector generalization of the velocity would then be

$$\underline{u} \equiv \frac{d\underline{x}(\tau)}{d\tau} = \frac{d}{d\tau}\begin{pmatrix} ct \\ \boldsymbol{x} \end{pmatrix} = \gamma \frac{d}{dt}\begin{pmatrix} ct \\ \boldsymbol{x} \end{pmatrix} = \gamma \begin{pmatrix} c \\ \boldsymbol{v} \end{pmatrix} . \tag{2.6}$$

For this definition to be consistent with our principle of Lorentz-invariant four-vector products, we can verify that $\underline{u} \cdot \underline{u} = \gamma^2(c^2 - v^2) = c^2\gamma^2(1 - \beta^2) = c^2$. We see that $\underline{u} \cdot \underline{u} = c^2$, independent of the choice of a reference frame in which we measure the particle's three-velocity $\boldsymbol{v}$. We note that, although $v < c$, the spatial component of $\underline{u}$, $\gamma \boldsymbol{v}$, can exceed the speed of light.

Using the definition of the four-velocity $\underline{u}$, it is straightforward to generalize the Newtonian expression for the momentum vector $\boldsymbol{p}_{\mathrm{Newt}} = m\boldsymbol{v}$, where m is the particle's mass. We find

$$\underline{p} \equiv m\underline{u} = m\gamma \begin{pmatrix} c \\ \boldsymbol{v} \end{pmatrix} . \tag{2.7}$$

The four-vector defined above is called the *energy-momentum four-vector*. We can rewrite its components as

$$\underline{p} = \begin{pmatrix} E/c \\ \boldsymbol{p} \end{pmatrix} , \tag{2.8}$$

where we have introduced the relativistic expression for the particle's energy, $E = \gamma mc^2$, and the relativistic three-momentum, $\boldsymbol{p} = \gamma m\boldsymbol{v}$. We can, again, show that the four-vector scalar product of $\underline{p}$ with itself is Lorentz invariant:

$$\underline{p} \cdot \underline{p} = \gamma^2 m^2 c^2 (1 - \beta^2) \equiv m^2 c^2 . \tag{2.9}$$

On the other hand, evaluating this scalar product using the representation (2.8), we find

$$E^2 = (mc^2)^2 + (|\boldsymbol{p}|c)^2 . \tag{2.10}$$

For a particle at rest ($\boldsymbol{p} = 0$), this reduces to Einstein's famous formula $E = mc^2$, illustrating the equivalence of mass and energy resulting from his special theory of relativity. To show that the interpretation of γmc^2 as the particle's energy makes sense, let us look at the limiting case of nonrelativistic velocities, $v \ll c$. In this case, we might expect to recover the Newtonian expression for the kinetic energy, $K = (1/2)mv^2$. We can recover this result using a Taylor expansion of the expression (2.5) for the Lorentz factor:

$$\gamma mc^2 = \frac{mc^2}{\sqrt{1 - (v/c)^2}} \approx mc^2 \left(1 + \frac{1}{2}\frac{v^2}{c^2}\right) = mc^2 + \frac{1}{2}mv^2 . \tag{2.11}$$

We see that in the nonrelativistic limit, the particle energy reduces to a term representing the rest-mass energy $E_{\mathrm{rest}} = mc^2$ and the classical kinetic energy K. The energy-momentum four-vector representation (2.8) also allows us to generalize this concept to massless particles like photons. Even for photons ($m = 0$), we may still write $\underline{p}_{\text{ph}} = (E_{\text{ph}}/c, \boldsymbol{p}_{\text{ph}})$, and evaluating the Lorentz-invariant scalar product $\underline{p}_{\text{ph}} \cdot \underline{p}_{\text{ph}} = (mc)^2 = 0$ then yields the relation

$$E_{\text{ph}} = |\boldsymbol{p}_{\text{ph}}| c \,, \tag{2.12}$$

or $p_{\text{ph}} = E_{\text{ph}}/c$, and hence,

$$\underline{p}_{\text{ph}} = \frac{E_{\text{ph}}}{c} \begin{pmatrix} 1 \\ \hat{k} \end{pmatrix} , \tag{2.13}$$

where $\hat{k}$ denotes a unit three-vector in the direction of propagation of the photon.

2.1.1
Interaction Thresholds

The Lorentz-invariance of four-vector scalar products allows us to evaluate the kinematics of elementary-particle interactions and, in particular, to calculate threshold energies required for interactions. The expressions we derive in this section will be used later when discussing the individual processes in more detail in Chapter 3.

2.1.1.1 $\gamma\gamma$ Pair Production

Just as an electron and its antiparticle, the positron, can annihilate to produce two photons, the inverse process can happen: two photons with sufficiently high energies can annihilate and produce an electron–positron pair. Obviously, the total available energy carried by the two photons has to be enough to come up with at least the rest-mass energy of the electron and positron for this interaction to be possible. Let us therefore consider the interaction of a high-energy (γ-ray) photon of energy E_1 with a target photon of energy E_2 and ask what is the minimum energy E_2 required for a $\gamma\gamma$ pair production event. We write the four-momenta of the two photons as

$$\underline{p}_{\text{ph}}^{(i)} = \frac{E_i}{c} \begin{pmatrix} 1 \\ \hat{k}^{(i)} \end{pmatrix} \quad (i = 1, 2) \,. \tag{2.14}$$

If a $\gamma\gamma$ pair production event happens, we write the four-momenta of the produced electron–positron pair as

$$\underline{p}_{\text{e}^\pm} = \gamma_\pm m_{\text{e}} c \begin{pmatrix} 1 \\ \boldsymbol{\beta}_\pm \end{pmatrix} \tag{2.15}$$

and note that energy and momentum conservation dictates that

$$\underline{p}_{\text{tot}}^{\text{in}} \equiv \underline{p}_{\text{ph}}^{(1)} + \underline{p}_{\text{ph}}^{(2)} = \underline{p}_{\text{tot}}^{\text{out}} \equiv \underline{p}_{\text{e}^+} + \underline{p}_{\text{e}^-} \,. \tag{2.16}$$

Now, we can take the four-vector scalar products of both sides of (2.16) with themselves, $\underline{P}_{\text{tot}} \cdot \underline{P}_{\text{tot}}$. But we know that for evaluating those four-vector scalar products, it does not matter which reference frame we use, since they are invariant under Lorentz transformation. Since we are interested in finding the threshold condition for the pair production process, the obvious choice for evaluating the product on the right hand side (i.e., involving the electron and positron) in the center-of-momentum (cm) frame of the interaction. If the interaction happens right at threshold, then the newly produced electron and positron will just be at rest in the cm frame. In other words, in the cm frame, the incoming photons have just enough energy to produce the rest-mass energy of the electron–positron pair. In the cm frame, the electron and positron will have equal and opposite velocities β_{cm}, and hence equal Lorentz factors γ_{cm}, so that $\underline{p}_{\text{e}^+} \cdot \underline{p}_{\text{e}^-} = (m_\text{e} c)^2 \gamma_{\text{cm}}^2 (1 + \beta_{\text{cm}}^2)$. Now, we note that $\underline{p}_{\text{ph}} \cdot \underline{p}_{\text{ph}} = 0$ and $\underline{p}_{\text{e}^\pm} \cdot \underline{p}_{\text{e}^\pm} = (m_\text{e} c)^2$. Furthermore writing the cosine of the interaction angle between the two photons as $\hat{k}^{(1)} \cdot \hat{k}^{(2)} = \cos\theta \equiv \mu$, we find

$$2\frac{E_1 E_2}{c^2}(1-\mu) = 2(m_\text{e} c)^2 + 2(m_\text{e} c)^2 \gamma_{\text{cm}}^2 \left(1 + \beta_{\text{cm}}^2\right) = 4(m_\text{e} c)^2 \gamma_{\text{cm}}^2 \,, \tag{2.17}$$

where we have used the identity $\gamma^2\beta^2 = \gamma^2 - 1$. We can thus write the cm electron and positron Lorentz factor as

$$\gamma_{\text{cm}} = \sqrt{\frac{E_1 E_2 (1-\mu)}{2(m_\text{e} c^2)^2}} \,. \tag{2.18}$$

We can invert the second equation in (2.5) to obtain $\beta = \sqrt{1 - 1/\gamma^2}$ and therefore

$$\beta_{\text{cm}} = \sqrt{1 - \frac{2\,(m_\text{e} c^2)^2}{E_1 E_2 (1-\mu)}} \,. \tag{2.19}$$

Now, the Lorentz factor γ_{cm} must be ≥ 1 for a physical solution. This leads to the threshold condition for $\gamma\gamma$ pair production of

$$E_2 \geq E_{\text{thr}} \equiv \frac{2\left(m_\text{e} c^2\right)^2}{E_1(1-\mu)} \,. \tag{2.20}$$

This threshold becomes smallest in the case of the two photons colliding head-on, that is, $\mu = -1$, where the threshold energy reduces to $E_{\text{thr}} = \left(m_\text{e} c^2\right)^2 / E_1$. The $\gamma\gamma$ absorption process and its relevance for the high-energy emission from extragalactic jet sources will be discussed in detail in Section 3.2.3.

2.1.1.2 **Photo-Pion Production**

It is often speculated that the relativistic jets of active galactic nuclei and gamma-ray bursts may be the site of the acceleration of the highest-energy cosmic rays. Those are elementary particles which have been observed on Earth up to energies of $E \gtrsim 10^{20}$ eV. If these ultra-high-energy cosmic rays are protons, they must have Lorentz factors of $\gamma \gtrsim 10^{11}$, which means they travel at 99.999999999999999999995% (i.e., $1 - 5 \times 10^{-23}$) of the speed of light. At these extremely high energies, those protons can interact with low-energy photons to produce neutral or charged pions, for example $p + \gamma \rightarrow p + \pi^o$. This process is called photo-pion production and may play an important role in the formation of the high-energy spectra observed from some active galactic nuclei.

We will now calculate the energy that a proton needs to be able to undergo a photo-pion production reaction with a photon of given energy E_{ph}. As in the case of $\gamma\gamma$ pair production above, we write the energy-momentum conservation equation as

$$\underline{P}_{tot}^{in} = \underline{p}_{p}^{(in)} + \underline{p}_{ph} = \underline{P}_{tot}^{out} = \underline{p}_{p}^{(out)} + \underline{p}_{\pi} \tag{2.21}$$

with $\underline{p}_{p/\pi} = \gamma_{p/\pi} m_{p/\pi} c(1, \boldsymbol{\beta}_{p/\pi})$. Again, we evaluate the Lorentz-invariant scalar products $\underline{P}_{tot} \cdot \underline{P}_{tot}$ on both sides, noting that $\underline{p} \cdot \underline{p} = (mc)^2$, which is 0 for photons. The left hand side then yields

$$\left(\underline{p}_{p}^{(in)} + \underline{p}_{ph}\right) \cdot \left(\underline{p}_{p}^{(in)} + \underline{p}_{ph}\right) = (m_p c)^2 + 2 E_{ph} \gamma_p m_p (1 - \beta_p \mu) , \tag{2.22}$$

where we have defined μ as the cosine of the interaction angle between the incoming proton and the photon, $\boldsymbol{\beta}_p \cdot \hat{k} = \beta_p \cos\theta \equiv \beta_p \mu$. For the right hand side, of (2.21), we choose, again, the center-of-momentum frame. At threshold, the incoming particles have just enough total energy to produce the pion in the interaction, so both proton and pion will be at rest. Hence, in the cm frame at threshold,

$$\left(\underline{p}_{p}^{(out)} + \underline{p}_{\pi}\right) \cdot \left(\underline{p}_{p}^{(out)} + \underline{p}_{\pi}\right) = (m_p + m_\pi)^2 c^2 . \tag{2.23}$$

We also realize that the energetically most favorable case for the interaction to happen, is a head-on collision, $\mu = -1$. For that case, we find for the threshold proton energy

$$\gamma_p m_p c^2 (1 + \beta_p) = \frac{m_\pi m_p c^4}{2 E_{ph}} \left(2 + \frac{m_\pi}{m_p}\right) . \tag{2.24}$$

Now, for target photons of energies much less than the pion rest-mass energy of $m_\pi c^2 \approx 140$ MeV, the proton needs to be highly relativistic in order to reach the threshold for pion production. Therefore, we may approximate $\beta_p \approx 1$ in (2.24) to find

$$E_p^{thr} = \frac{m_p m_\pi c^4}{2 E_{ph}} \left(1 + \frac{m_\pi}{2 m_p}\right) . \tag{2.25}$$

Ultra-high-energy cosmic rays reaching us from sources at cosmological distances, have to contend with the cosmic microwave background (CMB) radiation, which pervades all space with a blackbody radiation spectrum at a temperature of $T_{CMB} \approx 2.7$ K. The characteristic photon energy of this radiation field is

$E_{\rm CMB} \approx 6.6 \times 10^{-4}$ eV. Using this in (2.25) yields a threshold energy for cosmic rays to produce pions in the CMB, at $E_{\rm p}^{\rm thr} \approx 10^{20}$ eV. In 1966, Greisen [7] and Zatsepin and Kuz'min [8] predicted the existence of a cutoff (called the *GZK cutoff*) in the cosmic ray spectrum around this energy due to photo-pion production on the CMB. Recent results from the High-Resolution Fly's Eye (HiRes) and the Auger Cosmic Ray Experiments have now conclusively confirmed the existence of the GZK cutoff [9].

The photo-pion production process will be discussed in more detail in the context of emission processes in relativistic jets in Chapter 3.2.4.

2.2 Lorentz Transformations

It is rather well established that the radio through γ-ray emission we observe from extragalactic jets in AGN and GRBs emerges from small ($R \lesssim 10^{16}$ cm) emission regions which are moving at relativistic speeds. In this section, we will use the formalism derived above to derive several relativistic transformation laws between quantities measured in the comoving frame of those relativistically moving emission regions, and us, the (presumed "stationary") observers on Earth. In the rest of the book, we need to distinguish between the macroscopic motion of an entire emission region, and the motion of an individual, relativistic particle which is responsible for the observed radiation in the rest frame of the emission region. We will denote by β and γ the normalized speed and Lorentz factor of individual particles (electrons, protons, positrons, ...), while we will use Γ for the bulk Lorentz factor of the entire emission region, and $\beta_\Gamma = \sqrt{1 - 1/\Gamma^2}$ for its speed normalized to the speed of light:

$$
\begin{aligned}
\gamma &= \frac{1}{\sqrt{1-\beta^2}} = \text{Lorentz factor of an individual particle}\,, \\
\beta &= \frac{v}{c} = \text{Normalized speed of an individual particle}\,, \\
\Gamma &= \text{Bulk Lorentz factor of the emission region}\,, \\
\beta_\Gamma &= \sqrt{1-\frac{1}{\Gamma^2}} = \text{Normalized speed of the emission region}\,. \qquad (2.26)
\end{aligned}
$$

In the following, we will call the rest frame of the emission region the "em" frame, while our stationary observer's frame (in which we receive the emission produced in the "em" frame) is called the "rec" frame. All quantities will be labeled by superscripts "em" or "rec" depending on the frame in which they are measured. We define the direction of motion of the emission region (the jet axis) as the z-axis of our coordinate system. We define angles $\theta^{\rm em/rec}$ with respect to the z-axis, as shown in Figure 2.1. We define the cosines of those angles as $\mu^{\rm em/rec} \equiv \cos\theta^{\rm em/rec}$, and the azimuthal angles ϕ around the z-axis, choosing $\phi = 0$ in the direction of the observer. From symmetry considerations, it is obvious that the angle ϕ must be

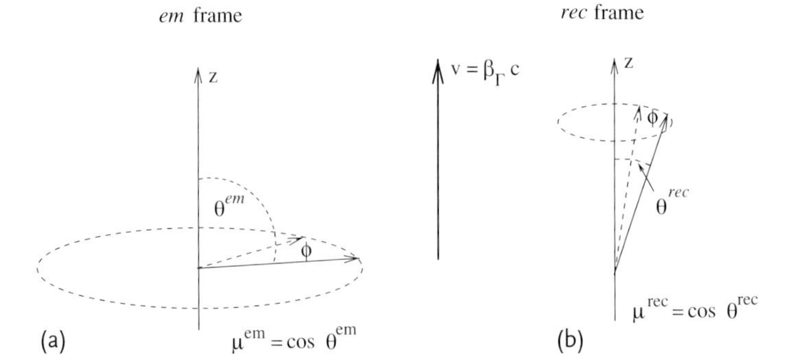

Figure 2.1 Definition of the geometry for transformations between the (a) emission region (em) and the (b) stationary observer (rec).

identical for both reference frames. For the transformations which we investigate in this section, we will need the four-vectors describing the velocities of the two frames as measured in either frame. We call $\underline{u}_{\rm em}^{\rm rec}$ the four-velocity of the "em" frame as measured in the "rec" frame, and so on. Thus, we have

$$\begin{aligned} \underline{u}_{\rm em}^{\rm rec} &= c\Gamma(1,0,0,\beta_\Gamma)\,, \\ \underline{u}_{\rm em}^{\rm em} &= c(1,0,0,0)\,, \\ \underline{u}_{\rm rec}^{\rm rec} &= c(1,0,0,0)\,, \\ \underline{u}_{\rm rec}^{\rm em} &= c\Gamma(1,0,0,-\beta_\Gamma)\,. \end{aligned} \tag{2.27}$$

Let us first consider how the energy and direction of propagation of photons will be altered as a result of the relativistic motion. Specifically, we consider a photon of energy $E^{\rm em}$ whose energy-momentum four-vector (in the "em" frame) is, as usual, $\underline{p}_{\rm ph}^{\rm em} = (E^{\rm em}/c)(1, \hat{k}^{\rm em})$. Now, to calculate what the observed photon energy $E^{\rm rec}$ of that same photon will be, we consider the Lorentz-invariant scalar products

$$\underline{p}_{\rm ph}^{\rm rec} \cdot \underline{u}_{\rm em}^{\rm rec} = \underline{p}_{\rm ph}^{\rm em} \cdot \underline{u}_{\rm em}^{\rm em} \tag{2.28}$$

to obtain

$$E^{\rm rec} = \frac{E^{\rm em}}{\Gamma(1-\beta_\Gamma \mu^{\rm rec})}\,, \tag{2.29}$$

where we have used $\hat{k}^{\rm rec} \cdot \boldsymbol{\beta}_{\rm em}^{\rm rec} = \beta_\Gamma \mu^{\rm rec}$. The factor determining the Lorentz boost of photon energies in (2.29) is called the *Doppler factor*, δ, and will continue to reappear throughout the discussions in this book:

$$\delta \equiv \frac{1}{\Gamma\,(1-\beta_\Gamma \mu^{\rm rec})}\,. \tag{2.30}$$

With this definition, we may write the photon energy boost as $E^{\rm rec} = \delta\, E^{\rm em}$. From (2.29), we see that photons received in the backward direction ($\mu^{\rm rec} = -1$), will be deboosted (redshifted) by a factor of

$$\delta_{\text{backward}} = \frac{1}{\Gamma(1+\beta_\Gamma)} \approx \frac{1}{2\Gamma} , \tag{2.31}$$

where the last equality holds for highly relativistic motion with $\Gamma \gg 1$. In the forward direction ($\mu_{\text{rec}} = +1$), we have

$$\delta_{\text{forward}} = \frac{1}{\Gamma(1-\beta_\Gamma)} = \frac{(1+\beta_\Gamma)}{\Gamma\left(1-\beta_\Gamma^2\right)} = (1+\beta_\Gamma)\Gamma \approx 2\Gamma , \tag{2.32}$$

where, again, the last equality holds for $\Gamma \gg 1$. Thus, photons emitted (and received) along the direction of the relativistic motion will be boosted to higher energies (blueshifted) by a factor of $\approx 2\Gamma$.

We can derive an alternative expression for the Lorentz boost of photon energies by using the four-vector scalar products

$$\underline{p}_{\text{ph}}^{\text{rec}} \cdot \underline{u}_{\text{rec}}^{\text{rec}} = \underline{p}_{\text{ph}}^{\text{em}} \cdot \underline{u}_{\text{rec}}^{\text{em}} . \tag{2.33}$$

This results in

$$E^{\text{rec}} = E^{\text{em}}\Gamma(1+\beta_\Gamma\mu^{\text{em}}) , \tag{2.34}$$

where we have used $\hat{k}^{\text{em}} \cdot \boldsymbol{\beta}_{\text{rec}}^{\text{em}} = -\beta_\Gamma\mu^{\text{em}}$. Now the Lorentz boost relations (2.29) and (2.34) must be equivalent. Therefore, we can combine them to find a relation between the two angles of propagation with respect to the jet axis, given by μ^{em} and μ^{rec}:

$$\mu^{\text{em}} = \frac{\mu^{\text{rec}} - \beta_\Gamma}{1 - \beta_\Gamma\mu^{\text{rec}}} \quad \text{and} \quad \mu^{\text{rec}} = \frac{\mu^{\text{em}} + \beta_\Gamma}{1 + \beta_\Gamma\mu^{\text{em}}} . \tag{2.35}$$

From (2.35), we see, first of all, that any photon emitted in the forward or backward direction ($\mu^{\text{em}} = \pm 1$) will be received in that same direction ($\mu^{\text{rec}} = \pm 1$). However, let us consider photons emitted at a right angle to the jet axis in the "em" frame ($\mu^{\text{em}} = 0$). Those will be received in the "rec" frame at an angle of $\mu^{\text{rec}} = \beta_\Gamma$. If $\Gamma \gg 1$, then β_Γ is close to one, and the corresponding angle $\theta^{\text{rec}} = \cos^{-1}(\beta_\Gamma)$ is small. We may therefore use a Taylor expansion for both β_Γ and $\cos\theta^{\text{rec}}$ to find

$$\begin{aligned} \beta_\Gamma &= \sqrt{1 - \frac{1}{\Gamma^2}} \approx 1 - \frac{1}{2\Gamma^2} \\ \mu^{\text{rec}} &= \cos\theta^{\text{rec}} \approx 1 - \frac{(\theta^{\text{rec}})^2}{2} . \end{aligned} \tag{2.36}$$

Comparing the two expressions, we find that photons emitted at a right angle to the jet axis in the "em" frame are received at an angle $\theta_{\text{rec}} \approx 1/\Gamma$ in the "rec" frame. Now, let us assume that photons are produced isotropically in the "em" frame, then our result implies that half of all photons will be beamed into a narrow cone of opening angle $\theta^{\text{rec}} = 1/\Gamma$ (see Figure 2.2).

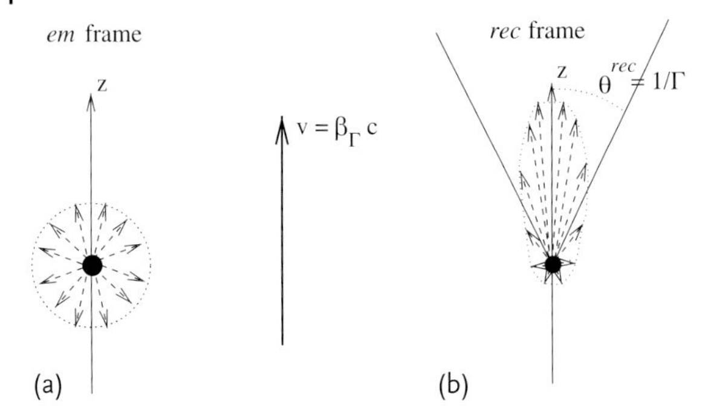

Figure 2.2 Light emitted from a relativistically moving source is beamed in the forward direction into a cone of opening angle $\theta^{\text{rec}} = 1/\Gamma$. In this figure, the length of the arrows represents the intensity of radiation in a given direction.

The distortion of the directions of propagation of photons described by (2.35) and illustrated in Figure 2.2 is known as *relativistic aberration*. Because of this effect, when we are looking at a relativistically moving object, we actually observe it from a different angle than our direct line-of-sight direction would suggest. For example, we can read from (2.35) that, when viewing an object at a right angle (in the "rec" frame) to the direction of motion, $\mu_{\text{rec}} = 0$, we observe photons that have actually been emitted, in the "em" frame, in the backward direction with $\mu_{\text{em}} = -\beta_\Gamma$, that is, at an angle $\theta_{\text{em}} \approx \pi - 1/\Gamma$. The effect combines with the relativistic length contraction and light travel time effects to distort the observed geometrical shape of an object moving at relativistic speed. The net effect is a rotated appearance of the object, which is also known as *Penrose rotation.*

We can use the photon energy transformation (2.29) also to find the transformation between time intervals in the "em" and "rec" frames, by noting that a photon energy corresponds to a frequency through $E_{\text{ph}} = h\nu$. Hence, the inverse of a photon energy (or frequency) describes a time interval measured in the respective frame. Therefore, we find

$$\frac{\Delta t^{\text{rec}}}{\Delta t^{\text{em}}} = \frac{\nu^{\text{em}}}{\nu^{\text{rec}}} = \frac{E^{\text{em}}}{E^{\text{rec}}} = \delta^{-1} . \tag{2.37}$$

This implies that any time interval in the emission frame "em" will be shortened by one factor of δ when observed in the "rec" frame.

Let us now consider how the observed fluxes of radiation received from a relativistically moving object are affected by the Lorentz transformations derived above. The energy flux in electromagnetic radiation is defined as

$$F_\nu \equiv \frac{h\nu d^3 N_{\text{ph}}}{dt dA d\nu} , \tag{2.38}$$

where dN_{ph} is the number of photons in a frequency interval $[\nu, \nu + d\nu]$. As this is an integer number, it will be the same in any reference frame. The term dA is a unit area interval. Introducing the distance D between the emitter and the receiver, we may write this as $dA = D^2 d\Omega$, where $d\Omega = d\mu d\phi$ is a solid angle element.

To find the transformation between fluxes perceived in the “em” and “rec” frame, we need to calculate

$$F^{\rm rec}_{\nu^{\rm rec}} = F^{\rm em}_{\nu^{\rm em}} \frac{\nu^{\rm rec}}{\nu^{\rm em}} \frac{d\nu^{\rm em}}{d\nu^{\rm rec}} \frac{d\Omega^{\rm em}}{d\Omega^{\rm rec}} \frac{dt^{\rm em}}{dt^{\rm rec}} . \tag{2.39}$$

The frequency transformations cancel out, while we see from (2.37) that $dt^{\rm em}/dt^{\rm rec} = \delta$. Since $\phi^{\rm em} = \phi^{\rm rec}$, the Jacobian $d\Omega^{\rm em}/d\Omega^{\rm rec}$ reduces to the derivative of (2.35):

$$\frac{d\Omega^{\rm em}}{d\Omega^{\rm rec}} = \frac{d\mu^{\rm em}}{d\mu^{\rm rec}} = \frac{1}{\Gamma^2(1-\beta_\Gamma \mu^{\rm rec})^2} = \delta^2 . \tag{2.40}$$

Therefore, we have

$$F^{\rm rec}_{\nu^{\rm rec}} = F^{\rm em}_{\nu^{\rm em}} \delta^3 . \tag{2.41}$$

As we will see in the following chapters, the radiation spectra of astrophysical jet sources are often dominated by nonthermal emission, well described by power-law spectra,

$$F_\nu \propto \nu^{-\alpha} , \tag{2.42}$$

over substantial frequency ranges. If we restrict our consideration to a range of frequencies over which (2.42) holds, we can further simplify the flux transformation (2.41), by expressing the flux in the emitting frame, $F^{\rm em}$ evaluated at the observed frequency, using

$$F^{\rm em}_{\nu^{\rm em}} = F^{\rm em}_{\nu^{\rm rec}} \left(\frac{\nu^{\rm rec}}{\nu^{\rm em}}\right)^\alpha = F^{\rm em}_{\nu^{\rm rec}} \delta^\alpha , \tag{2.43}$$

and hence,

$$F^{\rm rec}_{\nu^{\rm rec}} = F^{\rm em}_{\nu^{\rm rec}} \delta^{3+\alpha} . \tag{2.44}$$

This boosting effect is illustrated in Figure 2.3. A Lorentz boosting factor δ^3 results when comparing emission at $\nu^{\rm em}$ in the “em” frame with emission at $\nu^{\rm rec}$ in the observer’s frame. Comparing the fluxes at the same frequency in both frames, yields the boosting factor of $\delta^{3+\alpha}$ for a power-law radiation spectrum.

In applications to radio observations, it is often convenient to parameterize the spectral flux F_ν in terms of a *brightness temperature*, $T_{\rm b}$. For this purpose, one can assign an observed intensity

$$I_\nu = \frac{F_\nu}{\Delta\Omega} \tag{2.45}$$

corresponding to the intensity radiated by a blackbody emitter. In (2.45), the $\Delta\Omega$ is the solid angle over which the source extends as observed from Earth. This solid angle is related to the size of the emission region R and its angular distance from

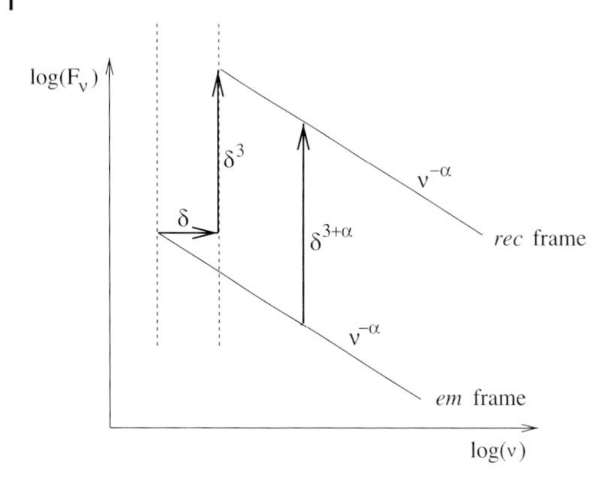

Figure 2.3 Illustration of the Lorentz boosting of radiative fluxes due to relativistic motion. Photon frequencies (energies) are boosted by one factor of δ, while the reception time contraction and relativistic aberration yield a boost of fluxes by δ^3.

Earth, d_{A}, through

$$\Delta\Omega = \frac{\pi R^2}{d_{\mathrm{A}}^2} . \tag{2.46}$$

The size of the emission region can be constrained either through direct measurement with high-resolution interferometry (e.g., VLBI, see Chapter 8), or through variability arguments (see Section 2.3.1). In the radio regime, the photon energy $h\nu \ll k_{\mathrm{B}} T$ for any temperature $T \gtrsim 100$ K. In that case, the blackbody intensity can be approximated in the Rayleigh–Jeans limit:

$$I_\nu = B_\nu(T_{\mathrm{b}}) = \frac{2h\nu^3}{c^2}\left(e^{h\nu/(k_{\mathrm{B}} T_{\mathrm{b}})} - 1\right)^{-1} \approx 2\left(\frac{\nu}{c}\right)^2 k_{\mathrm{B}} T_{\mathrm{b}} . \tag{2.47}$$

Setting the observed intensity I_ν equal to this limit yields the *brightness temperature*:

$$T_{\mathrm{b}} = \frac{2c^2 d_{\mathrm{A}}^2}{k_{\mathrm{B}} R^2} \frac{F_\nu}{\nu^2} . \tag{2.48}$$

Relativistic motion of the emission region will affect the perceived brightness temperature. If the size of the emission region is obtained by direct imaging, the observed brightness temperature, $T_{\mathrm{b}}^{\mathrm{rec}}$ can be related to the brightness temperature intrinsic to the source, $T_{\mathrm{b}}^{\mathrm{em}}$ by using (2.41) and (2.29):

$$T_{\mathrm{b}}^{\mathrm{rec}}(\mathrm{im}) = \delta\, T_{\mathrm{b}}^{\mathrm{em}}(\mathrm{im}) . \tag{2.49}$$

However, if the size of the emission region is obtained through the variability time scale as $R = c\Delta t_{\mathrm{var}}^{\mathrm{rec}}\delta$ (see Section 2.3.1), the factor R^2 in (2.48) yields two additional factors of δ so that

$$T_{\mathrm{b}}^{\mathrm{rec}}(\mathrm{var}) = \delta^3\, T_{\mathrm{b}}^{\mathrm{em}}(\mathrm{var}) . \tag{2.50}$$

Finally, we need to derive an expression to transform radiation energy densities between different moving frames. This will be important for applications in Chapters 3 and 8, where radiation fields from outside the emission region are being Compton scattered in the emission region to produce high-energy (X-ray and γ-ray) radiation.

In general, we can write a differential radiation energy density as

$$u_\nu(\Omega) = \frac{h\nu d^3 N}{dV d\Omega d\nu} \tag{2.51}$$

and the total radiation energy density as

$$u = \int_0^\infty d\nu \int_{4\pi} d\Omega \, u_\nu(\Omega) \,. \tag{2.52}$$

In contrast to the previous consideration of radiative flux, we will now be interested in transforming a radiation energy density from the stationary ("rec") frame to the comoving ("em") frame, and our calculations will be done in the "em" frame. Comparing (2.29) and (2.34), we note that we can express the Doppler factor δ as

$$\delta = \Gamma(1 + \beta_\Gamma \mu^{\mathrm{em}}) \,, \tag{2.53}$$

which we can use in the transformation laws $d\nu^{\mathrm{em}}/d\nu^{\mathrm{rec}} = \delta^{-1}$, $dV^{\mathrm{em}}/dV^{\mathrm{rec}} = dt^{\mathrm{em}}/dt^{\mathrm{rec}} = \delta$, $d\Omega^{\mathrm{em}}/d\Omega^{\mathrm{rec}} = \delta^2$. Hence, we find

$$u^{\mathrm{em}}_{\nu^{\mathrm{em}}}(\Omega^{\mathrm{em}}) = u^{\mathrm{rec}}_{\nu^{\mathrm{rec}}}(\Omega^{\mathrm{rec}}) \frac{dV^{\mathrm{rec}}}{dV^{\mathrm{em}}} \frac{d\Omega^{\mathrm{rec}}}{d\Omega^{\mathrm{em}}} = \frac{u^{\mathrm{rec}}_{\nu^{\mathrm{rec}}}(\Omega^{\mathrm{rec}})}{\Gamma^3(1 + \beta_\Gamma \mu^{\mathrm{em}})^3} \,. \tag{2.54}$$

To find the total radiation energy density in the comoving "em" frame, we have to integrate over photon frequency and solid angle:

$$\begin{aligned} u^{\mathrm{em}} &= \int_0^\infty d\nu^{\mathrm{em}} \int_{4\pi} d\Omega^{\mathrm{em}} u^{\mathrm{em}}_{\nu^{\mathrm{em}}}(\Omega^{\mathrm{em}}) \\ &= \int_0^\infty d\nu^{\mathrm{rec}} \int_{4\pi} d\Omega^{\mathrm{em}} \frac{u^{\mathrm{rec}}_{\nu^{\mathrm{rec}}}(\Omega^{\mathrm{rec}})}{\Gamma^4(1 + \beta_\Gamma \mu^{\mathrm{em}})^4} \,. \end{aligned} \tag{2.55}$$

The integration becomes particularly easy if the radiation field in the stationary ("rec") frame is isotropic, that is, if

$$u^{\mathrm{rec}}_{\nu^{\mathrm{rec}}}(\Omega^{\mathrm{rec}}) = \frac{u^{\mathrm{rec}}_{\nu^{\mathrm{rec}}}}{4\pi} \,. \tag{2.56}$$

In that case, writing $\int d\nu^{\mathrm{rec}} u^{\mathrm{rec}}_{\nu^{\mathrm{rec}}} \equiv u^{\mathrm{rec}}$, the integration in (2.55) reduces to

$$u^{\mathrm{em}} = \frac{u^{\mathrm{rec}}}{2\Gamma^4} \int_{-1}^{1} \frac{d\mu^{\mathrm{em}}}{(1 + \beta_\Gamma \mu^{\mathrm{em}})^4} = u^{\mathrm{rec}} \Gamma^2 \frac{3 + \beta_\Gamma^2}{3} \approx \frac{4}{3} u^{\mathrm{rec}} \Gamma^2 \,, \tag{2.57}$$

where the last equation holds for highly relativistic motion with $\beta_\Gamma \approx 1$. Hence, an isotropic radiation field in the stationary frame will be boosted by a factor $(4/3)\Gamma^2$ into the comoving frame of the emission region. Due to relativistic aberration, the external radiation field will appear to come from within a narrow cone of opening angle $1/\Gamma$ from the forward direction.

2.3
Relativistic Jet Diagnostics

In this section, we will see how the relativistic effects introduced in the previous sections can be used to diagnose the properties of the relativistic jets seen in AGN and GRBs.

2.3.1
Size Constraint from Variability

The most straightforward constraint one can deduce exploits the rapid variability seen in the optical through γ-ray emission in relativistic jet sources. If an extended source is transparent to the radiation we see (i.e., not optically thick to synchrotron self-absorption or γ-ray absorption through $\gamma\gamma$ pair production, as discussed in detail in Chapter 3), any fast variation of the emitted radiation throughout that source will be diluted by the finite light travel time through the source, which is of the order $\Delta t^{\mathrm{em}} \sim R/c$, where R is the characteristic size scale of the source. However, (2.37) shows that if the emission region is moving relativistically, any time interval in the comoving frame of the emission region will be observed as contracted by a factor $\Delta t^{\mathrm{rec}} = \Delta t^{\mathrm{em}}/\delta$. Hence, the smallest variability time scale which we expect from a relativistically moving source of size R, is $\Delta t^{\mathrm{rec}}_{\mathrm{var,\,min}} \geq \delta^{-1} R/c$. Hence, if we observe variability on a time scale $\Delta t^{\mathrm{obs}}_{\mathrm{var}}$, this yields a constraint on the size of the emission region as

$$R \leq c\Delta t^{\mathrm{obs}}_{\mathrm{var}}\,\delta\ . \tag{2.58}$$

Of course, the effective use of this constraint requires some knowledge of the Doppler factor δ. As we will see below, in the jets of AGN several arguments point towards Doppler factors of the order of $\delta \sim 10$. Blazars, the most extreme class of AGN, routinely exhibit variability on time scales of $\Delta t^{\mathrm{obs}}_{\mathrm{var}} \sim 1$ day. With these numbers, (2.58) yields a typical size constraint for the emission regions in these objects of $R \lesssim 2.6 \times 10^{16} (\Delta t^{\mathrm{obs}}_{\mathrm{var}}/1\,\mathrm{day})(\delta/10)\,\mathrm{cm}$. In some extreme cases, variability on time scales less than 1 h has been observed, which will then yield $R \lesssim 1.1 \times 10^{15} (\Delta t^{\mathrm{obs}}_{\mathrm{var}}/1\,\mathrm{h})(\delta/10)\,\mathrm{cm}$.

2.3.2
Superluminal Motion

High-resolution radio observations (see Chapter 5) allow us to resolve individual emission components ("knots") within the relativistic jets of radio-loud AGN. In many of these objects, these components appear to move across the sky with speeds faster than the speed of light, in some extreme cases with apparent transverse (i.e., in the plane of the sky) speeds of $v_{\perp,\text{app}} \sim 50c$. The accepted explanation for this phenomenon was first proposed by [10]. It turns out that this is merely a consequence of the finite speed of light, and does not pose any contradiction to special relativity.

Let us assume, as illustrated in Figure 2.4, that the radio knot moves with a relativistic speed $v = c\beta_\Gamma$ along the jet, which is directed at a small angle θ with respect to our line-of-sight. Now, consider the apparent speed at which an observer on Earth would see this knot move across the sky between points A and B in Figure 2.4. Let l be the distance between A and B, then the knot will move from A to B in a time (measured in the rest frame of the AGN) $\Delta t = l/(\beta_\Gamma c)$. The projected distance which it moves across the sky, is given by $s = l \sin\theta = v\Delta t \sin\theta$. Now, consider that the light emitted from the knot at point A has to travel an extra distance $x = v\Delta t \cos\theta$ to reach the observer, compared to light emitted at point B. Hence, the observer will see the knot at point B not a time Δt later, but only a time $\Delta t_{\text{obs}} = \Delta t - x/c = \Delta t(1 - \beta_\Gamma \cos\theta)$. Hence, the apparent motion across the sky which the observer measures, is given by

$$v_{\perp,\text{app}} = \frac{s}{\Delta t_{\text{obs}}} = \frac{v \sin\theta}{(1 - \beta_\Gamma \cos\theta)} . \tag{2.59}$$

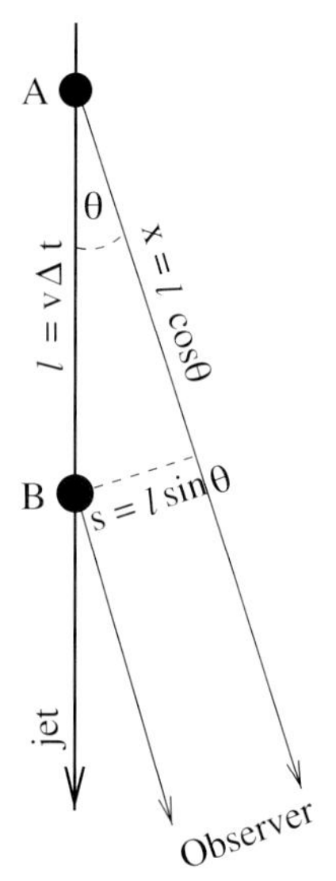

Figure 2.4 Illustration of the superluminal motion of radio knots in AGN.

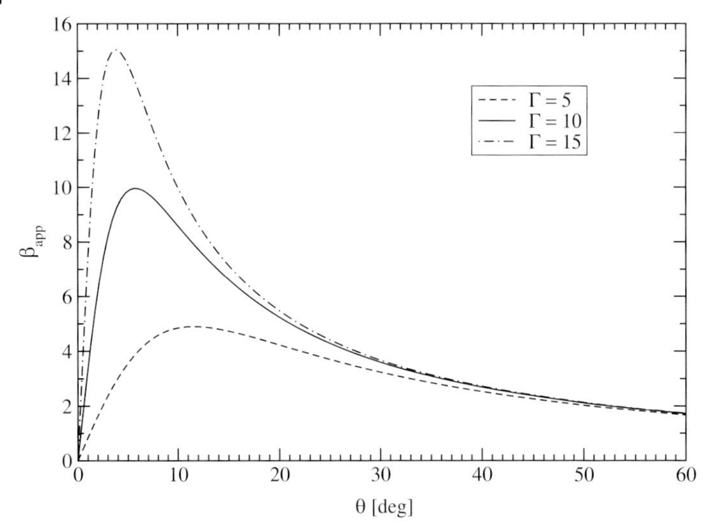

Figure 2.5 Dependence of the apparent transverse speed on the viewing angle θ.

Figure 2.5 shows the dependence of the apparent speed $v_{\perp,\text{app}}$ on the viewing angle θ for a few values of the bulk Lorentz factor Γ. For any given value of Γ, there is a preferred angle at which the apparent speed is maximum. It is straightforward to show that this happens (i.e., $dv_{\perp,\text{app}}/d\cos\theta = 0$) for

$$\cos\theta_{\text{sl}} = \beta_\Gamma \,. \tag{2.60}$$

This value of θ_{sl} is called the *superluminal angle* or *critical angle*. At this angle, $\sin\theta = \sqrt{1-\cos^2\theta_{\text{sl}}} = \sqrt{1-\beta_\Gamma^2} = 1/\Gamma$, and the apparent transverse speed is therefore

$$v_{\perp,\text{app}}(\theta_{\text{sl}}) = \frac{v/\Gamma}{(1-\beta_\Gamma^2)} = \Gamma v = c\beta_\Gamma\Gamma = c\sqrt{\Gamma^2-1}\,. \tag{2.61}$$

Hence, for any given value of Γ, the apparent superluminal speed can reach a value of $v_{\perp,\text{app}} \leq c\sqrt{\Gamma^2-1}$. This immediately yields a limit on the bulk Lorentz factor: if a jet knot with a superluminal speed of $\beta_{\perp,\text{app}}$ is observed, the bulk Lorentz factor must be at least

$$\Gamma \geq \sqrt{\beta_{\perp,\text{app}}^2+1}\,. \tag{2.62}$$

Conversely, one can find a limit on the viewing angle by realizing from (2.59) that for any given θ, the maximum apparent superluminal motion speed will be reached in the limit of $\beta_\Gamma \to 1$, that is, $\Gamma \to \infty$, as $\beta_{\perp,\text{app}}^{\max}(\theta) = \sin\theta/(1-\cos\theta)$. Setting this equal to the observed superluminal speed, β_{app}, we find a maximum viewing angle for which this superluminal speed can theoretically be achieved:

$$\cos\theta_{\max} = \frac{\beta_{\perp,\text{app}}^2-1}{\beta_{\perp,\text{app}}^2+1}\,. \tag{2.63}$$

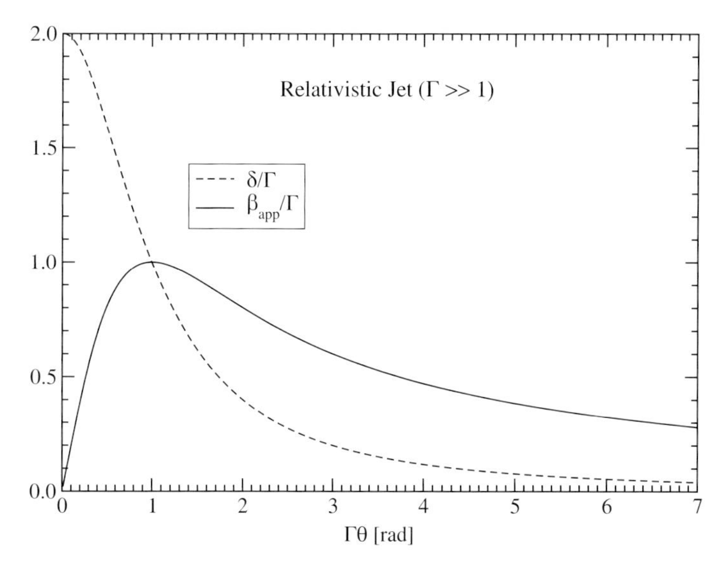

Figure 2.6 Dependence of the apparent transverse speed and the Doppler factor on the viewing angle θ for highly relativistic motion ($\Gamma \gg 1$). In this limit, the ratios δ/Γ and $\beta_{\perp,\mathrm{app}}/\Gamma$ depend only on the combination $\Gamma\theta$ (adapted from [11]).

For example, an apparent transverse superluminal speed of $\beta_{\perp,\mathrm{app}} = 10$ implies not only $\Gamma \geq 10.05$, but also a maximum observing angle of $\theta \leq 11.4°$.

For highly relativistic motion ($\Gamma \gg 1$), the apparent transverse speed $\beta_{\perp,\mathrm{app}}$ and the Doppler factor δ depend essentially only on the combination $\Gamma\theta$. This is illustrated in Figure 2.6, which shows that the maximum apparent speed occurs at $\Gamma\theta = 1$, and the maximum Doppler factor (for $\theta = 0$) is 2Γ.

2.3.3
Lorentz Factor and Viewing Angle Estimates

In the previous section, we derived lower and upper limits, respectively, on the bulk Lorentz factor and the viewing angle from observations of superluminal motion in relativistic jets, assuming no prior knowledge of the Lorentz factor Γ or the Doppler factor δ. In this section, we will derive actual estimates (instead of just limits) of the Lorentz factor and viewing angle, based either on the observation of both an approaching and a receding component in the jet, or independent knowledge of the Doppler boosting factor.

2.3.3.1 **Two-Sided Jets**

Here, we discuss the case in which both sides of a relativistic jet can be imaged in high-resolution radio observations. This is typically the case in sources classified as radio galaxies, in which we observe the radio jet at a relatively large viewing angle. Let us assume that we see an approaching jet component with a radio flux (at a fixed frequency) F_a, exhibiting apparent transverse (possibly superluminal) motion at a

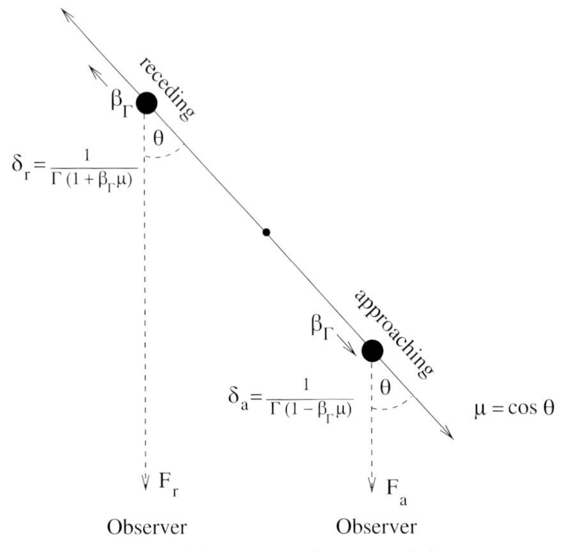

Figure 2.7 Sketch of the geometry for a two-sided jet.

speed $\beta_{\perp,\text{app}}$, as well as a component in the receding jet, from which we observe a flux F_r. The situation is illustrated in Figure 2.7.

A critical assumption for the following derivation is that the components are emitting the same radio luminosity isotropically in their respective rest frames, and they are moving at identical speeds in opposite directions along the jet. If that is the case, then we can write the fluxes in terms of the fluxes F_0 perceived in the comoving frames as

$$F_{\text{a,r}} = F_0 \delta_{\text{a,r}}^{3+\alpha} , \tag{2.64}$$

where α is the spectral index of the radio spectrum (see (2.42)), and the Doppler boosting factors are given by

$$\delta_\text{a} = \frac{1}{\Gamma(1-\beta_\Gamma \mu)} ,$$
$$\delta_\text{r} = \frac{1}{\Gamma(1+\beta_\Gamma \mu)} , \tag{2.65}$$

where μ is the cosine of the angle between the approaching jet and the line-of-sight. We can then combine the two boosting characteristics (2.64) to

$$\frac{F_\text{a}}{F_\text{r}} = \left(\frac{1+\beta_\Gamma \mu}{1-\beta_\Gamma \mu}\right)^{3+\alpha} \tag{2.66}$$

or

$$\frac{1+\beta_\Gamma \mu}{1-\beta_\Gamma \mu} = r , \tag{2.67}$$

where we define $r \equiv (F_a/F_r)^{1/(3+\alpha)}$. Combining (2.67) with (2.59) for the superluminal speed, we find, after some elementary algebra,

$$\beta_\Gamma = \frac{\sqrt{4\beta_{\perp,\mathrm{app}}^2 + (r-1)^2}}{r+1} \tag{2.68}$$

or

$$\Gamma = \frac{r+1}{2\sqrt{r - \beta_{\perp,\mathrm{app}}^2}} \tag{2.69}$$

and

$$\mu = \cos\theta = \frac{r-1}{\sqrt{4\beta_{\perp,\mathrm{app}}^2 + (r-1)^2}} \, . \tag{2.70}$$

These solutions agree with the expected outcome for a jet which we observe exactly at a right angle $\mu = 0$. In that case, the Doppler boosting effect will be identical for both knots, and therefore, $r = 1$. Under this viewing angle, the observed apparent transverse motion is identical to the actual speed of the knot, that is, $\beta_\Gamma = \beta_{\perp,\mathrm{app}}$.

We need to point out the caveat that the underlying assumption of equal intrinsic luminosity and speed of the approaching and receding component may be oversimplifying the problem. The superluminal components in extragalactic radio jets are typically resolved on physical scales of pc to kpc. Therefore, the difference between the light travel times from the approaching and the receding component, may be expected to be of the order of tens to thousands of years. In other words, we see the approaching component at a stage of its development that may be many years later than what we observe from the receding component. The assumption of identical speeds and intrinsic luminosities of both components can therefore, at best, only serve as a first guess.

2.3.3.2 One-Sided Jets

Due to the strong Doppler boosting effect, in many extragalactic radio jet sources only the approaching jet can be detected, because the counter-jet is strongly boosted out of our line-of-sight. However, even in those cases, under certain assumptions, Doppler factors can be estimated based on various arguments. For example, if the size of the emission region can be constrained through high-resolution radio imaging and/or variability arguments, one can estimate a limit on the Doppler factor under the assumption that the radiative energy output of the radiating particles in the emission region is not strongly dominated by Compton scattering their own synchrotron radiation. These concepts will be discussed in more detail in Chapters 3 and 5.

Let us therefore assume that we have independent knowledge of the Doppler factor δ, and observe superluminal motion as discussed in the previous section. In that case, we have two equations (2.59) and (2.30) with two unknowns (Γ and θ).

With some algebra, this system of equations can be solved to give

$$\Gamma = \frac{\beta^2_{\perp,\mathrm{app}} + \delta^2 + 1}{2\delta} \tag{2.71}$$

and

$$\theta = \arctan\left(\frac{2\beta_{\perp,\mathrm{app}}}{\beta^2_{\perp,\mathrm{app}} + \delta^2 - 1}\right) . \tag{2.72}$$

Applications of these estimates based on radio observations will be discussed in detail in Chapter 5.

This concludes our overview of special relativity and its effect on the observable properties of relativistic jets. We now have the necessary tools to understand the following introduction to radiation mechanisms in relativistic plasmas (Chapter 3), and the interpretation of observations of extragalactic jet sources in the following chapters.

References

1 Jackson, J.D. (1962) *Classical Electrodynamics*, Wiley & Sons, Inc.
2 Griffiths, D.J. (1981) *Introduction to Electrodynamics*, Pearson/Addison-Wesley.
3 Stephani, H. (1982) *Relativity*, Cambridge University Press.
4 Hartle, J.B. (2003) *Gravity: An Introduction to Einstein's General Relativity*, Addison Wesley.
5 Rindler, W. (2006) *Relativity*, Oxford University Press.
6 Narlikar, J. (2010) *An Introduction to Relativity*, Cambridge University Press.
7 Greisen, K. (1966) *Phys. Rev. Lett.*, **16**, 748.
8 Zatsepin, G.T. and Kuzmin, V.A. (1966) *JETP Lett.*, **4**, 78.
9 Sokolsky, P. (2008) *Mod. Phys. Lett. A*, **23**, 17.
10 Rees, M.J. (1966) *Nature*, **211**, 468.
11 Cohen, M. *et al.* (2007) *Astrophys. J.*, **658**, 232.

3
Radiation Processes

Markus Böttcher and Anita Reimer

In this chapter, we will introduce the fundamentals of radiation theory. We begin with some basic definitions of quantities needed to describe radiative transfer and introduce the radiative transfer equation and its solutions for a few special cases. The following sections will discuss the expressions for emission and absorption coefficients pertaining to individual radiation processes, the related energy-loss rates of the radiating particles, and the characteristics of the radiation spectra for nonthermal, relativistic particle distributions. We will then elaborate on the description of electromagnetic cascades, which are initiated by $\gamma\gamma$ pair production interactions of high-energy γ-rays or by photo-pion production interactions of ultrarelativistic protons with soft radiation fields. In our discussion of radiation mechanisms, we will focus on nonthermal radiation mechanisms relevant in the relativistic jets of AGN and GRBs. As the particle densities in those environments are believed to be very low ($n \lesssim 10^3\,\mathrm{cm}^{-3}$), any effects involving binary particle collisions (e.g., Coulomb scattering, bremsstrahlung, electron–positron pair annihilation) will not be discussed in this chapter. Also, as the emission regions in astrophysical jets are optically thin throughout most of the electromagnetic spectrum, we will not discuss any aspects of radiative diffusion in this chapter. The interested reader may find a more complete discussion of thermal and nonthermal emission mechanisms and radiative transfer phenomena in [1]. An excellent, in-depth discussion of various nonthermal radiation processes in relativistic outflows can also be found in [2].

3.1
Radiative Transfer: Definitions

This section introduces the basic definitions of quantities needed in the discussion of radiation mechanisms and radiative transfer relevant in the context of extragalactic jets.

Relativistic Jets from Active Galactic Nuclei, First Edition. Edited by M. Böttcher, D.E. Harris, H. Krawczynski.

3.1.1 Radiative Flux, Intensity, Energy Density

The *radiative flux* F was already introduced in the last chapter (2.38). More generally, it is defined as the amount of energy flowing through an area element dA in a time dt in the form of electromagnetic radiation,

$$F = \frac{d^2 E}{dA dt} \tag{3.1}$$

measured in $\mathrm{erg\,cm^{-2}\,s^{-1}}$. As the radiation usually consists of a continuous spectrum of frequencies, we define the differential flux per unit frequency as

$$F_\nu = \frac{d^3 E}{dA dt d\nu} . \tag{3.2}$$

A commonly used unit for the differential flux F_ν is Jy ("Jansky"), defined as

$$1\,\mathrm{Jy} = 10^{-23}\,\mathrm{erg\,cm^{-2} s^{-1}\,Hz^{-1}} . \tag{3.3}$$

In radio astronomy this is often also referred to as the *flux density*, S_ν. The radiative flux introduced above measures only the total net flux of energy through the area dA carried by photons traveling in *any direction*. However, the flux of photons may generally be different in different directions. A more specific quantity to measure the energy throughput per time *per solid angle* $d\Omega$ of arrival directions of photons, is called the *specific intensity* I_ν, defined as

$$I_\nu = \frac{d^4 E}{dA dt d\Omega d\nu} \tag{3.4}$$

measured in $\mathrm{erg\,cm^{-2}\,s^{-1}\,Hz^{-1}\,sr^{-1}}$. An average of the intensity over all solid angles is defined as the *mean intensity*, J_ν:

$$J_\nu = \frac{1}{4\pi} \int_{4\pi} I_\nu d\Omega . \tag{3.5}$$

We can relate radiative flux and specific intensity by integrating the specific intensity at the surface element dA over the solid angle of photon directions. However, we have to keep in mind that photons traveling at an angle θ with respect to the normal to dA, will travel at a reduced speed $c \cos\theta$ perpendicular to the surface area dA and will therefore contribute a correspondingly smaller fraction to the radiative flux:

$$F_\nu = \int_{4\pi} I_\nu \cos\theta \, d\Omega . \tag{3.6}$$

In the previous chapter, we also introduced the radiation energy density $u_\nu(\Omega)$ as

$$u_\nu(\Omega) = \frac{d^3 E}{dV d\nu d\Omega} . \tag{3.7}$$

Now, for any light ray moving in a direction specified by a solid angle Ω, we can define an area element dA perpendicular to that direction, and realize that in a time dt the light ray travels a distance cdt. Hence, the volume occupied by the bundle of light rays over the area dA is $dV = cdAdt$. Inserting this in (3.7), we see that

$$u_\nu(\Omega) = \frac{I_\nu}{c} \tag{3.8}$$

and

$$u_\nu = \frac{1}{c} \int\limits_{4\pi} I_\nu d\Omega = \frac{4\pi}{c} J_\nu . \tag{3.9}$$

Finally, we can relate the radiation pressure to the quantities introduced above. Recall that pressure P is force per unit area, force F is momentum change per time, and the momentum of photons is $p = E/c$, Therefore we may interpret the flux of momentum through a unit area dA as the radiation pressure of a photon flux in a direction perpendicular to dA, given by $dP_\nu = dF_\nu/c$. Now, considering an area dA' inclined at an angle θ with respect to the direction of the flux F_ν, then the component of the force normal to dA' (introducing an additional factor of $\cos\theta$) describes the contribution to the total radiation pressure, which is therefore given by

$$P_\nu = \frac{1}{c} \int\limits_{4\pi} I_\nu \cos^2\theta \, d\Omega . \tag{3.10}$$

The total radiation pressure is then $P = \int_0^\infty P_\nu d\nu$.

3.1.2
The Radiative Transfer Equation

The processes governing the transport of radiation through an astrophysical medium are, in general, emission, absorption, and scattering processes. In the context of the emission regions in relativistic jets, the probability of a photon to be scattered before escaping is, however, typically very small ($P \lesssim 10^{-6}$).[1] We will therefore focus on systems in which radiation transfer is governed by emission and absorption only.

The emission of photons by the various elementary processes discussed in the next section, is characterized by an *emission coefficient* j_ν, defined as the radiative energy produced per unit time, per unit volume, per unit frequency interval and per unit solid angle:

$$j_\nu = \frac{d^4 E}{dV dt d\nu d\Omega} . \tag{3.11}$$

1) This does not conflict with the hypothesis that Compton scattering might play a dominant role in the formation of the high-energy spectra of AGN and possibly also GRBs. While the scattered high-energy photons can carry a substantial (even dominant) fraction of the total energy output (transferred from relativistic electrons), their number is negligible compared to the number of available low-energy photons in the emission region.

A volume element of cross-section dA, so that $dV = dAds$, in which an emission process with emission coefficient j_ν is active, will then add to the intensity in a direction perpendicular to dA as

$$dI_\nu^{\text{em}} = j_\nu ds\ . \tag{3.12}$$

Radiation can also be absorbed by various processes. We describe the absorption of photons through an *absorption coefficient* α_ν. Obviously, the amount of absorbed radiation in length element ds depends on the amount of radiation passing through. α_ν is defined as the fraction of incident intensity which is absorbed per length element ds, and has a unit of cm^{-1}:

$$dI_\nu^{\text{abs}} = -I_\nu \alpha_\nu ds\ . \tag{3.13}$$

In the special case where we have a background radiation source with intensity I_ν^0, and a cloud of purely absorbing material of column length l between us and the source, the radiation transfer is described by

$$\frac{dI_\nu(s)}{ds} = -I_\nu \alpha_\nu(s) \tag{3.14}$$

with the boundary condition $I_\nu(0) = I_\nu^0$. Defining the *optical depth* τ_ν of the cloud as

$$\tau_\nu \equiv \int_0^l \alpha_\nu(s) ds\ , \tag{3.15}$$

(3.14) is solved by

$$I_\nu(l) = I_\nu^0 e^{-\tau_\nu}\ . \tag{3.16}$$

In the general case, the transfer of radiation is described by the *radiative transfer equation*

$$\frac{dI_\nu}{ds} = j_\nu - I_\nu \alpha_\nu\ . \tag{3.17}$$

Using $d\tau_\nu \equiv \alpha_\nu ds$, we can rewrite (3.17) as

$$\frac{dI_\nu}{d\tau_\nu} = S_\nu - I_\nu\ , \tag{3.18}$$

where we have defined the *source function* S_ν as

$$S_\nu \equiv \frac{j_\nu}{\alpha_\nu}\ . \tag{3.19}$$

The radiative transfer equation (3.18) has a rather simple solution if the source function S_ν is constant throughout the region under consideration. Let us denote the flux incident on the region (at $s = \tau_\nu = 0$) as I_ν^0, then (3.18) is solved by

$$I_\nu(\tau_\nu) = I_\nu^0 e^{-\tau_\nu} + S_\nu \left(1 - e^{-\tau_\nu}\right)\ . \tag{3.20}$$

In the applications in the remainder of this book, we will be mostly interested in well-localized emission regions with no significant background radiation source ($I_\nu^0 \approx 0$). Let us characterize those regions by an average column length l along our line-of-sight through the region and a corresponding optical depth τ_ν. Then, we can distinguish two limiting cases, depending on the optical depth, namely (i) for the optically thin regime, $\tau_\nu \ll 1$,

$$I_\nu(\tau_\nu \ll 1) = S_\nu \left(1 - e^{-\tau_\nu}\right) \approx S_\nu \tau_\nu = \frac{j_\nu}{\alpha_\nu} \alpha_\nu l = j_\nu l \,, \tag{3.21}$$

and (ii) for the optically thick regime, $\tau_\nu \gg 1$,

$$I_\nu(\tau_\nu \gg 1) = S_\nu \left(1 - e^{-\tau_\nu}\right) \approx S_\nu \,. \tag{3.22}$$

The functional form of the emission and absorption coefficients, j_ν and α_ν, will depend on the dominant emission processes, the energy distribution of radiating particles and other parameters characterizing the emission region, such as the magnetic field and the characteristics of the photon field that may be scattered by the radiating particles. In the following section, we will describe the evaluation of the emission and absorption coefficients pertaining to individual emission mechanisms, the rate of energy loss suffered by the radiating particles as a result of the emission of radiation, as well as the characteristics of the emanating radiation spectra.

3.2 Nonthermal Emission Processes

In this section, we will derive the emission and absorption coefficients and particle energy-loss rates. For illustrative purposes, we will show examples of characteristic photon spectra due to the individual emission processes, assuming a simple power-law spectrum of the nonthermal particles with power-law index p and low- and high-energy cutoffs, γ_1 and γ_2:

$$n(\gamma) = n_0 \gamma^{-p} \,, \quad \gamma_1 \le \gamma \le \gamma_2 \,. \tag{3.23}$$

Here $n(\gamma)$ is the number of particles per unit volume in a Lorentz-factor interval $[\gamma, \gamma + d\gamma]$. Such spectra may result from various acceleration mechanisms, as discussed in Chapter 9, although it is now well established that the particle spectra in astrophysical high-energy sources are more complicated than (3.23), most likely exhibiting spectral breaks, quasi-exponential cutoffs and a nonnegligible fraction of thermal particles in addition to the nonthermal distribution. In this section, we use the power-law representation (3.23) only to illustrate the generic shape of the resulting radiation spectra.

In most cases, we will only quote the expressions for the calculation of emission and absorption coefficients as their detailed derivation is beyond the scope of this book. Derivations of most of the expressions for emissivities, absorption coefficients, and particle energy-loss rates introduced here, can be found, for example, in [1, 2].

3.2.1 Synchrotron Radiation

Any particle moving through a magnetized region with magnetic field B, will follow a spiral trajectory. As this is an accelerated motion, classical electrodynamics predicts that radiation is emitted as a consequence. In the case of nonrelativistic particles, the resulting radiation is called *cyclotron* radiation, concentrated around the characteristic cyclotron frequency (and its harmonics). In the case of relativistic particles, which is of interest in the context of relativistic jets, a broad spectrum of *synchrotron radiation* results. In this section, we will discuss the particle energy-loss rates due to synchrotron emission as well as various expressions and approximations for the synchrotron emissivity and the synchrotron self-absorption coefficient. Our treatment will be based on results from classical electrodynamics only, and not consider any quantum mechanical effects.

3.2.1.1 Particle Energy-Loss Rate

Consider a particle with mass m, electric charge q and Lorentz factor γ (and, hence, total energy $E = \gamma m c^2$) and speed βc, moving at a pitch angle ψ with respect to an ordered magnetic field B (i.e., $\boldsymbol{B} \cdot \boldsymbol{\beta} = B\beta \cos\psi$). Classical electrodynamics (e.g., [3, 4]; see also [2]) tells us that this particle will lose energy due to its gyrational motion in the magnetic field as

$$\left(\frac{dE}{dt}\right)_{\rm sy}(\psi) = -\frac{16\pi c}{3}\left(\frac{q^2}{mc^2}\right)^2 u_B \beta^2\gamma^2 \sin^2\psi\,, \tag{3.24}$$

where $u_B = B^2/(8\pi)$ is the energy density in the magnetic field. In most application to radiation models in relativistic jets (and, in fact, most astrophysical applications), it is often assumed that there exists a mechanism which rapidly scatters the radiating particles in random directions. As a result, the distribution of particles will be randomly distributed in pitch angles with respect to the magnetic field. If this pitch-angle scattering happens on a time scale much shorter than the synchrotron energy-loss time scale, it is useful to average the energy-loss rate (3.24) over pitch angles, that is,

$$\left(\frac{dE}{dt}\right)_{\rm sy} = \frac{1}{4\pi}\int_{4\pi} d\Omega \left(\frac{dE}{dt}\right)_{\rm sy}(\psi) = -\frac{32\pi c}{9}\left(\frac{q^2}{mc^2}\right)^2 u_B\beta^2\gamma^2\,. \tag{3.25}$$

Introducing the Thomson cross-section

$$\sigma_{\rm T} = \frac{8\pi}{3} r_{\rm e}^2 = \frac{8\pi}{3}\left(\frac{e^2}{m_{\rm e}c^2}\right)^2 \tag{3.26}$$

we may write the energy-loss rate in terms of a change in Lorentz factor ($dE/dt = mc^2 d\gamma/dt$) as

$$\left(\frac{d\gamma}{dt}\right)_{\text{sy}} = -\frac{4}{3} c\sigma_{\text{T}} \frac{u_B}{m_e c^2} Z^4 \left(\frac{m_e}{m}\right)^3 \beta^2 \gamma^2 , \tag{3.27}$$

where Z is the particle's charge in units of the elementary charge e.

Equation (3.27) shows that the synchrotron energy loss of relativistic particles scales with mass as $dE/dt \propto -m^{-2}$ or $d\gamma/dt \propto -m^{-3}$. This implies, for example, that for a proton to suffer the same energy loss dE/dt as an electron with Lorentz factor γ_e, the proton would have to have a Lorentz factor of $(1836)^2 \gamma_e$, and its energy would have to be a factor $(1836)^3 \approx 6.2 \times 10^9$ higher. For this reason, electrons are the most efficient radiators. However, in order to radiate this efficiently, they need to be accelerated to ultrarelativistic energies, which becomes problematic just because of their very rapid radiative energy losses (see Chapter 9). Therefore, heavier particles like protons can be accelerated much more easily, but they need to acquire extreme energies (see Section 3.2.4) in order to produce an appreciable radiative output.

3.2.1.2 **Synchrotron Emissivity**

The synchrotron emission coefficient j_ν will (like most other emission coefficients discussed in this chapter) be written in terms of the radiative output of a single radiating particle, $P_\nu(\gamma)$. Given this single-particle radiative power and an arbitrary particle distribution $n(\gamma)$ (e.g., (3.23)), we find the emission coefficient as

$$j_\nu = \frac{1}{4\pi} \int_1^\infty d\gamma\, n(\gamma) P_\nu(\gamma) . \tag{3.28}$$

A detailed derivation of the synchrotron radiative power of a particle with energy γmc^2 in a magnetic field B, moving at a pitch angle ψ, summed over polarizations, is given by [1]:

$$P_\nu^{\text{sy}}(\gamma; \psi) = \frac{\sqrt{3} q^3 B}{mc^2} \sin\psi\, F(x) , \tag{3.29}$$

where $x = \nu/\nu_\psi$ with

$$\nu_\psi = \frac{3qB}{4\pi mc} \gamma^2 \sin\psi \tag{3.30}$$

and

$$F(x) = x \int_x^\infty K_{5/3}(\xi) d\xi . \tag{3.31}$$

Here, $K_{5/3}$ is the modified Bessel function of the second kind (a.k.a. McDonald function) of order 5/3.

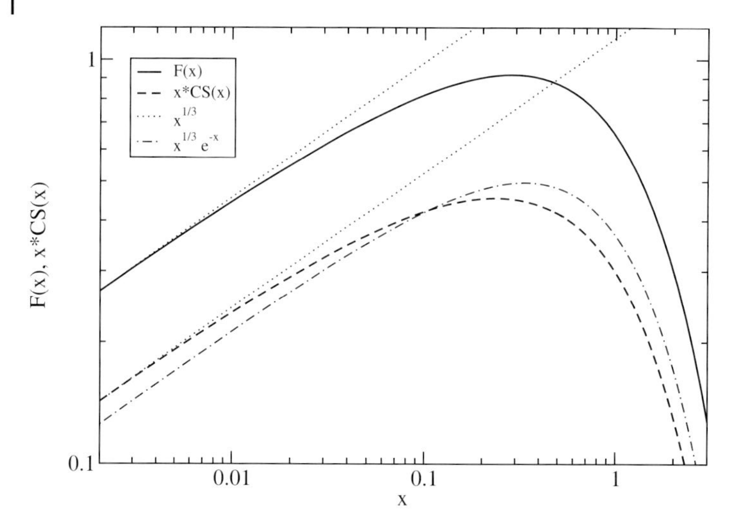

Figure 3.1 The functions $F(x)$ and $x\,C\,S(x)$ describing the spectral synchrotron power of a single particle. Also plotted is the low-frequency asymptote $x^{1/3}$ and the global approximation $x^{1/3}e^{-x}$.

Figure 3.1 illustrates that the spectral power $F(x)$ has a peak at $x \approx 0.29$, and asymptotes as

$$F(x) \propto \begin{cases} x^{1/3} & \text{for} \quad x \ll 1 \\ x^{1/2}e^{-x} & \text{for} \quad x \gg 1 \,. \end{cases} \tag{3.32}$$

In the case of an isotropic pitch-angle distribution of the radiating particles, one can evaluate the angle-averaged radiative synchrotron output by integrating (3.29) over all angles,

$$P_\nu^{\text{sy}}(\gamma) = \frac{1}{4\pi}\int\limits_{4\pi} P_\nu^{\text{sy}}(\gamma;\psi)\,d\Omega \,. \tag{3.33}$$

This integration has been evaluated by [5] and yields

$$P_\nu^{\text{sy}}(\gamma) = \frac{\sqrt{3}\pi q^3 B}{2mc^2}\,x\,C\,S(x)\,, \tag{3.34}$$

where $x = \nu/\nu_{\text{c}}$ with the critical frequency

$$\nu_{\text{c}} = \frac{3qB}{4\pi mc}\gamma^2 = 4.2\times 10^6\,B_G\,\gamma^2 Z\frac{m_{\text{e}}}{m}\,\text{Hz}\,. \tag{3.35}$$

The function $C\,S(x)$ is given in terms of Whittaker functions:

$$C\,S(x) = W_{0,\frac{4}{3}}(x)\,W_{0,\frac{1}{3}}(x) - W_{\frac{1}{2},\frac{5}{6}}(x)\,W_{-\frac{1}{2},\frac{5}{6}}(x)\,. \tag{3.36}$$

The function $x\,C\,S(x)$, which describes the spectral synchrotron power of a single electron, is plotted along with $F(x)$ in Figure 3.1. Just like $F(x)$, it exhibits a low-frequency asymptote $\propto x^{1/3}$ and an exponential cutoff at high frequencies. The peak spectral output occurs around the critical frequency ν_{c}, given by (3.35).

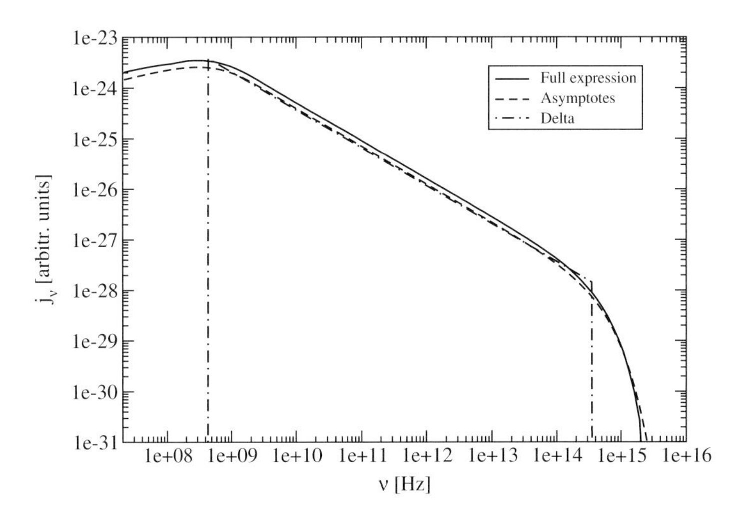

Figure 3.2 Optically thin synchrotron spectra from a power-law distribution of electrons with index $p = 2.5$ and cutoffs $\gamma_1 = 10$ and $\gamma_2 = 10^4$ for a magnetic field of $B = 1\,\text{G}$. The figure compares the full expression (3.34), the $\nu^{1/3} e^{-\nu/\nu_c}$ approximation (3.38), and the δ function approximation (3.40).

Figure 3.2 shows an optically thin synchrotron spectrum for a power-law distribution of electrons with index $p = 2.5$ and low- and high-energy cutoffs $\gamma_1 = 10$ and $\gamma_2 = 10^4$. The magnetic field is $B = 1\,\text{G}$. The power-law index α_{sy} of the photon spectrum is related to the index p of the electron spectrum through

$$\alpha_{\text{sy}} = \frac{p-1}{2} . \tag{3.37}$$

The evaluation of the full synchrotron emissivity as determined by (3.34) can be quite time-consuming for large-scale calculations. For most practical applications, however, simple approximations of the synchrotron emissivity can be used. The simplest approximation is based on the δ-function representation of the emissivity at the maximum synchrotron photon output for any given electron energy γ at ν_c according to (3.35),

$$P_\nu^{\text{sy},\delta}(\gamma) = \frac{32\pi}{9}\left(\frac{q^2}{mc^2}\right)^2 u_B \beta^2 \gamma^2 \delta(\nu - \nu_c) . \tag{3.38}$$

Using this approximation in the expression for the synchrotron emission coefficient (3.28) reduces to

$$j_\nu^{\text{sy},\delta} = \frac{4}{9}\left(\frac{q^2}{mc^2}\right)^2 u_B \nu_0^{-3/2} \nu^{1/2} n\left(\sqrt{\frac{\nu}{\nu_0}}\right) , \tag{3.39}$$

where ν_0 is defined through $\nu_c = \nu_0 \gamma^2$. For a power-law electron distribution with index p this immediately reproduces the synchrotron photon power-law index

$\alpha_{sy} = (p-1)/2$. The delta function approximation is compared to the full expression for the synchrotron emissivity in Figure 3.2. It shows that the representation is quite accurate over most of the straight power-law portion of the spectrum, but it produces artificially sharp low- and high-energy cutoffs at $\nu_{cut\,1,2} = \nu_0 \gamma_{1,2}^2$.

A somewhat more sophisticated approximation is given by a simple multiplication of the asymptotic behaviors of P_ν^{sy} for $\nu \ll \nu_c$ and $\nu \gg \nu_c$, as $P_\nu^{sy}(\gamma) \propto \nu^{1/3} e^{-\nu/\nu_c}$. The proper normalization to the total energy-loss rate (3.25) then yields

$$P_\nu^{sy,\,asympt}(\gamma) = \frac{32\pi c}{9\Gamma(4/3)} \left(\frac{q^2}{mc^2}\right)^2 u_B \beta^2 \gamma^2 \frac{\nu^{1/3}}{\nu_c^{4/3}} e^{-\nu/\nu_c} \,. \tag{3.40}$$

This asymptotic approximation is also compared to the full expression and the δ function approximation in Figure 3.2, and provides a rather accurate description of the synchrotron spectrum at all frequencies.

3.2.1.3 **Synchrotron Self-Absorption**

The synchrotron process can also result in the absorption of photons by relativistic electrons in a magnetic field. Under the assumption that the energy of the absorbed photon is much smaller than the energy of the particle which it interacts with, $h\nu \ll E$, the absorption coefficient associated with any radiation process can be evaluated as (e.g., [1])

$$\alpha_\nu = -\frac{1}{8\pi m\nu^2} \int_1^\infty d\gamma\, P_\nu(\gamma) \gamma^2 \frac{\partial}{\partial\gamma} \left(\frac{n[\gamma]}{\gamma^2}\right) . \tag{3.41}$$

In the case of a power-law distribution of particles with index p, this reduces to

$$\alpha_\nu^{PL} = -\frac{(p+2)}{8\pi m\nu^2} \int_1^\infty d\gamma\, P_\nu(\gamma) \frac{n(\gamma)}{\gamma} \,. \tag{3.42}$$

For illustration purposes, one can easily carry out the integration in (3.42) using the δ function approximation for the synchrotron power $P_\nu^{sy,\delta}$ from (3.38) to obtain

$$\alpha_\nu^\delta = \frac{(p+2)}{18\pi m} \left(\frac{3q}{4\pi mc}\right)^{\frac{p-2}{2}} \left(\frac{q^2}{mc^2}\right)^2 B^{\frac{p+2}{2}} \nu^{-\frac{p+4}{2}} \,. \tag{3.43}$$

The functional dependence on frequency, $\nu^{-\frac{p+4}{2}}$, and magnetic field, $B^{\frac{p+2}{2}}$, is the same when using the full expression (3.34) for the synchrotron power. It shows that for any nonthermal electron distribution, the opacity to synchrotron self-absorption rapidly increases with decreasing frequency. Therefore, nonthermal synchrotron sources tend to be optically thick below a critical break frequency ν_{SSA}, also called the *synchrotron self-absorption frequency*, where $\tau_{\nu_{SSA}} = R\alpha_{\nu_{SSA}} = 1$. The optically thick spectrum observed below the synchrotron self-absorption frequency is given, according to (3.22), by the source function, $S_\nu = j_\nu/\alpha_\nu$. If ν_{SSA} lies within

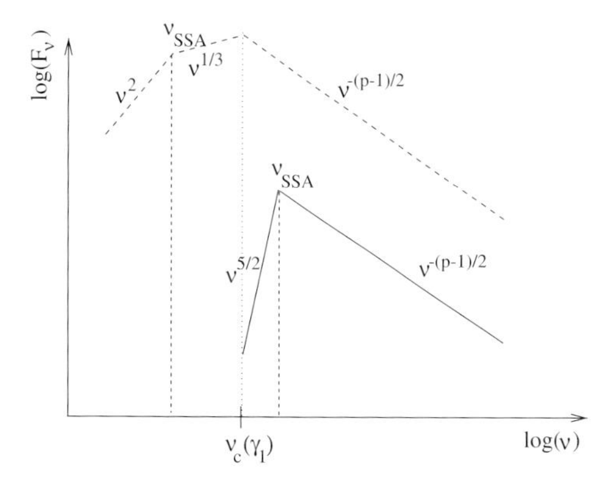

Figure 3.3 Synchrotron spectra from a power-law distribution of electrons. If the synchrotron self-absorption frequency is below the critical frequency of the lowest-energy electrons, $\nu_{\text{SSA}} < \nu_{\text{c}}(\gamma_1)$, the optically thick (low-frequency) branch of the spectrum has a ν^2 slope, turning into optically thin low-frequency synchrotron emission $\nu^{1/3}$ between ν_{SSA} and $\nu_{\text{c}}(\gamma_1)$. If $\nu_{\text{SSA}} > \nu_{\text{c}}(\gamma_1)$, the optically thick (low-frequency) part of the spectrum is a $\nu^{5/2}$ power law. In both cases, the optically thin high-frequency part of the spectrum is described by the standard slope $\nu^{-(p-1)/2}$.

the range of frequencies in which the power-law representation by $\alpha_{\text{sy}} = (p-1)/2$ holds (i.e., $\nu_{\text{SSA}} > \nu_{\text{c}}[\gamma_1]$), then the optically thick spectrum will be given by

$$S_\nu \Big|_{\nu_{\text{SSA}} > \nu_{\text{c}}(\gamma_1)} = \frac{j_\nu}{\alpha_\nu} \Big|_{\nu_{\text{SSA}} > \nu_{\text{c}}(\gamma_1)} \propto \nu^{5/2} . \tag{3.44}$$

However, if the synchrotron self-absorption frequency is below the $(p-1)/2$ power-law range of the synchrotron spectrum (i.e., $\nu_{\text{SSA}} < \nu_{\text{c}}(\gamma_1)$), the spectral power per electron is well approximated by the asymptotic $P_\nu(\gamma) \propto \nu^{1/3}$ for all γ with $n(\gamma) > 0$. Therefore, $j_\nu \propto \nu^{1/3}$. In this limit, $\alpha_\nu \propto \nu^{-5/3}$, and the source function is given by

$$S_\nu \Big|_{\nu_{\text{SSA}} < \nu_{\text{c}}(\gamma_1)} \propto \nu^2 . \tag{3.45}$$

The different cases of optically thick and thin synchrotron spectra are illustrated in Figure 3.3.

3.2.2 Compton Scattering

In this section we will discuss the effects of Compton scattering of various radiation fields off relativistic electrons in extragalactic jets. We will introduce several approximations to the Klein–Nishina cross-section, which are appropriate in different regimes, evaluate the electron energy-loss rates, and the spectral characteristics of the photon spectra resulting from Compton scattering. As mentioned at the beginning of this chapter, the Compton scattering optical depth of the emission regions in extragalactic jets is typically very small ($\tau_C = n R \sigma_T \lesssim 10^{-5}$). Therefore, multiple scatterings of any individual photon are very unlikely, and the radiative diffusion aspects related to multiple Compton scatterings will not be discussed in this chapter. The interested reader may find detailed discussions of these aspects in [1, 6].

In the discussion of high-energy emission processes, it will be convenient to normalize photon energies in units of the electron rest-mass energy $m_e c^2$. We therefore define

$$\epsilon \equiv \frac{h\nu}{m_e c^2} , \tag{3.46}$$

which will be used as our photon energy variable throughout the remainder of this chapter as well as Section 3.2.3 on $\gamma\gamma$ absorption and pair production.

3.2.2.1 The Compton Cross-Section

The Compton scattering between an electron of energy γ and a photon of energy ϵ is determined by the Compton cross-section, which is also known as the Klein–Nishina cross-section. The most convenient frame to evaluate the Klein–Nishina cross-section is in the rest frame of the electron before scattering. In the following, we denote quantities measured in the electron rest frame by primed symbols. Defining the energy of the scattered photon as ϵ'_s, and the scattering angle χ' (see Figure 3.4), the differential Klein–Nishina cross-section is calculated in quantum electrodynamics as [7]:

$$\frac{d\sigma_C}{d\Omega'_s d\epsilon'_s} = \frac{r_e^2}{2}\left(\frac{\epsilon'_s}{\epsilon'}\right)^2 \left(\frac{\epsilon'_s}{\epsilon'} + \frac{\epsilon'}{\epsilon'_s} - \sin^2\chi'\right) \delta\left(\epsilon'_s - \frac{\epsilon'}{1+\epsilon'[1-\cos\chi']}\right) , \tag{3.47}$$

where r_e is the classical electron radius (see (3.26)), $d\Omega'_s = d\cos\chi' d\phi'_s$ is a solid angle element of the direction of motion of the scattered photon, and the δ function results from energy and momentum conservation.

The total Compton cross-section can be found by integrating the Klein–Nishina cross-section (3.47) over all scattered photon energies ϵ'_s and directions corresponding to solid angles Ω'_s [8]:

$$\sigma_C(\epsilon') = \frac{\pi r_e^2}{\epsilon'^2}\left[4 + \frac{2\epsilon'^2(1+\epsilon')}{(1+2\epsilon')^2} + \frac{\epsilon'^2 - 2\epsilon' - 2}{\epsilon'} \ln(1+2\epsilon')\right] . \tag{3.48}$$

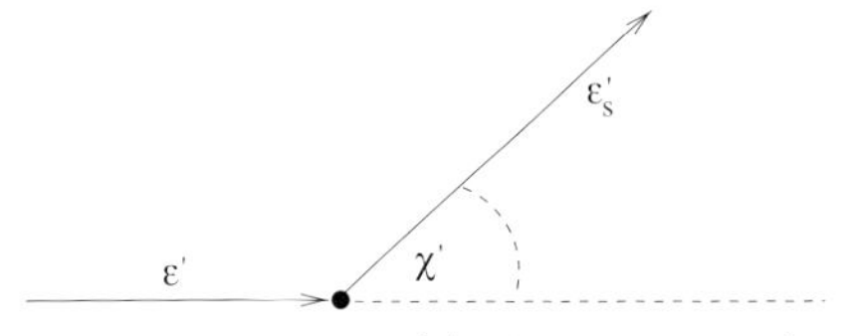

Figure 3.4 Geometry of the Compton scattering event in the electron rest frame.

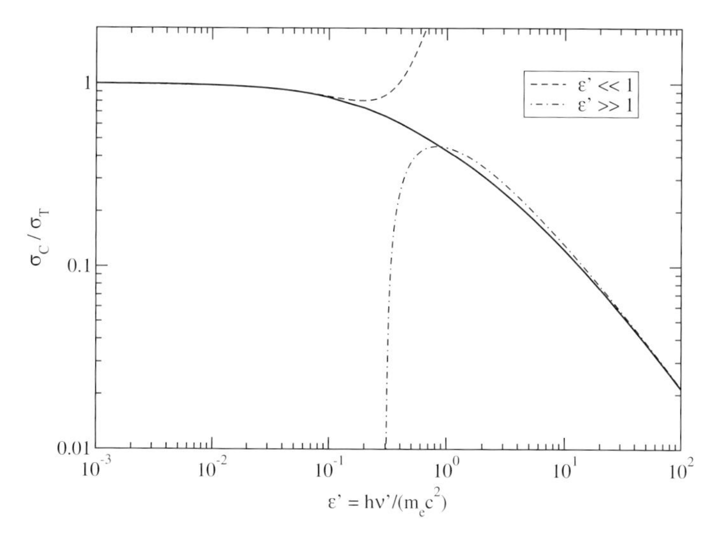

Figure 3.5 The Compton cross-section as a function of photon energy ϵ' in the electron rest frame. Also shown are the asymptotes for $\epsilon' \ll 1$ (dashed) and $\epsilon' \gg 1$ (dot-dashed).

The dependence of σ_{C} on ϵ' is illustrated in Figure 3.5. It shows that for small values of $\epsilon' \ll 1$, the cross-section assumes a constant value at the Thomson cross-section σ_{T} as introduced in (3.26). This energy regime is called the *Thomson regime*. The δ function in (3.47) then implies that in the rest frame of the electron, the energy of the scattered photon ϵ_s' is almost identical to the incident photon energy ϵ' in the Thomson regime. Therefore, Compton scattering in the Thomson regime can be considered elastic, and the recoil on the electron is negligible.

However, for large values of ϵ', the transfer of energy from the photon to the electron (in the electron rest frame) can become substantial. This goes along with a reduction of the cross-section. The limit $\epsilon' \gg 1$ is called the *Klein–Nishina limit*. In the two limiting cases for ϵ', the Compton cross-section can be approximated as

$$\sigma_{\mathrm{C}}(\epsilon') \approx \begin{cases} \sigma_{\mathrm{T}}\left(1 - 2\epsilon' + \frac{26}{5}\epsilon'^2\right) & \text{for} \quad \epsilon' \ll 1 \\ \frac{3}{8}\frac{\sigma_{\mathrm{T}}}{\epsilon'}\left(\ln[2\epsilon'] + \frac{1}{2}\right) & \text{for} \quad \epsilon' \gg 1\,. \end{cases} \tag{3.49}$$

Both asymptotes are compared to the full expression for the total Compton cross-section in Figure 3.5.

While the electron rest frame is the most convenient frame to evaluate the Compton cross-section, in the applications we are interested in, we will need to consider the scattering event between a relativistic electron with energy γ (and normalized velocity β) and a photon field, which are both specified in a certain reference frame, which we call the *laboratory frame*. In applications to high-energy emission from relativistic jets, this will usually be the rest frame of the emission region. Let us denote quantities in the laboratory frame by unprimed symbols.

Figure 3.6 illustrates the definition of angles and their cosines in the two reference frames. From those definitions and the Lorentz transformations of photon

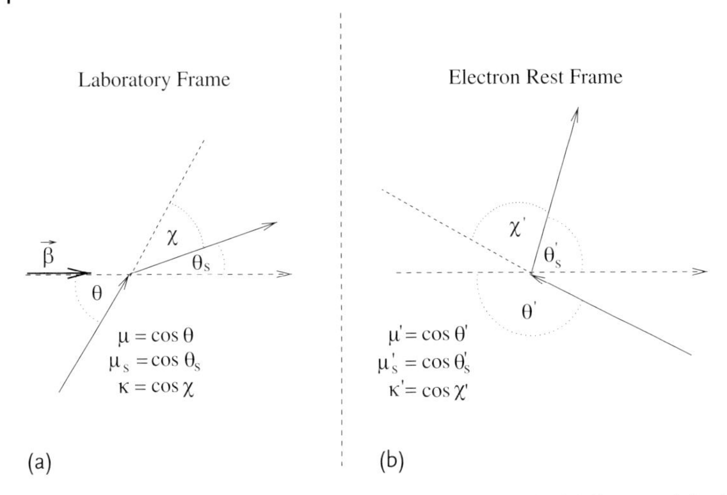

Figure 3.6 Geometry of a Compton scattering event in (a) the lab frame and (b) the electron rest frame.

energies derived in Section 2.2, we find that the electron rest frame photon energy ϵ' is related to the lab frame energy through

$$\epsilon' = \epsilon\gamma(1 - \beta\mu) \tag{3.50}$$

while the scattered photon energy ϵ_s in the lab frame is related to electron rest frame quantities by

$$\epsilon_s = \epsilon'_s\gamma(1 + \beta\mu'_s) \, . \tag{3.51}$$

Now, in the Thomson regime, the scattering will be elastic in the electron rest frame, that is, $\epsilon'_s \approx \epsilon'$. Therefore, in the Lab frame, the scattered electron will end up with an energy $\epsilon_s \sim \gamma^2\epsilon$. Thus, when relativistic electrons scatter low-energy photons, energy is transferred from the electron to the photon. This is often referred to as *inverse-Compton scattering.*

In order to calculate photon spectra and energy-loss rates due to Compton scattering off relativistic electrons, we will now need to transform the Klein–Nishina cross-section (3.47) from the electron rest frame to the lab frame. To do this, we first realize that the cross-section $d\sigma$ is an area perpendicular to the direction of motion of the electron and therefore remains invariant. Thus, we only need to account for the transformations of $d\Omega_s$ and $d\epsilon_s$. From Section 2.2, we see that

$$\frac{d\Omega'_s}{d\Omega_s} = \delta_s^2 \, , \tag{3.52}$$

and

$$\frac{d\epsilon'_s}{d\epsilon_s} = \frac{\epsilon'_s}{\epsilon_s} = \delta_s^{-1} \, , \tag{3.53}$$

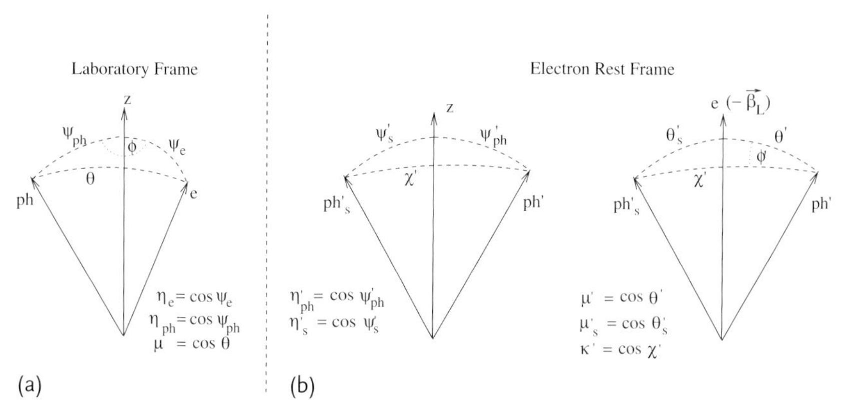

Figure 3.7 Definition of angles for transformation between (a) laboratory and (b) electron rest frames.

where $\delta_s \equiv 1/(\gamma[1-\beta\mu_s])$. Hence, $d\Omega'_s d\epsilon'_s = (\epsilon_s/\epsilon'_s) d\Omega_s d\epsilon_s$ and

$$\frac{d\sigma_C}{d\Omega_s d\epsilon_s} = \frac{\epsilon_s}{\epsilon'_s}\frac{d\sigma_C}{d\Omega'_s d\epsilon'_s} . \tag{3.54}$$

In order to evaluate the cross-section in the electron rest frame, we will, in addition to (3.50) and (3.51), also need the angle transformations

$$\mu'_s = \frac{\mu_s - \beta}{1-\beta\mu_s} ,$$
$$\mu' = \frac{\mu-\beta}{1-\beta\mu} . \tag{3.55}$$

Finally, we will want to evaluate the kinematics of the scattering event and the direction of the scattered radiation in the laboratory frame, in which the electron and the incoming photon have arbitrary directions of motion, specified by angles ψ_e and ψ_{ph} with respect to a z-axis in the laboratory frame, as illustrated in Figure 3.7. Therefore, we need to be able to calculate the angle cosines μ, μ_s, and κ (and their counterparts in the electron rest frame) from angles defined in the laboratory frame. Using spherical trigonometry, we find, for example,

$$\mu = \eta_{ph}\eta_e + \sin\psi_{ph}\sin\psi_e\cos\phi ,$$
$$\mu'_s = \kappa'\mu' + \sin\chi'\sin\theta'\cos\phi' . \tag{3.56}$$

In the case of Compton scattering by ultrarelativistic electrons ($\gamma \gg 1$, $\beta \to 1$), the relativistic aberration of photon paths between the laboratory and the electron rest frame becomes extreme. As discussed in Section 2.2, an electron traveling with a large Lorentz factor γ will "see" half of the photons of an isotropic radiation field coming from within a narrow cone of opening angle $1/\gamma$. Therefore, in the electron rest frame, $\mu' \approx -1$ for almost all photons. This is the basis for the *head-on approximation* to the Compton cross-section [2, 9, 10]. Since the Compton cross-section in the electron rest frame has only a weak dependence on the scattering angle κ' (see (3.47)), the angular characteristics of the scattered photons will also be dominated by relativistic aberration, resulting in a photon distribution strongly peaked along the direction of motion of the electron before scattering. This can also be seen in the dependence of the transformed Compton cross-section (3.54) which contains a factor $(\epsilon_s/\epsilon'_s) = \delta_s$, which is strongly peaked in the forward direction (see, e.g., Figure 2.6) for $\gamma \gg 1$. Hence, essentially all scattered photons will travel in the direction of the incoming electron, that is, $\Omega_s \approx \Omega_e$, and the Compton cross-section can be approximated as

$$\left.\frac{d\sigma_C}{d\Omega_s d\epsilon_s}\right|_{\text{head-on}} = \delta(\Omega_s - \Omega_e)\frac{d\sigma_C}{d\epsilon_s} , \tag{3.57}$$

where

$$\frac{d\sigma_C}{d\epsilon_s} = \frac{\pi r_e^2}{\gamma\epsilon'}\left[y + \frac{1}{y} - \frac{2\epsilon_s}{\gamma\epsilon' y} + \left(\frac{\epsilon_s}{\gamma\epsilon' y}\right)^2\right] H\left(\epsilon_s; \frac{\epsilon'}{2\gamma}, \frac{2\gamma\epsilon'}{1+2\epsilon'}\right) \tag{3.58}$$

with

$$y = 1 - \frac{\epsilon_s}{\gamma} \tag{3.59}$$

and $\epsilon' = \gamma\epsilon(1-\beta\mu)$ (see (3.50)), and the Heaviside function is defined as

$$H(x; a, b) = \begin{cases} 1 & \text{if} \quad a \le x \le b \\ 0 & \text{otherwise} . \end{cases} \tag{3.60}$$

If all scattering occurs in the Thomson regime, $\gamma\epsilon \ll 1$, while still all electrons are highly relativistic, $\gamma \gg 1$, then the scattering in the electron rest frame will be elastic, that is, $\epsilon'_s \approx \epsilon_s$, and relativistic aberration ($\mu_s \approx 1$) leads to $\epsilon_s \approx \gamma\epsilon'_s$. Therefore, $\epsilon_s \approx \gamma^2\epsilon(1-\beta\mu)$, and the Compton cross-section can be simplified further to yield a *delta function approximation* similar to (3.38) for synchrotron emission:

$$\left.\frac{d\sigma_C}{d\Omega_s d\epsilon_s}\right|_{\delta} = \sigma_T\delta(\Omega_s - \Omega_e)\delta(\epsilon_s - \gamma^2\epsilon[1-\beta\mu]) . \tag{3.61}$$

3.2.2.2 **Electron Energy Losses**

Generally, the energy-loss rate for an electron of energy γ traveling at an angle $\psi_e = \cos^{-1}\eta_e$ with respect to the z-axis and undergoing Compton scattering on a radiation field with photon distribution $n_{\text{ph}}(\epsilon, \Omega_{\text{ph}})$, can be calculated as

$$-\left(\frac{d\gamma}{dt}\right)_{\text{C}}(\eta_{\text{e}}) = c\int_{4\pi} d\Omega_{\text{ph}} \int_0^\infty d\epsilon (1-\beta\mu) n_{\text{ph}}(\epsilon, \Omega_{\text{ph}}) \times \int_{4\pi} d\Omega_{\text{s}} \int_0^\infty d\epsilon_{\text{s}} \frac{d\sigma}{d\epsilon_{\text{s}} d\Omega_{\text{s}}} (\epsilon_{\text{s}} - \epsilon)$$
$$= c\int_{4\pi} d\Omega_{\text{ph}} \int_0^\infty d\epsilon (1-\beta\mu) n_{\text{ph}}(\epsilon, \Omega_{\text{ph}}) \times \int_{4\pi} d\Omega'_{\text{s}} \int_0^\infty d\epsilon'_{\text{s}} \frac{d\sigma}{d\epsilon'_{\text{s}} d\Omega'_{\text{s}}} \left(\epsilon'_{\text{s}}\left[1+\beta\mu'_{\text{s}}\right] - \epsilon\right) . \tag{3.62}$$

In this integral, μ and μ'_{s} are given by (3.56). The ϵ'_{s} integration can be solved using the δ function in the Compton cross-section (3.47), and it is convenient to choose $d\Omega'_{\text{s}} = d\kappa' d\phi'$ (see Figure 3.7).

Analytical solutions to the angle integrations in (3.62) for general, azimuthally symmetric and isotropic radiation fields, can be found in [11].

To lowest order, the Compton energy loss on an isotropic radiation field in the Thomson limit reduces to

$$-\left(\frac{d\gamma}{dt}\right)_{\text{C, Thomson, iso}} \approx \frac{4}{3} c\sigma_{\text{T}}\gamma^2 \int_0^\infty d\epsilon\, \epsilon\, n_{\text{ph}}(\epsilon) = \frac{4}{3} c\sigma_{\text{T}} \frac{u_{\text{ph}}}{m_{\text{e}}c^2}\gamma^2 . \tag{3.63}$$

This expression is analogous to the electron energy loss due to synchrotron radiation (3.27), with u_{ph} replacing u_B. It indicates that energy losses through synchrotron and Compton radiation in the Thomson limit are related through $\dot{\gamma}_{\text{Thomson}}/\dot{\gamma}_{\text{sy}} = u_{\text{ph}}/u_B$. The total luminosity produced through any radiation mechanism can be written as

$$L_{\text{rad}} = m_{\text{e}}c^2 \int_{\text{Vol}} \text{dvol} \int_1^\infty d\gamma\, n_{\text{e}}(\gamma)\dot{\gamma} . \tag{3.64}$$

This indicates that the ratio of luminosities in synchrotron and Compton radiation in the Thomson limit is given by the ratio of energy densities in the radiation field and the magnetic field,

$$\frac{L_{\text{Thomson}}}{L_{\text{sy}}} = \frac{\dot{\gamma}_{\text{Thomson}}}{\dot{\gamma}_{\text{sy}}} = \frac{u_{\text{ph}}}{u_B} . \tag{3.65}$$

We note that the Thomson energy-loss rate (3.63) can also easily be recovered by using the δ function approximation for the Compton cross-section (3.61) in (3.62) and letting $\beta \to 1$ after the angle integration over $d\Omega_{\text{ph}} = d\mu d\phi$ has been performed. For large electron and photon energies ($\gamma\epsilon \gtrsim 1$), the Compton cross-section – and therefore also the Compton energy-loss rate – is suppressed with respect to the Thomson limit. This leads to a flatter energy dependence, $|\dot{\gamma}| \propto \gamma^a$ with $a < 2$, for high-energy electrons whose energy loss is dominated by Compton cooling in the Klein–Nishina regime.

3.2.2.3 **Compton Radiation Spectra**

Let us consider an electron distribution $n_e(\gamma, \Omega_e)$ and a photon distribution $n_{ph}(\Omega_{ph})$ with angles $\Omega_e = (\psi_e, \phi_e)$ and $\Omega_{ph} = (\psi_{ph}, \phi_{ph})$ as illustrated in Figure 3.7. In applications to relativistic jets, the natural choice for the z-axis will be the direction of bulk motion of the jet along $\boldsymbol{\beta}_\Gamma$. Most generally, the Compton scattered photon spectrum at frequency $\nu = \epsilon_s m_e c^2/h$ in the direction $\Omega_s = (\psi_s, \phi_s)$ can then be calculated as

$$j_\nu(\epsilon_s, \Omega_s) = h c \epsilon_s \int\limits_{4\pi} d\Omega_e \int\limits_1^\infty d\gamma\, n_e(\gamma, \Omega_e) \times \int\limits_{4\pi} d\Omega_{ph} \int\limits_0^\infty d\epsilon\, n_{ph}(\epsilon, \Omega_{ph})(1 - \beta\mu) \frac{d\sigma_C}{d\epsilon_s d\Omega_s} \tag{3.66}$$

with the Compton cross-section (3.47). The angles μ and μ_s and their counterparts in the electron rest frame can be found from (3.56) and (3.55). It becomes obvious that for arbitrary electron and photon distributions, the evaluation of the Compton emissivity becomes very cumbersome. In applications to relativistic jets, one can usually assume azimuthal symmetry, which will make one of the ϕ integrations trivial. It is also often assumed that the electron distribution is isotropic in the rest frame of the emission region, that is, $n_e(\gamma, \Omega_e) = n_e(\gamma)/(4\pi)$. Still, while the δ function in the Klein–Nishina cross-section (3.47) eliminates one integration, one is still left with a four-dimensional integral to be evaluated numerically. When doing so, special care has to be taken of the fact that the integrand is sharply peaked around $\Omega_e = \Omega_s$ for highly relativistic electron. As we recall, this is precisely the justification for the head-on approximation (3.57).

Even the use of the head-on approximation in (3.66) for an isotropic (in the emission region rest frame) electron distribution leaves us with four integrations to be done numerically:

$$j_\nu^{\text{head-on}}(\epsilon_s, \Omega_s) = \frac{h c \epsilon_s}{4\pi} \int\limits_1^\infty d\gamma\, n_e(\gamma) \int\limits_{-1}^1 d\eta_{ph} \int\limits_0^{2\pi} d\phi_{ph} \times \int\limits_0^\infty d\epsilon\, n_{ph}(\epsilon, \Omega_{ph})(1 - \beta\mu) \frac{d\sigma_C}{d\epsilon_s} \tag{3.67}$$

with $d\sigma_C/d\epsilon_s$ given by (3.58), and we can choose, without loss of generality, $\phi_s = 0$, so that $\phi = \phi_{ph}$ in (3.56) to evaluate μ.

For an isotropic distribution of electrons and photons, that is, $n_e(\gamma, \Omega_e) = n_e(\gamma)/(4\pi)$ and $n_{ph}(\epsilon, \Omega_{ph}) = n_{ph}(\epsilon)/(4\pi)$, the angle integrations in (3.67) can be performed analytically [10]:

$$j_\nu^{\text{head-on, iso}}(\epsilon_s) = \frac{h\epsilon_s}{4\pi} \int_1^\infty d\gamma\, n_e(\gamma) \int_0^\infty d\epsilon\, n_{ph}(\epsilon) g(\epsilon_s, \epsilon, \gamma)\,, \tag{3.68}$$

where

$$g(\epsilon_s, \epsilon, \gamma) = \frac{c\pi r_e^2}{2\gamma^4 \epsilon} \left(\frac{4\gamma^2 \epsilon_s}{\epsilon} - 1\right) \quad \text{if} \quad \frac{\epsilon}{4\gamma^2} \le \epsilon_s \le \epsilon\,, \tag{3.69}$$

and

$$g(\epsilon_s, \epsilon, \gamma) = \frac{2c\pi r_e^2}{\gamma^2 \epsilon} \left[2q \ln q + (1+2q)(1-q) + \frac{(4\epsilon\gamma q)^2}{(1+4\epsilon\gamma q)} \frac{(1-q)}{2}\right]$$
$$\text{if} \quad \epsilon \le \epsilon_s \le \frac{4\epsilon\gamma^2}{1+4\epsilon\gamma}\,, \tag{3.70}$$

where

$$q = \frac{\epsilon_s}{4\epsilon\gamma^2\left(1 - \frac{\epsilon_s}{\gamma}\right)}\,. \tag{3.71}$$

This is the expression most commonly used to evaluate the synchrotron self-Compton (SSC) spectrum from any system where the underlying electron distribution is assumed to be isotropic in the rest frame of the emission region, most notably, in AGN jets.

Using the δ function approximation in the Thomson regime to the Compton cross-section (3.61), the emissivity for isotropic photon and electron distributions simplifies to

$$j_\nu^{\delta,\text{iso}} = \frac{hc\sigma_T \epsilon_s^2}{8\pi} \int_{\epsilon_s}^\infty d\gamma \frac{n_e(\gamma)}{\gamma^4} \int_{\epsilon_s/(2\gamma^2)}^\infty d\epsilon \frac{n_{ph}(\epsilon)}{\epsilon^2}\,, \tag{3.72}$$

where the limits on the integrations arise from energy conservation in the Thomson limit through $\epsilon_s < \gamma$ and $\epsilon_s < 2\gamma^2\epsilon$. For illustrative purposes, we evaluate (3.72) for a monoenergetic photons, $n_{ph}(\epsilon) = n_{ph,0}\delta(\epsilon - \epsilon_0)$ and a power-law distribution of electrons as in (3.23) [2]:

$$j_\nu^{\delta,\text{iso, PL}} = \frac{hc\sigma_T n_0 n_{ph,0}}{8\pi(p+3)} \left(\frac{\epsilon_s}{\epsilon_0}\right)^2 \left\{\left(\max\left[\gamma_1, \epsilon_s, \sqrt{\frac{\epsilon_s}{2\epsilon_0}}\right]\right)^{-(p+3)} - \gamma_2^{-(p+3)}\right\}$$
$$\times H\left(\gamma_2 - \max\left[\epsilon_s, \sqrt{\frac{\epsilon_s}{2\epsilon_0}}\right]\right)\,, \tag{3.73}$$

where the Heaviside function is defined as

$$H(x) = \begin{cases} 1 & \text{if} \quad x \ge 0 \\ 0 & \text{else}\,. \end{cases} \tag{3.74}$$

One can see that in the ϵ_s range where the last term in the "max" expression dominates, that is, where $\epsilon_s > 2\epsilon_0\gamma_1^2$, and $\epsilon_s < 1/(2\epsilon_0)$, the δ function approximation predicts a power-law Compton spectrum with index

$$\alpha_C = \frac{p-1}{2} . \tag{3.75}$$

This is the exact same spectral shape as we found for synchrotron emission (see Section 3.2.1.2). It results from the $\epsilon_s \sim \gamma^2\epsilon_0$ scaling, which exhibits the same γ^2 dependence as the expression (3.35) for the critical synchrotron frequency $\nu_c(\gamma)$.

In Figure 3.8, we compare the Compton spectrum calculated with the delta approximation according to (3.73) with the corresponding expression using the Jones formula (3.68) for an electron distribution extending from $\gamma_1 = 10$ to $\gamma_2 = 10^6$ with a slope of $p = 2$, scattering an isotropic, monoenergetic photon field with energy $\epsilon_0 = 2 \times 10^{-5}$, corresponding to ultraviolet radiation with $E_0 \approx 10\,\text{eV}$.

The spectrum produced with the δ function approximation to the Compton cross-section shows excellent agreement with the full Compton cross-section within the power-law part of the Compton spectrum, as long as scattering is in the Thomson limit, that is, $\gamma\epsilon_0 \ll 1$, resulting in scattered photon energies $\epsilon_s \sim \gamma^2\epsilon_0 \ll 1/\epsilon_0$, corresponding to $\sim 3 \times 10^{24}$ Hz in our test case. The graph shows that substantial deviations from the Thomson power law are already evident at energies about two orders of magnitude below $1/\epsilon_0$. From (3.73), we see that the δ function approximation produces artificially sharp breaks at $\epsilon_s = \epsilon_0\gamma_1^2$ and $\epsilon_s = 1/(2\epsilon_0)$, with a hard ϵ_s^2 spectrum at low frequencies and a very soft $\epsilon_s^{-(1+p)}$ spectrum at high frequencies. This substantially deviates from the full Compton spectrum, which shows a more gradual turnover towards low frequencies, but eventually also reproduces an ϵ_s^2 spectrum, and exhibits a gradual softening towards high frequencies due to Klein–Nishina effects.

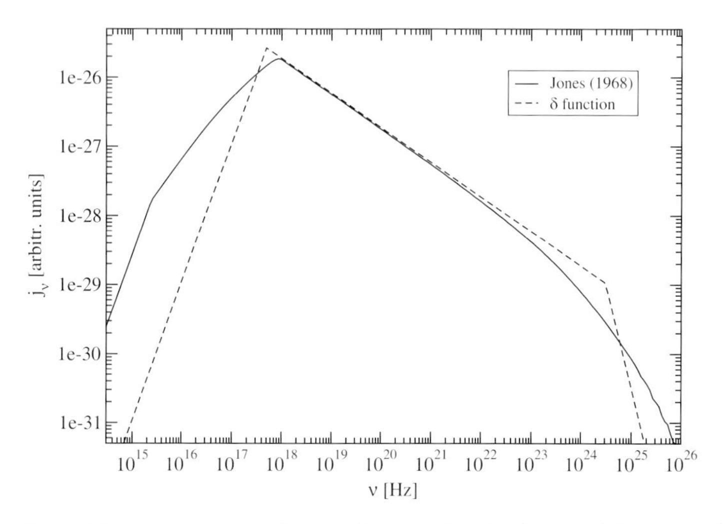

Figure 3.8 Compton scattered spectra for power-law distribution of electrons with $\gamma_1 = 10$, $\gamma_2 = 10^6$, $p = 2$ scattering a monoenergetic radiation field with $\epsilon_0 = 2 \times 10^{-5}$.

All calculations we have done so far, refer to the comoving frame of the emission region, which, in relativistic astrophysical jets, is moving with a bulk velocity $\beta_\Gamma c$ along the jet. Thus, the observer will see the emission Doppler boosted according to the transformation laws discussed in Section 2.2. There, we found that emission that is isotropically produced in the comoving frame of the emission region will be boosted in flux and intensity by a factor δ^3 into the observer's frame, along with a shift by a factor δ towards higher frequencies. In Chapter 8, we will discuss in more detail spectra resulting from Compton scattering of various radiation fields relevant to the high-energy emission from jets in AGN.

3.2.3
$\gamma\gamma$ Absorption and Pair Production

High-energy γ-ray photons can interact with each other (and with lower-energy photons) to produce an electron–positron pair. This is the only relevant absorption process for γ-rays in astrophysical environments. In Section 2.1.1.1, we have already calculated the threshold condition for $\gamma\gamma$ absorption. In this section, we will discuss the $\gamma\gamma$ absorption cross-section, the opacity of various astrophysical environments to high-energy γ-rays through $\gamma\gamma$ absorption, as well as the spectrum of e^+e^- pairs produced in the $\gamma\gamma$ absorption process. The results of this section will form the background for our later discussion of pair cascades in Section 3.3.

3.2.3.1 The $\gamma\gamma$ Absorption Cross-Section

Recall our discussion of Section 2.1.1.1. If a photon of energy ϵ_1 interacts with a photon of energy ϵ_2 under a collision angle $\theta = \cos^{-1}\mu$, producing an e^+e^- pair, in the center-of-momentum (cm) frame, the produced electron and positron will have equal Lorentz factors

$$\gamma_{\rm cm} = \sqrt{\frac{\epsilon_1\epsilon_2(1-\mu)}{2}} \tag{3.76}$$

and equal normalized velocities (in opposite directions) of

$$\beta_{\rm cm} = \sqrt{1 - \frac{2}{\epsilon_1\epsilon_2(1-\mu)}}\ . \tag{3.77}$$

The condition that (3.77) has a real-valued solution, yields the threshold condition on the energy ϵ_1 as

$$\epsilon_1 \geq \frac{2}{\epsilon_2(1-\mu)}\ . \tag{3.78}$$

For the most favorable interaction angle, $\mu = -1$, corresponding to a head-on collision of the two photons, the threshold reduces to $\epsilon_{1,\,\rm thr} = 1/\epsilon_2$.

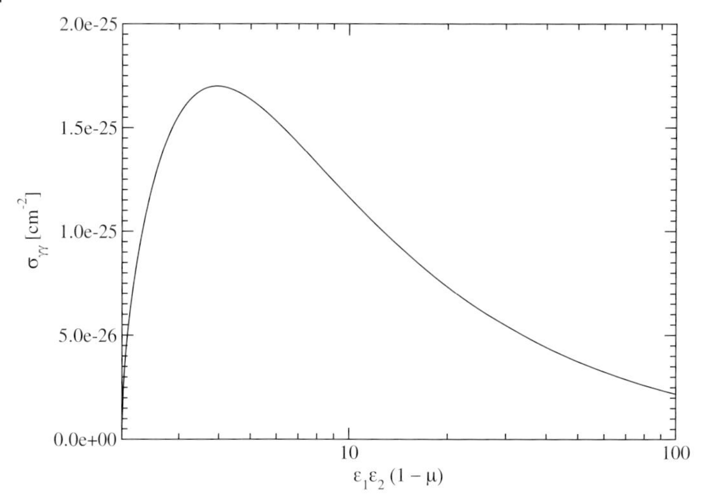

Figure 3.9 $\gamma\gamma$ absorption cross-section as a function of $x = \epsilon_1\epsilon_2(1-\mu)$.

The cross-section for $\gamma\gamma$ absorption is commonly written as a function of $\beta_{\rm cm}$ and is given by [7]:

$$\sigma_{\gamma\gamma}(\beta_{\rm cm}) = \frac{3}{16}\sigma_{\rm T}\left(1-\beta_{\rm cm}^2\right) \times \left(\left[3-\beta_{\rm cm}^4\right]\ln\left[\frac{1+\beta_{\rm cm}}{1-\beta_{\rm cm}}\right] - 2\beta_{\rm cm}\left[2-\beta_{\rm cm}^2\right]\right) . \tag{3.79}$$

Figure 3.9 shows the dependence of $\sigma_{\gamma\gamma}$ on the variable $x = \epsilon_1\epsilon_2(1-\mu)$. At threshold ($x = 2$), $\beta_{\rm cm} = 0$, and $\sigma_{\gamma\gamma} = 0$, as can be easily seen from (3.79). The cross-section has a strong peak at $x = 4$, where $\sigma_{\gamma\gamma}^{\rm peak} \approx \sigma_{\rm T}/4$. Interpreting this as a function of one of the photon energies, the peak is thus at $\epsilon = 2\epsilon_{\rm thr}$.

The strongly peaked shape of the $\gamma\gamma$ cross-section gives rise to a simple δ function approximation which is often useful for rough estimates of the $\gamma\gamma$ opacity of various astrophysical settings [12]:

$$\sigma_{\gamma\gamma}^{\delta}(\epsilon_1,\epsilon_2) = \frac{1}{3}\sigma_{\rm T}\epsilon_1\delta\left(\epsilon_1 - \frac{2}{\epsilon_2}\right) . \tag{3.80}$$

3.2.3.2 The $\gamma\gamma$ Opacity in Various Target Photon Fields

The opacity $\tau_{\gamma\gamma}(\epsilon_1)$ for $\gamma\gamma$ absorption of a γ-ray photon with energy ϵ_1 in a photon field $n_{\rm ph}(\epsilon_2,\Omega;x)$ over a path length l is given by

$$\tau_{\gamma\gamma} = \int_0^l dx \int_{4\pi} d\Omega\,(1-\mu) \int_{\frac{2}{\epsilon_1(1-\mu)}}^{\infty} d\epsilon_2\, n_{\rm ph}(\epsilon_2\Omega;x)\,\sigma_{\gamma\gamma}(\epsilon_1,\epsilon_2,\mu) . \tag{3.81}$$

If the absorbing radiation field is located outside the emission region, then $\gamma\gamma$ absorption simply leads to an exponential absorption term $e^{-\tau_{\gamma\gamma}}$ compared to the intrinsic emission $F_\nu^{\rm int}$ produced in the emission zone, that is,

$$F_\nu^{\rm obs}(\epsilon) = F_\nu^{\rm int}(\epsilon) e^{-\tau_{\gamma\gamma}(\epsilon)} . \tag{3.82}$$

In the application to AGN, such external $\gamma\gamma$ absorption effects can result, for example, from the direct accretion disk emission, the BLR line emission, and the infrared emission from the dust torus (e.g., [13–17]).

Figure 3.10 shows the results of such a calculation [19] for the prominent γ-ray bright quasar 3C 279. The figure illustrates the dependence of the $\gamma\gamma$ opacity of the BLR radiation field to high-energy γ-rays on γ-ray photon energy and on the location of the γ-ray emission region. It illustrates that if the emission region is located within the BLR, γ-rays above $\sim$ 100 GeV are likely to be heavily attenuated by $\gamma\gamma$ absorption.

Very-high-energy γ-rays from sources at cosmological distances will also suffer $\gamma\gamma$ absorption by the extragalactic (infrared and optical) background light (EBL: [20–26]). The EBL consists of infrared emission from dust, primarily in young, star-forming galaxies, as well as infrared and optical emission from stars in galaxies. Its spectrum and intensity depends on cosmological time and, hence, on redshift, and so does the $\gamma\gamma$ opacity for very-high-energy γ-rays. Figure 3.11 illustrates the energy and redshift dependence of $\tau_{\gamma\gamma}$ according to several currently discussed models of the EBL. The EBL is very difficult to measure, in particular, due to bright foreground emission from within the solar system and the Milky Way. However, if the intrinsic very-high-energy spectrum of an AGN can be in-

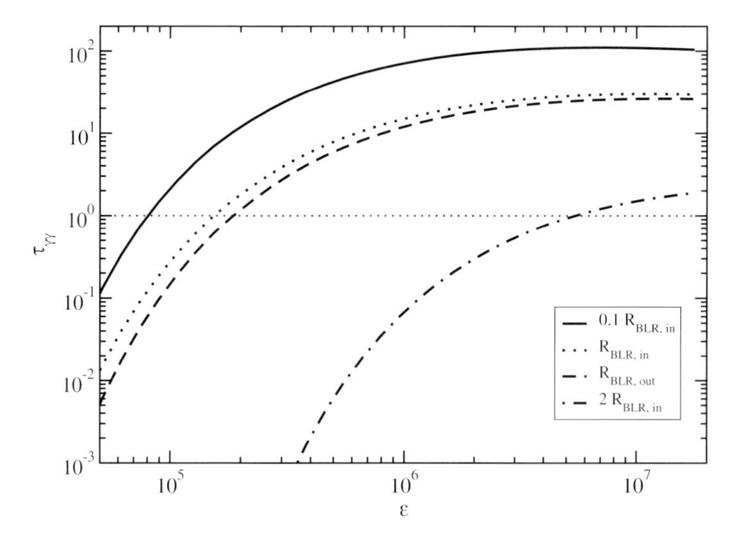

Figure 3.10 The $\gamma\gamma$ absorption opacity due to photons from the BLR in the γ-ray bright quasar 3C 279. Parameters as inferred from observations [18] include the accretion disk luminosity of $L_{\rm D} = 2 \times 10^{45}$ erg s^{-1}, a radially averaged Thomson depth of the BLR of $\tau_{\rm BLR} = 0.1$, and a distance of the inner edge of the BLR from the central engine of $R_{\rm BLR,\,in} = 0.03$ pc. The curves are labeled by the distance of the γ-ray emission region from the central engine. Reproduced from Böttcher *et al.* [19] with permission from the AAS.

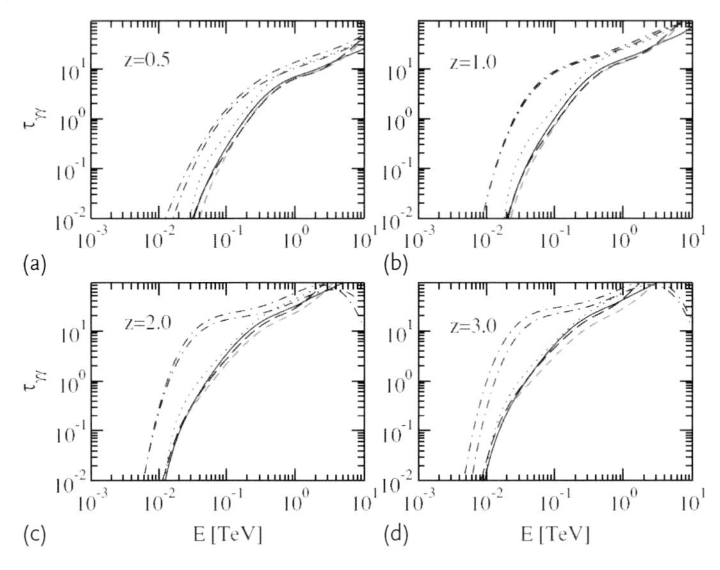

Figure 3.11 (a–d) The $\gamma\gamma$ absorption opacity of the Universe due to interaction with the Extragalactic Background Light for various models of the EBL. Reproduced from Finke *et al.* [26] with permission from the AAS.

ferred from modeling, constrained by the lower-energy emission, one can use the $\gamma\gamma$ absorption imprint of the EBL to diagnose the cosmological evolution of the EBL and, hence, the star-formation history of the Universe. For this reason, studies of the EBL and its $\gamma\gamma$ absorption effects are currently a very active field of research.

A γ-ray photon of energy ϵ_1 will preferentially interact with low-energy photons of energy $\epsilon_2 \sim 2/\epsilon_1$. This means, photons of energy E_1 will primarily probe the EBL at a wavelength λ_2 given by

$$\lambda_2 = 2.4 E_{1,\text{TeV}} \ \mu\text{m} . \tag{3.83}$$

Photons in the few 100 GeV energy range probe primarily the near-infrared regime of the EBL, while TeV photons probe the far-IR.

If $\gamma\gamma$ absorption happens within the emission region, its effect on the emerging spectrum will be governed by the radiative transfer equation (3.18). If the flux without $\gamma\gamma$ absorption would be F_ν^{int}, the physical conditions are homogeneous throughout the source, and there is no significant background source, the observed flux will be

$$F_\nu^{\text{obs}}(\epsilon) \approx F_\nu^{\text{int}}(\epsilon) \frac{1 - e^{-\tau_{\gamma\gamma}(\epsilon)}}{\tau_{\gamma\gamma}(\epsilon)} , \tag{3.84}$$

with the asymptotic behavior

$$F_\nu^{\text{obs}}(\epsilon) \approx \begin{cases} F_\nu^{\text{int}}(\epsilon) & \text{if} \quad \tau_{\gamma\gamma}(\epsilon) \ll 1 \\ \frac{F_\nu^{\text{int}}(\epsilon)}{\tau_{\gamma\gamma}(\epsilon)} & \text{if} \quad \tau_{\gamma\gamma}(\epsilon) \gg 1 . \end{cases} \tag{3.85}$$

The importance of internal $\gamma\gamma$ absorption is often diagnosed with the *compactness parameter* ℓ, which is defined as

$$\ell = \frac{L_\gamma \sigma_T}{4\pi R \langle\epsilon\rangle m_e c^3} . \tag{3.86}$$

This parameter is based on a rough estimate of the $\gamma\gamma$ opacity $\tau_{\gamma\gamma} \sim n_{ph}(\epsilon)\sigma_T R$. The number density of photons can be estimated from the γ-ray luminosity through $n_{ph}(\epsilon) \sim L_\gamma/(4\pi R^2 \langle\epsilon\rangle m_e c^3)$, where $\langle\epsilon\rangle$ is the average photon energy over which L_γ is measured. In environments where ℓ is large, high-energy photons are likely to be absorbed by $\gamma\gamma$ pair production, whereas they can easily escape from low-compactness ($\ell \ll 1$) environments.

If the soft target photons responsible for internal $\gamma\gamma$ absorption have a power law with energy index α_X, the absorption will result in a spectral break in the high-energy γ-ray spectrum by $\Delta\alpha = \alpha_X$. The break is expected to occur for the γ-ray energy where $\tau_{\gamma\gamma}(E_b) = 1$. Therefore, if a spectral break related to $\gamma\gamma$ absorption is observed at an energy $E_b = E_{GeV}$ GeV, one can infer an estimate of the Doppler boosting factor δ. Using the δ approximation for the $\gamma\gamma$ cross-section, the resulting Doppler factor estimate is

$$\delta \approx \left(10^{3\alpha_X} \frac{\sigma_T d_L^2}{3 m_e c^4 t_{var}^{obs}} F_{0-1} \frac{1-\alpha_X}{\epsilon_1^{1-\alpha_X} - \epsilon_0^{1-\alpha_X}} [1+z]^{2\alpha_X} E_{GeV}^{\alpha_X}\right)^{\frac{1}{4+2\alpha_X}} . \tag{3.87}$$

Here, d_L is the luminosity distance to the source, F_{0-1} is the soft photon (typically X-ray) flux between the dimensionless photon energies ϵ_0 and ϵ_1, and t_{var}^{obs} is the observed minimum variability time scale.

If no break is observed, (3.87) will yield a lower limit on δ by inserting the highest observed γ-ray energy for E_{GeV} [27]. This was one of the main arguments for relativistic Doppler boosting of the high-energy emission from blazars in the EGRET era. The Fermi Gamma-Ray Space Telescope has now observed spectral breaks with $\Delta\alpha \sim 1$ in the γ-ray spectra of several blazars [28]. However, the origin of those breaks is currently not understood, and there is no evidence that they can be related to internal $\gamma\gamma$ absorption.

3.2.3.3 The $\gamma\gamma$ Pair Production Spectrum

In the process of $\gamma\gamma$ absorption, an electron–positron pair is produced. We consider a spectrum of high-energy (γ-ray) photons given by $n_1(\epsilon_1)$, interacting with an isotropic target photon field $n_2(\epsilon_2)$. The spectrum of the produced pairs, $\dot{n}(\gamma)$, has been evaluated fully analytically in [29]. In this section, we describe various approximations to the exact $e^+ - e^-$ pair spectrum derived in [29].

The simplest approximation to the pair production spectrum was introduced by [30] and is based on the δ function approximation (3.80) for the $\gamma\gamma$ absorption cross-section and the realization that for interactions near threshold, $\gamma_+ \approx \gamma_- \approx (\epsilon_1 + \epsilon_2)/2$, where $\gamma_\pm$ are the Lorentz factors of the positron and electron, respectively. It is optimized for $\gamma\gamma$ absorption of high-energy ($\epsilon_1 \gg 1$) radiation by a power-law spectrum of low-energy emission given by $n_2(\epsilon_2) \propto \epsilon_2^{-\alpha}$, corresponding to an energy index $\alpha' = \alpha - 1$, and $\epsilon_1 \gg \epsilon_2$. Then the pair production spectrum can be approximated as

$$\dot{n}(\gamma)^{\mathrm{LZ}} \approx \eta(\alpha') c \sigma_{\mathrm{T}} \frac{n_1(\epsilon_0)}{\epsilon_0} n_2 \left(\frac{1}{\epsilon_0} \right) , \tag{3.88}$$

where $\epsilon_0 = \gamma + \sqrt{\gamma^2 - 1} \approx 2\gamma$ for $\gamma \gg 1$ and $\eta(\alpha')$ is a numerical factor dependent only on the spectral index of the low-energy photon spectrum. An example of the resulting pair spectrum for a γ-ray power-law spectrum interacting with a low-energy power-law photon spectrum is shown in Figure 3.12. It illustrates that the interaction of two power-law photon spectra results in a power-law pair spectrum over a substantial energy range, with gradual cutoffs at low and high energies determined by the minimum and maximum photon energy in the high-energy spectrum. Figure 3.12 illustrates that the δ approximation represents the straight power-law part of the pair spectrum very well, but predicts artificially hard cutoffs compared to the full expression of [29].

A somewhat more accurate approximation for the pair spectrum uses the full $\gamma\gamma$ absorption cross-section (3.79), but assumes $\epsilon_1 \gg \epsilon_2$ and $\gamma_+ = \gamma_- = (\epsilon_1 + \epsilon_2)/2 \approx \epsilon_1/2$. The resulting expression corresponds to a δ function approximation in the energy of the produced electron/positron, and can be written as

$$\dot{n}(\gamma)^{\delta_\gamma} = 2c \frac{n_1(2\gamma)}{(2\gamma)^2} \int_{1/2\gamma}^{1} d\epsilon_2 \frac{n_2(\epsilon_2)}{\epsilon_2^2} \int_{(\epsilon_{\mathrm{cm}}^L)^2}^{(\epsilon_{\mathrm{cm}}^U)^2} ds\, s\, \sigma_{\gamma\gamma}(s) , \tag{3.89}$$

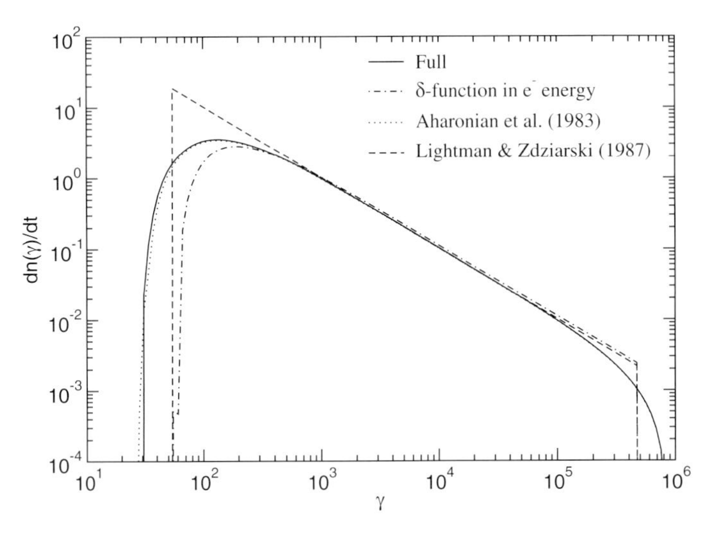

Figure 3.12 Pair production rate spectra. Comparison of the exact expression with different approximations for the $\gamma\gamma$ absorption interaction of two power-law photon spectra with indices $\alpha = 2.5$. The high-energy (γ-ray) spectrum extends from $\epsilon_1 = 10^2 - 10^6$, and the low-energy spectrum extends from $\epsilon_2 = 10^{-7} - 10^{-2}$.

where $s = \epsilon_{cm}^2$ (and hence $\beta_{cm} = \sqrt{1 - 1/s}$), and the limits $(\epsilon_{cm}^U)^2$ and $(\epsilon_{cm}^L)^2$ are the minimum and maximum kinetically allowed center-of-momentum photon energies. As illustrated in Figure 3.12, this approximation also reproduces the power-law portion of the pair spectrum very well, but fails to represent the end points of the spectrum as it obviously does not produce pairs at energies below half the minimum or above half the maximum energy of the high-energy photon spectrum.

Finally, a rather accurate approximation to the pair production spectrum, appropriate for various types of photon spectra, has been developed by [31]. It is based on a different representation of the pair production spectrum and is given by

$$\dot{n}(\gamma)^{\text{Ah}} = \frac{3}{32} c \sigma_{\text{T}} \int_{\gamma}^{\infty} d\epsilon_1 \frac{n_1(\epsilon_1)}{\epsilon_1^3} \int_{\frac{\epsilon_1}{4\gamma(\epsilon_1 - \gamma)}} d\epsilon_2 \frac{n_2(\epsilon_2)}{\epsilon_2^2}$$
$$\times \left\{ \frac{4\epsilon_1^2}{\gamma(\epsilon_1 - \gamma)} \ln \left(\frac{4\epsilon_2 \gamma(\epsilon_1 - \gamma)}{\epsilon_1} \right) - 8\epsilon_1\epsilon_2 \right.$$
$$\left. + \frac{2\epsilon_1^2(2\epsilon_1\epsilon_2 - 1)}{\gamma(\epsilon_1 - \gamma)} - \left(1 - \frac{1}{\epsilon_1\epsilon_2}\right) \frac{\epsilon_1^4}{\gamma^2(\epsilon_1 - \gamma)^2} \right\} . \quad (3.90)$$

The resulting pair production spectrum for the interaction of two power-law photon spectra is included in Figure 3.12 and shows excellent agreement with the full expression of [29].

3.2.4
γ-Hadron Interactions

Relativistic jets are among the most important candidate sources of the ultra-high-energy (UHE, $E > 10^{19}$ eV) cosmic rays. Furthermore, jet sources offer in general dense radiation fields. As a consequence, interactions between relativistic hadrons and photons are important in jet environments, and are mediated by either the electromagnetic or the strong force. In the following we shall consider the most important of these. Here we focus on power-law radiation fields with density $n_{\text{T}} \propto \epsilon^{-\alpha_\gamma}$ as the target for ultrarelativistic protons as the most common type of low-energy radiation in astrophysical jets.

3.2.4.1 Bethe–Heitler Pair Production

Electron–positron pair production by a nucleus on ambient photons

$$p + \gamma \longrightarrow p' + e^+ + e^-$$

takes place if the total center-of-momentum (cm) energy squared $s = m_{\text{p}}^2 c^4 + 2E_{\text{ph}} E_{\text{p}}(1 - \beta_{\text{p}} \cos\theta)$ of the initial particles is above the threshold energy of the interaction, $s_{\text{th}} = (m_{\text{p}}c^2 + 2m_{\text{e}}c^2)^2 \approx 0.882\,\text{GeV}^2$. Here $\beta_{\text{p}}c$ is the proton's velocity, $E_{\text{p}} = \gamma_{\text{p}} m_{\text{p}} c^2$ is the proton energy with γ_{p} the Lorentz factor of the proton, E_{ph} is the photon energy, and θ is the interaction angle between the initial proton and photon. The angle-averaged cross-section of this electromagnetic process, commonly known as "Bethe–Heitler pair production" [32],

$$\langle \sigma_{\mathrm{BH}}(\gamma_{\mathrm{p}}, x)\rangle = \frac{1}{2}\int_{-1}^{\cos\theta_{\mathrm{th}}} (1-\beta_{\mathrm{p}}\cos\theta)\sigma_{\mathrm{BH}}(s)d\cos\theta$$
$$= \frac{1}{8\beta_{\mathrm{p}}E_{\mathrm{p}}^2E_{\mathrm{ph}}^2}\int_{s_{\mathrm{th}}}^{s_{\mathrm{max}}} \sigma_{\mathrm{BH}}(s)\left(s - m_{\mathrm{p}}^2c^4\right)ds\,, \qquad (3.91)$$

where $s_{\max}$ denotes the cm frame energy squared for head-on collisions,

$$s_{\max} = m_{\mathrm{p}}^2c^4 + 2E_{\mathrm{p}}E_{\mathrm{ph}}(1+\beta_{\mathrm{p}})\,,$$

can be computed from standard quantum electrodynamics. It is shown in Figure 3.13 as a rising function of photon energy in the nuclear rest frame E'_{ph}. At the same time the mean fractional energy loss of the proton $\Delta E_{\mathrm{p}}/E_{\mathrm{p}} = \kappa_{\mathrm{BH}}$, or "mean inelasticity", decreases with E'_{ph} from $\kappa_{\mathrm{BH}} = 2m_{\mathrm{e}}/m_{\mathrm{p}}$ at threshold.

Due to the small inelasticity $\kappa_{\mathrm{BH}} \ll 1$ the interaction length

$$\frac{1}{\lambda_{\mathrm{BH}}(E_{\mathrm{p}})} = \frac{1}{8E_{\mathrm{p}}^2\beta}\int_{E_{\mathrm{ph,\,th}}}^{\infty} dE_{\mathrm{ph}}\frac{n(E_{\mathrm{ph}})}{E_{\mathrm{ph}}^2}\int_{s_{\min}}^{s_{\max}(E_{\mathrm{ph}},E)} ds(s-m_{\mathrm{p}}^2c^4)\sigma_{\mathrm{BH}}(s) \qquad (3.92)$$

with $E_{\mathrm{ph,\,th}} = (s_{\min} - m_{\mathrm{p}}^2c^4)/(2E(1+\beta))$, is much smaller than the mean energy-loss distance

$$x_{\mathrm{loss,BH}}(E_{\mathrm{p}}) = \frac{E_{\mathrm{p}}}{dE_{\mathrm{p}}/dx} = \frac{\lambda(E_{\mathrm{p}})}{\kappa_{\mathrm{BH}}(E_{\mathrm{p}})}\,. \qquad (3.93)$$

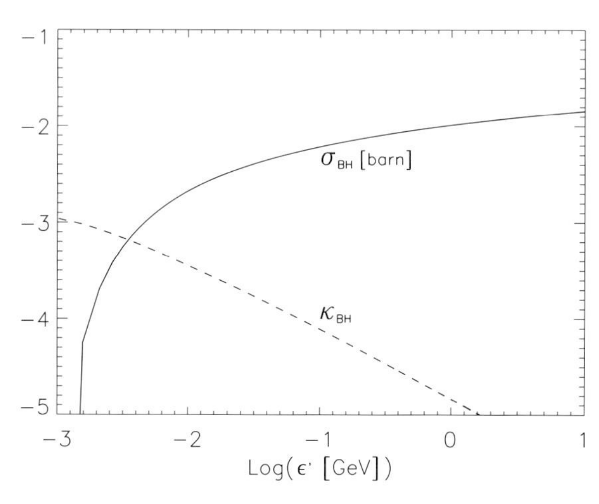

Figure 3.13 The angle-averaged Bethe–Heitler pair production cross-section $\langle\sigma_{\mathrm{BH}}\rangle$ (solid curve), and the angle-averaged Bethe–Heitler pair production inelasticity κ_{BH} (dashed curve) are plotted as a function of the photon's energy in the nuclear rest frame following the series expansion of [33, 34].

Depending on the target radiation field, energy-independent inelasticity approximations $\kappa_{\mathrm{BH}} \sim m_\mathrm{e}/m_\mathrm{p}$ may therefore overestimate the pair production rate:

$$q_{\mathrm{e,BH}}(\gamma_\mathrm{e}) = \int d\gamma_\mathrm{p} n_\mathrm{p}(\gamma_\mathrm{p}) \int d E_{\mathrm{ph}} n_\mathrm{T}(E_{\mathrm{ph}}) f(\gamma_\mathrm{p}, \gamma_\mathrm{e}, E_{\mathrm{ph}}) \langle \sigma_{\mathrm{BH}}(\gamma_\mathrm{p}, E_{\mathrm{ph}}) \rangle \tag{3.94}$$

with $f = d N_\mathrm{e}/d\gamma_\mathrm{e}$ the distribution of electrons with Lorentz factor γ_e, $n_\mathrm{T}(E_{\mathrm{ph}})$ the target photon distribution and $n_\mathrm{p}(\gamma_\mathrm{p})$ initial proton distribution.

Further reading and useful approximations may be found in, for example, [9, 34–37].

In (jet) radiation fields with a power-law ambient target photon field with index α_T this process is found to be less efficient in comparison to the hadronic interaction of a proton with the photons if $\alpha_\mathrm{T} < 2$ [38, 39]. This is a consequence of the small fractional proton energy loss $\leq 2m_\mathrm{e}/m_\mathrm{p}$, which is available for pair production, while in photohadronic interactions a significant fraction of the initial proton energy ($\geq$ 20%) is used for secondary particle production.

3.2.4.2 Photomeson Production

If the available cm frame energy exceeds the threshold energy of $s_{\mathrm{th}} = (m_\mathrm{p} c^2 + m_{\pi 0} c^2)^2 \approx 1.16\,\mathrm{GeV}^2$ ($m_{\pi 0}$ is the neutral pion rest mass) in the cm frame, equivalent to the photon energy in the proton's rest frame exceeding $m_\pi c^2(1 + m_\pi/2m_\mathrm{p}) \approx 145$ MeV (see Section 2.1.1.2), secondary particle production through photohadronic interactions becomes possible. While cross-section data of photomeson production are provided by fixed target collider experiments up to $\sqrt{s} \simeq$ a few GeV, interactions in astrophysical environments may reach up to $\sqrt{s} \simeq 10^3$ GeV. Therefore, models for photomeson production have been developed that adequately describe the interaction physics at low energies by comparing with accelerator data. The extension of such well-established phenomenological models that include all known symmetries of hadronic interactions, to high energies then allows predictions of hadronically initiated particle production in jet environments. Yet one has to keep in mind that predictions for photon and neutrino production in astrophysical sources still rely on a range of assumptions concerning the relevant interactions physics, apart from the various unknowns of the physics of the source itself. During the last ten years the SOPHIA Monte Carlo event generator [40] has established itself as a suitable and widely used software package to treat photomeson production in typical astrophysical environments. The following section is dominantly based on results from the SOPHIA code assuming unpolarized photons of an isotropic seed radiation field, and nucleons.

We describe photomeson production by dividing into partial cross-sections: (i) resonance excitation and decay, (ii) direct (nonresonant) pion production, (iii) diffractive scattering, and (iv) multipion production. Let's consider each channel in more detail. For the resonance region the nine most important resonances ($\Delta^+(1232)$, $N^+(1440)$, $N^+(1520)$, $N^+(1535)$, $N^+(1650)$, $N^+(1680)$, $\Delta^+(1700)$, $\Delta^+(1905)$ and $\Delta^+(1950)$) are considered. The Breit–Wigner formula together with their known properties of mass, width and decay branching ratios is used to determine their contribution to the individual interaction channels. Direct pion production ($p\gamma \rightarrow n\pi^+$, $p\gamma \rightarrow \Delta^{++}\pi^-$, $p\gamma \rightarrow \Delta^0\pi^+$), is found to adequately describe the residual cross-section at very low interaction energies. Noteworthy is that it produces exclusively charged pions. Diffractive scattering is due to the coupling of the photon to the vector mesons ρ^0 and ω, which are produced at very high interaction energies ($\sqrt{s} \geq 2\,\mathrm{GeV}$) with the ratio 9 : 1 and a cross-section proportional to the total cross-section. Finally, the residuals to the total cross-section are fitted by a simple model and treated as statistical multipion production which is simulated by a QCD string fragmentation model [41]. The total and partial cross-sections are shown in Figure 3.14. The so-called Δ approximation, σ_Δ, of this cross-section treats photomeson production as purely dominated by the largest resonance, the $\Delta(1232)$-resonance, and reads $\sigma_\Delta = 500\,\mu\mathrm{b}\,H(\sqrt{s} - m_\Delta + \Gamma_\Delta/2) \cdot H(m_\Delta + \Gamma_\Delta/2 - \sqrt{s})$, where $m_\Delta = 1.232\,\mathrm{GeV}$ is the mass and $\Gamma_\Delta = 0.115\,\mathrm{GeV}$ is the width of the $\Delta(1232)$-resonance, and H is the Heaviside step function. The Δ approximation uses the branching ratios of the $\Delta^+(1232)$-resonance to determine the number ratio π^0 to π^+ of 2 : 1. In order to illustrate the contribution of the different interaction processes in jet power-law target photon fields we convolve the cross-section with the proton and seed photon spectra. The resulting interaction probability distribution for a proton spectrum $n_E \sim E_\mathrm{p}^{-\alpha_\mathrm{p}}$ with index $\alpha_\mathrm{p} = 2$ (considered typical for shock accelerated particles) and various power-law seed photon spectra $n_\mathrm{T} \sim E_\mathrm{ph}^{-\alpha_\gamma}$ (typical for jet radiation fields) is shown in Figure 3.15. While in steep target photon fields the interaction occurs predominantly at low interaction energies (threshold and $\Delta(1232)$ reso-

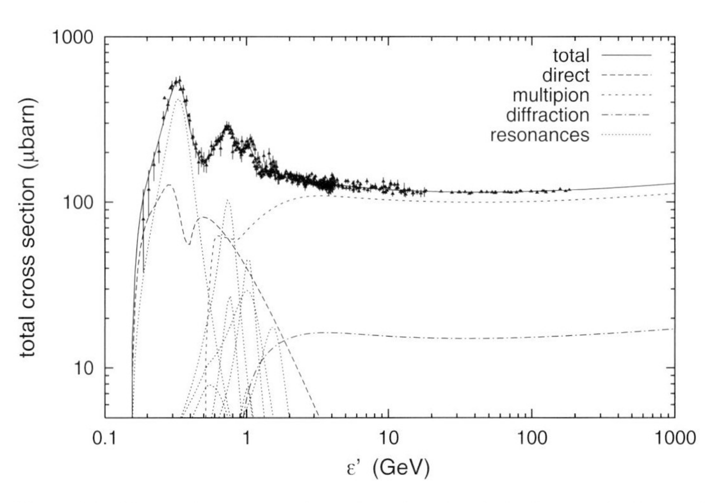

Figure 3.14 The total $p\gamma$ cross-section (based on SOPHIA version 1.4 [40]) with the contributions of the baryon resonances, the direct pion production, diffractive scattering, and the multipion production channel as a function of the photon's energy in the nuclear rest frame ($1\,\mu\mathrm{b} = 10^{-30}\,\mathrm{cm}^2$). Data are from [42]. Reprinted from Baldini *et al.* [42] with permission from Elsevier.

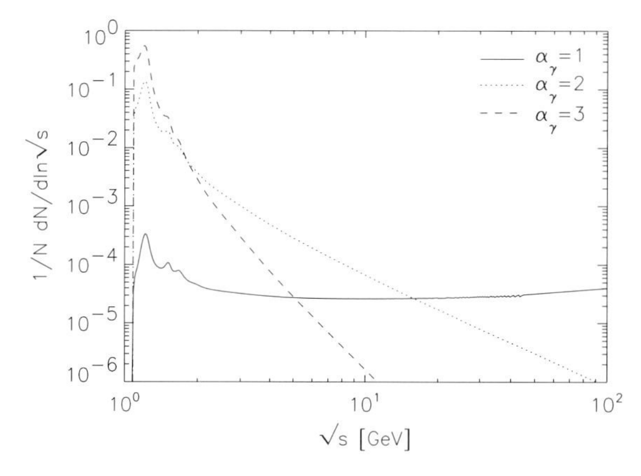

Figure 3.15 The interaction probability distribution as function of the cm frame energy for a power-law target photon field with index α_γ and proton spectrum $\sim E_p^{-2}$. Reprinted from Mücke *et al.* [43] with permission from the CSIRO.

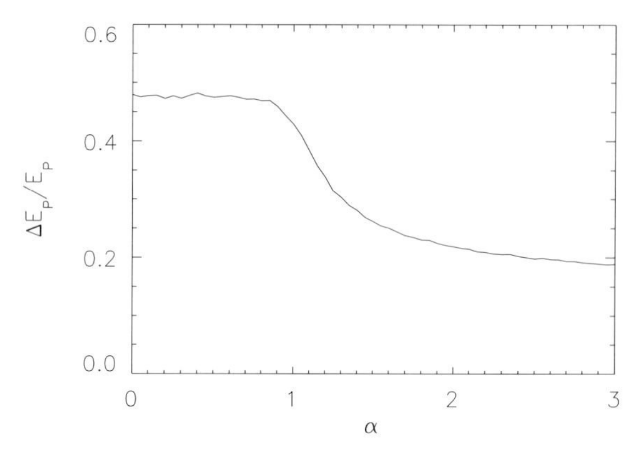

Figure 3.16 Fractional energy loss (inelasticity) for photomeson production in a power-law target photon field with index α.

nance region), interaction in flat target photon spectra (as can often been found in jets) occurs at all interaction energies nearly equally.

After the decay of all intermediate states, considering basic kinematical relations and accelerator data on rapidity distributions, the resulting distributions of protons, neutrons and pions in the source determine the measurable astrophysical quantities, for example, the cosmic ray, neutrino and γ-ray emission. Details are described in [40].

The "leading-baryon" of the final state particles carries the baryon quantum number of the incoming nucleon, and is degraded in energy, E_{lb}, with respect to the energy of the incoming nucleon, E_{in}, by the nucleon inelasticity $\kappa_{p\gamma} = (E_{lb} - E_{in})/E_{in}$. Unlike for $p\gamma$ pair production whose mean inelasticity is small $\lesssim m_e/m_p$, the fractional nucleon energy loss for photomeson production is appreciable, ≥ 0.2–0.5 (see Figure 3.16), and therefore can in general not be treated

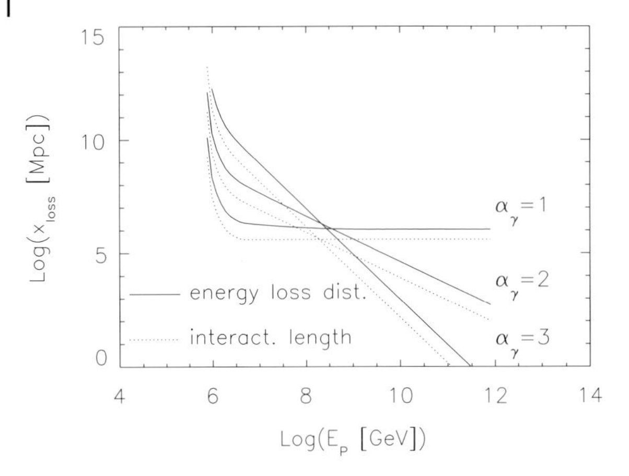

Figure 3.17 Mean energy-loss distance (solid lines) and mean interaction length (dotted lines) for photomeson production in power-law target photon field with index α_γ.

as a continuous loss process. For interaction in steep target photon fields (index ≥ 2) the nucleon mean inelasticity is $\kappa_{\mathrm{p}\gamma} \approx 0.2$, in agreement with expectations in the Δ approximation, while for flatter seed photon spectra the inelasticity increases to reach $\kappa_{\mathrm{p}\gamma} \approx 0.5$ for $\alpha \leq 1$. The mean interaction length,

$$\frac{1}{\lambda_{p\gamma}(E_\mathrm{p})} = \frac{1}{8E_\mathrm{p}^2\beta}\int\limits_{E_\mathrm{ph,\,th}}^{\infty} dE_\mathrm{ph}\frac{n(E_\mathrm{ph})}{E_\mathrm{ph}^2}\int\limits_{s_\mathrm{min}}^{s_\mathrm{max}(E_\mathrm{ph},E_\mathrm{p})} ds(s-m_\mathrm{p}^2c^4)\sigma_{p\gamma}(s) \tag{3.95}$$

with

$$s_\mathrm{min} = \left(m_\mathrm{p}c^2 + m_{\pi^0}c^2\right)^2 , \tag{3.96}$$

$$s_\mathrm{max}(E_\mathrm{ph}, E_\mathrm{p}) = m_\mathrm{p}^2c^4 + 2E_\mathrm{p}E_\mathrm{ph}(1+\beta) , \tag{3.97}$$

$$E_\mathrm{ph,\,th} = \frac{s_\mathrm{min} - m_\mathrm{p}^2c^4}{2E_\mathrm{p}(1+\beta)} , \quad \beta^2 = 1 - \frac{m_\mathrm{p}^2c^4}{E_\mathrm{p}^2} , \tag{3.98}$$

and E_p (E_ph) the proton (photon) energy, differs from the mean energy-loss distance, $x_\mathrm{loss}(E_\mathrm{p})$, by the mean inelasticity, $\kappa_{\mathrm{p}\gamma}(E)$:

$$x_\mathrm{loss}(E_\mathrm{p}) = \frac{E_\mathrm{p}}{dE_\mathrm{p}/dx} = \frac{\lambda(E_\mathrm{p})}{\kappa_{\mathrm{p}\gamma}(E_\mathrm{p})} . \tag{3.99}$$

Both are compared in Figure 3.17 for power-law seed photon spectra.

The dominant type of mesons produced through photohadronic interactions are pions. Neutral pions decay into two photons with half-life time $t_{1/2} \approx 8.4 \times 10^{-17}$ s while the charged pions decay following $\pi^\pm \longrightarrow \mu^\pm + \nu_\mu(\bar{\nu}_\mu)$ at half-life time of "only" $t_{1/2} \approx 2.6 \times 10^{-8}$ s. Pion multiplicities are therefore helpful to understand the dissipation of the initial energy into the gamma-ray and neutrino components. Neutral pion multiplicity is suppressed at threshold in favor of charged pions, reaching $\sim 2/3$ in the Δ-resonance region (following the branching ratio of the $\Delta(1232)$ resonance: $\Delta^+(1232) \rightarrow p + \pi^0$ in 2/3, and $\Delta^+(1232) \rightarrow n + \pi^+$ in 1/3 of all cases). In the multipion production region the multiplicities of pions of all charges follow approximately $\propto s^{1/4}$ in the cm frame, in good agreement with accelerator data (see [40]) and expectations from simple phenomenological models (e.g., Fermi gas model).

The solid angle integrated differential gamma-ray source function $q_\gamma(E_\gamma)$ can be calculated from the neutral pion source function $q_{\pi 0}(E_\pi)$ following the relations (e.g., [44]) of a two-body decay:

$$q_\gamma(E_\gamma) = 2 \int\limits_{E\pi\,\mathrm{min}}^{\infty} d E_\pi \frac{q_{\pi 0}(E_\pi)}{\sqrt{E_\pi^2 - m_\pi^2 c^4}} \tag{3.100}$$

with $E_{\pi,\mathrm{min}} = E_\gamma + (m_\pi c^2)^2/(4E_\gamma)$, E_π and E_γ the pion and photon energy, respectively. The decay of charged pions into muons and neutrinos is again a two-body decay, and can be simplified by noting that with $m_\nu \ll m_\mu$ one can neglect $m_\nu \approx 0$ here. Using conservation of four-momentum one finds for the Lorentz factor of the muon in the pion rest frame roughly unity. Thus the muon is close to rest in the pion rest frame, implying $\gamma_\mu \approx \gamma_\pi$ in the laboratory frame. This immediately allows us to simplify the muon source function $q_\mu(E_\mu)$ to

$$q_\mu(E_\mu) = \frac{m_\pi}{m_\mu} q_\pi(E_\pi) . \tag{3.101}$$

Note that the muons from pion decay are produced fully polarized (left-handed from π^+-decay, right-handed from π^--decay) due to parity non-conservation.

The decay of the muon into pairs and neutrinos $\mu^\pm \longrightarrow e^\pm + \nu_e(\bar{\nu}_e) + \bar{\nu}_\mu(\nu_\mu)$ is a three-body process (e.g., [44]) that is complicated by the muon's polarization properties. In the muon rest frame the neutrino distributions can be parametrized (e.g., [45]) as

$$q'_\nu(x) \propto f_0(x) \pm f_1(x) \cos\alpha \tag{3.102}$$

with $x = 2E'_\nu/m_\mu$, E'_ν the neutrino energy in the muon rest frame, α the angle between the direction of motion of the lepton and spin of the muon, and assuming zero neutrino mass. The functions f_0, f_1 are listed in [45]. After the Lorentz transformation into the laboratory frame the neutrino distributions can then be approximated by

$$q_\nu(y) \approx \beta_\mu^{-1}[g_0(y, \beta_\mu) - g_1(y, \beta_\mu) \cos\alpha_\pi] \tag{3.103}$$

in the relativistic limit $\beta_\mu \rightarrow 1$ with $y = E_\nu/E_\mu$ and α_π the angle between the pion and the reference axis, and functions g_0 and g_1 depending on the considered neutrino flavor [45]. Finally, for the derivation of the pair source function $q_e(\gamma_e)$ from the decay of the muon one finds [46]

$$q'_e \approx 1.78 \times 10^{-6} \gamma'_e \left(3 - \frac{\gamma'_e}{52}\right) \tag{3.104}$$

for the pair distribution in the muon rest frame with γ'_e the electron Lorentz factor in this frame, and after transformation into the laboratory frame

$$q_e(\gamma_e) \approx \int_1^{104} d\gamma'_e \frac{q'_e}{2\sqrt{\gamma'^2_e - 1}} \int_{\gamma_\mu^-}^{\gamma_\mu^+} \frac{q_\mu}{\sqrt{\gamma_\mu^2 - 1}} \tag{3.105}$$

where $\gamma_\mu^\pm = \gamma_e \gamma'_e (1 \pm \beta_e \beta'_e)$.

In addition to pions, also the production and decay of other mesons (in particular kaons and η mesons) contribute (up to 10–20%) to the overall photon, lepton and neutrino production. Convenient approximations of the corresponding SOPHIA simulations have been derived by [37].

In general one finds a maximum of each, $\sim$ 25% and 30%, of the total injected proton energy per interaction is converted into neutrinos and photons, respectively[2)] at high interaction energies. The ratio of total energy dissipated into photons and neutrinos, $\mathcal{E}_\gamma/\mathcal{E}_\nu$, however, remains approximately constant (unity) independent of whether one considers steep or flat target photon fields, and unlike the photon-to-neutrino ratio for interactions in the $\Delta(1232)$-resonance ($\mathcal{E}_\gamma/\mathcal{E}_\nu = 3$). This is because of the contribution from the secondary resonance and multipion production region in the case of flat seed photon spectra, and the contribution to the interaction from the direct channel at threshold if the ambient photon spectra are steep (see Figures 3.18 and 3.19).

The average energy of the produced neutrinos and photons, however, turns out to be sensitive to the steepness of the ambient photon field (Figure 3.18): low values for the case of flat target photon spectra where a large portion of the secondary particles is produced at high interaction energies, and larger values ($\bar{E}_\nu/E_p \approx 0.04$, $\bar{E}_\gamma/E_p \approx 0.06$) in steep seed photon fields, with most interactions produced in the resonance region. Only when most of the interactions occur near the $\Delta(1232)$-resonance is the Δ approximation for the cross-section a good description.

Note that the derivation of all source functions and results above is based on the assumption of prompt decay, that is, any loss processes prior to decay are not considered. This may, however, be relevant in highly magnetized jet environments where the synchrotron loss time scale can become shorter than the decay time scale. In this case, $\pi^\pm$ and $\mu^\pm$ synchrotron radiation and cooling prior to their respective decay will have its imprint not only in the expected photon component (see Section 8.3.2.2) but also the corresponding neutrino emission. For example, in a 1 kG magnetic field where relativistic charged pions and muons are located the

2) Assuming all pairs have 100% radiative efficiency, justified in dense radiative and/or highly magnetized environments.

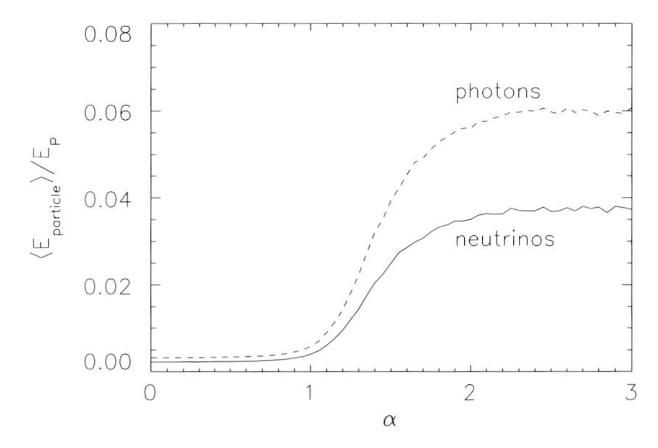

Figure 3.18 The average energy of the secondary neutrinos (solid line) and photons (dashed lines; assuming 100% radiative efficiency of the produced pairs) normalized to the proton input energy E_p as a function of the photon index α of the ambient photon field.

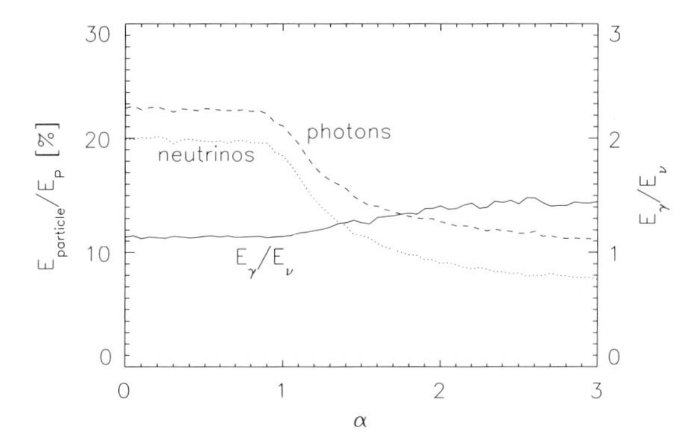

Figure 3.19 The total photon (dashed line) and ν-power (dotted line) normalized to the proton input energy E_p, and their ratio (solid line) emerging from a source due to photomeson production of protons of energy E_p in an ambient power-law target photon field with photon index α.

cooling break due to muon synchrotron radiation is then expected at

$$\gamma_\mu \approx \left[\frac{3 m_e c^2 m_\mu^3}{4 \sigma_T c u_B m_e^3 Z^4 t_{\mu,1/2}} \right]^{1/2} \approx 5.6 \times 10^7 B_{1000}^{-2}$$

with $B_{1000} = B/1\,\text{kG}$, corresponding to $\sim 10^8$ GeV in the neutrino spectrum, and particularly relevant for electron neutrinos since those are solely produced through muon decay. The pion synchrotron cooling break is apparent in the muon neutrino component slightly below 10^{10} GeV [47].

The decay chains of a charged pion produces mostly two muon neutrinos and one electron neutrino. Electron neutrinos are only produced from the decay of the muon here. In particular, the production of $\bar{\nu}_e$ (produced mostly in the $\pi^- \to \mu^- + \bar{\nu}_\mu \to e^- + \nu_\mu + \bar{\nu}_e + \bar{\nu}_\mu$ decay chain) is strongly suppressed for $p\gamma$ interactions $\sqrt{s} \leq 2\,\text{GeV}$ due to the small $p + \gamma \to p + \pi^+ + \pi^-$ cross-section. The nonzero mass of the neutrino allows them to oscillate between different flavor neutrino eigenstates ν_α. An approximation for the maximum oscillation length is given by (e.g., [48])

$$L_{\mathrm{osc,max}} \approx 120\, E_{20} \frac{7 \times 10^{-5}\mathrm{eV}}{\Delta m^2}\,\mathrm{pc}$$

with $E = 10^{20}\, E_{20}$ eV the neutrino energy and Δm^2 the neutrino mass difference squared. To calculate the observed flavor composition after propagation of a length L from the flavor composition at the source we shall consider the flavor neutrino eigenstate ν_α, $\alpha = e, \mu, \tau$, as a mix of mass eigenstates ν_i, $i = 1, 2, 3$: $\nu_\alpha = U_{\alpha i}\nu_i$ with $U_{\alpha i}$ the 3×3 mixing matrix whose values are observationally limited. For $L \gg L_{\mathrm{osc,max}}$ the probability for $\nu_\alpha \rightarrow \nu_\beta$ oscillation is then given by $P_{\alpha\beta} = \sum |U_{\alpha i}|^2 |U_{\beta i}|^2$. In particular, for sources at extragalactic distances the expected neutrino flavor ratio $\nu_\mu + \bar{\nu}_\mu : \nu_\mathrm{e} + \bar{\nu}_\mathrm{e} : \nu_\tau + \bar{\nu}_\tau = 2 : 1 : 0$ at the source is then transformed during propagation into a ratio $\nu_\mu : \nu_\mathrm{e} : \nu_\tau = 1 : 1 : 1$.

A property of photomeson production relevant to cosmic ray physics (and acceleration processes) is the possibility of proton-to-neutron (p/n) conversion (isospin-flip). Protons are magnetically confined to the accelerator while gaining energy. Unlike protons, neutrons are not magnetically confined and therefore can be ejected from the accelerator (unless $n\gamma$ interactions prevent their escape). Such neutrons decay on a mean distance of $x_{\mathrm{mean}} = c t_{n,1/2}\gamma \approx 8.6\gamma_9$ kpc ($\gamma_9 = \gamma/10^9$) to appear as "cosmic rays" propagating through the Universe. The microphysics of photomeson production predicts a p/n-ratio of ~ 0.3–1.3 in the threshold and resonance region, and ~ 2 in the multipion production region [40].

Photomeson production by nuclei of mass number $A > 1$ is in general considered to obey the Glauber rule $\sigma_{A\gamma} \simeq A^{2/3}\sigma_{p\gamma}$. However, in typical jet radiation fields high-mass nuclei are rather destroyed by photodisintegration than through pion production.

3.2.4.3 **Photodisintegration**

In photodisintegration interactions a photon reacts with an atomic nucleus of mass number A resulting in a higher excitation state A^* for this nucleus. The subsequent immediate deexcitation goes along with the emission of one or more protons, neutrons or α particles, and one or more photons. The interaction cross-section shows a marked resonance (the giant dipole resonance) with height $\sigma_{A\gamma} \approx A$ mb. A threshold energy of $E_{\mathrm{ph,\,thr}} > 0.5 A m_\mathrm{p} c^2 E_\mathrm{b}/E_A$ for the initial photon is necessary for this reaction, with E_A the energy of the incident nucleus and $E_\mathrm{b} \sim 10$–30 MeV the minimum energy to excite the giant dipole resonance (nuclear rest frame). An example for photodisintegration of ^{56}Fe during propagation in extragalactic low energy background radiation fields is given in, for example [49]. Noteworthy is that the energy-loss time is nearly independent of A: $E^{-1} dE/dt = E_A^{-1} dE_A/dt$. While photodisintegration is found to be more efficient than pion production on nuclei, it is interesting to note that Bethe–Heitler pair production on nuclei with charge number Z (cross-section $\sigma_{A\gamma,\mathrm{BH}} \approx Z^2$ mb) is superior by a factor Z^2/A to photodisintegration for heavy nuclei [50]. For protons, on the other hand, losses from photomeson production dominate over Bethe–Heitler pair production if above pion production threshold. In the environment of extragalactic jets, photodisintegration can therefore be considered negligible.

Another view of this process allows the question of the absorption probability of a high-energy gamma-ray by sufficiently dense ambient material, known as nuclear resonance absorption. For a target material of given metallicity several resonances may contribute, notably again the giant dipole resonance as the most prominent among them. If absorption columns along the line-of-sight of order $> 10^{26}\,\mathrm{cm}^{-2}$ exist, corresponding absorption troughs in the spectra of gamma-ray sources are expected. Details may be looked up in [51].

3.2.4.4 Neutron Decay

The neutron decays on a time scale of $t_{n,1/2} \approx 886.7\,\mathrm{s}$ in its rest frame into $n \rightarrow p + e^- + \bar{\nu}_\mathrm{e}$. The protons from this decay process carry most of the energy. The average energy of the $\bar{\nu}_\mathrm{e}$ is only $\approx 5\times10^{-4}$ times the original neutron energy. Those $\bar{\nu}_\mathrm{e}$ appear therefore at several orders of magnitude lower energies than those from the muon decay.

3.3 Electromagnetic Cascades

In photohadronic sources the opacity for the primary gamma-rays emerging from π^0-decay, or for the immediate synchrotron/Compton produced photons from pairs resulting in $\pi^\pm$-decay must be large, since otherwise the efficiency for the photohadronic interactions themselves would be too low to produce observable fluxes. Such primary photons or pairs initiate electromagnetic pair cascades which reprocess the injected power to lower photon energies where they eventually can escape the emission region, and be observed. In electromagnetic pair cascades photons are destroyed and produced in alternating processes such as $\gamma\gamma$ pair production and Compton scattering, synchrotron radiation, bremsstrahlung, and so on:

$$\gamma\gamma \rightarrow e^\pm \rightarrow \gamma\gamma \rightarrow e^\pm \rightarrow \ldots$$

A description of pair cascades has to include catastrophic particle and photon losses/production, continuous particle losses and particle and photon transport in a well-defined medium. The latter includes its particle, field and photon content (possibly including the self-produced cascade photons), and dielectric constant. At least two coupled differential equations are therefore necessary for an adequate pair cascade description, one Boltzmann kinetic equation for the pairs, and one for the photons:

$$\frac{\partial n_\mathrm{e}(\gamma_\mathrm{e}, t)}{\partial t} - \frac{\partial}{\partial \gamma_\mathrm{e}}[\dot{\gamma}_\mathrm{e} n_\mathrm{e}(\gamma_\mathrm{e}, t)] + \frac{n_\mathrm{e}(\gamma_\mathrm{e}, t)}{T_\mathrm{c}} = q_\mathrm{e}(\gamma_\mathrm{e}, t) + 4q_{\gamma\gamma}(n_\mathrm{ph}; 2\gamma_\mathrm{e}, t)\,, \tag{3.106}$$

$$\frac{\partial n_{\text{ph}}(E_{\text{ph}}, t)}{\partial t} + \frac{n_{\text{ph}}(E_{\text{ph}}, t)}{T_c} = q_\gamma(E_{\text{ph}}, t) + q_{\text{rad}}(n_e; E_{\text{ph}}, t) - q_{\gamma\gamma}(n_{\text{ph}}; E_{\text{ph}}, t) \tag{3.107}$$

with $n_e(\gamma_e, t)$ being the pair distribution and $n_{\text{ph}}(E_{\text{ph}}, t)$ the photon distribution, T_c the loss time scale of any catastrophic losses (including escape), $\dot{\gamma}_e$ the energy-loss rate for any continuous particle energy losses, q_e and q_γ the electron and photon source function, respectively, describing injection, q_{rad} the photon source from radiative particle losses (e.g., inverse-Compton, synchrotron radiation), and $q_{\gamma\gamma}$ describing the electron source and photon sink due to $\gamma\gamma$ pair production.[3] The radiative transfer has been expressed here by means of a photon escape probability. Examples for cascades in dielectric media include cosmic ray-induced air showers in the Earth's atmosphere and neutrino-induced showers in Moon regolith (via the Askaryan effect). In the following we consider only isotropic cascades in vacuum as the most relevant in jet environments. The isotropy of the cascade particles is ensured by pitch angle scattering in the magnetic field of the jets. Furthermore, thermal Comptonization shall be neglected, and we shall also follow the cascade in a spherical region of size R ("emission region").

A necessary condition to initiate pair cascades is a sufficiently high optical depth for photon photon pair production: $\tau_{\gamma\gamma} \sim \ell > 1$ [52] with the compactness parameter ℓ as defined in (3.86). This requires that the initial photon has sufficiently high energy to overcome the threshold for pair production (see Section 3.2.3), and the environment offers a sufficiently high density of ambient photons. The former may be produced, for example, in the course of photomeson production by ultrarelativistic protons. To describe the production of the initial photon or electron[4] requires a third kinetic equation that is coupled to the previous two cascade equations. For example, if the initial electron is produced by proton initiated processes where the proton losses can be considered as continuous,[5] the corresponding kinetic equation for protons with distribution n_p reads:[6]

$$\frac{\partial n_p(\gamma_p, t)}{\partial t} - \frac{\partial}{\partial \gamma_p}[\dot{\gamma}_p n_p(\gamma_p, t)] + \frac{n_p(\gamma_p, t)}{T_c} = q_p(\gamma_p, t) , \tag{3.108}$$

where $q_p(\gamma_p, t)$ is the proton source. If the cascade condition is met, pairs will be produced upon the initial photon/electron energy injection. Those pairs will lose their energy via Compton, synchrotron or bremsstrahlung, depending on the environment of the cascade. The so-produced photons possess less energy than the

3) The factor 4 associated with $q_{\gamma\gamma}$ is due to the pair injection term being proportional to the $\gamma\gamma$ pair production term and two particles (electron–positron) being produced with energy $\gamma_e \approx E_{\text{ph}}/2$ [12, 52].

4) Since high-energy electrons are close to 100% radiatively efficient in typical jet environments, we can treat here photons and electrons as equal.

5) For example, Bethe–Heitler pair production or synchrotron emission.

6) Note that photomeson production is *not* a continuous loss process.

initial photon, but possibly still enough to produce pairs via $\gamma\gamma$ pair production. This process continues, thereby degrading the energy of the initial photon, until pair production ceases to be possible, and the photons can escape the emission region. If this reprocessing of the initial photon from high to low energies occurs predominantly through the synchrotron (Compton) plus pair production channel, the corresponding cascades are called "synchrotron (Compton)-supported" cascades.

As an example let us follow the development of a series of generations of a Compton cascade in the Thomson regime, $\gamma\epsilon_s < 1$ with $\epsilon_s = E_s/m_e c^2 \ll 1$ ("Thomson cascades" if at least the first-order Compton scatterings (Section 3.2.2) take place in the Thomson regime, as opposed to "Klein–Nishina cascades"; [52]). The emission region of size R shall be filled with soft photons of energy ϵ_s and luminosity L_s, fulfilling $\tau_{\gamma\gamma} \gg 1$, and relativistic electrons with Lorentz factor $\gamma \gg 1$ and luminosity L_e with corresponding particle compactness parameter $\ell_e = L_e\sigma_T/(4\pi R m_e c^3)$. The first generation of photons after inverse-Compton scattering then has energy $\epsilon_1 \approx 4/3\gamma^2\epsilon_s$, pair production on the ambient photon field provides electrons of energy $\gamma_1 \sim \epsilon_1/2$. These electrons are upscattered again in the soft photon field to $\epsilon_2 \sim 4/3\gamma_1^2\epsilon_s \sim 2\gamma^4(2/3\epsilon_s)^3$, and those photons pair produce to give electrons of energy $\gamma_2 \sim \gamma^4(2/3\epsilon_s)^3$. The second generation has developed. Inverse-Compton scattering in the kth generation then delivers photons of energy $\epsilon_k \sim 2\gamma^\kappa(2/3\epsilon_s)^{\kappa-1}$ with $\kappa = 2^k$, and those photons pair produce to give electrons of energy $\gamma_k \sim \gamma^\kappa(2/3\epsilon_s)^{\kappa-1}$. Thus the condition for k pair generations and $k+1$ photon generations is $\gamma > (2/3\epsilon_s)^{1/\kappa-1}$, with $\mathrm{Int}[\ln(\ln[2/3\epsilon_s]/\ln[2/3\epsilon_s\gamma])/\ln 2]$ the number of pair generations[7] for given γ and ϵ_s. Similar considerations can be made for the case of synchrotron-supported pair cascades. One finds $\gamma > (3/16\epsilon_B)^{1/\kappa-1}$ with $\epsilon_B = B/B_{CR}$ and B_{CR} the critical magnetic field (Section 3.2.1) and $\mathrm{Int}[\ln(\ln[3/16\epsilon_B]/\ln[3/16\epsilon_s\gamma])/\ln 2]$ the number of pair generations for given γ and ϵ_B. In jet environments or intergalactic space one finds typically $\epsilon_B \ll \epsilon_s$. Thus by comparing with the case of a Compton-supported pair cascade one finds that in synchrotron-supported cascades the initial energy can be degraded efficiently by fewer generations than through Compton-supported cascades.

The following classification scheme follows closely [52]. So far we have only considered the case where the relevant soft photon field for the cascading does not change, the photons produced from the cascading have no significant impact on the cascading procedure. These cascades generally occur if $\ell_e < 1$ [52] and are classified as "completely linear cascades". If $\ell_e > 1$ and $L_e/L_s > 1$ the cascading takes place predominantly on those photons that are produced by the cascade themselves (cascade photons), while in the case $\ell_e > 1$ and $L_e/L_s < 1$ the cascade photons have only partly an impact on the cascading procedure. Those cascades are generally classified as "completely nonlinear" and "partly nonlinear cascades", respectively. It is straightforward to show that Compton-supported Thomson pair

7) Clearly, this requirement can be softened by including higher-order Compton scattering. This, however, will not be considered in this book.

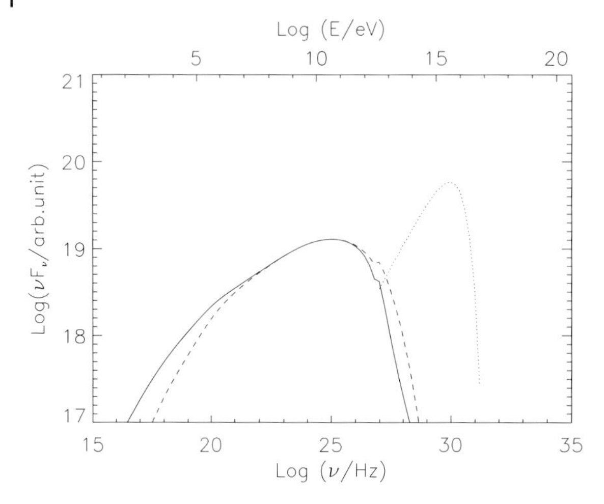

Figure 3.20 The emerging (solid line) and the first (dotted line) and second (dashed line) pair generation of synchrotron photons on production upon injection of a $\gamma_e = 10^{12}$ electron in a 1 G magnetic field and 10^{10} eV cm^{-3} target photon field between 0.01 and 100 eV.

cascades posses in general nonlinear properties, while Compton-supported Klein–Nishina pair cascades can be considered as linear cascades.

In a next step we include the effects of a limited size R of the emission region, in other words the escape probability of the photons $T_{\text{esc}} \propto R/c$. If far more photon pairs are produced than escape, $T_{\text{esc}}^{-1} \ll t_{\gamma\gamma}^{-1}(E)$, the cascade is called "saturated". It is a well-known property that the overall cascade spectrum from such cascades tends to $\alpha_\gamma \to 2$ with increasing number of generations. On the other side, "non-saturated cascades" develop if photons escape from the emission region rather than pair produce, and the cascade terminates after a few cycles. The latter is common in AGN jets where the emission region is size-limited, while saturated cascades develop in general during photon propagation in the diffuse background.

The cascade equations (3.107) can be solved analytically in only a very few limited and simplified cases, for example, fully saturated Compton pair cascades (e.g., [52]). In general they have to be solved numerically (e.g., [38]), through Monte Carlo simulations (e.g., [53]), or other methods like the matrix multiplication method [35, 54–57]. An example for the build-up of an unsaturated synchrotron-supported cascade is shown in Figure 3.20. One can clearly see the total injected power redistributed to lower energies in only 3–4 pair generations.

References

1 Rybicki, G.B. and Lightman, A.P. (1979) *Radiative Processes in Astrophysics*, Wiley-VCH Verlag GmbH.
2 Dermer, C.D. and Menon, G. (2009) *High Energy Radiation from Black Holes*, Princeton University Press.
3 Jackson, J.D. (1962) *Classical Electrodynamics*, John Wiley & Sons, Inc.
4 Griffiths, D.J. (1981) *Introduction to Electrodynamics*, Pearson/Addison-Wesley.
5 Crusius, A. and Schlickeiser, R. (1986) *Astron. Astrophys.*, **164**, L16.
6 Pozdnyakov, L.A., Sobol, I.M., and Sunyaev, R.A. (1983) *Sov. Sci. Rev. E*, **2**, 189.
7 Jauch, J.M. and Rohrlich, R. (1976) *The Theory of Photons and Electrons*, Springer-Verlag, New York.
8 Heitler, W. (1954) *The Quantum Theory of Radiation*, Dover, New York.
9 Blumenthal, G.R. and Gould, R.J. (1970) *Rev. Mod. Phys.*, **42**, 237.
10 Jones, F.C. (1968) *Phys. Rev.*, **167**, 1159.
11 Böttcher, M., Mause, H., and Schlickeiser, R. (1997) *Astron. Astrophys.*, **324**, 395.
12 Zdziarski, A.A. and Lightman, A.P. (1985) *Astrophys. J.*, **294**, L79.
13 Böttcher, M. and Dermer, C.D. (1995) *Astron. Astrophys.*, **302**, 37.
14 Donea, A.C. and Protheroe, R.J. (2003) *Astropart. Phys.*, **18**, 377.
15 Reimer, A. (2007) *Astrophys. J.*, **665**, 1023.
16 Liu, H.T., Bai, J.M., and Ma, L. (2008) *Astrophys. J.*, **688**, 148.
17 Sitarek, J. and Bednarek, W. (2008) *Mon. Not. R. Astron. Soc.*, **391**, 624.
18 Pian, E., Falomo, R., and Treves, A. (2005) *Mon. Not. R. Astron. Soc.*, **361**, 919.
19 Böttcher, M., Reimer, A., and Marscher, A.P. (2009) *Astrophys. J.*, **703**, 1168.
20 Stecker, F.W., de Jager, O.C., and Salamon, M.H. (1992) *Astrophys. J.*, **390**, L49.
21 MacMinn, D. and Primack, J.R. (1996) *Space Sci. Rev.*, **75**, 413.
22 Stecker, F.W., Malkan, M.A., and Scully, S.T. (2006) *Astrophys. J.*, **648**, 774.
23 Dwek, E. and Krennrich, F. (2005) *Astrophys. J.*, **618**, 657.
24 Franceschini, A., Rodighiero, G., and Vaccari, M. (2008) *Astron. Astrophys.*, **487**, 837.
25 Gilmore, R.C. *et al.* (2009) *Mon. Not. R. Astron. Soc.*, **399**, 1694.
26 Finke, J.D., Razzaque, S., and Dermer, C.D. (2010) *Astrophys. J.*, **712**, 238.
27 Dondi, L. and Ghisellini, G. (1995) *Mon. Not. R. Astron. Soc.*, **273**, 583.
28 Abdo, A.A. *et al.* (2010) *Astrophys. J.*, **710**, 1271.
29 Böttcher, M. and Schlickeiser, R. (1997) *Astron. Astrophys.*, **325**, 866.
30 Lightman, A.P. and Zdziarski, A.A. (1987) *Astrophys. J.*, **319**, 643.
31 Aharonian, F.A., Atoyan, A.M., and Nagapetyan, A.M. (1983) *Astrophysics*, **19**, 187.
32 Bethe, H. and Heitler, W. (1934) *Proc. R. Soc. Lond. A*, **146**, 83.
33 Maximom, L.C. (1968) *J. Res. Natl. Bur. Stand. Sect. B*, **72**, 79.
34 Chodorowski, M.J., Zdziarski, A.A., and Sikora, M. (1992) *Astrophys. J.*, **400**, 181.
35 Protheroe, R.J. and Johnson, P.A. (1996) *Astropart. Phys.*, **4**, 253, Erratum: *Astropart. Phys.*, **5**, 215.
36 Mastichiadis, A., Protheroe, R.J., and Kirk, J.G. (2005) *Astron. Astrophys.*, **433**, 765.
37 Kel'ner, S.R. and Aharonian, F.A. (2008) *Phys. Rev. D*, **78**, 034013.
38 Mannheim, K., Biermann, P.L., and Krülls, W.M. (1991) *Astron. Astrophys.*, **251**, 723.
39 Sikora, M., Kirk, J.G., Begelman, M.C., and Schneider, P. (1987) *Astrophys. J.*, **321**, 81.
40 Mücke, A., Engel, R., Rachen, J.P., Protheroe, R.J., and Stanev, T. (2000) *Comput. Phys. Commun.*, **124**, 290. photohadronic processes in astrophysics.
41 Andersson, B. *et al.* (1983) *Phys. Rep.*, **97**, 31.
42 Baldini, A., Flaminio, V., Moorhead, W.G., and Morrison, D.R.O. (1988), in *Landold–Börnstein, New Ser.*, vol. I/12b (ed. H. Schopper), Springer, Berlin, and references therein.

43 Mücke, A., Rachen, J.P., Engel, R., Protheroe, R.J., and Stanev, T. (1999) *Publ. Astron. Soc. Aust.*, **16**, 160. astrophysical environments.

44 Hagedorn, R. (1973) *Relativistic Kinematics*, Benjamin Reading, UK.

45 Gaisser, T.K. (1990) *Cosmic Rays and Particle Physics*, Cambridge University Press.

46 Ramaty, R. (1974) in *High Energy Particles and Quanta in Astrophysics* (eds F.B. McDonald and C.E. Fichtel), MIT Press, Cambridge, p. 122.

47 Mücke, A., Engel, R., Rachen, J.P., Protheroe, R.J., and Stanev, T. (1999) Proc. 26th Int. Cosm. Ray Conf. (Salt Lake City), vol. 2 (eds D. Kieda, M. Salamon, and B. Dingus), AIP Conf. Proc., vol. 516, p. 236.

48 Berezinsky, V. (2006) *Nucl. Phys. B (Proc. Suppl.)*, **151**, 260.

49 Stecker, F.W. and Salamon, M.H. (1999) *Astrophys. J.*, **512**, 521.

50 Aharonian, F. and Taylor, A.M. (2010) *Astropart. Phys.*, **34**, 258, arXiv:1005.3230.

51 Iyudin, A.F., Reimer, O., Burwitz, V., Greiner, J., and Reimer, A. (2005) *Astron. Astrophys.*, **436**, 763.

52 Svensson, R. (1987) *Mon. Not. R. Astron. Soc.*, **227**, 403.

53 Eungwanichayapant, A. (2003) PhD thesis, University of Heidelberg.

54 Protheroe, R.J. (1986) *Mon. Not. R. Astron. Soc.*, **221**, 769.

55 Protheroe, R.J. and Stanev, T. (1993) *Mon. Not. R. Astron. Soc.*, **264**, 191.

56 Mücke, A. and Protheroe, R.J. (2001) *Astropart. Phys.*, **15**, 121.

57 Mücke, A., Protheroe, R.J., Engel, R., Rachen, J.P., and Stanev, T. (2003) *Astropart. Phys.*, **18**, 593. proton blazar model.

4
Central Engines: Acceleration, Collimation and Confinement of Jets

Serguei Komissarov

4.1
Central Engine

Collimated flows, or jets, are observed in a variety of astrophysical systems, for example young stars, X-ray binaries, gamma-ray bursts, and AGN. In all these examples, the observations reveal clear evidence of ongoing accretion onto the central compact object, a star or a black hole. This suggests that disk accretion is an essential element of cosmic jet engines, providing either a source of power or collimation, or both.[1]

The physics of accretion is rich and complex. Many key aspects of accretion are the subject of ongoing investigation. Here we only outline a number of key and well-established facts and consider only the case of accretion onto supermassive black holes of AGN. A more detailed description can be found in the excellent textbook by Frank, King and Raine [3].

4.1.1
Bondi Flow

The simplest case of accretion onto a black hole is the spherical accretion from a uniform reservoir of gas at rest at asymptotically large distances from the black hole. In the Newtonian approximation this problem was solved by Bondi [4]. Although, close to the black hole the effects of general relativity become significant, most of the key properties of the flow are established at much larger distances, where the Newtonian solution is highly accurate. The Bondi solution describes a flow that becomes supersonic at the distance

$$r_{\rm acc} = \frac{5-3\gamma}{4}\frac{GM}{a_\infty^2} , \qquad (4.1)$$

1) The only exceptions are the jets of pulsar wind nebulae as there are no indications of accretion disks around radio pulsars. These jets are most likely produced not directly by the pulsars but instead form downstream of the termination shock of pulsar winds [1, 2].

Relativistic Jets from Active Galactic Nuclei, First Edition. Edited by M. Böttcher, D.E. Harris, H. Krawczynski.

where M is the black hole mass, a_∞ is the sound speed at infinity, and γ is the parameter of the assumed polytropic relation between the gas pressure and density, $p = K\rho^\gamma$, where K is constant. (This way the model captures both the adiabatic case, where $1 < \gamma < 2$ equals the ratio of specific heats, and, to a degree, cases with radiative losses or heating. For example, for isothermal flows $\gamma = 1$.) The term r_{acc} is often called the *accretion radius*. The mass accretion rate of this solution is

$$\dot{M} = 4\pi f(\gamma) r_{\text{acc}}^2 a_\infty \rho_\infty = g(\gamma) G^2 M^2 \frac{\rho_\infty}{a_\infty^3} , \tag{4.2}$$

where ρ_∞ is the gas density at infinity and

$$f(\gamma) = \left(\frac{2}{5-3\gamma}\right)^{\frac{1}{2}\frac{\gamma+1}{\gamma-1}} , \quad g(\gamma) = \pi \left(\frac{2}{5-3\gamma}\right)^{\frac{1}{2}\frac{5-3\gamma}{\gamma-1}} .$$

For $\gamma = 5/3$ we have $r_{\text{acc}} = 0$ and $g(\gamma) = \pi$, and for $\gamma = 1$ we have $g(\gamma) = \pi e^{3/2}$.

The remarkable property of adiabatic Bondi flow is that the accreted gas is essentially unbound – the *Bernoulli constant*

$$\text{Be} = \frac{v^2}{2} + h + \Psi = h_\infty , \tag{4.3}$$

where $h = [\gamma/(\gamma-1)]p/\rho$ is the enthalpy per unit mass and $\Psi = -GM/r$ is the gravitational potential of the black hole, is positive. This means that the gas has enough kinetic and thermal energy to escape from the gravitational potential well of the central mass. It is prevented from doing so by the symmetry of the flow allowing only strictly radial motion.[2] When the symmetry is broken, gas with $\text{Be} > 0$ may find routes of escaping back to infinity. In such cases, energy losses, for example via radiative cooling, become important for allowing accretion.

The symmetry can be broken when the accreting gas is not at rest at infinity but moves with speed $v_\infty \gg a_\infty$. Obviously, only the motion of fluid elements whose trajectories come sufficiently close to the black hole will be significantly effected by the hole gravity. The condition for significantly gravitational deflection of trajectories is $v_\infty^2/2 \leq GM/r_\text{b}$, where r_b is the *target parameter*, that is the distance of the undeflected trajectory from the black hole. The deflection causes shock collision of the oppositely directed streams of gas behind the hole, leading to dissipation of its kinetic energy, and subsequent cooling may result in the gas becoming gravitationally bound and capable of accreting onto the black hole. The radius of gravitational cross-section is then

$$r_{\text{acc}} = \frac{2GM}{v_\infty^2} \tag{4.4}$$

and the maximum mass accretion rate is

$$\dot{M} = \pi r_{\text{acc}}^2 v_\infty \rho_\infty = 4\pi G^2 M^2 \frac{\rho_\infty}{v_\infty^3} . \tag{4.5}$$

2) This is similar to the case of a test particle pushed radially towards the black hole with speed exceeding the local escape speed.

4.1.2 Disk Accretion

In reality the distribution of accreting gas is likely to be nonuniform at infinity and as the result the gravitationally captured gas will have net angular momentum. In this case, the centrifugal force will halt the Bondi type accretion around the so-called *circularization radius*

$$r_{\text{circ}} = \frac{l^2}{GM} \,, \tag{4.6}$$

where l is the angular momentum per unit mass (obviously, at this radius l equals the Keplerian value). The result is a rotationally supported configuration in the form of a disk or a torus. In order for the accretion to proceed further, a fraction of the trapped gas must lose its angular momentum.

In fact, the angular momentum can be redistributed between different parts of the disk in such a way that the inner parts lose angular momentum and move inwards whereas the outer parts gain angular momentum and move outwards. This can be realized via some sort of friction as the angular velocity of Keplerian motion decreases with radius and the friction between different parts of the disk would lead to outward transport of angular momentum. However, for the typical conditions of accretion disks the molecular viscosity, ν_{mol}, originated from Coulomb collisions between gas particles, is too small to have any noticeable effect. The corresponding *Reynolds number*, $\text{Re}_{\text{mol}} = v_{\text{k}} r / \nu_{\text{mol}}$, where $v_{\text{k}} = (GM/r)^{1/2}$ is the *Keplerian velocity* at the radius r, is very large, exceeding by many orders of magnitude the critical value $\text{Re}_{\text{crit}} \sim 10^3$ above which most flows become unstable and turbulent. This fact prompted the suggestion that the dynamics of accretion disks is driven by *turbulent viscosity*, $\nu_{\text{t}} \sim \lambda_{\text{t}} v_{\text{t}}$, where λ_{t} is the largest length scale of the turbulent motion and v_{t} is the corresponding velocity. Since λ_{t} cannot exceed the disk thickness H and normally v_{t} does not exceed the sound speed a_{s}, one has

$$\nu_{\text{t}} = \alpha a_{\text{s}} H \,, \tag{4.7}$$

where $\alpha \leq 1$. This argument does not preclude α to be different in different locations in the disk but without a good theory of accretion disk turbulence the most natural decision is to deal with constant α. Disk models utilizing this approach are known as *α-disks* [5]. However, it has been known for a long time that Keplerian disks, are in fact linearly stable, no matter how high their Reynolds number is. They satisfy the *Rayleigh stability criterion*, which requires the specific angular momentum to increase outwards for stability. This problem has been overshadowing the theory of accretion disks for years until in 1991 Balbus and Hawley [6] have shown that even very weak magnetic field can lead to instability in accretion disks. It turns out that this instability, now known as the *magnetorotational instability*, was already discovered by Velikhov [7] and Chandrasekhar [8] around 1960. The nonlinear phases of this instability and its ultimate effect on the accretion disk transport are the subject of intensive ongoing investigation, with great emphasis on advanced numerical simulations. It is still not entirely clear whether the α-disk model is a reasonable approximation or if another simple model can be developed to replace it. Given this state of affairs the α-prescription remains widely used in astrophysical applications and the numerical simulations are often utilized to determine the value of α. This value can also be deduced from astrophysical observations and there seems to be a contradiction between the observations and simulations, with the observational data being best fit by $\alpha \simeq 0.1$–0.4 and the numerical simulations suggesting much lower values [9].

As the gas of α-disk drifts towards smaller radii, the kinetic energy of its orbital motion dissipates and turns into heat. Provided the radiative cooling of the gas is very effective the disk is geometrically thin, with $\delta = H/r \ll 1$, otherwise it inflates and forms a thick disk or torus. The dynamics of a thin α-disk is particularly simple, as the problem of its vertical structure separates from the problem of its radial structure [5]. The condition of hydrostatic equilibrium in the direction normal to the disk plane yields the estimate for the sound speed

$$a_\mathrm{s} \simeq \delta v_\mathrm{k} \ll v_\mathrm{k} \,. \tag{4.8}$$

Thus, the orbital motion is highly supersonic. The radial velocity is given by

$$v_\mathrm{r} \simeq \frac{v_\mathrm{t}}{r} = \alpha \delta a_\mathrm{s} \simeq \alpha \delta^2 v_\mathrm{k} \,, \tag{4.9}$$

which is much less than the sound speed. The corresponding *viscous time scale* of thin disks can then be found simply as $t_\mathrm{v} = r/v_\mathrm{r}$.

The thin disk terminates at the *radius of the last stable circular orbit*, inside which the disk gas plunges into the black hole. The characteristic time scale of the plunging flow is very short and one can safely assume that the remaining energy and angular momentum of the accreting gas is swallowed by the black hole. The radius of the *last stable circular orbit* around a *Kerr black hole* is

$$r_\mathrm{ms} = r_\mathrm{g}\left[3 + Z_2 \mp \sqrt{(3 - Z_1)(3 + Z_1 + 2Z_2)}\right] \,, \tag{4.10}$$

where

$$r_\mathrm{g} = \frac{GM}{c^2} \simeq 1.5 \times 10^{13} \left(\frac{M}{10^8\, M_\odot}\right) \mathrm{cm} \tag{4.11}$$

is the *gravitational radius,*

$$Z_1 \equiv 1 + (1 - a^2)^{1/3}[(1 + a)^{1/3} + (1 - a)^{1/3}] \,,$$
$$Z_2 \equiv (3a^2 + Z_1^2)^{1/2} \,,$$

and $a = Jc/GM^2$ is the dimensionless spin of the black hole [10]. In (4.10) the upper sign corresponds to a prograde orbit and the lower sign to a retrograde orbit (we agree to use frames where $J > 0$ and $a > 0$). One can show that the absolute value of a cannot exceed unity, which corresponds to the so-called *maximally rotating black hole.* At large distances from the black hole the angular momentum vectors of

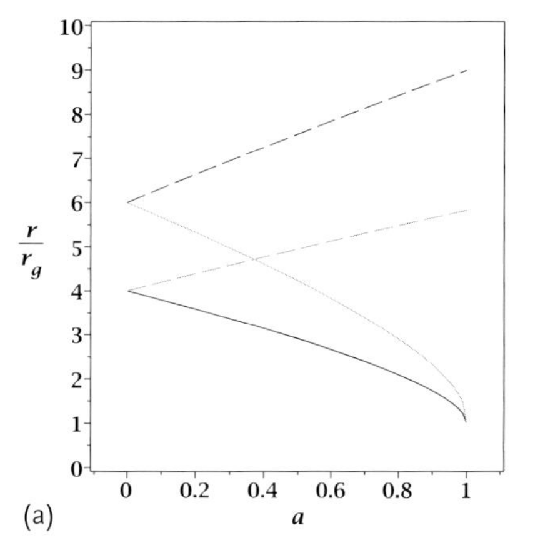

(a)

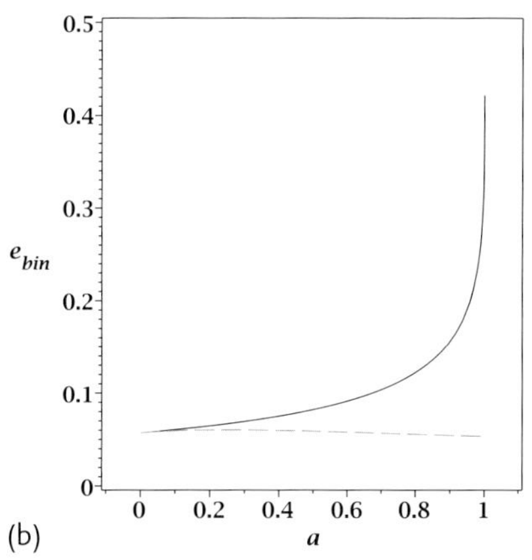

(b)

Figure 4.1 (a) Radii of last stable orbits (solid lines) and marginally bound orbits (dashed lines) for both prograde (the curves originating from (0,6)) and retrograde orbits (the curves originating from (0,4)). (b) Binding energy of the last stable prograde (solid line) and retrograde (dashed line) orbits.

accreting matter and black hole may not be parallel. However, a combination of the *Lense–Thirring precession* and viscous dumping forces the thin disk vector either to align or to counter-align with that of the black hole at smaller distances [11]. Both the prograde and the retrograde accretion disks appear to be stable [12]. The above equations give $r_{\rm ms} = r_{\rm g}$ for $a = 1$ and a prograde disk, $r_{\rm ms} = 9r_{\rm g}$ for $a = 1$ and a retrograde disk, whereas for $a = 0$, the limiting case known as *Schwarzschild black hole*, $r_{\rm ms} = 6r_{\rm g}$. The binding energy per unit rest mass of a particle on a circular orbit around a Kerr black hole is given by

$$e_{\rm bin} = c^2(1 - f(a,x)) \,, \quad \text{where} \quad f(a,x) = \frac{x^2 - 2x \pm a\sqrt{x}}{x(x^2 - 3x \pm 2a\sqrt{x})^{1/2}} \,, \tag{4.12}$$

and $x = r/r_{\rm g}$. In this equation the plus sign corresponds to a corotating orbit whereas the minus sign to a counterrotating one. According to this equation, the binding energy of the last stable orbit increases with a, yielding $e_{\rm bin}/c^2 \simeq 0.057$ for $a = 0$ and $e_{\rm bin}/c^2 \simeq 0.42$ for $a = 1$ and a prograde disk, and $e_{\rm bin}/c^2 \simeq 0.038$ for $a = 1$ and a retrograde disk (see Figure 4.1). Provided the disk remains radiatively efficient, and hence thin, all the way down to the last stable orbit, the binding energy gives us the radiated energy per unit accreted rest mass. Otherwise, some of the released energy will be advected with the flow into the black hole, thus lowering the bolometric luminosity of the disk. Introducing the *energy efficiency parameter* η, defined as the radiated energy per unit accreted rest mass, the bolometric luminosity of accretion disk can be written as

$$L_{\rm acc} = \eta \dot{M} c^2 \,, \tag{4.13}$$

where $\eta \leq e_{\rm bin}$. During the thermonuclear burning of H into He, the rest mass is converted into heat with the efficiency $\eta \sim 0.007$. Thus, the process of accretion onto black holes provides an extremely efficient source of energy.

4.1.3
The Eddington Limit

The radiation emitted from inner parts of accretion flow exerts pressure on its outer parts. The most relevant to AGN mechanism is the *Thomson scattering* on free electrons of fully ionized hydrogen. This leads to outward force per unit mass $L_{\rm acc}\sigma_{\rm T}/4\pi m_{\rm p} c r^2$, where $m_{\rm p}$ is the proton mass. For a sufficiently large $L_{\rm acc}$ this exceeds the gravity force GM/r^2 for all values of r. The critical value is known as the *Eddington luminosity*

$$L_{\rm Edd} = \frac{4\pi G m_{\rm p} c}{\sigma_{\rm T}} M \simeq 1.3 \times 10^{47} \left(\frac{M}{10^9\, M_\odot}\right) \mathrm{erg\,s^{-1}} . \tag{4.14}$$

Combining (4.13) and (4.14) we find the corresponding mass accretion rate

$$\dot{M} = \frac{4\pi G m_{\rm p}}{c\sigma_{\rm T}} \frac{M}{\eta} \simeq 2.3 \left(\frac{\eta}{0.1}\right)^{-1} \left(\frac{M}{10^8\, M_\odot}\right) M_\odot\,\mathrm{year}^{-1} . \tag{4.15}$$

The argument suggests that if the mass supply, as determined for example by (4.2) and (4.5), exceeds the Eddington mass accretion rate, the increased radiation pressure will try to repel some of the supplied gas. This may result in complex inflow-outflow flow pattern and/or in nonstationary flow dynamics, with an overall effect of reducing the mass accretion rate down to the Eddington value. In fact, the order of magnitude estimate given in (4.14) is similar to the bolometric luminosity of quasars, suggesting that their supermassive black holes are fed at the Eddington rate. The corresponding time scale, $M/\dot{M} \simeq 10^8$ years, suggests that the quasar phase of AGN is relatively short compared to the lifetime of their host galaxies.

However, the situation is much more complicated. Inside the radius

$$r_{\rm rd} \simeq 600 r_{\rm g} \alpha^{2/21} \left(\frac{M}{10^8 M_\odot}\right)^{2/21} \left(\frac{\dot{M} c^2}{L_{\rm Edd}}\right)^{16/21} , \tag{4.16}$$

where the pressure of thin accretion disks becomes dominated by radiation, they have constant height [5]

$$H \simeq \frac{3}{2} \left(\frac{\dot{M} c^2}{L_{\rm Edd}}\right) r_{\rm g} . \tag{4.17}$$

The last equation shows that although for $\dot{M} < \dot{M}_{\rm Edd}$ the disk is geometrically thin everywhere, for $\dot{M} > \dot{M}_{\rm Edd}$ the thin disk approximation breaks down near the black hole. This suggests that the radiative cooling becomes inefficient and, indeed, the inward advection of radiation wins against the outward diffusion at $r < r_{\rm tr}$, where

$$r_{\rm tr} = \left(\frac{\dot{M} c^2}{L_{\rm Edd}}\right) r_{\rm g} \tag{4.18}$$

is the so-called *trapping radius* [13]. The trapped radiation is advected into the black hole, thus reducing the overall energy efficiency of accretion below the standard values of the thin disk theory, and hence increasing the critical mass accretion rate. Moreover, the hydrostatic models of fat disks, or tori, show that the inner boundary of accreting tori is pushed closer to the black hole, towards the *marginally bound* circular orbit

$$r_{mb} = r_g(2 \mp a + 2\sqrt{1 \mp a}) , \tag{4.19}$$

where the binding energy vanishes (Figure 4.1), thus confirming the reduction of energy efficiency for super-Eddington accretion [14]. The bolometric luminosity of a radiationally supported torus with constant specific angular momentum is

$$L_{fd} \simeq L_{Edd} \log\left(\frac{r_{out}}{r_{in}}\right) , \tag{4.20}$$

where r_{out} and r_{in} are respectively the outer and the inner radii of the torus [15]. This exceeds the Eddington value but not by much. Taking r_{tr} as the outer radius and r_g as the inner radius, one then obtains the estimate

$$L_{fd} \simeq L_{Edd} \log\left(\frac{\dot{M}}{\dot{M}_{Edd}}\right) , \tag{4.21}$$

where

$$\dot{M}_{Edd} = \frac{L_{Edd}}{c^2} . \tag{4.22}$$

This result suggests that the mass accretion rate may significantly exceed $\dot{M}_{Edd}$, yet causing only moderate increase of luminosity above the Eddington value. The radiation from tori is expected to be significantly anisotropic, with the polar axis as preferred direction. Thus, even with super-Eddington luminosity the radiation pressure may not affect the gas supply to the disk in the equatorial plane.

Accretion flows with inefficient radiative cooling are often called *advection-dominated*. The discovery of almost dormant supermassive black holes, which show very low level of activity, in nuclei of most massive galaxies raised interest to the accretion regimes with $\dot{M} \ll \dot{M}_{Edd}$, as in this regime the radiation cooling may become inefficient because of the low mass density in the disk. Indeed, the observed luminosity of such dormant black holes is much lower than expected from (4.13) with $\eta \simeq 0.1$. Attempts have been made to build a self-similar model of such ADAF (*advection-dominated accretion flow*), and not without success. In fact, the model allows to explain the observed spectra of radiation from dormant black holes [16–18]. However, like the Bondi solution, it has positive Bernoulli function (see (4.3)) [18], indicating that the disk matter has sufficient energy to escape from the potential well of the black hole. This suggests that a strong wind can be relatively easily launched from the disk surface [19]. This idea has been developed further in the ADIOS (*advection-dominated inflow-outflow solution*) self-similar model, which combines the accretion flow with an outflow that carries away the angular momentum, mass, and energy from the radiatively inefficient disk [20]. This multiparametric model includes both the cases with insignificant mass loss from the disk and the cases where most of the gas supplied to the disk at its outer radius is ejected with the wind. Such a heavy mass loss is another factor which helps to explain the low observed luminosity of dormant black holes in this model.

4.1.4 Fuel Supply

The radial extent of a stationary α-disk is limited by the condition of dominating gravity of the black hole. The transverse (to the disk plane) components of gravitational force due to the black hole and the self-gravity of the disk become comparable at the distance

$$r_{\text{out}} \simeq 10^{17} \alpha^{2/3} \left(\frac{\delta}{10^{-2}}\right)^{8/3} \left(\frac{\dot{M}}{\dot{M}_{\text{Edd}}}\right)^{-3/2} \left(\frac{M}{10^8 \, M_\odot}\right)^{1/3} \text{cm} \,. \tag{4.23}$$

This is easily calculated using the expression for the mass accretion rate,

$$\dot{M} = 2\pi r H \rho v_{\text{r}} \tag{4.24}$$

and (4.9). The estimate $\delta \sim 0.01$ follows from the expressions for the thickness of the disk in the radiatively dominated inner region (see (4.17)), the radial extension of this region (see (4.16)), and the fact that δ grows very slowly with radius in the gas pressure-dominated outer region [5]. At the distance larger than $\delta^{-4/3} r_{\text{out}}$, where the disk mass equals the black hole mass, the disk becomes self-gravitating in the radial direction as well. Such self-gravitating disks are expected to become gravitationally unstable and fragment into clumps. Even if the interaction between clumps can lead to some sort of effective viscosity [21], the corresponding viscous time scale is too large. Using (4.9) one finds that

$$t_{\text{v}} \simeq 1.5 \times 10^9 \left(\frac{\alpha}{0.1}\right)^{-1} \left(\frac{\delta}{10^{-2}}\right)^{-2} \left(\frac{M}{10^8 \, M_\odot}\right)^{1/3} \left(\frac{r_{\text{out}}}{1\,\text{pc}}\right)^{3/2} \text{years} \,. \tag{4.25}$$

Thus, whatever the mechanism of feeding the supermassive black holes in AGNs, it must be capable of supplying gas into the inner parsec where the usual viscous disk transport can take over.

The actual mechanism of gas supply in AGN is still an open issue. Generally, the gas can come either from the galactic interstellar medium (ISM) or from stars. For example, stars can be tidally disrupted near the black hole. For this to occur the black hole mass has to be $\lesssim 10^8 \, M_\odot$ [22, 23]. More massive black holes swallow normal stars without disruption and substantial energy release. For dense central clusters, direct stellar collisions can also liberate mass comparable to the total mass of colliding stars. For these processes to be responsible for the quasar activity the disruption rate should be around one star per year, which is probably too high.

Alternatively, the gas can come from ISM, which is particularly plentiful in disk galaxies on kpc scales. The issue here is how to remove its angular momentum so it can flow towards the central black hole. One promising mechanism is the gravitational torque. For this to operate the gravitational potential has to be nonaxisymmetric, which can be a result of gravitational instabilities. Indeed, active galaxies exhibit various types of nonaxisymmetric perturbations on kiloparsec scales, such as "bars", "ovals", "rings", and other irregularities. However, not all irregular or barred spiral galaxies are active. In order to fuel the supermassive black holes, which, as we now believe, reside in the centers of all massive galaxies, nonaxisymmetry must be present on all scales from 10 kpc down to 1 pc. This suggests to look for hierarchical types of gravitational instabilities, such as the "bars-within-bars" [24]. Galaxies in the early Universe are particularly rich in gas. Moreover, strong interactions between galaxies – including galactic mergers, which are also frequent in the early Universe – are a source of strong perturbations capable of driving this gas into galactic nuclei. This could be the reason for the observed strong cosmological evolution of powerful AGN. Computer simulations that can follow the evolution of both the gas and the stellar components of galaxies are becoming an increasingly important tool for resolving this issue. Thus, the recent simulations of galactic mergers and bar-unstable galactic disks show that for gas-rich systems, hierarchical gravitational instabilities do indeed develop and are capable of driving up to 10 $M_\odot$ per year into the central 0.1 pc [25]. This is only a small fraction of gas driven inside the central 100 pc as most of this gas is converted into stars. The star formation and its feedback on the ISM gas are important ingredients of this process, in addition to the self-gravity. Since these processes cannot be included directly because of the limited computer power, they are incorporated phenomenologically and this is where the dominant uncertainties of such simulations reside.

The elliptical galaxies that host radio-loud AGN at low redshift are not gas-rich and the gas supply of their black holes has to be somewhat different. One possible channel is the mass loss experienced by stars in the cause of their natural evolution. For the typical stellar population of ellipticals, the mass-loss rate is expected to be in the range of 10^{-11}–10^{-12} year^{-1} per unit mass. Thus, only the combined mass loss from a substantial fraction of stars in the galaxy can provide the mass required to fuel a powerful AGN. Because of the slow rotation of elliptical galaxies this gas can accumulate within the central kiloparsec, without any need to lose its angular momentum, and form there a rotationally supported disk. The mass of this nuclear disk could grow until it becomes self-gravitating and gravitational instabilities trigger an inflow towards the central parsec.

4.2 Magnetic Fields

4.2.1 Basics

So far we have only mentioned the magnetic field in connection with the magnetorotational instability of accretion disks. However, they are also instrumental in many other aspects of the central engine dynamics, and first of all in the jet production [26, 31, 37].

Magnetic fields introduce a qualitative change in the dynamics of plasma. In gas dynamics the force associated with the internal energy of gas is reduced to isotropic pressure. In contrast, the *stress tensor* associated with the electromagnetic field is anisotropic

$$T^{ij} = -\frac{1}{4\pi}(E^i E^j + B^i B^j) + \frac{1}{8\pi}(E^2 + B^2)g^{ij} , \tag{4.26}$$

where E and B are the electric and magnetic fields, respectively, and g^{ij} is the metric tensor of space. The force per unit volume is given by the divergence of the stress tensor, $f^i = -\nabla_j T^{ij}$. In terms of the electric current density, $\boldsymbol{j}$, and the electric charge density, ρ_q, this reads as

$$\boldsymbol{f} = \rho_q \boldsymbol{E} + \frac{1}{c}\boldsymbol{j} \times \boldsymbol{B} . \tag{4.27}$$

The first term in this equation is known as the *electrostatic force*. The second term, often called the *Laplace force*, can be written as

$$\boldsymbol{j} \times \boldsymbol{B} = \boldsymbol{i}_n \frac{1}{R_c}\frac{B^2}{4\pi} + \boldsymbol{i}_t \frac{d}{ds}\left(\frac{B^2}{8\pi}\right) - \nabla\left(\frac{B^2}{8\pi}\right) , \tag{4.28}$$

where s is the distance along the magnetic field line, increasing in the direction of $\boldsymbol{B}$, $\boldsymbol{i}_t = d\boldsymbol{r}/ds$ is the unit tangent vector to the magnetic field line, $\boldsymbol{i}_n$ is the unit vector pointing towards the center of curvature of the magnetic field line, and R_c is its radius of curvature. The last term in this equation is the gradient of *magnetic pressure*, $p_m = B^2/8\pi$. The first two terms are often called the *magnetic tension* – it is these terms that make the dynamics of magnetized plasma qualitatively different from gas dynamics. In fact, the second term is completely balanced by the component of the magnetic pressure force along the magnetic field line, what can be anticipated already from the fact that the Laplace force is perpendicular to the magnetic field. The first term tends to reduce length of magnetic field lines. In axisymmetric configurations, it gives rise to the *hoop stress* associated with the azimuthal component of the magnetic field – the force that can lead to self-collimation of magnetized flows.

The electromagnetic field not only influences the dynamics of plasma via forces but also provides an additional channel for the energy transport. Now energy can be transported not only in the form of thermal and kinetic energies but also in the form of the electromagnetic energy. The density electromagnetic energy flux, or the *Poynting flux*, is

$$S = \frac{c}{4\pi} E \times B . \tag{4.29}$$

In the perfect conductivity approximation, the electric field in the fluid frame vanishes. In other frames, it is given by the Lorentz transformation, which yields

$$E = -\frac{1}{c} v \times B , \tag{4.30}$$

where v is the fluid velocity. In this approximation, the magnetic field lines are advected with the fluid as if they were *frozen* into it.

Sound waves are the only waves which can propagate through the fluid in hydrodynamics. In *magnetohydrodynamics* (MHD), the magnetic field results in the appearance of several new types of waves of this sort. First of all, this is the *Alfvén wave*. In the fluid frame this wave propagates with speed

$$c_a^2 = c^2 \frac{B^2}{B^2 + 4\pi w} \tag{4.31}$$

along the magnetic field lines. Here $w = p + e = \rho c^2 + p\gamma/(\gamma - 1)$ is the relativistic enthalpy, and γ is the ratio of specific heats. In the nonrelativistic limit, $p \ll \rho c^2$ and $B^2 \ll \rho c^2$, and thus $c_a^2 \to B^2/4\pi\rho$. This wave transports magnetic perturbations that keep the magnetic (and) total pressure unchanged. The second new mode is the *fast magnetosonic wave*. This wave is driven by perturbations of total pressure and it changes the gas pressure in the same sense as the magnetic pressure (e.g., if the gas pressure increases then so does the magnetic pressure). It can propagate in all directions with respect to the magnetic field and in the fluid frame its speed exceeds both the Alfvén speed and the *sound speed*,

$$a_s^2 = c^2 \gamma \frac{p}{w} . \tag{4.32}$$

In the nonrelativistic limit, this reduces to the familiar $a_s^2 = \gamma p/\rho$. The third new mode is the *slow magnetosonic wave*, which is also related to pressure perturbation but in a different way. In contrast to the fast magnetosonic mode this wave propagates mainly along the magnetic field lines and changes the gas pressure in the opposite sense to the magnetic pressure. Its speed is lower compared to both the sound speed and the Alfvén speed. Slow shock waves can be very efficient in dissipating magnetic energy and play important role in magnetic reconnection.

When the assumption of perfect conductivity is relaxed, the magnetic field provides an additional channel of plasma heating, the *magnetic dissipation*. In collisional plasma this is the *Ohmic dissipation*, where the electric currents, which support the magnetic field, decay as the result of mainly binary collisions between the charged particles carrying the current and the target particles. In collisionless plasma, the processes responsible for magnetic dissipation involve collective interaction of many charges. The result is not only plasma heating but also acceleration of nonthermal relativistic particles. The magnetic field is no longer perfectly frozen and the magnetic field lines can slip through the fluid and change their topology via the process of *reconnection*. In strongly magnetized configurations, where the magnetostatic equilibrium is supported via a fine balance between the magnetic pressure and tension, such reconnection can result in violent eruptions with dynamic conversion of magnetic energy into the kinetic energy of bulk motion.

The dynamics of magnetized plasma is a very rich and complex subject and all we can only afford within the scope of this book an outline of a number of key issues. Interested readers are directed to numerous textbooks on magnetohydrodynamics (MHD) available in all good science libraries. Most of them deal with Newtonian MHD. Although the relativistic MHD is not that well represented, mathematically well-advanced readers may find useful the monographs by Lichnerowicz [27] and Anile [28].

4.2.2
Powering Magnetic Winds and Jets

In order to understand the origin of magnetically driven outflows from the electrodynamic viewpoint it makes sense to consider the simple case of a perfectly conducting sphere with the *split-monopole* magnetic field, that is, a radial magnetic field that changes direction in the equatorial plane.[3] The magnitude of such a field depends only of the distance from the sphere, with $B \propto r^{-2}$. When such a sphere is set into rotation with the angular velocity Ω about its polar axis, free charges on its surface are subject to the Lorentz force, $f = q\boldsymbol{v} \times \boldsymbol{B}$, where q is the electric charge, $v = \Omega \varpi$ is the linear rotational velocity of the sphere, and ϖ is the cylindrical radius. Let us assume, without any loss of generality, that the magnetic moment vector is aligned with the spin vector, as shown in Figure 4.2a. In this case, the Lorentz force makes free positive charges, if they are available, to move towards the equator and the negative ones towards the poles. These separated electric charges will support a quadrupole electric field and maintain the difference between the electrostatic potentials of the equator and the poles. In other words, we have an electric battery. One only needs to connect it using sliding contacts and wires to some external load and the electric current will start flowing in the circuit. On the surface of the sphere the current will flow from the poles to the equator and the $\boldsymbol{j} \times \boldsymbol{B}$ force will give a net torque slowing down the sphere rotation. This is consistent with the fact that the rotational energy of the sphere is the only source of energy in this circuit. A rotating magnetized disk offers a similar configuration of electric currents (see Figure 4.2b).

In the astrophysical circumstances, there are no wires. Instead, the space around the magnetized rotator is filled with conducting plasma, so the electric current

3) In order to sustain such a field, a perfectly conducting plate has to reside in the equatorial plane. In astrophysical circumstances this role can be played by an accretion disk.

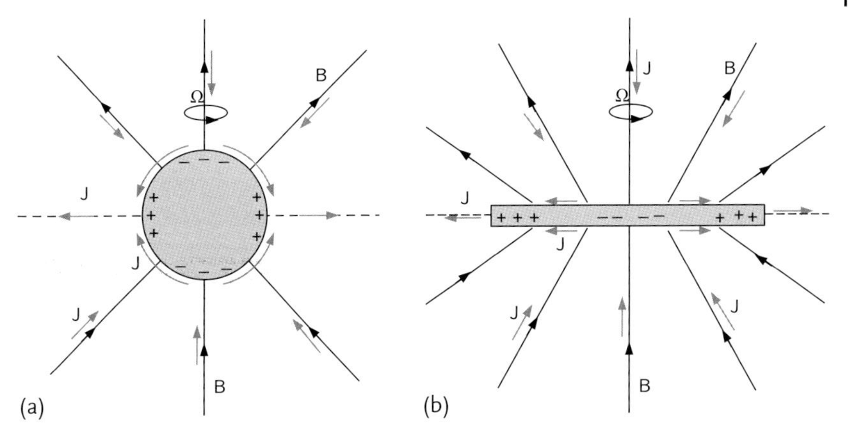

Figure 4.2 Electric charge separation and the current system of a rotating conducting sphere (a) and a rotating conducting disk (b). The azimuthal components of fields and currents are not shown.

of astrophysical circuits is volume distributed. Suppose that a perfectly conducting plasma exists around our rotating sphere and for simplicity consider only the steady-state axisymmetric case. Then the Faraday equation becomes $\nabla \times \boldsymbol{E} = 0$ and requires the azimuthal component of electric field to vanish, $E_\phi = 0$. Provided the magnetic field is so strong that to zero order one can ignore the inertial and gas pressure terms in the momentum equation, the magnetosphere is close to the so-called *force-free equilibrium*, where

$$\rho_{\mathrm{q}} \boldsymbol{E} + \frac{1}{c} \boldsymbol{j} \times \boldsymbol{B} = 0 \,. \tag{4.33}$$

This equation requires the poloidal component of electric current, j_{p}, to be parallel to the poloidal component of the magnetic field, B_{p}, and one may say that the magnetic field plays the role of wires in this case. When the inertial terms are retained the electromagnetic force is not exactly vanishing and it is responsible for the acceleration of the magnetospheric plasma. Equation (4.33) also yields $\boldsymbol{E} \cdot \boldsymbol{B} = 0$, and thus $\boldsymbol{E}_{\mathrm{p}} \cdot \boldsymbol{B}_{\mathrm{p}} = 0$, but this result is more general as it follows from the perfect conductivity condition, (4.30), as well.

The results $E_\phi = 0$ and $\boldsymbol{E} \cdot \boldsymbol{B} = 0$ allow us to write $\boldsymbol{E} = -k\varpi \boldsymbol{i}_\phi \times \boldsymbol{B}_{\mathrm{p}}$, where $\boldsymbol{i}_\phi$ is the unit vector of the azimuthal direction. Substituting this expression into the Faraday equation, $\nabla \times \boldsymbol{E} = 0$, one immediately finds that k is constant along magnetic field lines. Since on the surface of the sphere $E = -\boldsymbol{v} \times \boldsymbol{B}/c = -(\varpi \Omega/c)\boldsymbol{i}_\phi \times \boldsymbol{B}_{\mathrm{p}}$ we find that $k = \Omega$, and thus

$$\boldsymbol{E} = -\frac{\Omega \varpi}{c} \boldsymbol{i}_\phi \times \boldsymbol{B}_{\mathrm{p}} \,. \tag{4.34}$$

This is known as *Ferraro's law of isorotation* – the electric field is the same as if the magnetospheric plasma was in the state of pure rotation with the same angular velocity as the central body. However, in general (i) the magnetospheric plasma can flow along the poloidal field lines and (ii) the "azimuthal velocity", $\varpi \Omega$, may exceed the speed of light. Thus, this interpretation is not always accurate. The surface where $\varpi \Omega = c$ is called the *light cylinder*.

According to the Ampére law

$$\nabla \times \boldsymbol{B} = \frac{4\pi}{c} \boldsymbol{j} , \tag{4.35}$$

the poloidal electric current creates the toroidal component of magnetic field, B_t, and thus the magnetic field becomes helical. This result implies the nonvanishing poloidal electromagnetic fluxes of energy

$$S_e = \frac{c}{4\pi} E_p B_\phi \tag{4.36}$$

and angular momentum

$$S_{am} = \frac{1}{4\pi} B_\phi B_p . \tag{4.37}$$

In order to estimate the energy and the angular momentum fluxes one needs to know the azimuthal magnetic field and hence the electric current flowing in the magnetosphere. This can be estimated using the following simple understanding of the way the electromagnetic mechanism operates – the rotation at the base twists the magnetic field lines and the resultant helix propagates outwards with the speed of light. (In magnetospheres with low magnetization and hence low MHD wave speeds compared to the speed of light, the speed of the helix should also be lower.) In time dt the magnetic field line moves the distance $dl_\phi = \varpi \Omega \, dt$ in the azimuthal direction and the distance $dl_p = c dt$ in the poloidal direction (see Figure 4.3). Thus,

$$\left| \frac{B_\phi}{B_p} \right| = \frac{dl_\phi}{dl_p} = \frac{\varpi \Omega}{c} .$$

Taking into account that the magnetic field line is twisted in the direction opposite to its rotation, this gives us

$$B_\phi = -\frac{\varpi \Omega}{c} B_p . \tag{4.38}$$

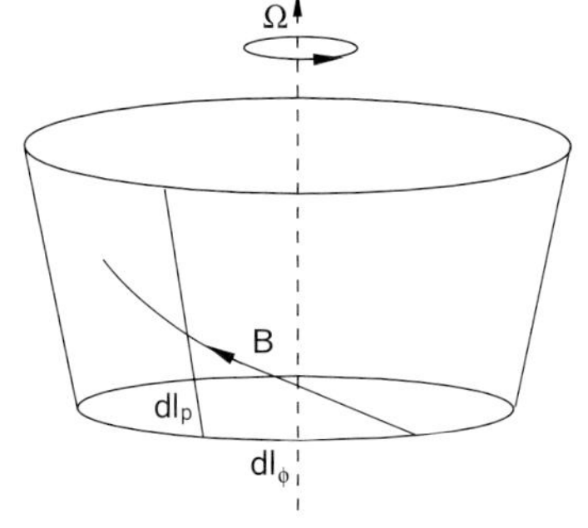

Figure 4.3 Magnetic helix created by the rotation at the base of the magnetosphere.

It has the same magnitude as the electric field (see (4.34)). For the split-monopole configuration of the poloidal magnetic field, $B_{\rm p} = \Psi/2\pi$, where Ψ is the total magnetic flux through the northern hemispheres, and thus

$$B_\phi = \frac{\Psi \Omega \sin\theta}{2\pi r} . \tag{4.39}$$

Now we can use the integral form of the Ampére law (4.35) in order to find the poloidal electric current flowing inside the conical flux surface of opening angle θ,

$$I(\theta) = \frac{\varpi B_\phi c}{2} = I_{\rm GJ} \sin^2\theta , \tag{4.40}$$

where $I_{\rm GJ} = \Psi\Omega/4\pi$ is called the *Goldreich–Julian* current.

Substituting (4.34) and (4.39) into (4.36) and integrating over the solid angle we find that the total electromagnetic luminosity is

$$L_{\rm e} = \frac{8}{3c}\left(\frac{\Psi\Omega}{4\pi}\right)^2 . \tag{4.41}$$

The corresponding total flux of angular momentum is

$$L_{\rm am} = \frac{1}{\Omega} L_{\rm e} . \tag{4.42}$$

This loss of angular momentum of the magnetized rotator provides an additional mechanism, different from the viscous friction discussed in Section 4.1.2, of driving accretion in accretion disks. This mechanism does not require dissipation of the orbital kinetic energy, which is generic in the viscous models, and may result in rather cold, and hence geometrically thin, dissipationless disks.

The inertia of magnetospheric plasma leads to a deviation from the force-free condition (4.33). As a result, the poloidal current is no longer exactly parallel to the poloidal magnetic field. Instead, it crosses the poloidal field lines and forms a closed circuit inside the magnetosphere, as shown in Figure 4.4. Now, the Laplace force has a component which is parallel to the poloidal magnetic field and accelerates the magnetospheric plasma, converting the Poynting flux into the kinetic energy of bulk motion. This is the origin of astrophysical winds and jets in the theory of magnetically driven outflows. The plasma has to be constantly resupplied into the magnetosphere from the surface of the rotator. In the case of accretion disks, the mechanisms of plasma supply can be similar to those operating on the Sun – *coronal mass ejection* and other phenomena related to the Sun's magnetospheric activity. In the case of black hole (and neutron star) magnetospheres, different mechanisms, which are quantum in nature, come into play.

The detailed mathematical theory of MHD winds is rather complicated and is beyond the scope of this book – interested readers will find useful the comprehensive analysis of this problem in [29, 30]. We only mention that the steady-state wind solutions describe flows that become faster with the distance and one after another cross three critical surfaces, slow magnetosonic, Alfvénic, and fast magnetosonic.

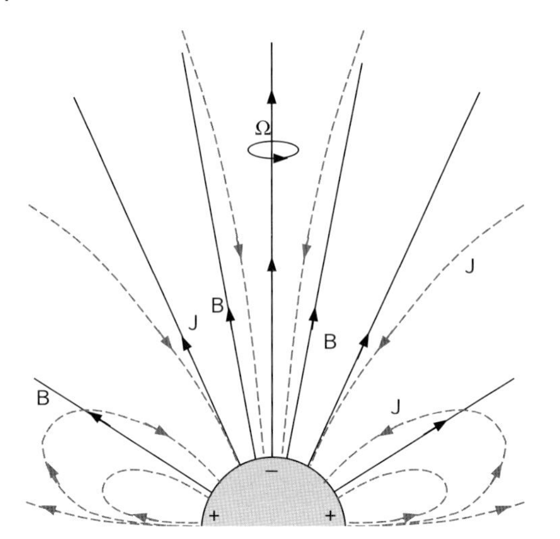

Figure 4.4 Current closure in non-force-free rotating magnetosphere. The solid lines show the poloidal component of the magnetic field and the dashed lines show the poloidal electric current (the azimuthal components are not shown).

An MHD wave created downstream of the corresponding critical surface cannot move upstream across this surface, for example the slow magnetosonic wave cannot cross the slow magnetosonic surface, and so on. Thus, each such surface plays the role of event horizon for the relevant wave and separates causally disconnected parts of the flow. As a result, the solution upstream of the fast magnetosonic surface is completely independent from the downstream solution.

Generally, the MHD wind equations are too complex to yield an analytic solution. Therefore, the attention of theorists has been focused on particularly simple cases, where the self-similarity or perturbative approach can be utilized. For example, Blandford and Payne [31] developed the self-similar model of nonrelativistic magnetized wind from a thin accretion disk, which was generalized later to the relativistic regime [32]. Nowadays, numerical simulations can be used to complement the theoretical studies of this sort and to overcome their limitations.

4.2.3 The Blandford–Znajek Mechanism

In general relativity the total energy of a particle in a gravitational field, which combines its rest-mass energy, kinetic energy, and the gravitational energy, is called the *redshifted energy*, or the *energy at infinity*. It turns out that near a rotating black hole there are orbits which correspond to negative redshifted energy. They exist only inside the closed surface called the *ergosphere*, which lies outside the event horizon – in the equatorial plane its radius is $2r_g$ and in the polar direction it has the same radius as the event horizon. Particles entering (or leaving) the ergosphere always have positive redshifted energy, and thus in order to obtain negative redshifted energy a particle has to interact with other particles inside the ergosphere. If, for example, two particles with initial redshifted energies $e_1 > 0$ and $e_2 > 0$ come inside the ergosphere and collide in such a way that the first one is pushed into orbit with negative redshifted energy, $e_1' < 0$, the redshifted energy of the second one increases by the amount $e_1 - e_1' > e_1$. If this particle then escapes, to infinity it carries away the energy that exceeds the total initial energy of the two particles by the amount $-e_1'$. The black hole energy decreases by exactly the same amount when it eventually swallows the trapped first particle. This means that it must be possible, at least in principle, to extract some, but not all, of the mass-energy of a rotating black hole. The total energy extractable via this mechanism, named after it's author R. Penrose [33], is

$$\mathcal{E}_{\rm rot} = f_{\mathcal{E}}(a) M c^2 \simeq 1.8 \times 10^{63} f_{\mathcal{E}}(a) \left(\frac{M}{10^9\, M_\odot} \right) \text{erg} , \tag{4.43}$$

where $f_{\mathcal{E}}(a) = 1 - \sqrt{r_+/2r_{\rm g}} \le 0.29$, $r_+ = r_{\rm g} f_+(a)$ is the radius of the event horizon, and $f_+(a) = 1 + \sqrt{1-a^2}$. Provided the black hole is rapidly rotating, $a \sim 1$, this is sufficient to sustain the quasar-scale power for a million years. Unfortunately, the Penrose mechanism in its original mechanical form is unlikely to operate in astrophysical conditions. However, it has an electromagnetic version, which is much more promising.

In 1974, Wald discovered that if a rotating black hole is placed in a vacuum with a uniform magnetic field, which is aligned with the black hole's rotational axis, then a stationary quadrupole electric field, similar to that around a rotating magnetized sphere, is generated as well [34]. This is a pure GR effect as the problem does not involve electric charges, and hence the electric field is not related to any kind of electric charge separation. In fact, one can write the tensor equations of black hole electrodynamics in exactly the same vector form as the Maxwell equations for an electromagnetic field in matter. From these equations it follows that the curved space of black holes acts as a medium, whose electromagnetic properties satisfy the constitutive equations

$$\boldsymbol{E} = \alpha \boldsymbol{D} + \boldsymbol{\beta} \times \boldsymbol{B} , \quad \boldsymbol{H} = \alpha \boldsymbol{B} - \boldsymbol{\beta} \times \boldsymbol{D} , \tag{4.44}$$

where α and β are functions describing the space-time metric (see (10.5)). One can attribute Wald's electric field to the electromagnetic action of this "medium" [35].

The similarity between the vacuum electromagnetic field of a rotating black hole and that of a rotating conducting sphere prompted the investigation of whether the rotational energy of black holes can be extracted electromagnetically in the same fashion as we described in Section 4.2.2, and whether this can be important in the astrophysics of black holes. One important issue is the supply of charged particles. They cannot be extracted from the black hole as the event horizon can only act as a sink of particles. However, electron–positron pairs can be created in the black hole magnetosphere via collisions of high energy photons emitted by the accretion disk (see Chapters 2 and 3 on $\gamma\gamma$ pair production). In addition, electron–positron pairs can be created in the potential gaps in the same fashion as in radio pulsars [36].

These processes create very tenuous plasma and like the magnetospheres of pulsars the black hole magnetospheres are expected to be extremely magnetized, with $\rho c^2 \ll B^2$ and $c_a \simeq c$.

The second issue is the development of electric currents. Blandford and Znajek [37] analyzed this problem using the force-free approximation and the perturbative approach for the case of a slowly rotating black hole, $|a| \ll 1$. They considered two stationary magnetic configurations, one with the split-monopole and one with the parabolic geometry. For the split-monopole they found that the asymptotic solution is the same as that for a rotating sphere with the angular velocity $\Omega = 0.5\Omega_h$, where

$$\Omega_h = f_\Omega(a)\frac{c^3}{2GM} \simeq 10^{-4} f_\Omega(a)\left(\frac{M}{10^9 M_\odot}\right)^{-1} s^{-1} , \tag{4.45}$$

where $f_\Omega(a) = a/(1+\sqrt{1-a^2})$, is the angular velocity of the black hole.[4] Later, the computer simulations of black hole magnetospheres confirmed this result and revealed that it is quite accurate even for rapidly rotating black holes [38, 39]. Combining this result with (4.41), we obtain the following expression for the electromagnetic luminosity of the black hole

$$L_{BZ} = \frac{2}{3c}\left(\frac{\Psi\Omega_h}{4\pi}\right)^2 . \tag{4.46}$$

The structure of the plasma flow, which develops in black hole magnetospheres, can be described as a paired wind. It involves not only the outflow, which as before crosses all three critical surfaces and becomes superfast magnetosonic on its way to infinity, but also an inflow, which becomes superfast magnetosonic before crossing the black hole horizon, and thus making the event horizon causally disconnected from the outflow. The total outgoing energy flux however is constant all the way from the black hole horizon to "infinity". This apparent paradox is resolved in a similar fashion as in the Penrose mechanism. Roughly speaking, the inflow carries into the black hole negative redshifted electromagnetic energy. More precisely, inside the ergosphere the flux of redshifted energy can be directed outwards and at the same time any local physical observer orbiting the black hole would measure an ingoing flux of electromagnetic energy [40].

In order to estimate the electromagnetic power of black holes and accretion disks one needs to know the magnetic flux they can accumulate. Kerr black holes cannot have their own magnetic field, so the question is how much magnetic flux can be squeezed onto the black hole by its accretion disk. This depends on the poorly understood dynamics of magnetic fields in accretion disks. We are not even sure on its origin – it may be dragged in along with gas from the ISM or generated *in situ* via magnetic dynamo. Given the observed power of the AGN jets and the estimated masses of their black holes one can estimate the magnetic flux required

4) For an observer at infinity this is the angular velocity of particles as they approach the event horizon.

by the BZ-mechanism

$$\Psi_{\rm h} \simeq 1.4 \times 10^{33} \frac{1}{f_\Omega(a)} \left(\frac{L}{10^{46}\,{\rm erg\,s^{-1}}}\right)^{1/2} \left(\frac{M}{10^9\,M_\odot}\right) {\rm G\,cm^2}\,. \tag{4.47}$$

Dividing this by the half of the event horizon area, $S = 4\pi r_{\rm g}^2 f_+(a)$, one obtains the corresponding strength of the magnetic field

$$B_{\rm h} \simeq 5 \times 10^3 \frac{1}{f_\Omega(a) f_+(a)} \left(\frac{L}{10^{46}\,{\rm erg\,s^{-1}}}\right)^{1/2} \left(\frac{M}{10^9\,M_\odot}\right)^{-1} {\rm G}\,. \tag{4.48}$$

The usual assumption is that the magnetic pressure is a fraction of the total thermodynamic disk pressure near the inner edge of the disk, $B^2/8\pi = \beta_{\rm m} p$. For example, consider a radiation pressure-dominated disk with $\dot{M} > L_{\rm Edd}/c^2$. Then according to (4.17) this disk is geometrically thick near the black hole and its pressure can be estimated as $p = GM\rho/H$. The gas density can be found from the mass conservation as $\rho = \dot{M}/2\pi r H v_{\rm r}$, where $v_{\rm r} = \alpha a_{\rm s} H/r$ and $a_{\rm s}^2 \simeq p/\rho$. Combining these equations we find

$$p = \frac{\dot{M}(GM)^{1/2}}{2\pi\alpha H^{5/2}}\,. \tag{4.49}$$

The corresponding magnetic field is

$$B_{\rm h} \simeq 1.7 \times 10^4 \left(\frac{\beta_{\rm m}}{\alpha}\right)^{1/2} \left(\frac{M}{10^9\,M_\odot}\right)^{-1/2} \left(\frac{\dot{M}c^2}{L_{\rm Edd}}\right)^{-3/4} {\rm G}\,, \tag{4.50}$$

and the BZ-power is

$$L_{\rm BZ} \simeq L_{\rm Edd} \frac{f_+^2(a) f_\Omega^2(a)}{4} \left(\frac{\beta_{\rm m}}{\alpha}\right) \left(\frac{\dot{M}c^2}{L_{\rm Edd}}\right)^{-3/2}\,. \tag{4.51}$$

This magnetic field is sufficiently strong and the BZ-power is comparable with the Eddington luminosity. Somewhat surprisingly, the power actually decreases with the mass accretion rate. This is because a higher mass accretion rate leads to a higher disk thickness and a lower radiation pressure, $p \propto \dot{M}^{-3/2}$. A significantly higher magnetic field is probably not sustainable as the black hole magnetic field would begin to strongly upset the force balance and would be pushed back into the disk via interchange instability. The actual disk magnetic field can be significantly lower compared to that given by (4.50) as the flow of gas in the plunging region proceeds at high speed, supersonic and super-Alfvénic, so the magnetic field becomes trapped in this flow and eventually accreted by the black hole. One may argue that strong magnetospheric magnetic field would diffuse back into the turbulent disk [41, 42]. However, it is usually the other way around as the magnetic field tends to escape from turbulent domains and strong magnetic field tends to dump turbulence, thus reducing magnetic diffusivity [43–46]. Indeed, recent GR MHD numerical simulations show that the poloidal magnetic field accumulated by the black hole can be substantially higher compared to that of the inner disk [47–50].

Most theoretical models, as well as numerical simulations, of magnetically driven wind simply assume that the disk has strong regular poloidal magnetic field from the start. However, the exact nature of such field is not very clear. The two obvious options are accumulation of magnetic field from the interstellar medium (ISM), and the accretion disk dynamo. The mean value of the galactic ISM field in clouds with densities below $10^3\ \mathrm{cm}^{-3}$ is $\sim 6\ \mu\mathrm{G}$ [51]. For larger densities, it scales as $\propto n^{0.65}$ [52]. Provided the ISM magnetic field has a regular structure on the length scale r, the corresponding magnetic flux is

$$\Psi \simeq 3 \times 10^{32} \left(\frac{B}{10\ \mu\mathrm{G}}\right) \left(\frac{r}{1\ \mathrm{pc}}\right)^2 \mathrm{G\ cm}^2 . \tag{4.52}$$

Comparing this result with (4.47), one can see this option requires to accumulate the ISM magnetic field from the region exceeding only by one or two orders of magnitude the outer radius of the accretion disk (see Section 4.1.4). How exactly this field could be transported towards the inner part of the disk is an open issue. In a turbulent disk, the field will be a subject of outward diffusion with the same rate ν_t/r as the inward advection (see (4.9)), which makes accumulation of magnetic flux in the central part of the disk rather problematic [41]. This argument, however, does not apply if the accretion is driven not by the “friction” associated with turbulence, and magnetorotational instability, but with magnetic torques associated with the advected magnetic field. A number of possible realizations of this idea have been investigated recently [46, 53, 54].

The disk dynamo may well produce strong regular magnetic field or alter the advected ISM magnetic field. The dynamo magnetic field is predominantly azimuthal and contained within the disk. The magnetic buoyancy may well result in a poloidal magnetic field above the disk surface but mainly in the form of loops. Thus, the net magnetic flux of such a poloidal field is zero. In principle, these coronal loops may reconnect and form larger loops, so that a nonvanishing magnetic flux may exist over a relatively large area on the disk surface. However, the efficiency of such an inverse cascade is not very high [55]. In any case, the most relevant characteristic length scale in the dynamo problem seems to be the disk thickness H, particularly in the case of radiation-supported disks. It certainly limits the length scale of turbulent motion in the vertical direction. So one would expect the characteristic length scale of the regular poloidal magnetic field not to exceed H by much.

Violent variability in many radio-loud AGN, in the form of flares accompanied by ejection of superluminal “blobs”, suggests strong nonstationarity of the central engine on the time scale around one year. Since neither the black hole mass nor its spin can change in such a short time, this variability can only be related to the change in the black hole magnetic field. In spite of this clear message, the theory of variability is not yet developed and we can only speculate on the possible mechanisms. If the disk magnetic field is regular on scales much larger than H, which implies that the magnetic field is accreted from ISM, then the magnetic flux of a black hole can change significantly only if from time to time the inner part of the accretion disk collapses, for some reason, and the black hole magnetic field escapes into the created opening between the horizon an the inner edge of the truncated disk, which should exceed the radius of the last stable orbit. If, on the other hand, the poloidal magnetic field of the disk alternates on a relatively short length scale along the disk radius, which is favored by the disk magnetic dynamo theory, then the black hole flux will be fluctuating even if the disk itself is steady. Assuming that the poloidal disk flux alternates on the scale $r_a = \kappa H$, with $\kappa > 1$, we can use the thin disk theory and estimate the variability time scale as $t_{var} = r_a/v_r$, where $v_r = \alpha a_s H/r_a$. This yields a very reasonable result

$$t_{var} \simeq 9 \left(\frac{\alpha}{0.1}\right)^{-1} \left(\frac{\kappa}{10}\right)^{7/2} \left(\frac{M}{10^9 M_\odot}\right) \left(\frac{\dot{M} c^2}{L_{Edd}}\right)^{3/2} \text{years} , \tag{4.53}$$

provided $\kappa \sim 10$. This model assumes that the maximum magnetic flux accumulated by the black hole around the time of flare is on the order of the disk flux, which is spread over the radius r_a. Then the disk magnetic field can be estimated as $B_d \simeq \Psi_h/2\pi r_a^2$, where the black hole flux Ψ_h is determined by (4.50). The corresponding magnetic energy density can then be compared with the energy density of the turbulent motions $\rho v_t^2/2$, where $v_t \leq \alpha a_s$. The ratio of magnetic and turbulent energies is

$$\left(\frac{E_m}{E_t}\right) \simeq 6 \frac{f_+^2(a)\beta_m}{\alpha^2 \kappa^{5/2}} \left(\frac{\dot{M} c^2}{L_{Edd}}\right)^{-4} . \tag{4.54}$$

For $\alpha = \beta_m = 0.1$, $\kappa = 10$, and $a = 1$ the product of the first terms on the right hand side of this equation $\simeq 0.2$, showing that the magnetic energy is below the energy of turbulent motion inside the disk, as expected in the theory of a magnetic dynamo. For the development of a magnetorotational instability, the Alfvén speed has to be below the sound speed and this condition is also satisfied. It remains to be seen if the magnetic dynamo can actually generate a regular poloidal field on the scale ~ 10 H.

Another open issue is the division of AGN on radio-loud and radio-quiet members. Indeed, for some reason, otherwise very similar AGN may or may not be powerful sources of radio emission, which is associated with AGN jets. Orientation effects can play an important role in determining the observed ratio of radio and UV/optical emission, in particular the jet emission is strongly Doppler-beamed. But these factors alone do not seem to explain the observation, though the modern data seem to be more consistent with a smooth single distribution of radio loudness rather than a bimodal one. If the relativistic AGN jets originate from black hole magnetospheres, then (4.46) tells us that there should be a significant dispersion in either the black hole magnetic flux, or spin, or both of them. It is hard to see, why the magnetic disk dynamo would yield very different strengths of magnetic field in similar black hole-accretion disk systems. However, the ISM magnetic field can vary significantly from cloud to cloud, and so the magnetic explanation model cannot be ruled out at present.

In principle, the black hole spin can also vary significantly from one AGN to another. The Blandford–Znajek process tends to reduce the spin, but this is a rather slow process. Disk accretion seems more important in determining the spin. A prograde accretion can spin the black hole up to the maximal rotation rate on the time scale $\simeq M/\dot{M}$ [56]. A retrograde accretion spins it down on an even shorter time scale [57, 58]. If the accretion consists of a large number of episodes with random orientation of the gas angular momentum, then the black hole mass grows systematically, whereas its mean angular momentum grows in a random walk fashion, yielding a decreasing black hole spin parameter a [57]. If this is normally the case, then most black holes are slow rotators, and only a few "fortunate" ones rotate rapidly. This may explain why only around 10% of quasars are radio-loud. The fortunate black holes could by chance experience very few retrograde accretion events. Moreover, they can be born as the result of coalescence of a binary supermassive black hole system. In such a case the black hole spin can increase at the expense of the orbital angular momentum of the binary. Such binaries are expected to form during galactic mergers, which are expected to be rather frequent in the early Universe [59]. This may explain, why the fraction of radio-loud AGN strongly increases with the cosmological redshift.

Another possibility has been proposed recently, where radio-loud AGN are associated with the retrograde accretion [60]. This possibility arises from the observation that the rapid inflow in the plunging region allows the black hole to accumulate much larger magnetic flux compared to the case where this region is considered as empty, allowing the flux to spread over a larger area (see Figure 4.5). This so-called *flux-trapping* effect [61] is strongest for the retrograde orbit around the black hole with $a = 1$, and weakest for the prograde orbit with $a = 1$. Although the disk pressure, and presumably the disk magnetic field strength, are weaker at the inner edge of the retrograde disk, the flux-trapping may still result in a significantly stronger black hole magnetic field compared to the prograde case.

In general, the "antialignment paradigm" has a number of attractive features and has the potential to resolve several long-standing issues in the astrophysics of AGN. In particular, it provides a very plausible alternative explanation to the ob-

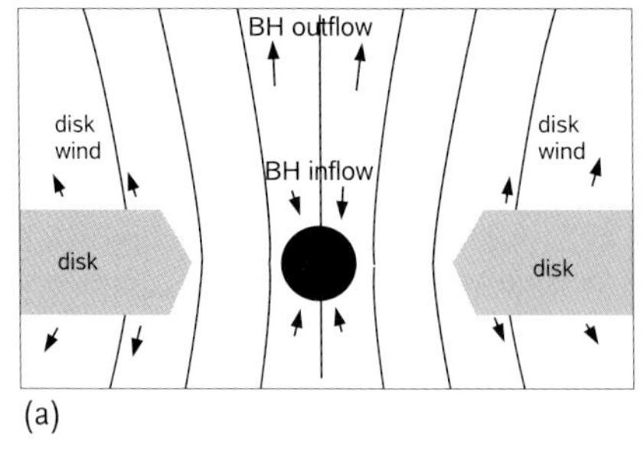

(a)

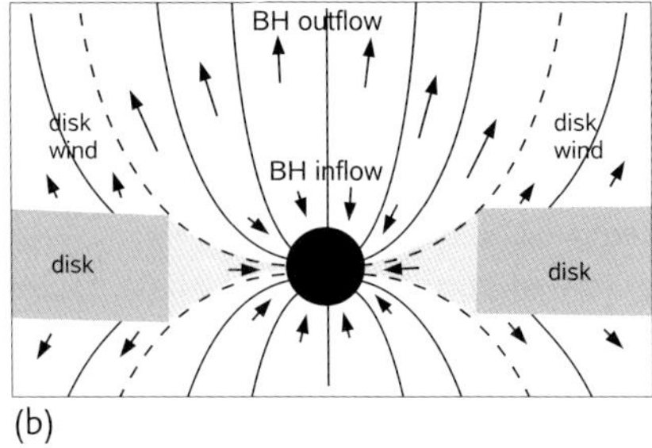

(b)

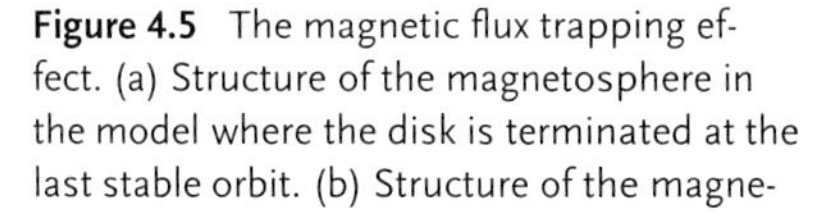

Figure 4.5 The magnetic flux trapping effect. (a) Structure of the magnetosphere in the model where the disk is terminated at the last stable orbit. (b) Structure of the magnetosphere with included accretion flow in the plunging region (light shadow of gray). Notice that in this case the magnetic flux threading the black hole is higher.

served rarity of radio-loud AGN and their cosmological evolution. Indeed, prograde systems are self-sustained, whereas in a retrograde system the black hole spin decreases until the system becomes prograde. A prograde system may also become retrograde, but only if the angular momentum of accreting gas changes dramatically. This, however, may require a major event like a galactic merger. In the violent early Universe such mergers were quite common and hence one would expect a large fraction of retrograde systems. At present, however, the galactic mergers are much less frequent, resulting in a much smaller fraction of retrograde systems. This agrees, at least qualitatively, with the fact that in the early Universe powerful radio sources were much more common.

Since in any version of a magnetic central engine the accretion disk is magnetized, we expect a wind to be driven from its surface. It should be most powerful from the inner part of the disk, where the magnetic field is strongest and the angular velocity is comparable to that of the black hole. It is even possible that the disk wind power is comparable or even exceeds the BZ-power. However, the mass-loading of disk winds is expected to be too high, ruling out an ultrarelativistic terminal velocity, and hence the AGN jets are much more likely to originate from black hole magnetospheres. Recent computer simulations support this conclusion [47, 48, 50].

4.3 Confinement, Collimation, and Acceleration of Jets

The AGN jets are highly collimated, with intrinsic opening angles as small as a few degrees already on pc scales [62]. The M 87 jet, one of the closest and best-studied examples, shows gradually increasing collimation from $\sim 60°$ on the scales ~ 0.01 pc ($\sim 50 r_g$) to $\sim 5°$ on a kpc scale [63]. On the other hand, the split-monopole solution of Blandford–Znajek [37] shows no collimation, suggesting that the process of self-collimation via the magnetic hoop stress is not particularly efficient for ultrarelativistic flows. Thus, the observed collimation of AGN jets indicates that the central engine must have a component which limits the volume occupied by the relativistic outflow. This confining component could be the disk wind, as already demonstrated in the parabolic solution found in [37]. Further out, this can be ISM of the host galaxy, possibly in combination with the jet cocoon. When jets become superfast magnetosonic their confinement may no longer be required, as free sideway expansion of such jets cannot increase their opening angle by more than the local Mach angle associated with fast magnetosonic waves (see below). However, the external confinement, and the gradually increasing collimation that stems from it, are important for the process of magnetic acceleration as well.

In fact, relativistic jets can remain collimated even in the subsonic regime, and even when they expand in a vacuum. The sideways expansion increases the total flow speed but in the relativistic regime it has to remain below the speed of light. This limits the jet spreading to the angle $\sim 1/\Gamma_j$, where Γ_j is the initial jet Lorentz factor. For most AGN jets this angle is still too large compared to their observed collimation.

One of the most remarkable facts about AGN jets is that their length exceeds the size of the central black hole by up to ten orders of magnitude. Along the jets, the specific volume of plasma increases enormously, and the corresponding adiabatic losses, in combination with various radiative losses, ensure that the plasma particles lose essentially all their random motion energy, which they might have had inside the central engine, very quickly. Yet, the observations show that the jet brightness does not decline so rapidly. This suggests that most of the jet energy is in a different form, and that the observed emission is the result of its conversion into the energy of emitting particles.

In principle, the model of magnetically driven outflow can explain the observations by local dissipation of magnetic energy in magnetic reconnection events, though the details, including the dissipation rates and the properties of emerging radiation, are yet to be worked out (e.g., [64, 65]). According to this interpretation, the energy of AGN jets may still be mainly in the electromagnetic form. In such *Poynting jets,* the Poynting flux dominates over the kinetic and thermal energy fluxes.

Another possibility is that the observed AGN jets are already kinetic energy-dominated flows [66, 67]. Indeed, a number of factors make this idea very attractive. First, such flows do not require external support in order to preserve their collimation as they are also superfast magnetosonic. Second, they are much more stable and can propagate large distances without significant energy losses, in an essentially ballistic regime. Third, when they interact with the external medium, the result is shocks, which dissipate kinetic energy locally and thus can produce bright compact emission sites, reminiscent of the knots and hot spots of AGN jets. Given this, and thanks to the success of the shock model in many other areas of astrophysics, the current magnetic paradigm assumes that most of the jet Poynting flux is eventually converted into its bulk kinetic energy, which then dissipates at shocks.

The problem of magnetic acceleration is rich and mathematically complex. This complexity is the reason why the attention of theorists has been directed mostly towards the development of the *standard model*, which deals with steady-state axisymmetric flows of perfect fluid. Although this is the simplest case, it is still impossible to give a comprehensive and mathematically rigorous review within the scope of this book. Instead, I will focus more on physical arguments.

4.3.1 Acceleration in Supersonic Regime

One clear difference between nonrelativistic and relativistic flows is that in the nonrelativistic limit most of the energy conversion occurs already in the subfast magnetosonic regime whereas in the relativistic limit this occurs in the superfast magnetosonic regime. This is true both for the thermal and magnetic mechanisms (in the case of unmagnetized plasma the fast magnetosonic speed equals the sound speed).

Consider first the nonrelativistic limit. For a hot gas with polytropic equation of state, the thermal energy density $e = p/(\gamma - 1)$ and the sound speed $a_s^2 = \gamma p/\rho$, where γ is the ratio of specific heats, p and ρ are the gas pressure and density, respectively. From this we immediately find that when the flow speed equals the sound speed, $\rho v^2 = \gamma(\gamma - 1)e$. Thus, the kinetic energy is already comparable with the thermal energy. For a cold magnetized flow the fast magnetosonic speed is $c_f^2 = B^2/4\pi\rho$ and at the sonic point $\rho v^2 = B^2/4\pi$. Thus, the kinetic energy is already comparable with the magnetic energy.

The relativistic expression for the sound speed is $a_s^2 = (\gamma p/w)c^2$, where $w = \rho c^2 + p\gamma/(\gamma - 1)$ is the gas enthalpy, ρ and p are measured in the fluid frame. The condition of highly relativistic asymptotic speed requires relativistically hot gas (initially), that is the Lorentz factor of thermal motion $\Gamma_{th} \gg 1$, $p \gg \rho c^2$, and $\gamma \approx 4/3$. This means that at the sonic point $v^2 \approx c^2/3$ and the corresponding Lorentz factor is only $\Gamma^2 \approx 3/2 \ll \Gamma_{th}^2$. Thus, most of the energy is still in the thermal form.

The relativistic expression for the fast magnetosonic speed in cold gas is

$$c_f^2 = \frac{c^2 B'^2}{B'^2 + 4\pi\rho c^2} , \tag{4.55}$$

where B' is the magnetic field as measured in the fluid frame. Thus, at the sonic point $\Gamma^2 = 1 + B'^2/4\pi\rho c^2$. Now one can see that large asymptotic Lorentz factors imply $B'^2/4\pi\rho c^2 \gg 1$ at the sonic point. Indeed, if $B'^2/4\pi\rho c^2 \lesssim 1$ then there is not enough magnetic energy for further acceleration and yet the Lorentz factor is still rather low (see also (4.62) below).

In the subfast magnetosonic regime, different fluid elements can communicate with each other by means of fast magnetosonic waves both along and across the direction of motion. In the superfast magnetosonic regime, the causal connectivity is limited to the interior of the Mach cone, and this has important implications for the efficiency of magnetic acceleration.

4.3.2
Acceleration and Differential Collimation

Consider first the thermal acceleration of relativistic flows in the supersonic regime. The mass and energy conservation laws for steady-state flows imply that both the mass and the total energy fluxes are constant along the jet:

$$\rho\Gamma c A = \text{const.} , \tag{4.56}$$

$$(\rho c^2 + 4p)\Gamma^2 c A = \text{const.} , \tag{4.57}$$

where $A \propto \varpi_j^2$ is the jet cross-section area, ϖ_j is the jet radius, the jet speed $v \approx c$, and for simplicity we assume $\gamma = 4/3$. From these equations it follows that

$$\left(1 + \frac{4p}{\rho c^2}\right)\Gamma = \Gamma_{max} \tag{4.58}$$

(this is known as the Bernoulli equation). $\Gamma_{\max}$ is the constant that equals the Lorentz factor of the flow after complete conversion of its thermal energy into the kinetic one. From mass conservation we find that $\rho \propto \Gamma^{-1} \varpi_{\mathrm{j}}^{-2}$ and thus

$$\frac{p}{\rho} \propto \rho^{1/3} \propto \Gamma^{-1/3} \varpi_{\mathrm{j}}^{-2/3} . \tag{4.59}$$

When $p \gg \rho$, allowing plenty of thermal energy to be spent on continued plasma acceleration, this equation and the Bernoulli equation yield

$$\Gamma \propto \varpi_{\mathrm{j}} . \tag{4.60}$$

Thus, the sideways (or transverse) jet expansion is followed by rapid acceleration. In particular, for freely expanding conical jets $\Gamma \propto z$, there z is the distance along the jet.

For cold magnetized flows the energy equation can be written as

$$\left(\rho c^2 + \frac{B_\phi'^2}{4\pi}\right) \Gamma^2 c A = \text{const.} , \tag{4.61}$$

where we ignore the contribution due to the small poloidal component of magnetic field.[5] Then the Bernoulli equation reads

$$\left(1 + \frac{B_\phi'^2}{4\pi\rho c^2}\right) \Gamma = \Gamma_{\max} , \tag{4.62}$$

where $\Gamma_{\max}$ is the Lorentz factor after complete conversion of the magnetic energy into the kinetic one. Assuming that the radii of all streamlines evolves like ϖ_{j} and using the magnetic flux-freezing condition, one finds the familiar law for the evolution of transverse magnetic field $B_\phi \propto \varpi_{\mathrm{j}}^{-1}$ and $B_\phi' = B_\phi/\Gamma \propto \Gamma^{-1}\varpi_{\mathrm{j}}^{-1}$. This gives us $B_\phi'^2/\rho = \varpi_{\mathrm{j}}^0 \Gamma^{-1}$ and the Bernoulli equation yields the uncomfortable result

$$\Gamma = \text{const.} \tag{4.63}$$

The same conclusion can be reached in a slightly different way. When a fluid element expands only sideways, its volume grows as $V \propto \varpi_{\mathrm{j}}^2$ and its magnetic energy $e_{\mathrm{m}} \propto B_\phi^2 V \propto \varpi_{\mathrm{j}}^0$ remains unchanged, implying no conversion of the magnetic energy and no acceleration. Thus, in contrast to the thermal case, the sideways expansion of a cold magnetized flow is not sufficient for its acceleration.

For the magnetic mechanism to work a special condition, which can be described as *differential collimation*, has to be satisfied. In order to see this, we refine our analysis and consider the flow between two axisymmetric flow surfaces with cylindrical radii $\varpi(z)$ and $\varpi(z) + \delta\varpi(z)$ (Figure 4.6).

5) This is sufficiently accurate when $\varpi_{\mathrm{j}} \gg \varpi_{\mathrm{LC}}$, the light cylinder radius, the condition which is normally satisfied in the supersonic regime.

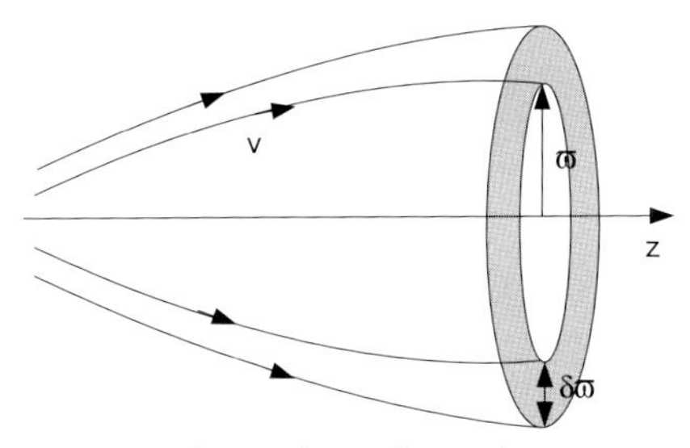

Figure 4.6 Flow surfaces of a steady-state axisymmetric jet.

Now $A \propto \varpi\,\delta\varpi$, $\rho \propto \Gamma^{-1}(\varpi\,\delta\varpi)^{-1}$, $B_\phi \propto \delta\varpi^{-1}$, $B'^2_\phi/\rho \propto (\varpi/\delta\varpi)\Gamma^{-1}$ and the magnetic Bernoulli equation yields

$$\Gamma \simeq \Gamma_{\max}\left(1 - \frac{\varpi}{\delta\varpi}\frac{\delta\varpi_0}{\varpi_0}\right), \tag{4.64}$$

where ϖ_0 and $\delta\varpi_0$ are the initial surface parameters, and we assume that the initial Lorentz factor $\Gamma_0 \ll \Gamma_{\max}$. This result shows that the magnetic acceleration requires $\varpi/\delta\varpi$ to decrease with the distance along the jet. In other words, the separation between neighboring flow surfaces should increase faster than their radius. For example, consider a flow with parabolic flow surfaces, $z = z_0(\varpi/\varpi_0)^\kappa$, where the power index varies from surface to surface, $\kappa = \kappa(\varpi_0)$. Then

$$\frac{\varpi}{\delta\varpi} = \frac{\varpi_0}{\delta\varpi_0}\left[1 - \frac{d\kappa}{d\varpi_0}\frac{\varpi_0}{\kappa^2}\ln\frac{z}{z_0}\right]^{-1}. \tag{4.65}$$

Thus, if $d\kappa/d\varpi_0 = 0$, and hence all flow surfaces are "uniformly" collimated, then $\varpi/\delta\varpi$ =const. and the magnetic acceleration fails. This includes the important case of ballistic conical flow with radial streamlines. If, however, $d\kappa/d\varpi_0 < 0$, and thus the inner flow surfaces are collimated faster compared to the outer ones, then $\varpi/\delta\varpi$ decreases along the jet and the flow accelerates.

Whether such differential collimation can arise naturally depends on the details of the force balance across the jet. Such balance is described by the Grad–Shafranov equation (e.g., [30]), which is notoriously difficult to solve – only very few analytic or semianalytic solutions for rather simple cases have been found so far (e.g., [68–70]). Recently, this issue has been studied using time-dependent numerical simulations [71–75]. The results show that the differential collimation can develop under the self-collimating action of magnetic hoop stress associated with the azimuthal magnetic field, but only in cases where efficient externally imposed confinement helps to keep the jet sufficiently narrow. For example, the asymptotic Lorentz factor of conical jets decreases with the opening angle and, unless the angle is small enough, the jet remains Poynting-dominated. In the case of parabolic jets, the opening angle decreases with distance and the acceleration continues until the kinetic energy becomes comparable with the magnetic one.

When the external confinement is provided by medium with the pressure distribution $p_{\rm ext} \propto z^{-s}$, where $s < 2$, the jet shape is indeed parabolic $\varpi_{\rm j} \propto z^{s/4}$ and its

Lorentz factor grows as

$$\Gamma \simeq \frac{\varpi_{\rm j}}{\varpi_{\rm LC}} \simeq \left(\frac{z}{\varpi_{\rm LC}}\right)^{s/4} \tag{4.66}$$

until the energy equipartition is reached [72, 76]. As far as the dependence on $\varpi_{\rm j}$ is concerned, this is as fast as in the thermal mechanism. However, the Lorentz factor is a slower functions of z compared to the case of thermally accelerated conical jet. For $s > 2$, the external confinement is insufficient – the jets eventually develop conical streamlines and do not accelerate efficiently afterwards. Various components of the Lorentz force, the hoop stress, magnetic pressure, and electric force, finely balance each other. This is in contrast to the thermal acceleration, which remains efficient for jets with conical geometry.

Although the detailed analysis of this issue is rather involved, one can get a good grasp of it via the causality argument. Indeed, the favorable differential self-collimation can only be arranged if the flow surfaces "know" what other flow surfaces do. This information is propagated most effectively by fast magnetosonic waves.[6] In the subsonic regime, these waves can propagate in all directions and have no problem in establishing causal communication across the jet. In the supersonic regime, they are confined to the Mach cone, which points in the direction of motion. When the characteristic opening angle of the Mach cone,

$$\sin\theta_{\rm M} = \frac{\Gamma_{\rm f} c_{\rm f}}{\Gamma v}\,, \tag{4.67}$$

where $c_{\rm f}$ and $\Gamma_{\rm f}$ are the fast magnetosonic speed and the corresponding Lorentz factor, respectively, becomes smaller than the jet opening angle, $\theta_{\rm j}$, the communication across the jet is disrupted. Thus, the condition for effective magnetic acceleration is $\theta_{\rm j} < \theta_{\rm M}$. Using (4.67), (4.55), (4.62), one can write this condition as

$$\Gamma < \left(\frac{\Gamma_{\rm max}}{\sin^2\theta}\right)^{1/3}\,. \tag{4.68}$$

For a spherical wind with $\Gamma_{\rm max} \gg 1$, this condition reads as $\Gamma < \Gamma_{\rm max}^{1/3} \ll \Gamma_{\rm max}$. Thus, the magnetic acceleration of relativistic winds is highly inefficient. Higher efficiency can be reached for collimated flows. For $\Gamma \geq 0.5\Gamma_{\rm max}$, this condition requires

$$\theta_{\rm j}\Gamma \leq 1\,, \tag{4.69}$$

where we used the small angle approximation. Condition (4.69) is satisfied by AGN jets, where $\langle\theta_{\rm j}\Gamma\rangle \simeq 0.26$, with significant spread around this value [62].

Equation (4.66) can be used to estimate the distance where the jet reaches the values of Lorentz factor inferred from the VLBI observations. This is important as

6) Alfvén waves transport information only along the magnetic field lines. Slow magnetosonic waves can transport information only up to some angle to the fields and at a smaller speed compared to fast waves.

the absence of bulk-Comptonization spectral signatures in blazars has been used to conclude that the AGN jet Lorentz factors $\Gamma > 10$ are only attained on scales $> 10^{17}$ cm, suggesting that at smaller scales the jets are still Poynting-dominated [77]. First, we need to know the radius of the light cylinder. Assuming that the jet originates from the central black hole, we find from (4.45) that

$$\varpi_{\rm LC} \simeq 2\frac{c}{\Omega_{\rm h}} = 4 f_{\Omega}^{-1}(a) r_{\rm g} . \tag{4.70}$$

For $a = 1$ this gives $\varpi_{\rm LC} = 4r_{\rm g}$. Next, (4.66) can be used to show that

$$z \simeq 4 r_{\rm g} \Gamma^{4/s} \quad \text{and} \quad \theta_{\rm j} \simeq 2\Gamma^{(s-4)/s} . \tag{4.71}$$

With $s = 3/2$ as an example, these equations yield

$$z \simeq 0.06 \left(\frac{M}{10^8\, M_\odot}\right) \left(\frac{\Gamma}{20}\right)^{8/3} \text{pc} , \tag{4.72}$$

$$\theta_{\rm j} \simeq 0.^\circ 8 \left(\frac{\Gamma}{20}\right)^{-5/3} , \quad \text{and} \quad (\theta\Gamma) \simeq 0.3 \left(\frac{\Gamma}{20}\right)^{-2/3} . \tag{4.73}$$

Thus, the slow magnetic acceleration of relativistic jets may well fit the observational constraints on the jet speed and geometry.

However, the standard model has a number of shortcomings. First of all, it does not allow energy dissipation and hence does not explain the observed jet emission. In order to include dissipation one would have to break the imposed high degree of symmetry of this model, and perhaps introduce variability of the central engine. It remains to be seen whether or not such modifications can be introduced without making a strong impact on the overall jet dynamics.

The observed linear polarization angles (EVPA) of AGN jets also seem to present a problem for the standard model. In approximately half of all cases, the electric field vector is normal to the jet direction [78]. This means that in the comoving jet frame the longitudinal component of magnetic field is at least comparable to the transverse one [79]. On the other hand, the standard model predicts that beyond the light cylinder, $B_\phi / B_{\rm p} \simeq (\varpi_{\rm j}/\varpi_{\rm LC})$, where $B_{\rm p}$ is the poloidal component (predominantly longitudinal) and B_ϕ is the azimuthal components of magnetic field as measured in the observer's frame. In the comoving jet frame this leads to $B'_\phi / B'_{\rm p} \simeq \Gamma^{-1}\varpi_{\rm j}/\varpi_{\rm LC}$. For rapidly rotating black holes ($a \simeq 1$), this leads to

$$\frac{B'_\phi}{B'_{\rm p}} \simeq \frac{10^3}{\Gamma} \left(\frac{\theta_{\rm j}}{1^\circ}\right) \left(\frac{l_{\rm j}}{1\,\text{pc}}\right) \left(\frac{M}{10^8\, M_\odot}\right)^{-1} , \tag{4.74}$$

where l_j is the distance from the black hole. Thus, unless the AGN jets are produced by very slowly rotating black holes (so that $\varpi_{\rm LC} \gg 4r_{\rm g}$), which is highly unlikely, the idealized standard model seems to be inconsistent with the polarization data.

Current-driven instabilities may randomize the magnetic field, transferring energy from the slowly decaying transverse component, $B'_\perp \propto \Gamma^{-1}\varpi_{\rm j}^{-1}$, to the rapidly decaying longitudinal component, $B'_{\parallel} \propto \varpi_{\rm j}^{-2}$ [80]. As a result, the magnetic field strength evolves as $B' \propto \varpi_{\rm j}^{-2}$, and $B'^2/\rho \propto \varpi_{\rm j}^{-2}\Gamma$, and, in the magnetically dominated regime, the Bernoulli equation (4.62) yields $\Gamma \propto \varpi_{\rm j}$. Thus, a randomized magnetic field behaves as an ultrarelativistic gas with $\gamma = 4/3$, providing as rapid an acceleration as the thermal mechanism. The observed polarization of AGN jets then reflects the level of turbulence and its anisotropy. Such instabilities are also likely to be followed by magnetic dissipation and plasma heating. This may also facilitate bulk acceleration of jets [81, 82]. In particular, heat is easily converted into kinetic energy during sideways expansion. Nonstationary central engine may generate jets in an impulsive fashion and this may also facilitate jet acceleration by allowing the plasma expansion along the jet axis [83–86].

Since the initial magnetization σ_0 determines the asymptotic speed of both continuous and impulsive flows, the magnetic theory of AGN jets will not be complete until we know the processes which determine this key parameter. This issue is closely related with the issue of jet composition. Winds emerging from accretion disks are most certainly dominated by protons. On the other hand, relativistic jets emerging from black hole magnetospheres can be made of a pure electron–positron pair plasma. However, it is unlikely that this initial composition is preserved all the way to pc and kpc scales, at which the AGN jets can be studied in details. The very fact that the observed jet Lorentz factors are much lower compared to what they could have been if the jets consisted only of pairs, required the screening of the component of the electric field parallel to the magnetic field; this tells us that these jets are likely to be "contaminated" with protons.

The gamma-ray observations of AGN show that they can be very powerful emitters of gamma-rays. Combined with the data in lower frequency bands these data can be used to estimate the jet energy density in magnetic form and in synchrotron electrons and positrons. Although the calculations are not very straightforward and model-dependent, the results suggest that the gamma-ray luminosity often exceeds, and sometimes significantly, the jet luminosity both in the magnetic form and in the form of hot pairs at the distance of 10^{17}–10^{18} cm from the central black hole [87]. The fact that the jet is not completely destroyed in the process of gamma-ray emission means that the radiation reaction deceleration was not too strong, which implies the existence of another nonradiating jet component that carries bulk of the energy. The only candidates are cold pairs and protons. In the case of 3C 454.3, at least one proton per 10 hot pairs is required, assuming that there are no cold pairs [88]. The corresponding jet power then exceeds the bolometric disk luminosity by a factor of a few, with the jet kinetic energy being about the same as its Poynting flux.

AGN jets can be enriched with heavy particles in many different ways. For example, they can be entrained from the surrounding gas via boundary instabilities. In the scenario where the black hole magnetosphere is constantly rebuilt with alternating poloidal magnetic field of the disk, the newly added field lines are already loaded with protons. Finally, protons can be supplied by the stars which happen to occupy the same volume as the jets [89].

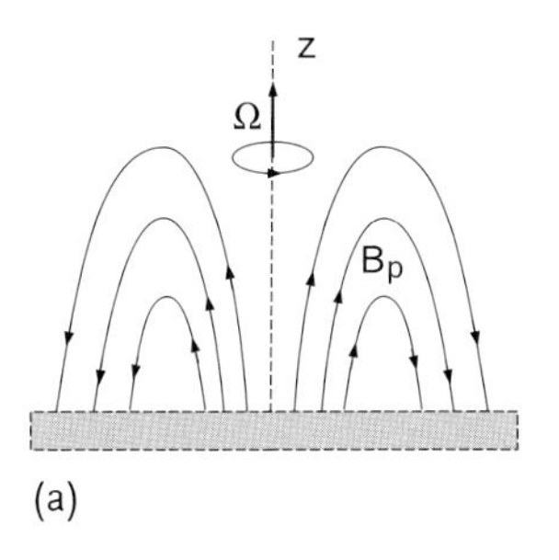

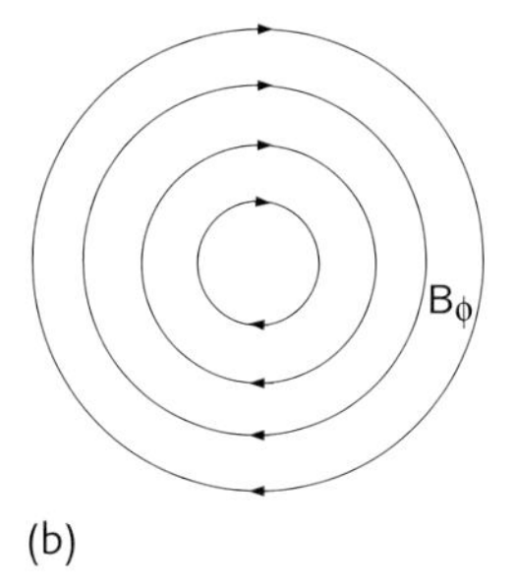

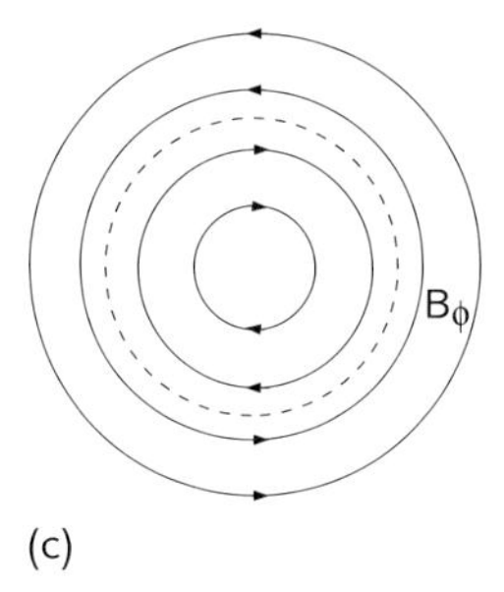

Figure 4.7 (a) The dipolar initial poloidal magnetic field of the magnetic tower model. (b) The azimuthal magnetic field generated in the magnetic tower. (c) The azimuthal magnetic field generated in a superfast magnetosonic outflow with the poloidal field of the same topology. The dashed line shows the location of the current sheet where the azimuthal, as well as the poloidal, magnetic field changes direction.

4.3.3
Jets and Magnetic Towers

Consider an axisymmetric dipolar-like magnetosphere arching within a small volume above the inner part of an accretion disk, as shown in Figure 4.7a, and surrounded by gas with constant pressure. The two foot points of any field line are anchored to the disk at different radii. Because of the differential rotation of the disk, the field line is twisted and an azimuthal component of magnetic field is generated (Figure 4.7b). The increased magnetic pressure forces the magnetosphere to expand. Assuming that the gas pressure in the magnetosphere is small compared to the magnetic one and that the expansion speed is low compared to the Alfvén speed, the expansion can be treated as a sequence of magnetostatic configurations, each corresponding to the number of twists produced by the time. The result is a “magnetic tower” growing higher and higher above the disk surface [90]. The anisotropy of this expansion stems from the hoop stress of the generated azimuthal magnetic field. It causes compression of the tower so that in the tower cross-section the total pressure peaks at the axis. Thus, at the tip of the tower the magnetic pressure is higher compared to that at its walls and this causes faster axial expansion. In fact, the sideways expansion is soon brought to a halt. Similar configurations have been observed in several axisymmetric numerical simulations of magnetized disks and the possibility that the astrophysical jets can be associated with magnetic towers has been discussed.

In order to decide whether the magnetic tower model can be used to interpret the astronomical observations and the results of numerical simulations one has to check whether the assumptions of the model are satisfied. The most constraining is the assumption of magnetostatic equilibrium along the tower. It requires the expansion speed to remain well below the Alfvén speed, which is the same as the fast magnetosonic speed in the limit of low gas pressure assumed in the magnetic tower model. In astrophysical conditions, the external pressure declines with the height z above the disk, and this results in accelerated expansion of the tower. For

$p_{\text{ext}} \propto z^{-s}$ one has $v_{\text{tower}} \propto z^{s/4}$ [90]. Thus, one has to worry if this speed eventually exceeds the Alfvén speed. In the magnetic tower model, this issue is avoided as the Alfvén speed is assumed to be infinitely large and the inertia of plasma is completely ignored. However in reality, both the observed and the simulated flows have finite Alfvén speed. An outflow may begin as a magnetic tower but eventually become a superfast magnetosonic flow, this transition can be described as a jet breakout. In this flow the twists generated at the inner foot point of a field line cannot propagate all the way to the other foot point and dictate the sign of the azimuthal component there. Instead, at both foot points the field line is swept back and the magnetic configuration is as shown in Figure 4.7c.

References

1 Lyubarsky, Y.E. (2002) *Mon. Not. R. Astron. Soc.*, **329**, L34.

2 Komissarov, S.S. and Lyubarsky, Y.E. (2004) *Mon. Not. R. Astron. Soc.*, **349**, 779.

3 Frank, J., King, A.R., and Raine, D.J. (2002) *Accretion Power in Astrophysics*, Cambridge University Press, Cambridge.

4 Bondi, H. (1952) *Mon. Not. R. Astron. Soc.*, **112**, 195.

5 Shakura, N.I. and Sunyaev R.A. (1973) *Astrophys. J.*, **24**, 337.

6 Balbus, S.A. and Hawley, J.F. (1991) *Astrophys. J.*, **376**, 214.

7 Velikhov, E.P. (1959) *J. Exp. Theor. Phys.*, **36**, 1398.

8 Chandrasekhar, S. (1960) *Proc. Natl. Acad. Sci. USA*, **46**, 253.

9 King, A.R., Pringle J.E., and Livio M. (2007) *Mon. Not. R. Astron. Soc.*, **376**, 1740.

10 Bardeen, J.M., Press, W.H., and Teukolsky, S.A. (1972) *Astrophys. J.*, **178**, 347.

11 Bardeen, J.M. and Petterson, J.A. (1975) *Astrophys. J.*, **195**, L65.

12 King, A.R., Lubow, S.H., Ogilvie, G.I., and Pringle, J.E. (2005) *Mon. Not. R. Astron. Soc.*, **363**, 49.

13 Begelman, M.C. (1979) *Mon. Not. R. Astron. Soc.*, **187**, 237.

14 Abramowicz, M., Jaroszynski, M., and Sikora M. (1978) *Astron. Astrophys.*, **63**, 221.

15 Abramowicz, M., Calvani, M., and Nobili, L. (1980) *Astrophys. J.*, **242**, 772.

16 Ichimaru, S. (1977) *Astrophys. J.*, **214**, 840.

17 Narayan, R. and Yi, I. (1994) *Astrophys. J.*, **428**, L13.

18 Narayan, R. and Yi, I. (1995) *Astrophys. J.*, **452**, 710.

19 Narayan, R. and Yi, I. (1995) *Astrophys. J.*, **444**, 231.

20 Blandford, R.D. and Begelman, M.C. (1999) *Mon. Not. R. Astron. Soc.*, 303, L1.

21 Paczynski, B. (1978) *Acta Astron.*, **28**, 91.

22 Hills, J.G. (1975) *Nature*, **254**, 295.

23 Hills, J.G. (1978) *Mon. Not. R. Astron. Soc.*, **182**, 517.

24 Shlosman, I., Frank, J., and Begelman M.C. (1989) *Nature*, **338**, 45.

25 Hopkins, P.F. and Quataert, E. (2010) *Mon. Not. R. Astron. Soc.*, **407**, 1529.

26 Lovelace, R.V.E (1976) *Nature*, **262**, 649.

27 Lichnerowicz, A. (1967) *Relativistic Hydrodynamics and Magnetohydrodynamics*, Benjamin, New York.

28 Anile, A.M. (1989) *Relativistic Fluids and Magneto-Fluids*, Cambridge University Press, Cambridge.

29 Mestel, L. (1999) *Stellar Magnetism*, Clarendon, Oxford.

30 Beskin, V.S. (2009) *MHD Flows in Compact Astrophysical Objects: Accretion, Winds and Jets*, Springer.

31 Blandford, R.D. and Payne, D.G. (1982) *Mon. Not. R. Astron. Soc.*, **199**, 883.

32 Li, Z.-Y., Chiueh, T., and Begelman, M.C. (1992) *Astrophys. J.*, **394**, 459.

33 Penrose, R. (1969) *Rev. Nuovo Cim.*, **1**, 252.

34 Wald, R.M. (1974) *Phys. Rev D*, **10**(6), 1680.
35 Komissarov, S.S. (2004) *Mon. Not. R. Astron. Soc.*, **350**, 427.
36 Cheng, K.S., Ho, C., Ruderman, M.A. (1986) *Astrophys. J.*, **300**, 500.
37 Blandford, R.D. and Znajek, R.L. (1977) *Mon. Not. R. Astron. Soc.*, **179**, 433.
38 Komissarov, S.S. (2001) *Mon. Not. R. Astron. Soc.*, **326**, L41.
39 Tchekhovskoy, A., Narayan, R., and McKinney, J.C. (2010) *Astrophys. J.*, **711**, 50.
40 Komissarov, S.S. (2009) *J. Korean Phys. Soc.*, **54**, 2503.
41 Lubow, S.H., Papaloizou, J.C.B., and Pringle, J.E. (1994) *Mon. Not. R. Astron. Soc.*, **267**, 235.
42 Livio, M., Ogilvie, G.I., and Pringle, J.E. (1999) *Astrophys. J.*, **512**, 100.
43 Zel'dovich, Ya.B. (1957) *Sov. Phys. JETP*, **4**, 460.
44 Rädler, K.-H. (1968) *Z. Naturforsch. A*, **23**, 1851.
45 Tao, L., Proctor, M.R.E., and Weiss, N.O. (1998) *Mon. Not. R. Astron. Soc.*, **300**, 907.
46 Rothstein, D.M., Lovelace, R.V.E. (2008) *Astrophys. J.*, **677**, 1221.
47 McKinney, J.C. (2006) *Mon. Not. R. Astron. Soc.*, **368**, 1561.
48 Barkov, M.V. and Komissarov, S.S. (2008) *Mon. Not. R. Astron. Soc.*, **385**, L28.
49 Komissarov, S.S. and Barkov, M.V. (2009) *Mon. Not. R. Astron. Soc.*, **397**, 1153.
50 McKinney, J.C. and Blandford, R.D. (2009) *Mon. Not. R. Astron. Soc.*, **394**, L126.
51 Heiles, C. and Troland, T.H. (2005) *Astrophys. J.*, **624**, 773.
52 Crutcher, R.M. (2007) in *Magnetic fields in the Non-Masing ISM*, (eds J.M. Chapmen and W.A. Baan), Cambridge University Press, Cambridge, p. 47.
53 Spruit, H.C. and Uzdensky, D.A. (2005) *Astrophys. J.*, **629**, 960.
54 Beckwith, K., Hawley, J.F., and Krolik, J.H. (2009) *Astrophys. J.*, **707**, 428.
55 Tout, C.A. and Pringle, J.E. (1996) *Mon. Not. R. Astron. Soc.*, **281**, 219.
56 Page, D.N. and Thorne, K.S. (1974) *Astrophys. J.*, **191**, 499.
57 Moderski, R., Sikora, M., and Lasota, J.-P. (1998) *Mon. Not. R. Astron. Soc.*, **301**, 142.
58 King, A.R. and Pringle J.E. (2006) *Mon. Not. R. Astron. Soc.*, **373**, L90.
59 Wilson, A.S. and Colbert E.J.M. (1995) *Astrophys. J.*, **438**, 62.
60 Garofalo, D., Evans, D.A., and Sambruna, R.M. (2010) *Mon. Not. R. Astron. Soc.*, **406**, 975.
61 Reynolds, C.S., Garofalo, D., and Begelman, M.C. (2006) *Astrophys. J.*, **651**, 1023.
62 Pushkarev, A.B., Kovalev, Y.Y., Lister, M.L., and Savolainen, T. (2009) *Astron. Astrophys.*, **507**, L33.
63 Biretta, J.A., Junor, W., and Livio, M. (2002) *New Astron. Rew.*, **46**, 239.
64 Lyutikov, M. and Blandford, R.D. (2003) arXiv1004.2429.
65 Lyubarsky, Y.E. (2005) *Mon. Not. R. Astron. Soc.*, **358**, 113.
66 Scheuer, P.A.G. (1974) *Mon. Not. R. Astron. Soc.*, **166**, 513.
67 Blandford, R.D. and Rees, M.J. (1974) *Mon. Not. R. Astron. Soc.*, **169**, 395.
68 Beskin, V.S., Kuznetsova, I.V., and Rafikov, R.R. (1998) *Mon. Not. R. Astron. Soc.*, **341**, 1998.
69 Beskin, V.S. and Nokhrina, E.E. (2006) *Mon. Not. R. Astron. Soc.*, **367**, 375.
70 Vlahakis, N. and Königl A. (2003) *Astrophys. J.*, **596**, 1080.
71 Komissarov, S.S., Barkov, M.V., Vlahakis, N., and Königl, A. (2007) *Mon. Not. R. Astron. Soc.*, **380**, 51.
72 Komissarov, S.S., Vlahakis, N., Königl, A., and Barkov, M.V. (2009) *Mon. Not. R. Astron. Soc.*, **394**, 1182.
73 Komissarov, S.S., Vlahakis, N., and Königl, A. (2010) *Mon. Not. R. Astron. Soc.*, **407**, 17, arXiv0912.0845.
74 Tchekhovskoy, A., McKinney, J.C., and Narayan, R. (2009) *Astrophys. J.*, **699**, 1789.
75 Tchekhovskoy, A., Narayan, R., and McKinney, J.C. (2010) *New Astron.*, **15**, 749.
76 Lyubarsky, Y.E. (2009) *Astrophys. J.*, **698**, 1570.
77 Sikora, M., Begelman, M.C., Madejski, G.M., and Lasota, J.-P. (2005) *Astrophys. J.*, **625**, 72.

78 Wardle, J.F.C. (1998) ASP Conference Series, vol. 144, Astronomical Society of the Pacific, p. 97.

79 Lyutikov, M., Pariev, V.I., and Gabuzda, D.C. (2005) *Mon. Not. R. Astron. Soc.*, **360**, 869.

80 Heinz, S. and Begelman, M.C. (2000) *Astrophys. J.*, **535**, 104.

81 Drenkhahn G. (2002) *Astron. Astrophys.*, **387**, 714.

82 Drenkhahn, G. and Spruit, H.C. (2002) *Astron. Astrophys.*, **391**, 1141.

83 Contopoulos, J. (1995) *Astrophys. J.*, **450**, 616.

84 Granot, J., Komissarov, S.S., and Spitkovsky A. (2011) *Mon. Not. R. Astron. Soc.*, **411**, 1323, arXiv1004.0959.

85 Lyutikov, M. and Lister, M. (2010) *Astrophys. J.*, **722**, 197, arXiv1004.2430.

86 Lyutikov, M. (2010) *Phys. Rev. E*, **82**, 6305, arXiv1004.2429.

87 Ghisellini, G., Tavecchio, F., and Ghirlanda, G. (2009) *Mon. Not. R. Astron. Soc.*, **399**, 2041.

88 Ghisellini, G. and Tavecchio, F. (2009) *Mon. Not. R. Astron. Soc.*, **409**, L79.

89 Komissarov, S.S. (1994) *Mon. Not. R. Astron. Soc.*, **266**, 649.

90 Lynden-Bell, D. (2003) *Mon. Not. R. Astron. Soc.*, **341**, 1360.

Part Three Phenomenology

5
Observational Details: Radio

A.H. Bridle and M.H. Cohen

5.1
Overall Structures of Radio Sources

In this chapter we discuss the observations of radio jets. After some introductory remarks, mainly describing the classic Fanaroff–Riley separation of double radio sources, we separate the topic into small-scale (parsec) and large-scale (kiloparsec) jets. These two types of objects have been studied with different telescopes and the two subfields have developed largely independently of each other. Very Long Baseline Interferometry (VLBI) with angular resolution of 1 milliarcsecond (mas) and better is used at parsec (pc) scales, and the Very Large Array (VLA) and similar instruments are used for larger scales. VLBI is sensitive only to high surface-brightness features ($\gtrsim 10^5$ K) and so typically studies only the core region of a source; whereas the VLA can image the faint outer plumes of a radio galaxy but its core region is subsumed into an unresolved bright spot. This instrumental distinction, however is not definitive, as some observations at certain frequencies cover the intermediate range, e.g., MERLIN and its connections to the European VLBI Network cover a wide range of scales. See Figure 1.1 for a montage of VLBI and VLA images that span the pc–kpc range.

Fanaroff and Riley [1] showed that the morphologies of extended double radio sources produced by AGN separate into two main classes in a way that correlates with the overall radio source luminosity. In Fanaroff–Riley Class I (FR I) sources, the separation between the regions of highest brightness is *less* than half the largest size of the source, while in Class II (FR II) sources this separation is *more* than half the largest size of the source. For $L_{1.4\ \mathrm{GHz}}$ below $\sim 5 \times 10^{25}\ \mathrm{W\ Hz^{-1}}$ most sources are FR Is, while above this power most are FR IIs. In detail, the break radio power increases with the optical luminosity of the host galaxy [2, 3]; when this dependence is accounted for, the Fanaroff–Riley transition is remarkably sharp (Figure 5.1).

Figure 5.2 shows 3C 31, an example of a "plumed" FR I source. An unresolved radio *core* at the AGN produces two collimated expanding *jets* that bend and widen into meandering *plumes*. The plumes widen and become fainter away from the core, so that the morphology of the entire source is an edge-darkened structure several

Relativistic Jets from Active Galactic Nuclei, First Edition. Edited by M. Böttcher, D.E. Harris, H. Krawczynski.

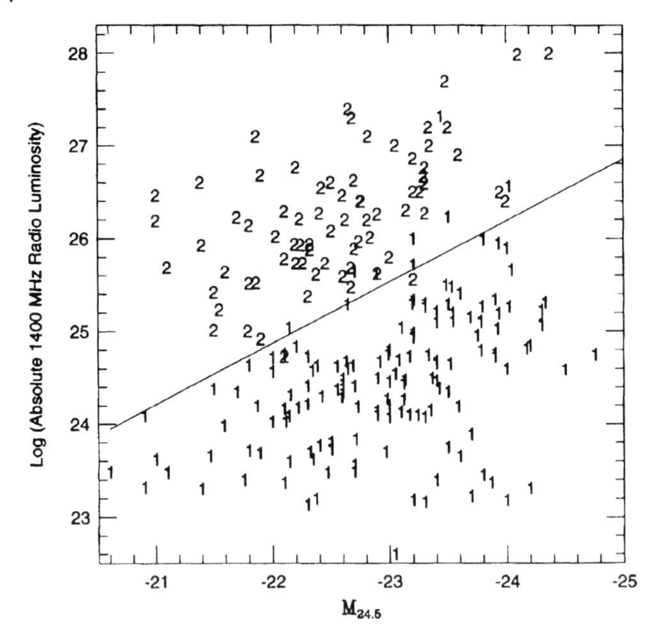

Figure 5.1 Fanaroff–Riley morphological classification vs. radio and optical luminosities. FR I sources are plotted with symbol "1", FR II sources with symbol "2". Reproduced from Ledlow and Owen [3] with permission from the AAS.

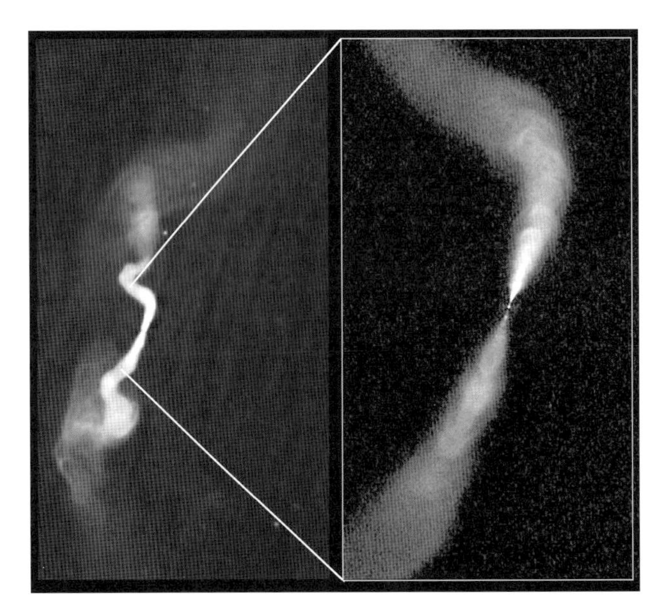

Figure 5.2 VLA images of the plumed Fanaroff–Riley Class I radio galaxy 3C 31: 1.4 GHz, 5.5″ FWHM resolution, 15′ (300 kpc) field of view, and (inset) 8.4 GHz, 0.3″ FWHM resolution, 2′ (40 kpc) field of view.

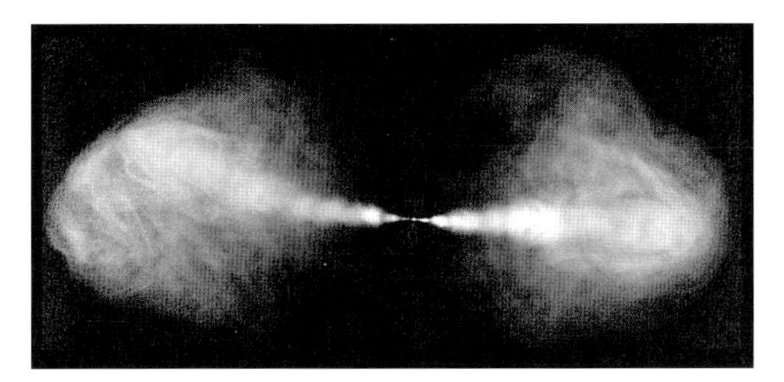

Figure 5.3 VLA image of the lobed Fanaroff–Riley Class I radio galaxy 3C 270: 4.9 GHz, 0.6″ FWHM resolution, 7.5′ (70 kpc) field of view.

hundred kpc in extent whose outer boundaries are ill-defined. At centimeter wavelengths, the spectral indices of such plumes generally increase away from the AGN towards the outer parts of the source, suggesting that the electron populations most affected by synchrotron aging in such structures are those farthest from the AGN.

Figure 5.3 shows 3C 270, an example of a "lobed" FR I source. The unresolved *core* produces two expanding *jets*, but the extended structure has a C-shaped outer envelope delineating two *lobes* of more extended emission apparently surrounding the jets. (In 3C 270, the locations of the highest brightness regions are somewhat resolution-dependent, so the classification of the eastern half of the structure as an FR I is less clear than for the western half.) The spectral indices of the extended lobes generally increase towards the center of such sources, suggesting that the electron populations most affected by synchrotron aging are those furthest from the C-shaped envelope. The spectral difference between the extended regions of lobed and plumed FR I sources, and the result that lobed FR I sources are in the majority in complete samples of low-luminosity radio galaxies [4] make it unlikely that the lobed sources are bent plumed sources seen in a confusing projection; that possibility cannot be ruled out for individual cases, however.

Figure 5.4 shows 3C 353, an example of a lobed FR II source associated with an elliptical galaxy at $z = 0.03$. The unresolved *core* sits between two well-collimated *jets* that extend towards bright *hot spots* in the outer half of the structure. The more extended emission forms well-bounded *lobes* with much fine structure in filaments, wisps and secondary hot spots. There is also weak diffuse emission in the center of the source. Compact hot spots are not *required* to be present on both sides of a lobed source for it to be classified as an FR II, but their presence is often a conspicuous difference between lobed FR II sources (like 3C 353) and lobed FR I sources (like 3C 270). There are also significant differences between the jets in the two types of lobed sources, see Section 5.3 where we describe two jet "flavors" whose characteristics correlate well with the Fanaroff–Riley classification of the extended structures that contain them.

Figure 5.5 shows 3C 204 and 3C 212, examples of lobed FR II sources associated with quasars at $z \sim 1$. The unresolved radio *cores* and the *jets* are more prominent relative to the extended structure, and the jets are more "one-sided", than in radio

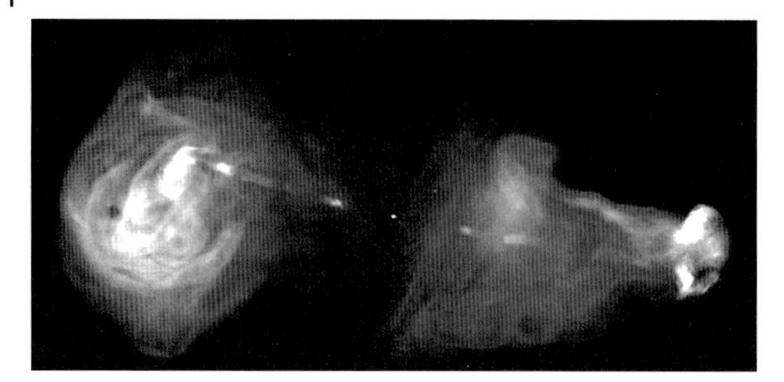

Figure 5.4 VLA image of the Fanaroff–Riley Class II radio galaxy 3C 353: 8.4 GHz, 0.44″ FWHM resolution, 5′ (180 kpc) field of view.

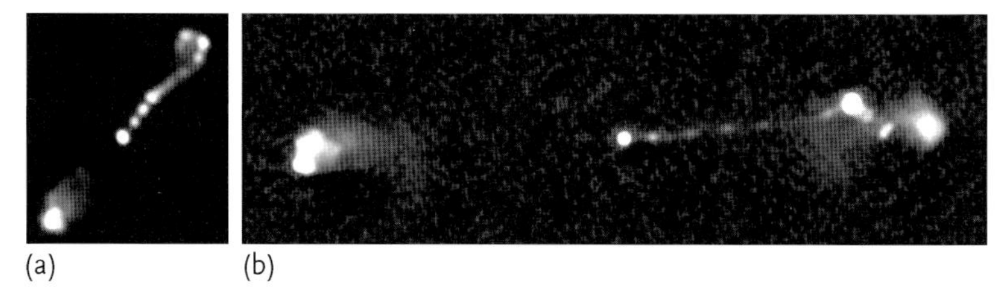

(a) (b)

Figure 5.5 VLA images of two Fanaroff–Riley Class II quasars. (a) 3C 212: 8.94 GHz, 0.24″ FWHM resolution, 12″ (∼ 100 kpc) field of view. (b) 3C 204: 4.9 GHz, 0.34″ FWHM resolution, 38″ (∼ 300 kpc) field of view.

galaxies such as 3C 353 (Figure 5.4). Despite the apparent one-sidedness of the jets in these sources, there are often well-defined hot spots in both lobes, suggesting ongoing activity on both sides of the AGN. The lobes of the two quasars shown are also more compact than those in the radio galaxy 3C 353, but this is not always the case – some FR II sources produced by quasars have more diffuse lobes than these, and a few have diffuse extensions resembling the plumes in FR I sources. The spectral indices of FR II source lobes generally increase away from the hot spots at centimeter wavelengths, and are highest in the diffuse emission near the centers of the sources, as in the lobed FR I cases.

5.1.1 Terminology

Several terms used above to describe features of double radio sources have been given precise definitions to assist more fine-grained classification of the radio structures.

Bridle [5] defined a radio "jet" as "a feature that is at least four times as long as it is wide, separable at high resolution from other extended structures (if any) either by brightness contrast or spatially (e.g., it should be a narrow ridge running through more diffuse emission, or a narrow feature in the inner part of a source entering more extended emission in the outer part), and aligned with the nucleus of the parent object where it is closest to it".

Leahy [6] defined a "lobe" as "an extended region of emission which is not a jet, showing billowy or filamentary substructure, whose perimeter is mostly well defined in the sense that the projected magnetic field is parallel to the edge, the intrinsic polarization is $> 40\%$, and the intensity tends to zero as the perimeter is approached".

Bridle *et al.* [7] defined a lobe "hot spot" as the brightest feature in the lobe having a surface brightness greater than four times that of the surrounding emission and a FWHM less than 5% of the largest diameter of the source, while being further from the AGN than the end of the jet if one is detected (to distinguish hot spots from jet knots – see Figure 5.5).

The spectral index, α, is defined by flux density, $S_\nu \propto \nu^{-\alpha}$.

5.2 Parsec-Scale Jets

Most jets are believed to be intrinsically two-sided, because the outer radio structures of the sources are generally symmetric, as in Figures 5.2–5.5. However, on a small scale, most jets appear to be one-sided. As explained in Chapter 2, this is due to relativistic boosting, which creates a strong front-to-back ratio when the geometry is propitious; that is, when $\theta \lesssim \Gamma^{-1}$, where θ is the angle between the jet and the line-of-sight (LOS), and Γ is the Lorentz factor. The boosting greatly raises the apparent luminosity of the approaching jet, so that surveys down to a fixed flux density level are strongly biased, and contain one-sided sources well out of proportion to their true numbers in the sky. One-sided jets are discussed here, and compact two-sided jets are considered in Section 5.2.2.

5.2.1 One-Sided Jets

Figure 5.6 shows a "stacked" 15-GHz image of the quasar 3C 345, from the MOJAVE website [9], showing a one-sided jet. No counterjet is seen. In this case the jet is a stable structure and individual "blobs" or "components" appear near the base and tend to move outward. The stack shows the weighted average of these components, as they move along the track. In other sources the jet is not stable, but can have a variable ejection angle as well as variations in the ridge line.

Figure 5.7 shows the BL Lac object 0003-066 at three epochs. The structure of the jet changes markedly over the ten-year period [8]. In another case, 3C 279 [10], the jet changed direction over a period of several years, but at a point well downstream of the base. In these cases, it is important to remember that we measure changes in the image as projected on the sky, and small intrinsic bends can be strongly amplified in the projection when θ is small.

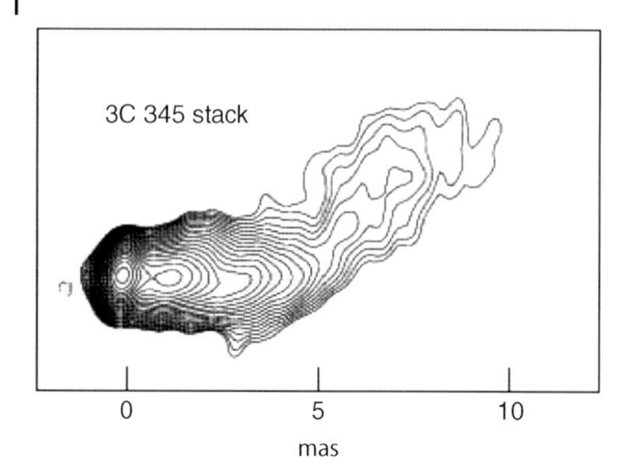

Figure 5.6 Stacked image of 3C 345 at 15 GHz. Contours change in steps of $\sqrt{2}$. Thirty-nine separate VLBA images made over the interval 1995–2005 were coadded for this stack [8], which has a dynamic range of about 2000:1. The linear scale is 6.6 pc mas^{-1} so the overall length of the jet seen here is about 60 pc, as projected on the sky.

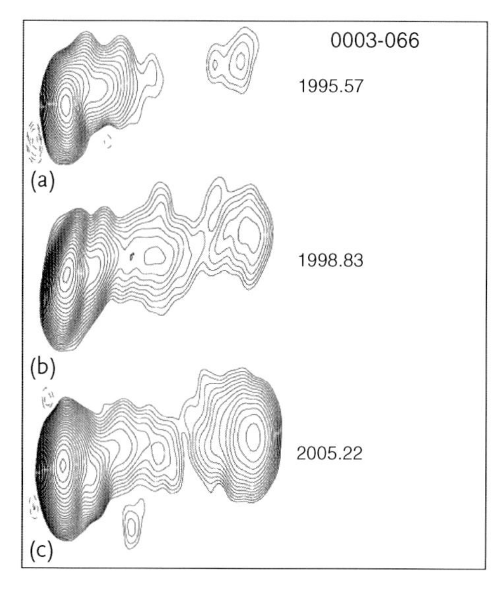

Figure 5.7 Images of quasar 0003-066 at 15 GHz at three epochs: (a) 1995.57, (b) 1998.83, and (c) 2005.22. Images at further epochs can be found at the MOJAVE website [9], along with a movie. Due to the low declination, the restoring beam is elongated NS. The projected position angle of the jet near the core changes from about −65 to −110° over the 10 years. The projected separation between the two strong peaks in the 2005.22 image is 7.0 mas, or 34 pc.

In most cases the beginning of the jet is marked by a bright compact component, called the core. It generally is regarded as the location in the jet where the optical depth to synchrotron emission, τ becomes unity, but Marscher [11] points out that in some cases it could be a shocked region downstream of $\tau = 1$. The jet generally has decreasing brightness along its length, although strong downstream components often stand out, as in Figure 5.7.

In 1–3% of the sources in large VLBI surveys (Section 5.2.3), the images show only a core. This could be due to insufficient dynamic range, or perhaps to intermittent jet flow.

5.2.1.1 Bends and Helices

Bends such as the one seen in Figure 5.6 are common in pc-scale jets, and they also are seen in large-scale jets, e.g., Figure 5.2. In Figure 5.6 the bend occurs at about 25 $\csc\theta$ pc from the center of the host galaxy. As discussed later, θ is probably only a few degrees, so the bend occurs about 500 pc from the center.

The apparent misalignment frequently seen between the ejection at the core and the direction of the large-scale jet was first investigated by Pearson and Readhead [12] who showed that the distribution of misalignment angles peaked around $0°$, with a secondary maximum near $90°$. In Figure 5.6, 3C 345 displays a bend of around $40°$, but larger-scale images show a total misalignment near $90°$. Conway and Murphy [13] suggested that this tendency for a $90°$ bend between the small- and large-scale structure could be due to low-pitch helical motion, combined with relativistic beaming. More recently, however, Kharb *et al.* [14], with higher-frequency data and more sources, found that the distribution of misalignment angles decreases rather smoothly from 0 to $180°$, with only a weak $90°$ peak.

On parsec scales, the shape of a jet projected on the sky is frequently sinuous, and some jets are edge-brightened. This has prompted analyses to describe it as the projection of a helix. Helical structures are expected to occur as a result of nozzle precession, or Kelvin–Helmholtz instabilities (Section 10.3.1), or via a "magnetic tower" (Section 4.3.3). Many jets have been analyzed in terms of helices: a few recent ones are for BL Lac [15], which yielded $\Gamma = 3$–6; 3C 273 [16] which was fit with a double helix; and 3C 120 [17] which was fit in detail, using a "pseudosynchrotron" analysis that included relativistic effects, and showed that a helical expanding jet could match the observations.

5.2.1.2 Wiggles, Wobbles, Kinks

The bends seen in the 3C 273 jet in Figure 5.8 are often called *wiggles*. The jet in 0003-066 (Figure 5.7; $z = 0.357$, scale $= 4.88\,\mathrm{pc\,mas^{-1}}$) is swinging rapidly, and this is sometimes called *wobble* [19]. The MOJAVE website [9] shows movies of many sources, and the wobble phenomenon can be seen there. The observed time scale for wobble is 2–20 years [19], although the upper limit probably only reflects the length of time the observations have been made. Wiggles and wobbles are often thought to trace a helical jet; in the latter case the jet is precessing on a short time scale. See Agudo *et al.* [20] and Britzen *et al.* [21] for detailed discussions of wobble in the sources 0355+508 and 0735+178, respectively.

The 3C 273 jet (Figure 5.8) shows a kink in the 15 GHz image, at about 10 mas (projected) from the core. This kink is visible episodically, and sometimes appears to move. See Section 5.2.4 and images on the MOJAVE website [9]. This structure may represent a kink instability; see Figure 11.9 for a 3D nonrelativistic MHD simulation, and Section 10.3.1.1.

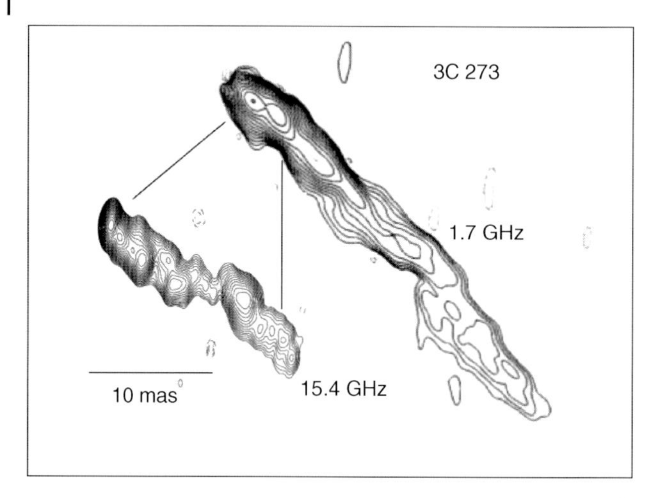

Figure 5.8 Images of 3C 273 at 1.7 GHz (epoch 1988.9 [18]), and at 15 GHz (epoch 2006.2 [8]). The images have dynamic range of about 5000:1. Redshift $z = 0.158$ and linear scale is 2.7 pc mas^{-1}. The projected length at 1.7 GHz is about 200 pc, and the true length is a few kpc. Reprinted from Davis *et al.* [18] with permission from Macmillan Publishers Ltd.

5.2.1.3 **Spectrum Along a Jet, Coreshift**

The spectrum along a jet generally shows a gradient, with the core having the flattest spectrum (see Section 7.6.1; this effect is called a "progression" by X-ray astronomers). This means that observations at lower frequencies usually emphasize the downstream portions of the jet; when combined with the generally lower angular resolution at lower frequencies, this can give a view of the core that is blended with the nearby, steeper-spectrum jet. At higher frequencies the core can be isolated from the downstream jet, and individual components can be seen emerging from the core. For example, in 3C 273 (Figure 5.8) the 1.7 GHz core is composed of a number of 15 GHz components that are blended together. Similarly, the core in the 15 GHz image becomes decomposed at 43 GHz [22].

The Blandford–Königl model [23, 24] for the synchrotron emission from an expanding jet shows that there should be a *coreshift* such that the core ($\tau = 1$ region) moves upstream as the observing frequency is raised. Hence, cores at different frequencies may refer to different locations, and comparison of fluxes or images of such cores should be treated with caution. However, the downstream components are optically thin, and their positions are thought to be stable as a function of frequency. Alignment of the optically thin components at different frequencies is a customary way to measure the coreshift [25–27]. Measured coreshifts are a few tenths of a milliarcsecond at cm wavelengths, and they vary with frequency in the way predicted by the theory. This shows that the expanding jet model, with the radio core at $\tau \approx 1$, is a reasonable representation of the inner jet.

5.2.2
Two-Sided Jets

Lister *et al.* [28] discuss five nearby galaxies that have two-sided parsec-scale jets in their nuclei. Their proximity makes them amenable to study with VLBI, even though they have low luminosity and subrelativistic motions. As an example, NGC 1052 is shown in Figure 5.9 at four frequencies [29]. The gap in the center is probably caused by absorption in obscuring clouds, possibly arranged in a torus; the absorption is frequency-dependent and at 5 GHz appears to hide component B. The NE jet is approaching. This source is mildly variable. The components are seen to move outward on both sides, with apparent transverse speed $v_{app} \sim 0.2\text{–}0.3c$ [30]. With the assumption that the two sides are intrinsically similar, $57° < \theta < 72°$ [29], and the true speed of the beam is $v \approx 0.25c$ (2.68) and (2.70). The beam in NGC 1052 is not relativistic.

Counterjets are thought to exist in the blazars and radio galaxies, but are rarely seen because they are relativistically deboosted. With sufficient sensitivity, however, the counterjets should be seen. The best cases for this are M 87 [31], NGC 1275 (3C 84) [32], and Cen A [33] where weak counterjets are seen with confidence.

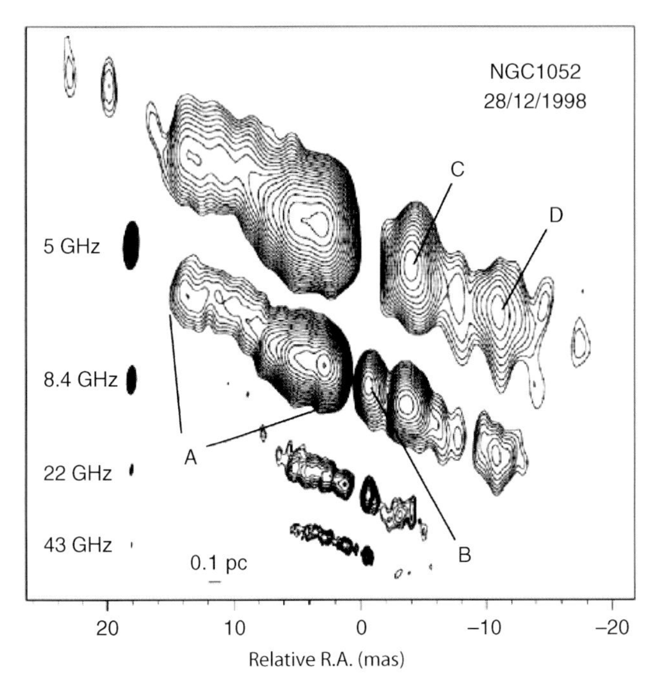

Figure 5.9 Images of NGC 1052 at four frequencies. This galaxy is 22 Mpc away so the linear resolution is high, and the overall length of the radio source is only about 4 pc. It is oriented near the plane of the sky, so its projected length is close to the true length. Taken from Kadler *et al.* [29] and reproduced with permission of Astronomy and Astrophysics.

Compact Symmetric Objects (CSO) are small but luminous. They have oppositely directed jets tens or hundreds of parsecs long, and morphologically are often miniature versions of the large extended radio sources. They are young, only thousands of years old, and are expanding but are only mildly relativistic. In some cases there also is large-scale outer radio structure, and the CSO may be part of a recurrent phenomenon [34]. CSOs often have the spectral characteristics of the Compact Steep Spectrum (CSS) and Gigahertz Peaked Spectrum (GPS) sources, which mainly differ by the frequency at the peak of the spectrum: GPS peak above 5 GHz and CSS peak below that, down to 100 MHz.

5.2.3 VLBI Surveys

Recent major VLBI surveys of compact sources are listed here. The sources are typically selected on the basis of flux density, compactness, and high-energy emission. There is substantial overlap in sources among the surveys. Several of them include subsamples that themselves form a statistically complete, flux density-limited sample. Most of the surveys are continuing, to increase sample size, to obtain more accurate kinematic data, and to study the relations between the radio and the high-energy emissions, especially gamma-rays measured with the Fermi Observatory.

VIPS, CJF (5 GHz) The VLBA Imaging and Polarimetry Survey (VIPS) [35] subsumes the 5-GHz Caltech–Jodrell Bank Flat-Spectrum Survey (CJF; [36]). VIPS includes 1127 sources, most of which are in the SDSS footprint. It comprises a large database that readily allows radio-optical comparisons, although the SDSS does not have spectral and morphological data for all the objects. The CJF contains 293 objects and is a complete flux density-limited subset of VIPS. At least three epochs of observations at 5 GHz are available for each CJF source, and images and kinematics are in [37, 38], respectively.

MOJAVE (15 GHz) MOJAVE (Monitoring Of Jets in Active galactic nuclei with VLBA Experiments) [8] is a 15-GHz monitoring survey that currently is being expanded to include about 280 sources. It includes a statistically complete subsample of 163 objects, the known compact sources stronger than 1.5 Jy at any epoch since 1994, and north of $\delta = -30^\circ$. Full polarization images of each source are made at each epoch. The cadence for observing the sources varies from three weeks for rapidly moving complex sources, to about two years for more stable objects. The median number of epochs per source is about 10; this is sufficient for accurate measurements of component motions.

MOJAVE is a continuation of the 2-cm VLBA survey that started when the VLBA became available in 1994 [39]. The main objectives of the program include studying the internal motions and polarizations of the sources, and correlating the high-resolution radio results with gamma-ray emissions as measured by the Fermi satellite. Data and results from MOJAVE are available at their website [9].

TANAMI (8.4, 22 GHz) The MOJAVE survey has a restricted declination range, and to cover the whole sky the southern hemisphere TANAMI program (Tracking Active galactic Nuclei with Austral Milliarcsecond Interferometry) was instituted. It is monitoring strong compact sources south of $\delta = -30°$ with VLBI at 8.4 and 22 GHz. It includes flat-spectrum radio sources with flux density above 2 Jy at 5 GHz plus a number of other sources of interest, including gamma-ray sources seen by the LAT on the Fermi Observatory. This program is still at an early stage, but has published first-epoch results at 8.4 GHz on 43 objects [40].

VCS1 (2, 8 GHz) The VLBA Calibrator Survey (VCS1) [41] includes 1332 compact sources observed with the VLBA at 2.3 and 8.4 GHz with astrometric accuracy. The objective is to provide a list for phase referenced VLBI observations, and for astrometry and geodesy. The data are also useful for astronomical studies of compact jets.

RRFID (2, 8 GHz) The RRFID (the US Naval Observatory's Radio Reference Frame Image Database) supports precision astrometry and geodesy by maintaining a large database[1)] of images of compact radio sources [42–44]. Observations are performed every other month at 2 and 8 GHz, using the VLBA together with up to 10 geodetic antennas. Currently the database contains 517 sources, chosen for their suitability for astrometry and geodesy. The *RRFID kinematic survey* [45] is a subset of the database comprising 87 sources with at least three epochs of observation at 8 GHz.

VSOP (5, 22 GHz) The VSOP (VLBI Space Observatory Program) included a prelaunch survey [46] that detected strong fringes from 136 sources at 22 GHz, and provided a guide for the main program between the HALCA (Highly Advanced Laboratory for Communications and Astronomy) satellite and the ground-based VLBI antennas. In the satellite program 242 sources were studied [47].

Measurement of brightness temperatures of AGN cores was a major part of the program. HALCA baselines were three times longer than those available between two telescopes on earth, and so in principle should have yielded brightness temperatures (or limits) an order of magnitude larger than those available from terrestrial measurements. However, noise and calibration difficulties on the long baselines to the satellite reduced the available T_b range in some cases. See Section 5.2.6.

43 GHZ survey Jorstad *et al.* [22] made VLBA observations of 15 AGN at roughly two-month intervals during 1998–2001, at 43 GHz. Both total and polarized images of the flux density were obtained. This is the largest published set of multiepoch VLBI observations that exists at millimeter wavelengths. The targets include objects that are bright and variable at 43 GHz, and the objective is to gain information relevant to jet collimation and acceleration, and to evaluate correlations between

1) http://rorf.usno.navy.mil/RRFID.shtml (accessed 31 August 2011).

gamma-ray emission and the radio structures. This work has continued, currently with monthly observation of about 30 blazars with the VLBA. Images and model details are available on their website.[2)]

GMVA (86 GHz) VLBI at 86 GHz is important but difficult. At this frequency the core of a jet may be optically thin, and so allow a view deep into the central region of the AGN. The resolution is high since a global array has $d/\lambda > 2 \times 10^9$; this corresponds to less than 60 Schwarzschild radii at a nearby object like M 87 [48].

Lee *et al.* [49] have made a survey of 127 sources with the GMVA (Global Millimeter VLBI Array) at 86 GHz, of which 109 yielded images. The dynamic range is low but some of the sources show several components and can be said to have a jet-like structure.

5.2.4
Motions in the Jet

The components seen along the jets in Figures 5.6–5.9 tend to move downstream, away from the core. Multiepoch images are used to track their motions, as in Figure 5.10, where the distance of the centroid of each component from the core is plotted. Figure 5.10 shows a simple case, where there is no ambiguity in following the components. Often, however, components are close together and some follow curved trajectories, and in complex cases it is necessary to make observations at close intervals to follow the components unambiguously. Many examples are in the literature, e.g., [28] at 15 GHz and [22] at 43 GHz. The slope of the trajectory gives a speed that is usually expressed as a multiple of the velocity of light; see Section 5.2.5.

The RRFID survey at 8.6 GHz is closely matched in resolution to the MOJAVE survey at 15 GHz, because RRFID has longer baselines that roughly compensate for the longer wavelength. Piner *et al.* [45] have used this fact to compare the motions at the two frequencies for the 36 sources they have in common. For a few components in certain sources the motions are in good agreement; but for most components they are not (see Figure 8 of [45]). The differences likely result from time-dependent spectral differences among the various components, combined with difficulties in identifying components in the images, and differences in modeling techniques. More work is needed to clarify the frequency dependence of component identifications.

A substantial fraction of the moving components are not moving radially out from the core (32% at 15 GHz in the MOJAVE survey). In these cases a plot like Figure 5.10 can be misleading. Several of the surveys described above present data on both radial and transverse components of the velocity. In addition, some components are seen to accelerate, either by changing direction or by changing speed with a constant direction.

2) http://www.bu.edu/blazars/VLBAproject.html (accessed 31 August 2011).

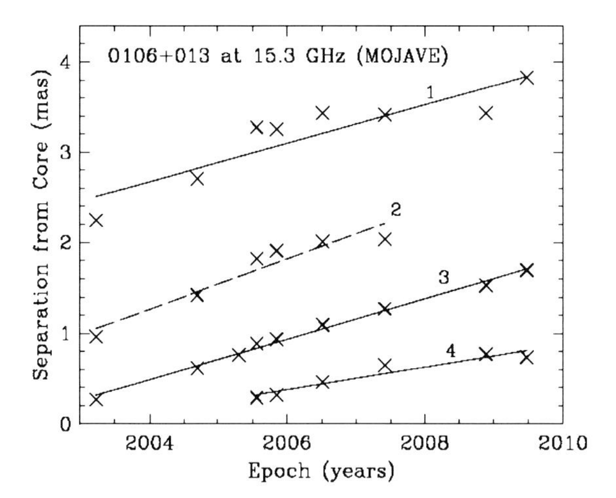

Figure 5.10 Expansion diagram for four components in 0106+013. Solid lines indicate uniform vector motion that is radial, to within the uncertainties, while the dashed line for component 2 indicates a significantly nonradial motion [28].

In all surveys, a small number of sources are found that have components moving towards the core. These appear in sources that otherwise are normal, and explanations for this behavior have been offered within the relativistic outflow model [28, 37]. The first is that as a component moves downstream around a bend, its projected position can move towards the core, and it could even move behind the core and emulate a counterjet. Secondly, as a new component begins to emerge from the core, the centroid of the enlarged core moves downstream, meaning that the separation of this centroid from the next component can get smaller [50].

Ballistic vs. streaming motions An early question was whether the bends are formed by dynamical processes, with the components streaming along the jet; or the flow is ballistic, with bends formed by rotations of the nozzle. It now is clear that this question is too simplistic, since both types of motion can occur in the same source. 3C 345, however, presents a clean case of streaming around a bend (Figure 5.6). The bend at 4 mas was stable for 14 years, while the speed of the individual blobs was about 0.56 mas/year [28]. Had the flow been ballistic, the bend would have been substantially broadened to the west. As it is, individual components were tracked going around the bend (see Lister *et al.* [28], Figure 3.340).

The motions in 3C 273 are more complex. The position angle (PA) at the nozzle was about 144° in 2006, as seen at 15.4 GHz in Figure 5.8. However, this angle is resolution-dependent; if 3C 273 had been farther away, or if the observations had been made at a longer wavelength, the PA would have appeared to be smaller. The kink near 10 mas connects regions that are offset but roughly parallel; this feature appears in 3C 273 at many epochs. During some intervals the kink appears to move out with the flow; see the sequence of images and the movie in the MOJAVE website [9], especially the interval 2004–2008. Another feature of 3C 273 is that the initial PA itself changes, e.g., from about 110° in 1995 to about 144° in 2006, at 15 GHz. Abraham and Romero [51] modeled 3C 273 with a precessing helix, with a period of about 16 years; while Lobanov and Zensus [16] used a double helix to allow for the transverse structure in the jet.

Pattern vs. beam Another early question was whether VLBI was measuring relativistically moving blobs of plasma, or patterns such as those formed on a screen, or by stationary but blinking lights. Several lines of evidence show that the radiating plasma is indeed moving. First, there is strong evidence for relativistic motion, as described in Chapters 1 and 2 and in Section 5.3.1. Secondly, the fact that 95% or more of the moving components are moving out, not in, eliminates random motions. Thirdly, statistical studies of strong radio jets show that the apparent luminosity is correlated with the apparent transverse speed, as expected for relativistic beams (Section 5.2.6). This would require an *ad hoc* explanation for screens.

However, there may well be a difference between the motion of the underlying flow and that of a visible moving radio component. The enhanced density responsible for the component may be due to a shock, whose apparent speed and direction will be different from that of the general flow (see Chapter 9). A stationary shock could produce a stationary component, and indeed many of these are seen. Low pattern speed (LPS) features in the MOJAVE sample have been studied by [28], who show that most LPS components are found within 6 pc (projected) of the core. They may be due to standing shocks, possibly generated at a bend in the jet. Some of them could simply reflect the geometry; for example, a helical beam could have its Doppler shift enhanced at a stationary point where the beam lies close to the LOS. See [52] for a discussion of a new stationary component that suddenly appeared in 3C 120.

Some components may be due to trailing shocks (e.g., Agudo *et al.* [53]) that are slower than the flow; use of such components will underestimate the flow velocity. The use of the fastest component to derive the Lorentz factor [28] will yield a velocity that will overestimate the flow speed if that component is a density enhancement in a downstream shock.

5.2.5 Relativistic Beams

Relativistic beams are discussed in Chapter 2, and the results developed there are used here. A moving component is characterized by its Lorentz factor, Γ, and the angle to the LOS, θ; from these the Doppler factor, δ, and the normalized apparent transverse speed, $\beta_{\mathrm{app}} = v_{\mathrm{app}}/c$ can be found (2.30 and 2.59). When $\Gamma^2 \gg 1$, the beam parameters are simply related as in Figure 2.6, which shows β_{app} and δ vs. θ, all normalized by Γ. Note that the peak Doppler factor is approximately twice the Lorentz factor and is attained when $\theta = 0°$. When $\sin\theta = \Gamma^{-1}$, $\delta = \Gamma$ and $\beta_{\mathrm{app}} = \beta_{\mathrm{app,\,max}} = \beta\Gamma$. This value of θ is called the "critical angle", θ_{crit}, or the "superluminal angle", θ_{sl} (see Section 2.3.2).

When $\Gamma > \sqrt{2}$, then $\beta_{app} > 1$ in some range of θ, in which case the apparent motion is said to be superluminal. Another relativistic effect is a reduction in apparent time intervals, proportional to δ. This increases the apparent flux density variability. The MOJAVE survey includes sources whose flux density has exceeded the lower limit at some time during the specified program interval. Thus the list is biased towards sources with rapid variability, i.e., a high Doppler factor, which implies a high Lorentz factor, because $\Gamma > \delta/2$.

The flux density received from an approaching relativistic beam is increased by a factor of δ^n over what a comoving observer would measure, where $n = p + \alpha$; α is the spectral index and p is the Doppler boost index, typically 2 or 3 (see Figure 2.3 and (2.44). The exponent $p = 3$ is appropriate for the radiation from a source that is smaller than the beam, but $p = 2$ for a continuous jet [54]; see also Section 5.2.7. Since δ can be 20 and more, it is clear that beams aimed close to the LOS can be greatly magnified in strength, and those aimed in other directions will be weakened. This biases the selection of beamed sources; a sample chosen on the basis of the beamed flux density will preferentially contain sources with small values of θ, even though the solid angle associated with powerful boosting is small [55]. With a simple model for a flux density-limited sample of beamed sources, Cohen *et al.* [56] showed that roughly 3/4 of the sample will have $\theta < \theta_{crit}$. A further implication is that the parent population of the observed relativistic sources must be huge; Lister and Marscher [57] estimate that there are millions of misdirected AGN to provide for the 293 objects in the CJF survey.

5.2.5.1 **Superluminal Motion**

A source will typically produce a new moving component every few years, and in some sources many components have been followed. In Figure 5.10 four components are labeled, and it is noticeable that their lines are roughly parallel; they have similar speeds. In MOJAVE [58] and RRFID [45] the variance of the speeds in a source is less than the variance of the speeds among all sources, so many sources can be said to have a characteristic speed. In fact most sources are not as clean as 0106+013 (Figure 5.10), and many have both stationary and moving features.

Although there is difficulty in following individual components at different frequencies, the statistics of the speeds can provide useful information. Lister *et al.* [28] use the “fastest” moving component for studying the MOJAVE sources, because that might best represent the speed of the underlying flow; slower components might be due to trailing shocks, which are slower than the flow [53]. Figure 5.11 shows the distribution of β_{app} for the MOJAVE sample. The open line is for all robust jet features, and the solid histogram shows the fastest robust component for each source. The distribution of the fastest components is roughly flat to $\beta_{app} \sim 15$ with a tail to $\beta_{app} = 50$. The distribution of all features has a strong peak at low speeds; 40% of all features have $\beta_{app} < 5$.

CJF is also a flux-limited survey, but a comparison with MOJAVE is difficult because the results are reported differently. Figure 7 in Britzen *et al.* [38] shows a histogram of speeds for all “quality 1” components, but this includes more than one point for some sources. The shape of the histogram is similar to the open line

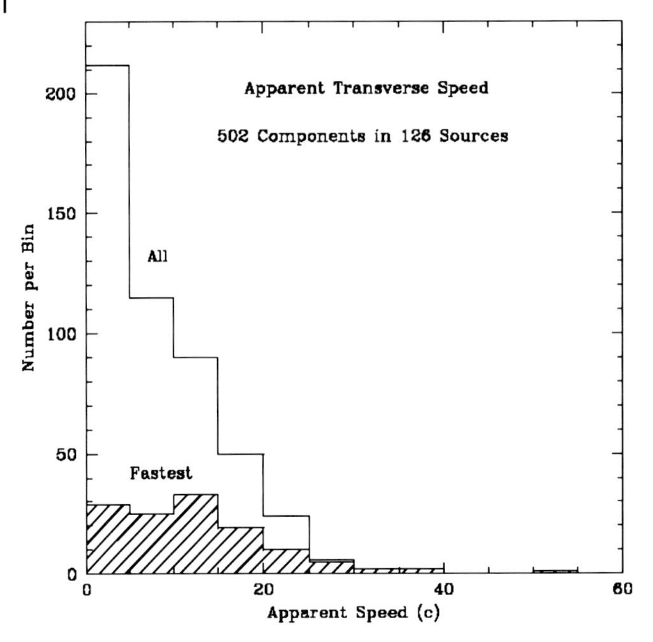

Figure 5.11 Histograms of β_{app} for MOJAVE; open for all robust components, and hatched for the fastest component in each soure. From Lister *et al.* [28], Figure 7.

in Figure 5.11, but is narrower, with a tail to $\beta_{app} \sim 25$. The 5 GHz speeds on average are lower than the 15 GHz speeds, very roughly by a factor of 2. See [38] for further discussion.

RRFID and the 43 GHz survey do not use complete samples, but rather concentrate on particularly bright and compact sources. However, CJF and MOJAVE are restricted to *flat-spectrum* sources, which in general are bright and compact, and there are many sources in common among these four surveys. At 8 GHz, RRFID shows a histogram of 54 speeds, for the fastest component in each source. It is similar to the 15 GHz plot, roughly flat to $\beta_{app} \sim 10$ with a tail to $\beta_{app} \sim 30$, but with a peak in the first bin (0–2c). Jorstad *et al.* [22] study only 15 sources and an equivalent histogram is not warranted. However, from their Table 5 we find that at 43 GHz the median of the fastest average speeds is $\beta_{app,med} \approx 17$. From these comparisons, it appears that the fastest speeds increase as the observing frequency increases from 5 to 43 GHz. Lister *et al.* [28] suggest that at the higher frequencies, superior resolution and a reduced sampling interval allow the faster components to be detected more easily. Furthermore, as seen in Figure 5.8, a component can be a blend of higher-frequency components, and if there is a distribution of speeds among the components it is only at the higher frequency that the fastest speed can be detected.

5.2.5.2 Lobe-Dominated Quasars

Most of the radio studies involving relativistic beams use objects whose jets lie close to the LOS; the cores and jets are strong and amenable to study. The more numerous jets lying close to the plane of the sky are strongly deboosted and invisible. But an intermediate angular range, roughly from 10 to 40°, is being studied, albeit with difficulty. Hough *et al.* [59, 60] are observing a complete sample of 25 weak-nucleus *Lobe-Dominated Quasars* (LDQ) from the 3CR catalog, with the VLBA. They have made images of all the objects, and have found that 22 have one-sided jets; the other three are barely resolved. For ten of the objects speeds have been determined, and they fall in the range $\beta_{\text{app}} \sim 0{-}10$. The LDQ appear to be physically the same as the strong core-dominated quasars, but at higher angles to the LOS.

5.2.6
Statistical Studies of Compact Jets with VLBI

The Lorentz factor Γ is an important physical quantity that characterizes a beam, but can be estimated only indirectly. The measured distribution of β_{app} (Figure 5.11) shows that there must be a distribution of Γ, for if all sources had the same Γ, then the distribution of β_{app} would have a sharp peak at $\beta_{\text{app}} \approx \Gamma$, which it does not have [57, 61]. A power-law distribution has generally been used in statistical studies involving Γ [57].

The observable quantities are flux density, brightness distribution, redshift and proper motion, and from the latter two the apparent transverse speed β_{app} is calculated. In a number of cases where both a jet and counterjet are seen (or where useful limits can be set) the Doppler factor has been estimated by assuming that the two jets are intrinsically identical, as discussed in Chapter 2. Once β_{app} and δ are in hand, then Γ and θ can be calculated with (2.71) and (2.72).

For a one-sided source, it is sometimes assumed that $\Gamma \approx \beta_{\text{app}}$ (i.e., that $\theta \approx \theta_{\text{crit}}$). However, this will be incorrect in many cases. From Monte Carlo simulations with a power-law distribution of Lorentz factors, Cohen *et al.* [56] showed that $\Gamma > 1.5\beta_{\text{app}}$ for roughly 30% of the sources with $\beta_{\text{app}} > 4$, in a flux density-limited sample of relativistic jets.

The Doppler factor δ may be estimated from the brightness temperature of the jet, T_{b}, which can be directly measured with VLBI. The Doppler factor is found by dividing T_{b} by an "intrinsic temperature", T_{int} (2.49), usually taken to be the "equipartition temperature", $T_{\text{eq}} \sim 5 \times 10^{10}$ K defined by Readhead [62]; or a limit can be found by using the inverse-Compton limit $T_{\text{IC}} \sim 10^{12}$ K [63]. δ and β_{app} then uniquely fix the value of Γ (2.71). Homan *et al.* [64] separate the MOJAVE observations into low- and high-flux density states, and show that in the low-state $T_{\text{int}} \sim 3 \times 10^{10}$ K, and in the high-state $T_{\text{int}} \geq 2 \times 10^{11}$ K. For sources in the high state the median of the peak values of T_{b} is $\sim 2 \times 10^{12}$ K, and the lower limits on T_{b} extend up to 5×10^{13} K [65]. If we take $T_{\text{int}} \sim 2 \times 10^{11}$ K then $\delta_{\text{med}} \sim 10$ and δ_{peak} is at least 250, although it may be the case that T_{int} is higher for sources with higher brightness temperatures, and δ_{peak} correspondingly lower. In most cases $\delta > \Gamma > \delta/2$ [56], so Γ_{med} is between 5 and 10.

The VSOP has T_{b} results [47] that are similar to those for MOJAVE. The measured values have a median of about $10^{12.1}$ K, while including the limits would probably increase the median. There is little difference in T_{b} between 5 and 15 GHz, as expected for optically thick synchrotron cores having the same Doppler factors at the two frequencies.

A brightness temperature $T_{\text{b,var}}$ can also be estimated from the time scale for flux-density variability. The cube root of the ratio of $T_{\text{b,var}}$ to T_{int} gives the "variability Doppler factor" D_{var} that, with β_{app}, gives the Lorentz factor. See (2.50); also Lähteenmäki and Valtaoja [66]. Jorstad *et al.* [22] developed a method that also uses the flux density variability but with the measured angular diameter, not an assumed intrinsic temperature. The various estimates for Γ agree reasonably well for the Jorstad sample, and lead to distributions of Γ that peak at $\Gamma \sim 20$ with a tail to $\Gamma \sim 40$ for quasars, and lower values for BL Lac objects and radio galaxies. However, since $\Gamma > \beta_{\text{app}}$ and the fastest speeds at both 15 and 43 GHz are near $\beta_{\text{app}} = 50$, it is clear that the distribution of Γ must extend up to at least $\Gamma = 50$.

Luminosity-speed correlation Figure 5.12 shows the peak apparent speed and median apparent luminosity for 121 jets in the MOJAVE sample. The line is an "aspect curve" [56], the run of $(\beta_{\text{app}}, L_{\text{app}})$ for a jet with $\Gamma = 50$ and an intrinsic luminosity $L_0 = 3 \times 10^{25}$ W Hz^{-1}, Doppler-boosted by a factor $\delta^{1.63}$, as θ increases from $10''$ at the right-hand end through $1.1°$ at the peak to $23°$ at the left end. Cohen *et al.* [56] and Lister *et al.* [28] show that the lack of sources to the left of the line is not a selection effect, but represents a real correlation between speed and luminosity:

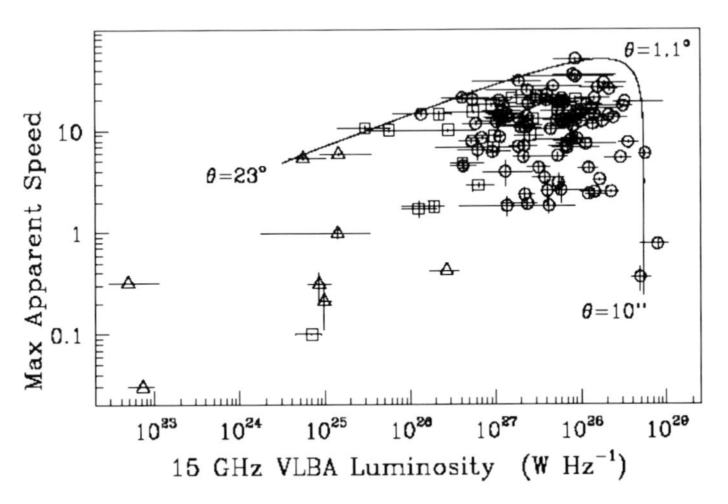

Figure 5.12 Maximum apparent jet speed vs. VLBA luminosity at 15 GHz for 121 jets in the MOJAVE sample. Circles indicate quasars, squares BL Lac objects, and triangles, galaxies. Horizontal bars indicate the range of luminosity during 1994.0–2004.0. The curve is parametric in θ, see text. Adapted from [28].

low-luminosity sources do not have high-speed components. This is seen also in the 5 GHz CJF sample [38].

It is tempting to regard Figure 5.12 as showing that the strongest jets in the sky have $\Gamma \approx 50$ and $L_0 \approx 3 \times 10^{25}$ W Hz^{-1} at 15 GHz. There are only a few of these objects and they are at various θ; the ones with $\theta \lesssim 2°$ lie near the line in Figure 5.12, at the right-hand side and at the top. The ones with higher θ, however, are increasingly unlikely to be in the survey, as they are deboosted. Cohen *et al.* [56], using probability arguments, showed that sources near the line but with $\theta \gtrsim 2°$ have lower values of Γ and L_0. These studies assumed that the distributions of both the intrinsic luminosity and Γ for a population of beamed jets have power-law shapes. The luminosity functions have been studied by [57, 67].

5.2.7 Spine-Sheath Configuration

The nearby galaxy M 87 has a long jet that can be studied in detail, and it displays notable transverse structure. It is *edge-brightened* (see, e.g., [31, 68] for images at 15 and 43 GHz, respectively). This is thought to be due to a transverse velocity gradient, with the jet velocity faster on-axis than at the edge, where there is a slowly moving "sheath". The viewing angle is greater than the critical angle for the central spine, $\theta > \theta_{\mathrm{crit,spine}}$ so that the spine is deboosted, but $\theta < \theta_{\mathrm{crit,sheath}}$ and the radiation from the limb is relativistically enhanced. Spine-sheath models in an X-ray context are discussed in Chapter 7, and numerical simulations are given in Chapters 10 and 11. See [69] for an analysis of 3C 31 that leads to a transverse velocity gradient in a large-scale jet.

The radio galaxy Cygnus A may be similar to M 87. Cohen *et al.* [56] suggest that its powerful radio lobes are indicative of a highly relativistic spine that is invisible because of its large angle, while it has a mildly relativistic sheath that produces the observed weak jet and counterjet.

The spine-sheath jet configuration is also useful in population studies. The unification of FR I radio galaxies with BL Lac objects, with the galaxies being off-axis BL Lacs, has difficulty that is eased if the Doppler factor that controls the galaxy luminosity is smaller than the one that controls the superluminal motion [70]. An object that has a spine-sheath configuration will have an angular dependence of flux density that varies with θ more slowly than the canonical δ^p, with $p = 2$ or 3.

5.3 Kiloparsec-Scale Jets

5.3.1 Correlations with Extended Structure and Luminosity

Many properties of kpc-scale radio jets produced by AGN correlate well with other features of the radio sources, as follows:

- *Correlations between jet properties and the Fanaroff–Riley structure classes.* The kpc-scale jets in low-luminosity FR I radio galaxies are generally prominent (often $> 10\%$ of the total extended radio power), rapidly spreading (FWHM opening angles $> 8°$), and "two-sided" (jet-counterjet intensity ratio $< 4:1$), whether the large-scale structure is plumed (e.g., Figure 5.2) or lobed (e.g., Figure 5.3). Those in higher-luminosity FR II sources are better collimated, more "one-sided", and are generally less prominent relative to the lobes [71] in radio galaxies (e.g., Figure 5.4) than in quasars (e.g., Figure 5.5).
- *Correlations between kiloparsec and parsec scales.* The brighter kpc-scale jet is always a plausible extension of the brighter parsec-scale jet and is always on the same side of the source as any superluminal motion (see Section 5.2.5.1). The kpc-scale jets are also well aligned with the parsec-scale jets in the FR I sources [72–74] and in lobe-dominated FR II sources. In core-dominated sources (whose jets may lie closer to the line-of-sight where *angular* relationships can be exaggerated by projection) even when the jets appear more bent the brighter large-scale jet is generally a plausible continuation of the brighter small-scale jet. These correlations imply that the initial brightness asymmetries of the jets on both parsec and kiloparsec scales, and the superluminal motions, all share a common explanation as consequences of bulk relativistic outflow from the AGN.
- *Correlations between jet sidedness and depolarization asymmetry.* In sources whose kpc-scale jets differ greatly in brightness, the brighter jet is on the side of the source that depolarizes less at long wavelengths [4, 75–78]. This is consistent with the brighter jet being on the side of the source that has the shorter path length to the observer through any intervening Faraday-rotating medium, hence with the apparently brighter jet always being on the side that is nearer the observer, as required if the brightness asymmetry is (primarily) due to relativistic aberration.
- *Correlations between magnetic field orientation and jet sidedness.* Well-resolved kpc-scale jets are often highly linearly polarized, so their magnetic fields must be partially ordered transverse to the line-of-sight. Sensitive, resolved polarimetry shows that the one-sided jet features (i.e., all of the jet lengths in FR II sources and the brighter regions near the bases of two-sided jets in FR I sources) are dominated by field components oriented *parallel* to the jet axes. Straight segments of kpc-scale jets in FR I sources usually become dominated by apparently a *perpendicular* field on the jet axes far from the AGN [71], while a parallel field may still prevail at the jet edges (see Section 5.4.2) and on the outside of pronounced bends in the jets.

5.3.2
The Two Jet "Flavors"

Kiloparsec-scale radio jets evidently come in two principal flavors [79, 80], whose properties are closely related to the Fanaroff–Riley structure classes:

- *Weak-flavor jets.* Prominent, mostly two-sided, rapidly spreading, dominated by a perpendicular magnetic field near the axis over much of their length, and terminating either in meandering plumes or in lobes without hot spots to form low-luminosity FR I sources.
- *Strong-flavor jets.* Less prominent in radio galaxies than in quasars, mostly one-sided, narrowly spreading, dominated by a parallel magnetic field over much of their length, terminating at hot spots to form luminous two-lobed FR II sources.

The side-to-side brightness asymmetries of the inner, straighter segments of weak-flavor jets decrease both with distance from the AGN and towards the outer edges of these jets [81, 82]. However, in both plumed and lobed FR I sources, the asymmetries of the most extended structures usually become *more* marked further from the AGN. The first trend suggests that the *initial* asymmetries of weak-flavor jets arise from relativistic aberration, but also that relativistic asymmetries decrease with distance from the AGN due to jet deceleration. In contrast, environmentally induced asymmetries grow with distance from the AGN and eventually dominate those due to aberration. The clear continuity of jet brightness asymmetries between the parsec and kiloparsec scales in individual sources supports the notion that the radio asymmetries are dominated on small scales by relativistic effects, but on large scales by environmental effects. In strong-flavor jets, however, it appears that significant effects of relativistic aberration persist at least as far as the terminal hot spots, and perhaps somewhat beyond them.

5.3.3 Internal Structures of Kiloparsec-Scale Radio Jets

Figure 5.13 illustrates typical structures found near the bases of weak-flavor jets using the well-resolved example of NGC 315 [83, 84].

Spreading rates Neither weak- nor strong-flavor jets appear to be freely expanding (i.e., conical) outflows on kpc scales. Rather, there is evidence that both flavors continually interact with, and adjust to their environment.

Weak-flavor jets that have been resolved transverse to their axes always begin well collimated and faint, then *brighten* and *flare*, i.e., spread much more rapidly, a few kpc from the nucleus. At high resolution, the brightening region often contains off-axis discrete knots, as well as a more general increase in intensity. *Brightening* usually precedes geometric *flaring* (as in Figure 5.13a) but the two effects can occur close together for example, in 3C 31 [69]. Most weak-flavor jets recollimate, that is, revert to less-rapid spreading, further from the AGN. Some exhibit several distinct episodes of fast spreading followed by recollimation (e.g., near the center of Figure 5.3 where the lobes are faint relative to the jets).

Strong-flavor jets spread more slowly than weak-flavor jets [71] and variations in their spreading rates are less evident far from the AGN. Note however that as FR II sources are generally more luminous than FR I sources, there are fewer nearby

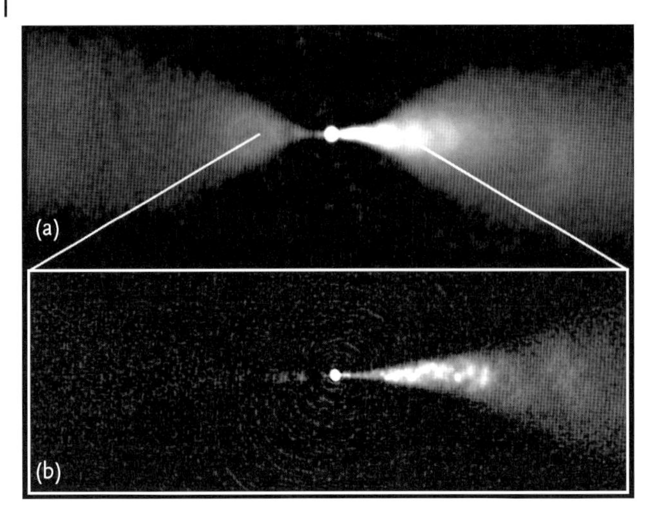

Figure 5.13 Structures at the kpc-scale base of the jet and counterjet in the FR I radio galaxy NGC 315. (a) VLA 4.9 GHz with 2.35″ FWHM resolution, 300″ (100 kpc) field of view. (b) VLA 4.9 GHz with 0.36″ FWHM, 60″ (18 kpc) field of view.

examples so their jets are harder to resolve; 3C 353 (Figure 5.4, [85]) is the best resolved.

In some lobed FR I sources where jets appear to be surrounded by more extended radio emission (at least in projection) the counterjets appear to spread more rapidly than the brighter main jets. Section 5.5 suggests a possible explanation for this.

Transverse brightness profiles The counterjets in weak-flavor jets are less strongly center-brightened than the main jets close to the AGN; some are noticeably center-darkened. The transverse profiles of total intensity are generally symmetric away from sharp bends or localized knots.

The few strong-flavor jets that have been well resolved in the transverse direction show limb-brightening or flat-topped emission profiles away from bright knots [85, 86]. The apparent suppression of emission near the jet axis implied by these profiles could have two causes: (i) differential Doppler boosting between a fast relativistic spine and a slower-moving boundary layer and/or (ii) intrinsically enhanced emissivity in the boundary layer (e.g., from *in situ* particle acceleration or magnetic field amplification by velocity shear). Whichever effect predominates, visualization of emission features in strong-flavor jets appears to be biased toward their (possibly slowest moving) outer layers. Higher angular resolution and sensitivity (e.g., the EVLA) are needed to explore this effect more thoroughly, and detections of strong-flavor counterjets might distinguish the two ways of explaining the transverse intensity profiles. Strong edge-brightening has also been seen in the well resolved, and notably one-sided pc-scale jet in M 87 [31], in which measured proper motions are small.

Polarization and magnetic field orientation There is a characteristic asymmetry across the AGN in the degree and direction (corrected for Faraday rotation) of linear polarization of weak-flavor jets, which correlates well with the asymmetry in total intensity: the base of the brighter jet has a polarization minimum on-axis, which always extends further from the nucleus than its equivalent in the counterjet. Before the polarization minimum, the apparent magnetic field on the axis is oriented *along* the jet. After the minimum, the apparent magnetic field is oriented *across* the jet. In both the main jet and the counterjet, the *transverse* profiles of the degree of polarization and total intensity are usually quite symmetric in regions away from bright local knots.

This strong coupling of the asymmetries in the linear polarization and intensity of weak-flavor jets is a key to modeling them as intrinsically symmetrical *decelerating* relativistic outflows (Section 5.4).

The faint inner regions of weak-flavor kpc-scale jets resemble strong-flavor jets in that they are well collimated and dominated by magnetic field components parallel to the jet axis. *All well-studied weak-flavor jets undergo a transition in which their collimation and magnetic structure change dramatically near the distance from the AGN where their sidedness also changes rapidly.* Elucidating the physics of this transition region in the weak-flavor jets is clearly important for understanding the relationship between the two jet flavors and the two Fanaroff–Riley structure classes.

Spectral index The bases of weak-flavor jets exhibit a remarkably narrow range of initial spectral indices $\alpha \sim 0.62$, corresponding to a radiating particle energy index of 2.24, in the regions where the jets brighten before they flare. The spectral index α then decreases, both with increasing distance from the AGN and towards the edges of the jets, usually falling to $\alpha \sim 0.5$–0.55, before it eventually increases with distance from the AGN (as expected from synchrotron aging). Strong-flavor jets have somewhat steeper spectra than those found at the bases of weak-flavor jets, with indices in the range $\alpha \sim 0.65$–0.9. Less is known about their internal spectral structure, due to generally poorer transverse resolution.

Proper motions The FR I sources in which weak-flavor jets are found are thought to be the side-on counterparts of the BL Lac objects, in which pc-scale relativistic motion is well established [87]. Superluminal motions have been seen directly on milliarcsecond scales in several weak-flavor jets [88] and on arcsecond [89] but *not* milliarcsecond [31] scales in the M 87 jet. Knots in the closest known AGN jet, in Cen A [90] show proper motions at $\sim 0.5c$ in the brighter jet and are much brighter than any counterparts in the counterjet.

Arcs, knots, and filaments Some weak-flavor jets are crossed by "arc"-like emission features in which the apparent magnetic field is tangent to the arc. The shapes of these arc-like features also exhibit systematic differences between the main and counterjets, which can be explained by differential relativistic aberration [91].

Some strong-flavor jets (e.g., Figure 5.5) exhibit semiperiodic strings of knots [7]. Phenomena that could produce this include nonlinear growth of Kelvin–Helmholtz

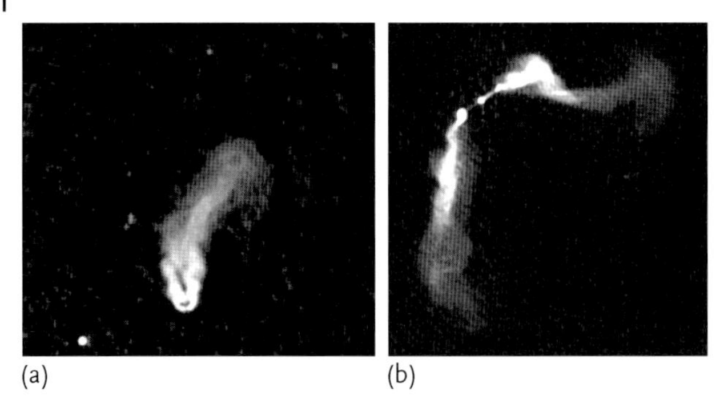

Figure 5.14 Bending of jets in cluster environments. (a) The narrow-angle tail source 3C 83.1 around NGC 1265 (VLA image at 1.4 GHz with 12″ FWHM resolution; the extended structure shown is ~ 300 kpc in extent). (b) The wide-angle tail source 3C 465 around NGC 7720 (VLA image at 1.6 GHz with 5″ FWHM resolution; the extended structure shown is ~ 360 kpc in extent).

instabilities, criss-crossing oblique shocks in confined jets, or regular perturbations of the flow velocity or direction at the AGN. High-resolution imaging and polarimetry with EVLA or e-MERLIN will help us explore these phenomena more fully.

5.3.4
Jet Bending on Kiloparsec Scales

Weak-flavor jets in dense environments can be bent into U-shapes by ram pressure of the "cross-winds" resulting from movement of their parent galaxy relative to the cluster intergalactic gas, as in the "narrow-angle tail" NGC 1265 [92]. A second class of bent sources appears to be bent by global flows from subcluster merger events [93], as in "wide-angle tail" sources such as NGC 7720 (Figure 5.14). Numerical simulations by [94] and lab experiments by [95] imply that jet bending by a cross-wind proceeds by driving oblique shocks into the flow, producing a series of deflections by continual contact with the wind. The jets in wide-angle tails [96] initially resemble strong-flavor jets, i.e., they are straight, narrow and relatively inconspicuous until they suddenly brighten, bend, and broaden into the meandering plumes. The location of the strongest bending in these sources may mark the transition from jet propagation in the ISM of the galaxy into propagation in the global flow in the intracluster medium, but the mechanism of this transition is unclear [93, 97, 98] because there is no clear X-ray evidence for discontinuities in the gas properties near the jet-plume transitions.

5.4
Modeling Jet Kinematics from Radio Data

The kinematics and magnetic field evolution of well-resolved weak-flavor jets can be inferred by fitting 3D models of their geometry, velocity fields, emissivity variation and magnetic field anisotropies to sensitive radio imaging and polarimetry. The input data needed are deep, well-resolved images of Stokes parameters I, Q, and U at several frequencies. An essential requirement is to measure the percentage and intrinsic E-vector directions of the linear polarization (corrected for foreground Faraday rotation) throughout the jets and the counterjets in the same source. The apparent brightness ratios and the polarization asymmetries between the main jet and the counterjet are then used *simultaneously* to constrain fits of parameterized 3D models to the data. The key assumption is that the jet and the counterjet are *intrinsically identical* axisymmetric outflows with relativistic velocities $v = \beta c$ and magnetic field anisotropies that vary both *across* (i.e., transverse to the jets) and *along* them. The spreading geometry is determined from the observed jet shapes. The parameters describing the flow velocity field, emissivity, and magnetic field geometry are then adjusted to minimize residuals between the observed intensity and polarization data and the jet and counterjet models integrated through the jets at every measured position, taking into account relativistic aberration and the resolution of the telescope.

5.4.1
Intensity Asymmetry Modeling: Velocity-Angle Degeneracy

The variations of the jet-counterjet brightness asymmetry (i.e., of the jet "sidedness") along and across both jets are assumed to arise from the dependence of observed intensity $S_\nu = \delta^{(2+\alpha)} S_0(\nu)$ on the Doppler factor $\delta = 1/\Gamma(1 \pm \beta \cos\theta)$ and angle θ to the line-of-sight via relativistic beaming. As the observed sidedness ratios constrain only $\beta \cos\theta$, however, fits to sidedness alone are degenerate in β and θ. As described in Section 2.3.3, this degeneracy can be resolved for parsec-scale jets using constraints from the apparent superluminal motion. For kpc-scale weak-flavor jets it can be resolved by taking into account the *systematic* jet-counterjet polarization asymmetry.

5.4.2
Polarization Asymmetry Modeling: Resolving the Degeneracy

Differential relativistic aberration makes the *observed* inclination angles θ' to the line-of-sight $\sin\theta' = \delta \sin\theta$ systematically different throughout the jet and the counterjet. For well-resolved relativistic jets with anisotropic magnetic fields, the observed degree and orientation of the net linear polarization differ between the two jets in ways that depend on the variation of the ratios of the radial, toroidal and longitudinal field components along the lines-of-sight through the jets, and on the jet velocity fields. This dependence lets us solve for the magnetic field structure, velocity fields and orientation of well-resolved, polarized jets and counterjets simultaneously, removing the degeneracy between β and θ.

Figure 5.15 shows the generic polarization structure *observed* in straight segments of weak-flavor jets just after the transition region: the apparent magnetic field is transverse to the jet near the jet axis but longitudinal near its edges

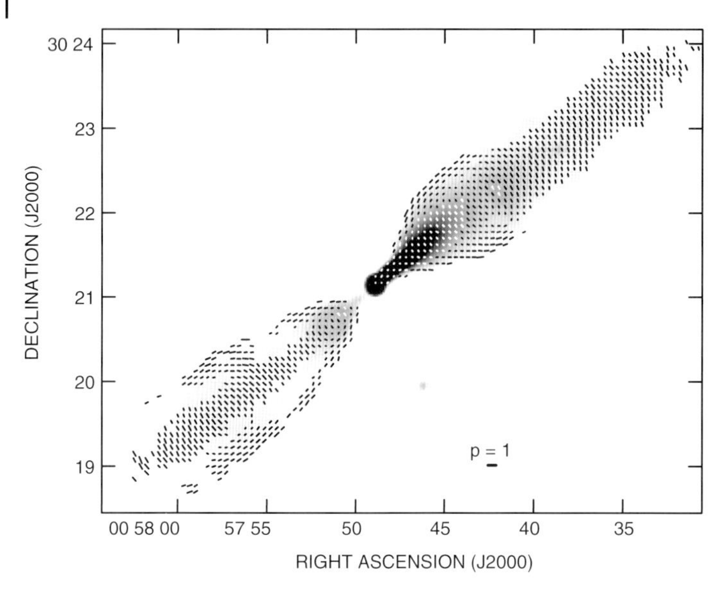

Figure 5.15 The magnetic field configuration in NGC 315 at 5.5″ FWHM resolution. Lengths of vectors are proportional to the degree of linear polarization p and their orientations are those of the apparent magnetic field. Reproduced from Laing *et al.* [83] with permission from the Royal Astronomical Society.

(Figure 5.15). A 3D ordered helical field can exhibit such a polarization pattern but well-ordered helical fields cannot reproduce the observed *transverse symmetry* of the total intensity and degree of polarization profiles [99]. Figure 5.16 sketches two ways in which *anisotropic* field configurations with significant longitudinal and toroidal components *can* in principle produce the observed transverse profiles. The first configuration has a *spine* of two-dimensional field tangled in the plane perpendicular to the jet (with little longitudinal component) surrounded by a *sheath* of purely longitudinal field. Such a configuration might arise if a fast-moving jet spine is surrounded by a slower-moving boundary layer where velocity shear makes the perpendicular fields become longitudinal. The second configuration (Model "B" of [100]) has concentric two-dimensional sheets of tangled longitudinal and toroidal field but little or no radial field, owing to velocity shear between the (fast) center and (slower) edges of the jet.

Figure 5.17 shows how the two anisotropic field configurations shown in Figure 5.16 can be distinguished in *decelerating* relativistic jets by the relative positions of the transition between apparently longitudinal and perpendicular fields on the axes of two intrinsically identical, conically expanding, jets. In the first case, this transition occurs further from the nucleus in the counterjet (and may not be seen at all in the main jet). In the second case, the converse is expected.

In *every* weak-flavor jet for which good data are available, the transition from parallel to perpendicular apparent field near the jet axis occurs further from the AGN in the brighter jet, that is, the actual 3D field configuration must be a closer match to the *wrapped-sheet* case. Detailed modeling of individual jets using parameterized

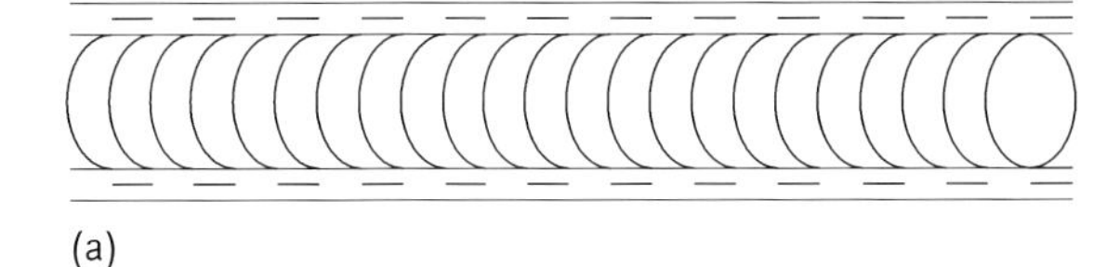

(a)

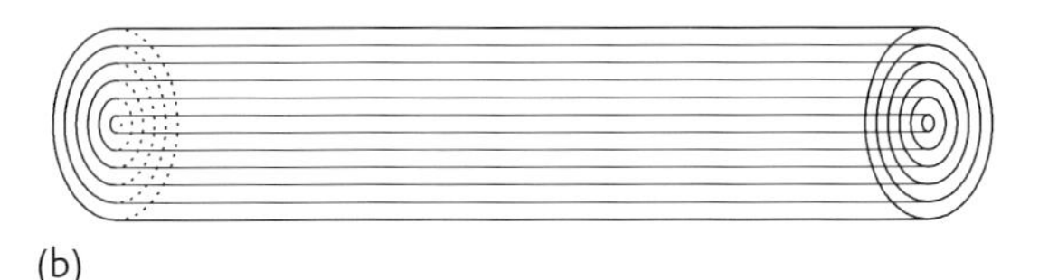

(b)

Figure 5.16 Two local geometries for tangled magnetic field configurations that might account for the generic polarization structure shown in Figure 5.15, ignoring the jet expansion for simplicity. (a) A "spine" in which the fields are tangled in the plane perpendicular to the jet axis, surrounded by a "sheath" dominated by longitudinal field – only sheath field is shown explicitly. (b) Sheets of tangled field with equal longitudinal and toroidal components wrapped around the jet axis. Only the concentric cylindrical geometry of the tangled field sheets is shown, not the tangled fields themselves.

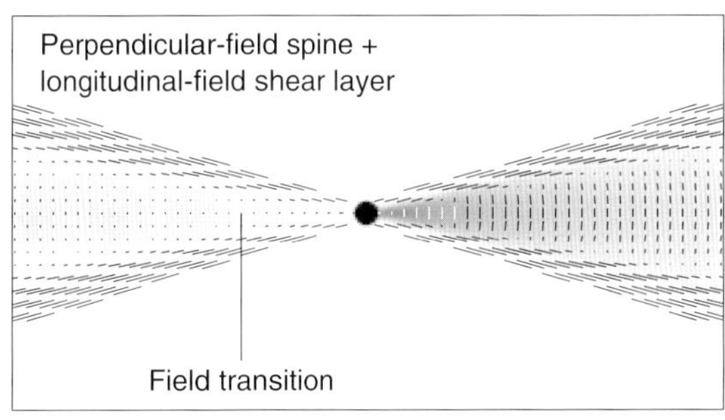

(a)

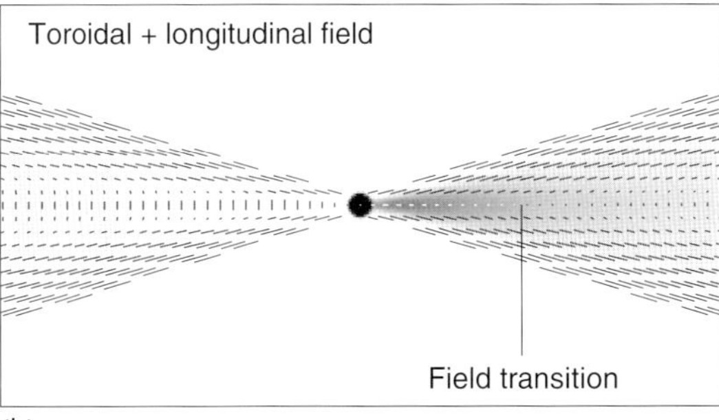

(b)

Figure 5.17 Examples of apparent magnetic field configurations resulting from two possible magnetic field configurations (a, b) in a relativistic jet decelerating with the parameters inferred for NGC 315 by [101]. Lengths of vectors are proportional to the degree of linear polarization p and their orientations are those of the apparent magnetic field.

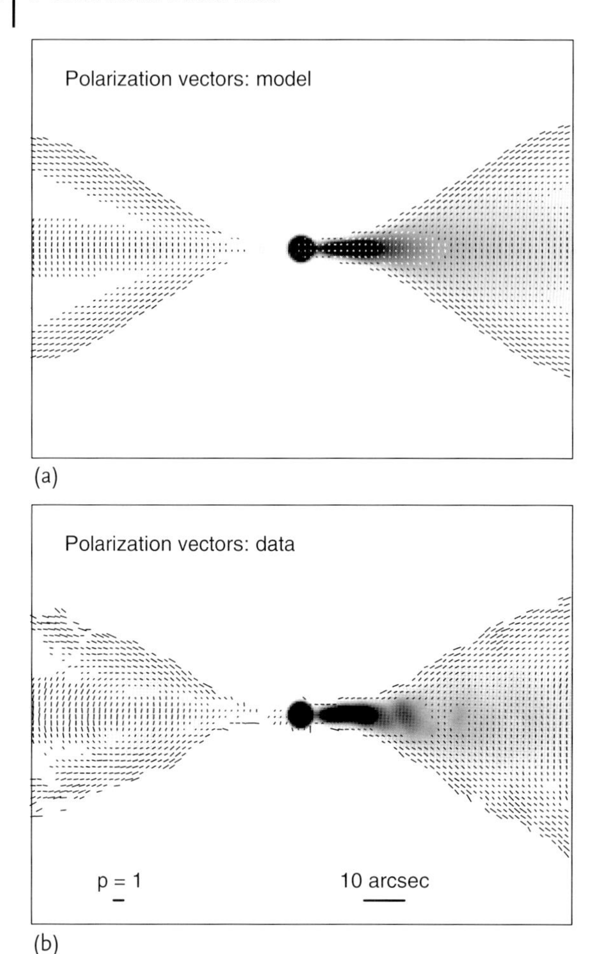

Figure 5.18 Modeled (a) and observed (b) polarization distributions for the inner jet and counterjet in the FR I radio galaxy NGC 315 [101]. Lengths of vectors are proportional to the degree of linear polarization p and their orientations are those of the apparent magnetic field. Reproduced from Canvin *et al.* [101] with permission from the Royal Astronomical Society.

variants of this configuration can then be used to quantify how the field components vary with distance from the AGN in those jets, as illustrated in Figure 5.18.

5.4.3
Velocity Fields in Weak-Flavor Jets

Laing and Bridle [69] used an iterative procedure to fit (in the χ^2 sense) a parametric jet model to the correlated jet-counterjet brightness and polarization asymmetries in 3C 31 (Figure 5.2) and to derive how the flow velocity and field components evolve along this jet. Applying this approach to straight segments of other well-resolved weak-flavor jets [91, 101, 102] has shown that their systematic asymmetries can be well accounted for if:

- β is about 0.8–0.9 where the jets first brighten.
- All of the jets decelerate abruptly in the "flaring" region where they begin to spread rapidly, but the deceleration and flaring take place at different distances from the nucleus in different sources.
- The jets decelerate to $\beta = 0.1-0.4$ further from the nucleus, sometimes achieving constant, low velocities before intrinsic asymmetries begin to dominate.
- The outflow near the edges of many weak-flavor jets is slower than the outflow on-axis, so there is significant velocity shear across them, although some regions have a top-hat velocity profile.

5.4.4 Magnetic Field Evolution in Weak-Flavor Jets

The magnetic field configurations inferred from fitting observed jet-counterjet linear polarization distributions and their asymmetries determine field component *ratios*, not the absolute field strengths or topologies. Some clear systematics of the ratios in weak-flavor jets have emerged:

- The longitudinal and toroidal components are generally similar to each other close to the nucleus, but the toroidal component dominates at large distances. This behavior is qualitative, but not quantitative, as expected from flux-freezing in an expanding flow wherein the longitudinal field component would be $\propto r_{\text{jet}}^{-2}$ while the toroidal field component would be $\propto (\Gamma \beta r_{\text{jet}})^{-1}$.
- The radial field components are generally weak compared to the other components. The jets in 3C 31, which show unusually clear evidence for ongoing deceleration at large distances from the AGN, also have unusually strong radial field components at their edges. This suggests a connection between ongoing deceleration of that jet and maintenance of some radial field by turbulent entrainment during ingestion of the ISM across its boundary. In other sources the radial field may be suppressed by the velocity shear.

We emphasize that these systematics are derived from studies of the initial segments of weak-flavor jets within which both the jet and counterjet remain sensibly straight and collinear. Far from the AGN, both the morphologies and magnetic field structures of the weak-flavor jets and their counterjets are modified by ongoing interactions with their environments, as evidenced by large-scale bending of the jets and the formation of plumes or lobes (Figures 5.2, 5.3, and 5.14).

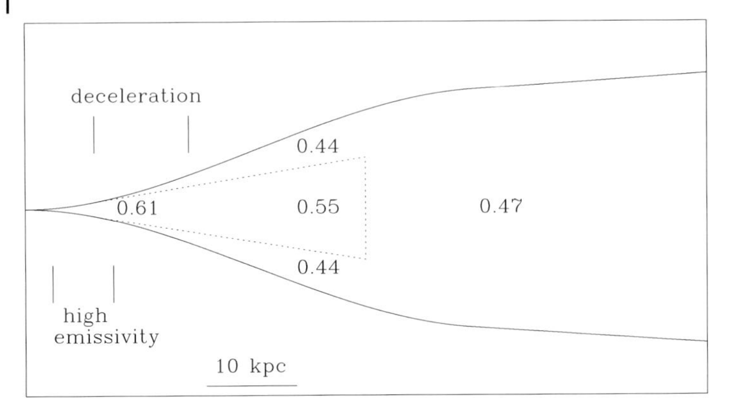

Figure 5.19 Jet deceleration, emissivity and spectral index (values shown) in the inner 80 kpc of the base of the jet in NGC 315 [83]. Reproduced from Laing *et al.* [83] with permission from the Royal Astronomical Society.

5.4.5 Emissivity Evolution in Weak-Flavor Jets

Far from the AGN, where jet deceleration inferred from model fitting again becomes slow, the emissivities of weak-flavor jets decline with distance roughly *adiabatically*, that is, the rate of decline is consistent with flux-freezing with the observed spreading rate, and no particle acceleration. In the initial fast-moving and rapid-deceleration regions, there are *two* distinct regions where particle acceleration is required. The first is associated with enhanced X-ray emission and with the appearance of nonaxisymmetric knots in the jets (e.g., [84]). The second appears to be associated with velocity gradients (i.e., with velocity shear) at the jet edges.

- Jet brightening always begins shortly before the most rapid deceleration, which is in turn complete before recollimation occurs.
- The regions of high X-ray to radio brightness ratio are also regions of high flow velocity.
- The initial energy index of 2.24 is close to that of 2.23 expected in the limit from Fermi I acceleration by ultrarelativistic shocks but the flow velocities inferred where this index occurs are less than expected from this interpretation.

Figure 5.19 sketches the relationship between regions of enhanced emissivity, strongest jet deceleration, and spectral index structure at the base of the well-resolved jet in NGC 315 [83].

5.4.6 Mass, Momentum and Energy Fluxes

Once the jet velocity field and spreading geometry are known, application of the relativistic conservation laws for mass, energy and momentum lets us deduce the global variations of density, pressure and entrainment rate along the jets when the confining media have been detected by X-ray observations. For such sources we can estimate the external pressure gradient within which the kinematics have been manifested, so we can obtain constraints on jet dynamical quantities. For 3C 31, Laing and Bridle [69] showed that well-constrained solutions exist, subject to two key assumptions:

- $\Phi = \Pi c$ where Φ is the energy flux (with rest-mass energy subtracted) and Π is the momentum flux. This must hold quite accurately if the jets have decelerated from bulk Lorentz factors on parsec scales.
- The jets reach pressure balance with the external medium in their outer regions, as suggested by their eventual recollimation.

Boundary-layer entrainment (*ingestion* of the ISM across the surface of the jets) and mass input (*injection*) from stars within the volume traversed by the jets are probably both important in slowing the weak-flavor jets in FR I sources. Conservation-law model fits (e.g., [103]) suggest that the jets are initially light enough to be slowed significantly by stellar mass injection, which may enable them to reach a transonic regime in which the ingestion of ambient ISM across the jet boundary is significant, as suggested by [104]. For example, the internal Mach number inferred for 3C 31 by [103] was ≈ 1.5 through the deceleration region, the mechanical luminosity of the jet was estimated at 1.1×10^{37} W, its mass flux $5 \times 10^{-4}\ M_\odot\ \mathrm{year}^{-1}$, and the entrainment rate 10 kpc from the AGN to be $1.2 \times 10^{10}\ \mathrm{kg\,kpc^{-1}\,s^{-1}}$.

5.4.7
Comparisons with Strong-Flavor Jets

The only strong-flavor jets which have been well enough resolved to allow detailed modeling of their 3D emissivity and field structures are those in 3C 353 [85]. Even there, full kinematic modeling along the lines used for weak-flavor jets has not yet been possible, owing to the difficulty of detecting the counterjets at all with good transverse resolution, let alone obtaining the sensitive counterjet polarimetry needed to apply these methods. There is some evidence however that the magnetic fields in the jets of 3C 353 are also dominated by their toroidal and longitudinal components [85].

The association of strong-flavor jets with the formation of compact hot spots in the lobes of FR II sources suggests that they maintain higher Mach numbers than weak-flavor jets to large distances from the AGN, and perhaps undergo less deceleration, consistent with preserving the jet collimation. Until counterjets can be detected and resolved in more FR II sources with strong-flavor jets, it will remain difficult to compare their properties directly with those inferred from modeling the deceleration of weak-flavor jets.

The prominent differences between strong flavor jets in radio galaxies and quasars probably arise from the systematic orientational difference between them, with the radio galaxy population closer to the plane of the sky than the quasars.

5.5
Backflow in Bilobed FR I Sources?

In a few of the "lobed" FR I sources whose structure suggests that their jets may be propagating through lobe material, the brighter jets are isophotally narrower than the counterjets, e.g., Figure 5.20. High-resolution imaging of two such sources, B2 0206+35 and B2 0755+37 [105] has shown that their counterjets contain features whose geometry matches that of the brighter "main" jets, but with brightness *minima* near the jet axis where the brighter jets have brightness maxima. Furthermore, the edges of these counterjets appear *brightness-enhanced* compared to the corresponding structure around the main jets, so that the counterjets appear wider at low resolution. No examples are known where the main jets are isophotally wider than the counterjets. The brightness and polarization asymmetries in these sources can be fitted with intrinsically symmetric relativistic jet models if both jets are surrounded by mildly relativistic symmetric *backflow*, which appears brighter around the counterjets and fainter around the edges of the main jets due to differential beaming (the backflow around the counterjet is *approaching* the observer, while that around the main jet is receding).

Such symmetric backflow has also been seen in some numerical simulations of overpressured, fast, light jets (where it may however be an artifact of forced axisymmetry, low resolution, or open boundary conditions). Symmetric backflow might be thought hard to set up in practice, but the radio data suggest that it may, nevertheless, occur with $\beta \sim 0.05$–0.4 and enhanced emissivity in both B2 0206+35

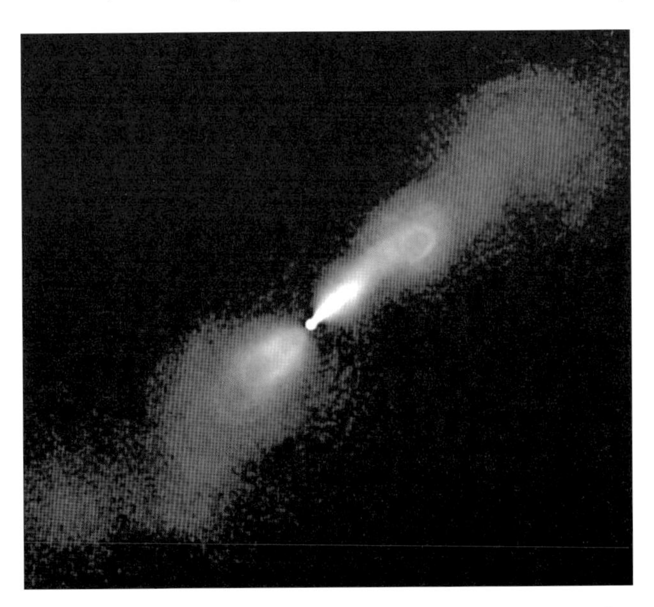

Figure 5.20 The central region of the FR I radio galaxy B2 0206+35 showing apparent structural *asymmetries* between the jet and counterjet that can be explained by *symmetric* backflow (VLA 4.9 GHz image, 0.35″ FWHM). The field of view is $\sim 80''$, corresponding to ~ 60 kpc.

and B2 0755+37. The dominant magnetic field component inferred in the putative backflow in these sources is, perhaps unsurprisingly (e.g., [106]), toroidal. Whether such toroidal fields (and associated return currents) in the backflow play a role in confining these jets is unclear because, as emphasized earlier, the field *strengths* are unknown without further assumptions.

Acknowledgment

AHB thanks Robert Laing for many stimulating discussions and insights about observations of radio jets and their interpretation. MHC thanks Matt Lister and Ken Kellermann for helpful comments on the manuscript.

References

1 Fanaroff, B.L. and Riley, J.M. (1974) *Mon. Not. R. Astron. Soc.*, **167**, 31.
2 Owen, F.N. and White, R.A. (1991) *Mon. Not. R. Astron. Soc.*, **249**, 164.
3 Ledlow, M.J. and Owen, F.N. (1996) *Astron. J.*, **112**, 9.
4 Parma, P., de Ruiter, H.R., and Fanti, R. (1996) in *Extragalactic Radio Sources* (eds R. Ekers, C. Fanti, and L. Padrielli), IAU Symposium 175, Kluwer Academic Publishers, p. 137.
5 Bridle, A.H. (1986) *Can. J. Phys.*, **64**, 353.
6 Leahy, J.P. (1993) in *Jets in Extragalactic Radio Sources* (eds H.-J. Röser and K. Meisenheimer), Lecture Notes in Physics 421, Springer-Verlag, Berlin, p. 1.
7 Bridle, A.H., Hough, D.H., Lonsdale, C.J., Burns, J.O., and Laing, R.A. (1994) *Astron. J.*, **108**, 766.
8 Lister, M.L. *et al.* (2009) *Astron. J.*, **137**, 3718.
9 MOJAVE (Monitoring Of Jets in Active galactic nuclei with VLBA Expe4riments), Perdue University, http://www.physics.purdue.edu/astro/MOJAVE/ (accessed 31 August 2011).
10 Homan, D.C., Lister, M.L., Kellermann, K.I., Cohen, M.H., Ros, E., Zensus, J.A., Kadler, M., and Vermeulen, R.C. (2003) *Astrophys. J.*, **589**, L9.
11 Marscher, A.P. (2008) in *Extragalactic Jets: Theory and Observation from Radio to Gamma Ray* (eds T.A. Rector and D.S. De Young), ASP Conference Series, vol. 386, Astronomical Society of the Pacific, p. 437.
12 Pearson, T.J. and Readhead, A.C.S. (1988) *Astrophys. J.*, **328**, 114.
13 Conway, J.E. and Murphy, D.W. (1993) *Astrophys. J.*, **411**, 89.
14 Kharb, P., Lister, M.L., and Cooper, N.J. (2010) *Astrophys. J.*, **710**, 764.
15 Denn, G.R., Mutel, R.L., and Marscher, A.P. (2000) *Astrophys. J. Suppl. Ser.*, **129**, 61.
16 Lobanov, A. and Zensus, A. (2003) *Science*, **294**, 128.
17 Hardee, P.E., Walker, R.C. and Gomez, J.L. (2005) *Astrophys. J.*, **620**, 646.
18 Davis, R.J., Unwin, S.C., and Muxlow, T.W.B. (1991) *Nature*, **354**, 374.
19 Agudo, I. (2009) in *Approaching Micro-Arcsecond Resolution with VSOP-2: Astrophysics and Technology* (eds Y. Hagiwara, E. Fomalont, M. Tsuboi, and Y. Murata), ASP Conference Series, vol. 402, Astronomical Society of the Pacific, p. 330.
20 Agudo, I. *et al.* (2007) *Astron. Astrophys.*, **476**, L17.
21 Britzen *et al.* (2010) *Astron. Astrophys.*, **515**, 105.
22 Jorstad, S.G. *et al.* (2005) *Astron. J.*, **130**, 1418.
23 Blandford, R.D. and Königl, A. (1979) *Astrophys. J.*, **232**, 34.
24 Königl, A. (1981) *Astrophys. J.*, **243**, 700.

25 Lobanov, A.P. (1998) *Astron. Astrophys.*, **330**, 79.

26 O'Sullivan, S.P. and Gabuzda, D.C. (2009) *Mon. Not. R. Astron. Soc.*, **400**, 26.

27 Sokolovsky, K.V., Kovalev, Y.Y., Pushkarev, A.B., and Lobanov, A.P. (2011) *Astron. Astrophys.*, **523**, A38.

28 Lister, M.L. *et al.* (2009) *Astron. J.*, **138**, 1874.

29 Kadler, M., Ros, E., Lobanov, A.P., Falcke, H., and Zensus, J.A. (2004) *Astron. Astrophys.*, **426**, 481.

30 Vermeulen, R.C., Ros, E., Kellermann, K.I., Cohen, M.H., Zensus, J.A., and van Langevelde, H.J. (2003) *Astron. Astrophys.*, **401**, 113.

31 Kovalev, Y.Y., Lister, M.L., Homan, D.C., and Kellermann, K.I. (2007) *Astrophys. J.*, **668**, L27.

32 Walker, R.C., Dhawan, V., Romney, J.D., Kellermann, K.I., and Vermeulen, R.C. (2000) *Astrophys. J.*, **530**, 233.

33 Horiuchi, S., Meier, D.L., Preston, R.A., and Tingay, S.J. (2006) *Publ. Astron. Soc. Japan*, **58**, 211.

34 O'Dea, C.P. (1998) *Publ. Astron. Soc. Pac.*, **110**, 493.

35 Helmboldt, J.F. *et al.* (2007) *Astrophys. J.*, **658**, 203.

36 Taylor, G.B., Vermeulen, R.C., Readhead, A.C.S., Pearson, T.J., Henstock, D.R., and Wilkinson, P.N. (1996) *Astrophys. J. Suppl. Ser.*, **107**, 37–68.

37 Britzen *et al.* (2007) *Astron. Astrophys.*, **472**, 763.

38 Britzen *et al.* (2008) *Astron. Astrophys.*, **84**, 119.

39 Kellermann, K.I., Vermeulen, R.C., Zensus, J.A., and Cohen, M.H. (1998) *Astron. J.*, **115**, 1295.

40 Ojha, R. *et al.* (2010) *Astron. Astrophys.*, **519**, 450.

41 Beasley, A.J., Gordon, D., Peck, A.B., Petrov, L., MacMillan, D.S., Fomalont, E.B., and Ma, C. (2002) *Astrophys. J. Suppl. Ser.*, **141**, 13.

42 Fey, A.L., Clegg, A.W., and Fomalont, E.B. (1997) *Astrophys. J. Suppl. Ser.*, **105**, 299.

43 Fey, A.L. and Charlot, P. (1997) *Astrophys. J. Suppl. Ser.*, **111**, 95.

44 Fey, A.L. and Charlot, P. (2000) *Astrophys. J. Suppl. Ser.*, **128**, 17.

45 Piner, B.G., Mahmud, M., Fey, A.L., and Gospodinova, K. (2007) *Astron. J.*, **133**, 2357.

46 Moellenbrock, G.A. *et al.* (1996) *Astron. J.*, **111**, 2174.

47 Dodson, R. *et al.* (2008) *Astrophys. J. Suppl. Ser.*, **175**, 314.

48 Krichbaum, T.P. *et al.* (2007) in *Extragalactic Jets: Theory and Observation from Radio to Gamma Ray* (eds T.A. Rector and D.S. De Young), ASP Conference Series, vol. 386, Astronomical Society of the Pacific, p. 186.

49 Lee, S.S. *et al.* (2008) *Astron. J.*, **136**, 159.

50 Orienti, M. and Dallacasa, D. (2010) *Mon. Not. R. Astron. Soc.*, **406**, 529.

51 Abraham, Z. and Romero, G.E. (1999) *Astron. Astrophys.*, **344**, 61.

52 Roca-Sogorb, M. Gomez, J.L., Agudo, I., Marscher, A.P., and Jorstad, S.G. (2010) *Astrophys. J.*, **712**, L160.

53 Agudo, I. *et al.* (2001) *Astrophys. J.*, **549**, L183.

54 Lind, K.R. and Blandford, R.D. (1985) *Astrophys. J.*, **295**, 358.

55 Scheuer, P.A.G. and Readhead, A.C.S. (1979) *Nature*, **277**, 182.

56 Cohen, M.H., Lister, M.L., Homan, D.C., Kadler, M., Kellermann, K.I., Kovalev, Y.Y., and Vermeulen, R.C. (2007) *Astrophys. J.*, **658**, 232.

57 Lister, M.L. and Marscher, A.P. (1997) *Astrophys. J.*, **476**, 572.

58 Kellermann, K.I. *et al.* (2004) *Astrophys. J.*, **609**, 539.

59 Hough, D.H. *et al.* (2002) *Astron. J.*, **123**, 1258.

60 Hough D. (2008) in *Extragalactic Jets: Theory and Observation from Radio to Gamma Ray* (eds T.A. Rector and D.S. De Young), ASP Conference Series, vol. 386, Astronomical Society of the Pacific, 274.

61 Vermeulen, R.C. and Cohen, M.H. (1994) *Astrophys. J.*, **430**, 467.

62 Readhead, A.C.S. (1994) *Astrophys. J.*, **426**, 51.

63 Kellermann, K.I. and Pauliny-Toth, I.I.K. (1969) *Astrophys. J.*, **155**, L71.

64 Homan, D.C., Kovalev, Y.Y., Lister, M.L., Ros, E., Kellermann, K.I., Cohen, M.H.,

Vermeulen, R.C., Zensus, J.A., and Kadler, M. (2006) *Astrophys. J.*, **642**, L115.
65 Kovalev, Y.Y. *et al.* (2005) *Astron. J.*, **130**, 2475.
66 Lähteenmäki, A. and Valtaoja, E. (1999) *Astrophys. J.*, **521**, 493.
67 Cara, A. and Lister, M.L. (2008) *Astrophys. J.*, **674**, 111.
68 Ly, C., Walker, R.C., and Junor, W. (2007) *Astrophys. J.*, **660**, 200.
69 Laing, R.A. and Bridle, A.H. (2002) *Mon. Not. R. Astron. Soc.*, **336**, 328.
70 Chiaberge, M., Celotti, A., Capetti, A., and Ghisellini, G. (2000) *Astron. Astrophys.*, **358**, 104.
71 Bridle, A.H. (1984) *Astron. J.*, **89**, 979.
72 Giovannini, G., Cotton, W.D., Lara, L., Marcaide, J., and Wehrle, A.E. (1994) in *Compact Extragalactic Radio Sources* (eds J.A. Zensus and K.I. Kellermann), NRAO, p. 61.
73 Venturi, T., Giovannini, G., Feretti, L., Cotton, W.D., Lara, L., Marcaide, J., and Wehrle, A. (1994) in *The Physics of Active Galaxies* (eds G.V. Bicknell, M.A. Dopita, and P.J. Quinn), ASP Conference Series, vol. 54, Astronomical Society of the Pacific, p. 241.
74 Venturi, T., Castaldini, C., Cotton, W.D., Feretti, L., Giovannini, G., Lara, L., Marcaide, J.M., and Wehrle, A.E. (1995) *Astrophys. J.*, **454**, 735.
75 Laing, R.A. (1988) *Nature*, **331**, 149.
76 Garrington, S.T., Leahy, J.P., Conway, R.G., and Laing, R.A. (1988) *Nature*, **341**, 147.
77 Garrington, S.T., Conway, R.G., and Leahy, J.P. (1991) *Mon. Not. R. Astron. Soc.*, **250**, 171.
78 Morganti, R., Parma, P., Capetti, A., Fanti, R., and de Ruiter H.R. (1997) *Astron. Astrophys.*, **326**, 919.
79 Bridle, A.H. (1991) in *Testing the AGN Paradigm* (eds S.S. Holt, S.G. Neff, and C.M. Urry), AIP Conference Proceedings 254, Springer, p. 386.
80 Laing, R.A. (1993) in *Astrophysical Jets* (eds D. Burgarella, M. Livio, and C.P. O'Dea) Space Telescope Science Symposium No. 6, Cambridge University Press, p. 95.
81 Laing, R.A. (1996) in *Energy Transport in Radio Galaxies and Quasars* (eds P.E. Hardee, A.H. Bridle, and J.A. Zensus), ASP Conference Series, vol. 100, Astronomical Society of the Pacific, p. 241.
82 Laing, R.A., Parma, P., de Ruiter, H.R., and Fanti, R. (1999) *Mon. Not. R. Astron. Soc.*, **306**, 513.
83 Laing, R.A., Canvin, J., Cotton, W.D., and Bridle, A.H. (2006) *Mon. Not. R. Astron. Soc.*, **368**, 48.
84 Worrall, D.M., Birkinshaw, M., Laing, R.A., Cotton, W.D., and Bridle, A.H. (2007) *Mon. Not. R. Astron. Soc.*, **380**, 2.
85 Swain, M.R., Bridle, A.H., and Baum, S.A. (1998) *Astrophys. J. Lett.*, **507**, L29.
86 Carilli, C.L., Perley, R.A., Bartel, N., and Dreher, J.W. (1996) in *Cygnus A – Study of a Radio Galaxy* (eds C.L. Carilli and D.E. Harris), Cambridge University Press, p. 76.
87 Urry, C.M. and Padovani, P. (1995) *Publ. Astron. Soc. Pac.*, **107**, 803.
88 Giovannini G., Cotton W.D., Feretti L., Lara L., and Venturi T. (2001) *Astrophys. J.*, **552**, 508.
89 Biretta, J.A., Zhou, F., and Owen, F.N. (1995) *Astrophys. J.*, **447**, 582.
90 Hardcastle, M.J., Worrall, D.M., Kraft, R.P., Forman, W.R., Jones, C., and Murray, S.A. (2003) *Astrophys. J.*, **593**, 169.
91 Laing, R.A., Canvin, J., Bridle, A.H., and Hardcastle, M.J. (2006) *Mon. Not. R. Astron. Soc.*, **372**, 510.
92 O'Dea, C.P. and Owen, F.N. (1986) *Astrophys. J.*, **301**, 841.
93 Loken, C., Roettiger, K., Burns, J.O., and Norman, M. (1995) *Astrophys. J.* **445**, 80.
94 Balsara, D. and Norman, M. (1992) *Astrophys. J.*, **393**, 631.
95 Lebedev, S.V., Ampleford, D., Ciardi, A., Bland, S.N., Chittenden, J.P., Haines, M.G., Frank, A., Blackman, E.G., and Cunningham, A. (2004) *Astrophys. J.*, **616**, 988.
96 O'Donoghue, A., Eilek, J.A., and Owen, F.N. (1993) *Astrophys. J.*, **408**, 428.
97 Hardcastle, M.J., Sakelliou, I., and Worrall, D.M. (2005) *Mon. Not. R. Astron. Soc.*, **359**, 1007.
98 Jetha, M.N., Sakelliou, I., Hardcastle, M.N., Ponman, T.J., and Stevens, J.R. (2005) *Mon. Not. R. Astron. Soc.*, **358**, 1394.

99 Laing, R.A., Canvin, J., and Bridle, A.H. (2006) *Astron. Nacht.*, 327, 523.

100 Laing, R.A. (1980) *Mon. Not. R. Astron. Soc.*, **193**, 439.

101 Canvin, J.R., Laing, R.A., Bridle, A.H., and Cotton, W.D. (2005) *Mon. Not. R. Astron. Soc.*, **363**, 1223.

102 Canvin, J.R. and Laing, R.A. (2004) *Mon. Not. R. Astron. Soc.*, **350**, 1342.

103 Laing, R.A. and Bridle, A.H. (2002) *Mon. Not. R. Astron. Soc.*, **336**, 1161.

104 Bicknell, G.V. (1995) *Astrophys. J. Suppl. Ser.*, **101**, 29.

105 Bondi M., Parma P., de Ruiter H.R., Laing R.A., and Fomalont E.B. (2000) *Mon. Not. R. Astron. Soc.*, **314**, 11.

106 Begelman, M.C., Blandford, R.D., and Rees,M.J. (1984) *Rev. Mod. Phys.*, **56**, 255.

6
Optical, Infrared and UV Observations

Eric Perlman

Jets are less common in the optical regime than in the radio regime. To exhibit detectable optical emissions, a synchrotron-emitting jet has to have highly energetic particles that must be accelerated *in situ* – for example, in a 100 μG magnetic field, emission at 4×10^{14} Hz (i.e., R band) requires particles with $\gamma = 1.2 \times 10^6$ and radiative lifetimes of a few hundred years ($\tau \propto \nu^{-1/2}$). Despite this, optical emission is an important facet of many relativistic jets. In this chapter, we will review the properties of jets in the ultraviolet, optical and infrared regimes, exploring the commonalities and differences to the characteristics found in other bands. We will also fully explore what we can learn about jets from their UV/optical/IR emission. The list of topics here is surprisingly diverse, as it turns out that UV/optical/IR emission can teach us not only about the highest energy particles and how they are accelerated in jets, but also about jet structure and dynamics – and perhaps (surprisingly) also about the lowest-energy particles.

6.1
A Historical Perspective

Twenty years ago, optical jets were a novelty. Prior to the launch of the Hubble Space Telescope (HST) in 1990, only five resolved optical jets were known: those of M 87 (Heber Curtis's original "curious straight ray"), 3C 273, 3C 66B, 3C 277.3 (Coma A), and PKS 0521-36. Beyond this, there were a number of jet-associated phenomena that were known: the extreme variability, power-law continua, and high polarizations observed in blazars (these phenomena will be discussed in Chapter 8). The rarity of the phenomenon was such that not only were their characteristics poorly characterized, but the term "optical jet emission" was not even well defined in the literature.

In the 1980s, many researchers were using the name "optical jet" to describe outflows of emission line gas from Seyfert galaxies such as NGC 4258 and NGC 1097 [1, 2], as well as a few radio galaxies (e.g., [3]). While the Seyfert outflows are of themselves highly interesting and are rather more common than optical

Relativistic Jets from Active Galactic Nuclei, First Edition. Edited by M. Böttcher, D.E. Harris, H. Krawczynski.

jets, it is important to stress that they are different phenomena, being both nonrelativistic as well as associated with line emission (and now known to be part of the narrow-line region). This terminology has persisted in the literature; however, we believe it is important to be precise, and for this reason we believe that another term should be adopted for these outflows. On the other hand, it is also important to point out that the two phenomena are not necessarily separate. For example, in some Seyferts, such as NGC 4258, there is an association with the radio outflow, but in others, such as NGC 1097, no such correlation is observed. The radio outflow in these objects is likely to be slower given that no evidence of relativistic motions and/or beaming have been found in these objects. In a number of radio galaxies with jets, a similar correlation is observed (e.g., [3, 4]).

This picture changed completely with the launch of HST in 1990. Within a few years, optical emission was discovered from the radio jets of several more objects, for example 3C 78, 3C 264, and 3C 371. However, the pace of discovery really quickened after the launch of the Chandra X-ray Observatory. As will be discussed in Chapter 7, Chandra began to discover X-ray emission from jets with almost its very first observation. The X-ray emission from some jet features was quickly interpreted as coming from inverse-Comptonization of the lowest-energy electrons. The fact that high-energy emission might come from a mandatory process sparked a number of surveys – as well as optical follow-ups. As a result, X-ray and optical emission was discovered from many more objects. Today, about 45 jets have confirmed optical emission, with possible emission having been observed in several others.

Searches for optical emission from microquasars are also underway. Theoretically, there is no reason why such emission should not be present; however, historically, it has been much more difficult to find optical emission from microquasars than from AGN. One of the reasons for this is that the jet flow is transient in microquasars, observed only for a few weeks to months after outbursts, whereas in AGN the jets are persistent on time scales much longer than a year. In addition, existing observations have yet to be deep enough to find resolved optical emission from a microquasar jet – although it should be noted that in two cases (GRO 1655-40 and Sco X-1 [5]), intrinsic polarization was detected in near-IR observations of the star itself, arguing for a significant (up to $\sim$ 5% in the case of GRO 1655-40) synchrotron contribution to the unresolved emission. Moreover, in SS 433 we see unresolved blue and redshifted emission lines [6] at speeds of $0.26c$, similar to those observed at radio wavelengths [7, 8]. For a review of the observational situation in stellar mass black hole systems, the reader is referred to Russell and Fender [9] and Corbel *et al.* [10], while for neutron star binaries, the reader is referred to Russell, Fender and Jonker [11]. In this chapter we concentrate on AGN jets; however, all the comments made here about nonthermal jet emission from AGN jets can be applied equally well to microquasars, should such emission be found from them in the future.

In Table 6.1, we list the jets where optical emission has been confirmed, along with basic information about the radio source. In column 3, the labels Q and RG refer to quasars and radio galaxies, respectively, while the symbols I and II refer to Fanaroff–Riley class I and II sources; see Section 5.1 for definitions and examples.

Table 6.1 Known optical/IR jet sources.

Object	Redshift	Cl.	log P_{tot}, 1.4 GHz	log P_{core}, 5 GHz	log P_{jet}, 1.4 GHz
3C 15	0.0730	RG I	25.41	23.48	24.59 (22.87)
PKS 0208-512	1.003	Q II	28.37	28.00	27.01
3C 17	0.22	RG II	26.93	26.30	25.91
3C 31	0.0169	RG I	24.21	22.45	23.48 (23.39)
Pictor A	0.035	RG II	25.93	24.01	
3C 66B	0.0215	RG I, Sey 1	24.69	22.59	23.90 (23.90)
3C 78	0.0289	RG I, Sey 1	24.83	23.77	... (...)
3C 120	0.0334	RG I, Sey 1	24.76	24.93	24.07
3C 133	0.2775	RG II	26.72	25.33	24.68
PKS 0521-365	0.061	Q I	25.83	24.75	24.87
PKS 0637-752	0.654	Q II	27.84	27.40	...
3C 179	0.846	Q II	27.39	26.62	26.21
B2 0755+37	0.0413	RG I	24.49	23.59	23.52 (...)
3C 200	0.458	RG II	26.73	24.96	25.72
3C 207	0.684	Q II	27.23	26.45	26.93
3C 212	1.049	Q II	27.61	25.38	...
3C 245	1.029	Q II	27.80	27.16	27.08
PKS 1136-135	0.554	Q II	27.15	26.27	...
3C 264	0.0208	RG I	24.48	23.09	23.67 (...)
4C 49.22	0.334	Q I	26.43	25.85	26.14
PKS 1202-262	0.789	Q II	27.59	27.11	27.15
3C 273	0.158	Q II	27.14	26.92	26.66
PKS 1229-021	1.045	Q II	28.05	27.53	27.48
M 87	0.0043	RG I, LINER	24.78	22.92	23.27
3C 277.3	0.0857	RG II	25.38	22.99	23.52
3C 279	0.536	Q II	27.56	27.56	26.66
Cen A	0.00012	RG I	24.62	22.12	21.97
3C 293	0.0452	RG I	25.01	22.70	... (...)
PKS 1354+195	0.720	Q II, Sey 1	27.15	26.92	26.45
3C 296	0.0237	RG I	24.43	22.69	... (...)
3C 303	0.141	RG II, Sey 1	25.75	24.53	24.33
B2 1553+24	0.0426	RG I	23.36	23.01	22.93 (22.69)
4C 69.21	0.751	Q II	27.07	27.08	26.25
3C 345	0.594	Q II	27.57	27.57	26.37 (24.90)
3C 346	0.16	RG I, Sey 2	26.05	24.81	25.42
B2 1658+30	0.0344	RG I	23.88	22.89	23.17
1745+624	3.889	Q II	28.85	28.37	28.11
3C 371	0.050	RG II	24.84	24.60	23.43
3C 380	0.692	Q II	27.94	27.30	27.73
4C 73.18	0.302	Q II	26.58	26.49	24.8 (24.0)
3C 401	0.201	FR II	26.37	24.09	25.24
3C 403	0.059	FR II	25.41	22.59	23.99
PKS 2201+044	0.028	RG I, Sey 1	24.10	23.41	...
3C 454.3	0.859	Q II	28.10	28.03	26.43
4C 13.85	0.673	Q II	26.88	26.51	...

Table 6.2 Candidate optical/IR jets.

Object	Redshift	Cl.	log P_{tot}, 1.4 GHz	log P_{core}, 5 GHz	log P_{jet}, 1.4 GHz
NGC 6251	0.0230	S2 II	24.14	23.66	24.11 (22.46)
PKS 0445+097	2.110		27.78	27.29	...
3C 20	0.174		26.52	23.07	24.42
3C 216	0.668		27.37	26.86	26.39
3C 388		Seyfert	25.73	23.76	23.71 (23.32)
3C 433	0.1016		26.15	22.76	24.24

This list is based on one originally compiled by Sebastian Jester[1], but has been updated substantially to include jets seen only in the IR (a class which will become increasingly important in the next decade with the launch of JWST) as well as others discovered within the last few years. An online version will be published that will include search tools and additional information. Table 6.2 gives candidate jet sources. We have been restrictive about what we term candidate jets, because many such candidates that have been found with ground-based data do not hold up when the source is imaged with the HST.

Optical and infrared emission can also be found in the terminal hot spots of jets. This topic is closely related to that of optical emission from the jets themselves, in the first place because the radiation mechanisms are the same and also because the radiating particles may have been in the jet's flow before traveling to the hot spots where they are accelerated to optically emitting energies. The known hot spots with optical or infrared emission are listed in Table 6.3. We distinguish hot spots from jets via the functional definition that the hot spots are usually terminal features.

With the vast majority of the known optical jets having been discovered in the last decade, it is safe to say we are only just learning about their morphologies. This is the subject of the next section. Before discussing the source morphologies, however, it is useful to go over some class properties.

6.2 Studies of Sample Properties

Table 6.1 allows us to discuss whether one can use sample properties to study the physics of optical jets. In Figure 6.1 we show histograms of the total and core luminosities at 1.4 and 5 GHz, following [12], as well as the ratio of core to extended radio luminosity R (normalized to 5 GHz, and using a spectral index of 0.7 ($F_\nu \propto \nu^{-\alpha}$) for the extended emission), for the optical jet sources. With the current data, there appears to be a preference for FR II type sources (Figure 6.1a); however, that population has been surveyed considerably more heavily for optical and X-ray jet emission (e.g., [13–15]; also note that the 3CR sample, which has been surveyed

1) http://home.fnal.gov/~jester/optjets/.

Table 6.3 Known optical/IR hot spots.

Object	Redshift	Cl.	log P_{tot}, 1.4 GHz	log P_{core}, 5 GHz
3C 20	0.174	RG II	26.52	23.07
3C 33	0.0567	RG II	26.38	23.29
PKS 0405-123	0.574	Q II	27.39	26.60
3C 105	0.089	RG II	26.24	23.36
3C 111	0.0485	RG II	26.29	24.47
Pictor A	0.035	RG II	25.93	24.01
3C 195	0.111	RG II	25.73	23.88
4C 29.30	0.0643	RG II?	24.73	23.86
3C 208	1.110	Q II	27.52	26.27
3C 263	0.646	RG II	27.90	25.98
3C 275.1	0.555	RG II	27.08	25.74
PKS 1421-490	0.663	Q II	29.15	$<$ 26.08
3C 303	0.141	RG II	25.75	23.44
3C 336	0.927	RG II	27.31	25.56
3C 351	0.371	RG II	26.65	24.14
3C 390.3	0.056	RG II	25.59	24.10
3C 403	0.059	RG II	25.41	22.59
Cygnus A	0.057	RG II	27.73	24.12
3C 445	0.057	RG II	25.30	23.51

by HST in snapshot mode, is 80–90% FR II), whereas only one considerably smaller sample of FR I radio galaxies (the B2 sample [16]) has been surveyed in a similar way. Thus despite the obvious imbalance in numbers it is difficult to compute a frequency of detection for any class.

Unfortunately we cannot yet calculate an optical luminosity function for jets. The value of studying this function would be hard to overstate. Given the wide range of object types (Figure 6.1) and source numbers, and the fact that 2–10 components are known per jet, there is a wide variety of structure types and luminosities. With the luminosity function, structures could be studied in groups not only by morphology but also by distance from the center, luminosity, and other parameters.

Moreover, such a study would allow us to discover biases that currently are unknown, by allowing comparisons to luminosity functions seen in other bands. Indeed, there are many factors that could affect whether or not a jet can be observed in the optical – including, but not limited to, its broadband spectral energy distribution, viewing angle, and Lorentz factor. Many of these parameters are hidden or not immediately obvious, so obtaining an unbiased luminosity function may be difficult – however, the difficulty should not dissuade us. Unfortunately, papers in the literature often do not include the requisite information. What can be seen, however, in Figure 6.1c, is that there is a definite preference among the optical jet sources for objects with high values of log R, where $R = L_{core}/L_{ext}$, the ratio of compact to extended flux at radio wavelengths. By comparison, more typical values of log R for unbeamed FR I sources are -2 to -1 [17, 18], while those of unbeamed FR II sources are even lower [17].

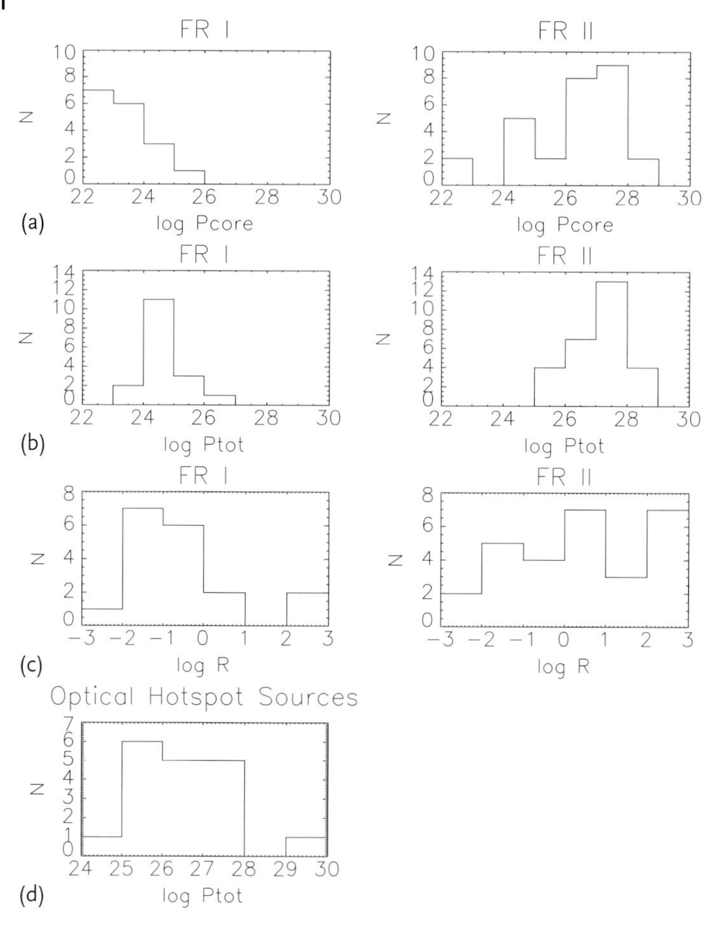

Figure 6.1 Population statistics for the confirmed optical jet and hot spot population (see Table 6.1). (a) We show the total power at 1.4 GHz for the sources with optical jets. (b) The core power at 5 GHz for the jet sources. (c) The ratio of core to extended luminosity for the optical jet sources, normalized at 5 GHz. Note that all 7 objects in the highest-R bin in (c) represent only lower limits. We have also broken up the FR I and FR II optical jet populations for easy reference. (d) We show the total power at 1.4 GHz for the sources with optical hot spots.

The high values of log R indicate that many of the sources are relativistically beamed, as log R is often taken as a proxy for θ [17]. The fact that no extragalactic optical counterjet has ever been seen also points towards beaming, with relatively small inclinations to the line-of-sight. Scarpa and Urry [19], based on a study of about 15 optical jets, suggested that the typical Lorentz factors are $\Gamma \approx 7.5$ with $\theta \approx 20°$. This gives a Doppler factor $\delta \approx 1.95$ and a relatively modest front-to-back ratio of order 10–20 (see (2.66)). However, this calculation is somewhat naive, because it assumes a uniform selection with a complete, flux-limited sample. As noted above, this is almost certainly not the case for the existing sample of optical jets, which have been selected in a very inhomogeneous way. They also assume a beaming factor that is the same for the optical and radio-emitting jet material, which as detailed later in this chapter is likely not true. Thus this computation, while tempting, really needs to wait for the computation of the optical luminosity function using a complete sample.

By comparison to the optical jet sources, the characteristics of the optical hot spot sources (Figure 6.1d) are much more tightly clustered, with all being of the FR II class except the lowest-power one, 4C 29.30, which has been called a "restarting" radio galaxy and is hard to characterize morphologically. All but two have total radio powers at 1.4 GHz between 10^{25} and 10^{28} W Hz^{-1}. This distribution is typical of FR II sources [20]. Unlike the optical jet sources, the distribution of core to extended flux ratios in the hot spot sources is fairly typical for FR IIs, so that there is no evidence that the population of radio sources with optically seen hot spots has jets seen closer to our line-of-sight than random chance would predict. This is sensible according to jet models, which predict that the jet hot spots represent working surfaces where the jet is already subrelativistic and interacting with the intergalactic medium.

6.3 Source Morphologies, Superluminal Motion and Variability

The first, and most basic study one can do in the optical, as in any other band, is to take an image, describe and characterize the morphology, and compare it with that seen in other bands. For the optical jets, this leads to some interesting conclusions, which can be illustrated by discussing the two best-known cases, the jets of M 87 and 3C 273. These two jets are the prototypes, the first ones discovered, as well as the brightest optical jets in (respectively) the FR I and FR II classes. As such, they give a wealth of information about what can be learned by comparing properties between the optical and other bands.

M 87 is the nearest optical jet source. (Cen A is nearer but while its jet can be seen in the mid-IR at distances $> 10''$ from the nucleus it is invisible at shorter wavelengths. Thus the best images of its jet are with the *Spitzer* Space Telescope, and have FWHM ~ 2–$3''$ [21].) The M 87 jet is unique in that it was discovered in the optical in 1918, decades before the existence of radio astronomy. The optical jet is seen only on one side, which is believed to be approaching us. With optical and near-IR observations from the ground, the jet of M 87 is essentially unresolved across its width. But with HST, it is very well resolved (Figure 6.2). The jet is revealed to have a knotty structure in the optical, with some of the knots having a "filamentary" structure, particularly knots E, A and B. Other knots have strong structures transverse to the jet axis (knots A and C), giving rise to the suggestion that they are perpendicular shocks, as also appears in the radio.

The high resolution of the HST images allows direct comparison of the morphology in the optical to that seen in high resolution radio images [22, 23]. The same components are seen in the radio and optical; however, detailed and careful study

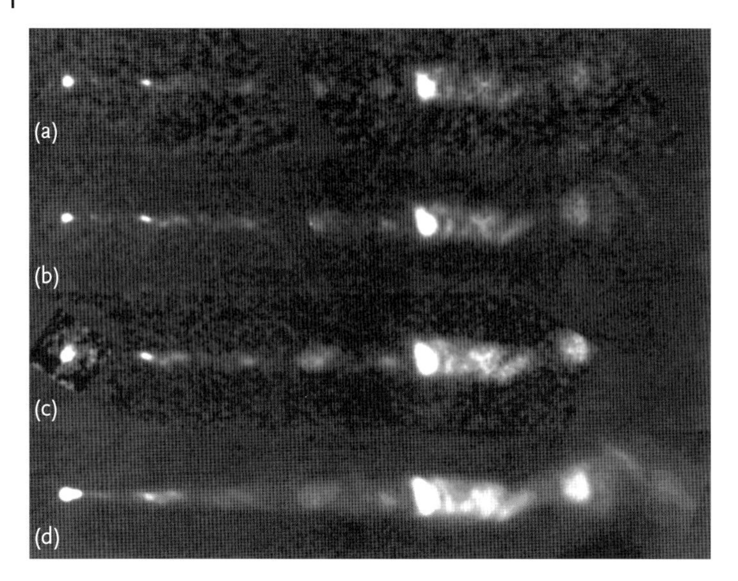

Figure 6.2 Images of the M 87 jet in three near-UV bands: taken in 1993 (a) $\lambda = 140$ nm, (b) $\lambda = 220$ nm, (c) $\lambda = 372$ nm, and (d) in the radio 1.3 cm, taken in 1985. The images have been rotated so that the jet lies along the X-axis, and the approximate length of the jet is 23 arcsec. Reproduced from Sparks *et al.* [22] with permission from the AAS.

of the HST images reveals that the optical emission is more concentrated in the knots and along the centerline of the jet than the radio emission. There are also indications that the trend of increasing "contrast" with decreasing wavelength continues from the optical into the UV. The comparison gets more interesting yet when one compares the optical morphology to that seen in the X-rays with Chandra. In Figure 6.3, we show the Chandra image of the M 87 jet, with contours overlaid from HST observations that are smoothed to the same resolution, corresponding to the capabilities of Chandra. As can be seen, the properties seen in the optical are somewhat intermediate between those seen in the radio and the X-rays: the jet is somewhat "knottier" in the X-rays than in the optical, which as already mentioned is knottier than the radio. The width of the knots also varies with energy, with the knots being widest in radio and narrowest in X-rays, and the optical width being intermediate between the two.

The jet of M 87 also displays apparent superluminal motion in the optical [25], Figure 6.4, with components observed to be moving as fast as $6c$. The fastest moving components are in the HST-1 knot region, some 60 pc (0.85 arcsec) projected distance from the nucleus. HST-1 has exhibited massive variability during the last 10 years ([26–28]; see Chapter 7 for more information). These variations have been interpreted in the same framework as blazar flares – namely as being due to enhanced particle acceleration within the flow. Superluminal motion was in fact observed with the HST out to 6.3 arcsec (500 pc projected) from the core. When the HST data are combined with earlier VLA radio observations [29], some evidence

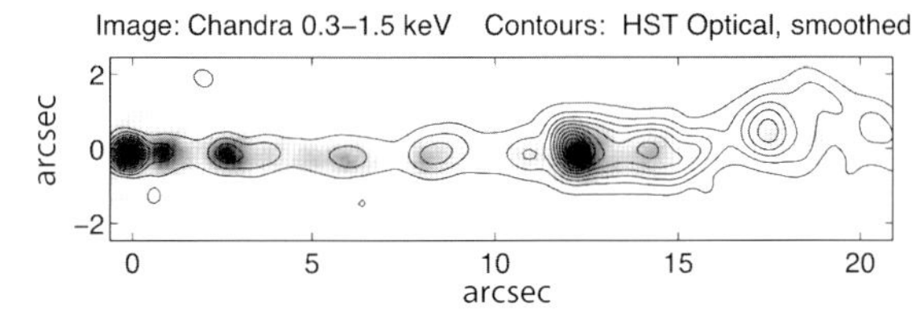

Figure 6.3 Images of the M 87 jet in optical (HST F814W) and X-rays (Chandra 0.3–1.5 keV). The optical image was smoothed with a Gaussian to a resolution of 0.5″ to match the FWHM of the PSF-deconvolved Chandra image, and the Richardson–Lucy deconvolution has been used on the Chandra data as discussed in Perlman and Wilson [24]. Note that while there is a close correspondence of many features, there are significant differences, as discussed in the text. Reproduced from Perlman and Wilson [24] with permission from the AAS.

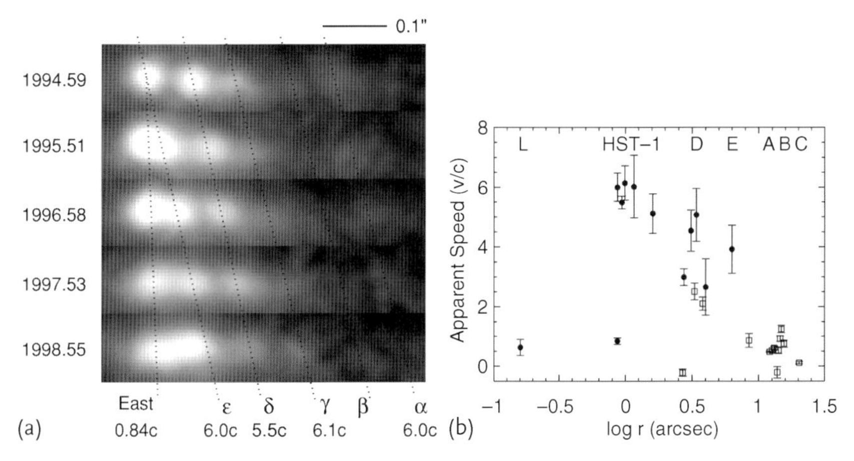

Figure 6.4 (a) Images of the knot HST-1 region at five epochs. Five superluminal components have been noted with dashed lines, along with one nearly stationary component at the upstream end of the complex. (b) A plot of measured apparent speed vs. distance from the nucleus. Speeds measured with HST are denoted by filled symbols, and speeds measured on VLA images are denoted by open symbols. Reproduced from Biretta *et al.* [25] with permission from the AAS.

is seen for deceleration on scales of hundreds of parsecs, although this has not been observed directly in any component to date. The tentative evidence for deceleration is consistent with models of FR I jets, which are thought to decelerate to subrelativistic speeds within the host galaxy (e.g., [30, 31]).

There is also evidence for faster speeds in the optical data than in the radio data for M 87, and in fact speeds as high as $6c$ have not yet been seen even with VLBA observations [32, 33]. If in fact the speeds seen in the optical represent bulk motions, as is believed, then constraints can be placed on the Lorentz factor of the jet's flow as well as its viewing angle. The data require a bulk Lorentz factor $\Gamma > 6$ and a viewing angle $\theta < 19^\circ$. Larger viewing angles require larger Lorentz factors; however, $\theta = 19^\circ$ requires $\Gamma = 40$, while $\theta = 10^\circ$ requires a much smaller value, $\Gamma = 6$. In any event, this means that the jet to counterjet ratio must be at least a few times 10^4, consistent with the observation that the counterjet is not seen at arc-second scales, in any band. However, deep VLBA observations at 15 GHz do reveal a short, weak counterjet [34].

The small viewing angle required by the superluminal motions is in conflict with the larger angle $\theta \sim 40°$ required by the appearance of the H_α disk; it is generally assumed that the jet is perpendicular to such a disk [35]. This discrepancy has yet to be resolved, although Lobanov *et al.* [36] proposed a way to reconcile it with helically wound filaments. The superluminal components move along the filaments, with the observed optical motions being representative of the pattern speed of normal mode disturbances and/or motion of the spots where the jet tangent is closest to our line-of-sight. The Lobanov *et al.* model requires a more modest Lorentz factor $\sim 3-3.5$; however, the predicted jet to counterjet ratio is then much smaller, ~ 200, which is hard to reconcile with observations.

The jet of 3C 273 is much more powerful (Table 6.1 as well as longer than the jet of M 87 (60 kpc as compared to 2 kpc in projected length), and it was the first FR II jet discovered in the optical, having been observed first in the early 1960s. Images of the 3C 273 jet in the optical are seen in Figures 6.5 and 6.6. The 3C 273 jet's optical morphology is knotty, with the same components being seen in the optical as in the radio. Differences appear only when the jet is observed with the subarcsecond resolution of HST, which reveals that (similar to M 87) the jet is knottier and the components are more compact in the optical than in the radio. Other differences are also present, particularly in the knot B region, where the optical maximum is displaced in the northern direction from the maximum seen in the radio, and in the hot spot region H3-H2, where the flux seen in the optical fades much more quickly than that seen in the radio. Unlike in M 87, the width of jet components within the optical-UV is nearly constant (Figure 6.6a). The optical and ultraviolet images show distinct "filamentary" structures, as in the M 87 jet, although the angular resolution in the HST images is not quite sufficient to fully resolve the two "strands". When the optical morphology is compared to that seen in the X-ray by Chandra, the picture can be filled out even more. The same components are seen in the X-rays as in the optical, and the ridge line in the knot B region matches that seen in the optical and not the radio. The trend of increased knottiness continues from the optical into the ultraviolet, and also the flux profile of the components is seen to change radically as one goes from the radio through to the X-rays. In the optical, the flux profile of different components is much more constant down the jet in the optical than it is in the radio (where the flux increases monotonically from knot A, $12''$ from the core to the hot spot H3, $22''$ away) or X-ray (where the flux decreases monotonically from knot A out to H3). This trend is seen within the optical band as well, as in the near-infrared we see that the hot spot is the brightest feature, with a monotonic increase from knot B1 to H3, while by the ultraviolet the trend has reversed itself.

Other FR II jets have not been studied to the same degree as that of 3C 273. Perhaps the next best-studied cases are the jets of PKS 1136-135 and 1150+497 [40, 41], with PKS 0637-752 [42] close behind. All three jets reveal a knotty structure in the optical, with some significant differences between their morphologies in

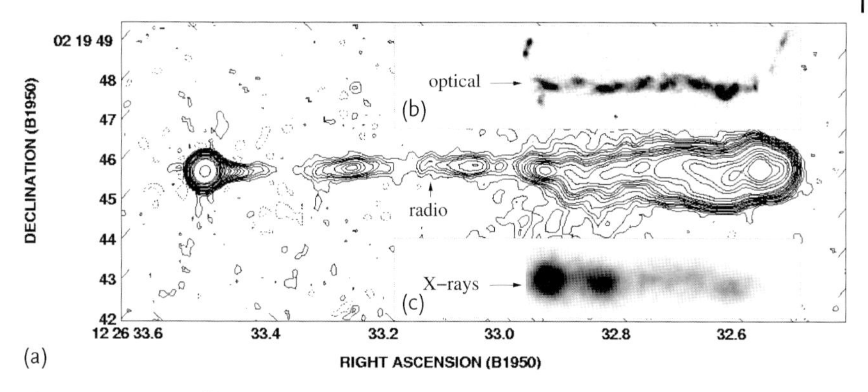

Figure 6.5 Images of the 3C 273 jet in radio contours (a), optical (b), and X-rays (c), as seen by the VLA, HST and Chandra, respectively. The radio image has been convolved to 0.5″ resolution to match that of the X-ray image, while the HST image is shown at full resolution. Reproduced from Perlman *et al.* [37] with permission from the Astronomical Society of the Pacific.

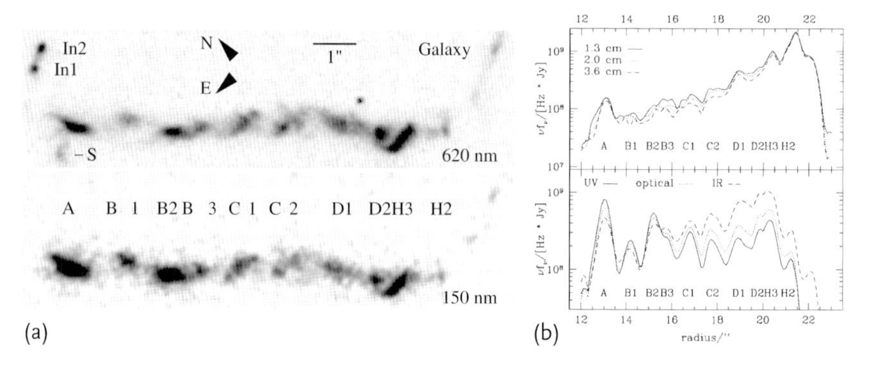

Figure 6.6 (a) Images of the 3C 273 jet in the optical and far-UV are compared to one another. Plot taken from [38]. (b) Normalized cuts through the jet profile at various distances from the quasar at 300 nm (solid line), 620 nm (short dash), 1.6 m (long dash), and 2.0 cm (dotted); resolution is 0.3″. Negative offsets are to the south of the jet. The "southern extension", due to a background galaxy, is visible in the short-wavelength profiles. However, the apparently corresponding feature in the radio profile is in fact unrelated, due instead to the radio cocoon around the jet. (a) is reproduced from Jester *et al.* [38] with permission from the Royal Astronomical Society; (b) is from [39] with permission from Astronomy and Astrophysics.

the optical, radio and X-ray. PKS 1136-135, in particular, shows some of the same morphological trends seen in the 3C 273 jet [40, 41, 43]. However, both of these objects are at higher redshifts, and the surface brightnesses of these objects are considerably lower than the jet of 3C 273, so that diffuse emission can only be seen in two of them. Thus deeper imaging is needed to discuss these objects further, although we will return to the subject of the spectra and polarization.

Much less is known about the morphology of optical hot spots. This is partially because the number of hot spots with known optical emission is only about 1/3 the number of known optical jets. In addition, in the R band the known hot spots tend to be 20th mag per square arcsecond and fainter [44–46], whereas the optical jets range up to 20th mag per square arcsecond for FR IIs, and 16th mag per square arcsecond for FR Is. The structures are typically fairly compact [46], with optical emission seen from only the regions of the hot spots that are brightest in the radio. An example, the southern hot spot of 3C 445, is shown in Figure 6.7. The optical/IR image has a knotty morphology, similar to what is seen in high-frequency radio observations. A similar knotty structure is seen in the western hot spot of Pic A, the only other hot spot imaged with HST (discussed further in Section 6.5), albeit with more extension along the jet direction.

From the examples given above, a number of trends become clear. All of our prototype objects have optical characteristics that differ somewhat from what is seen in other bands. The optical morphology tends to be knottier than that seen in the radio; and the width of components also varies across the IR-UV bands, such that components are observed to be narrower in the ultraviolet than in the infrared. In the optical, knots also tend to be more compact in the direction parallel to the flow than they are in the radio. This makes sense from the standpoint of radiative cooling times, which are a factor $\sim$ 5 shorter for particles emitting at a wavelength of 1000 Å than they are for particles emitting at 2.5 μm. Moreover, as already stated,

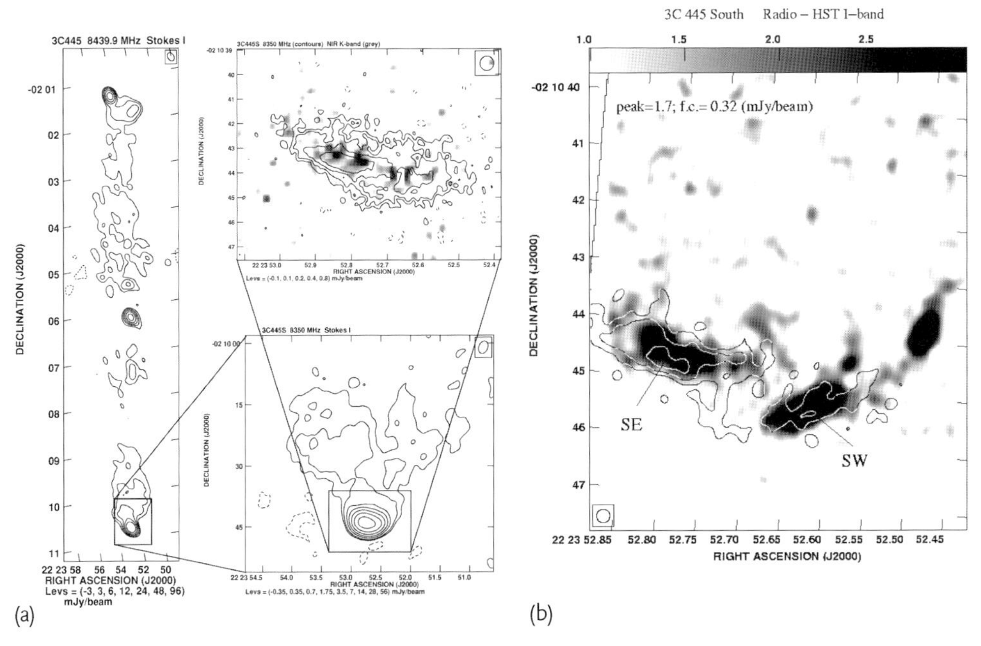

Figure 6.7 The contours are plotted for the radio and optical morphology of the southern hot spot in the 3C 445 jet, as viewed in the radio by the VLA, VLT in near-IR (a), and HST (b). The left three plots in (a) are reproduced from Mack *et al.* [46] with permission from the Royal Astronomical Society. (b) is reproduced from Orienti *et al.* [47] with permission from the Royal Astronomical Society.

the radiative cooling times are short compared to the light travel time down the jet. As already mentioned at the beginning of this chapter, the typical particle radiative lifetime of a few hundred years (which was calculated for magnetic fields in the 100 μG range, typical of an FR I jet), is a factor of 10–100 less than the light travel time down the optically visible part (1–10 kpc) of an FR I jet. A similar statement can be made for FR II jets, where the magnetic field tends to be a factor $\sim$ 10 smaller but the optically visible portion is at least a factor 10 larger. This point has two implications. First of all, optical synchrotron emission, under most conditions, requires *in situ* particle acceleration, as otherwise the particles could not have made it from the nucleus to the optical emitting knots. Secondly, particle lifetimes of hundreds to thousands of years can, on nearby objects, be translated into cooling lengths that are actually observable. For example, the proximity of the M 87 jet (d = 16 Mpc, at the center of the Virgo cluster) gives an image scale of 78 pc (projected) per arcsecond – which means that in most of the jet components, the resolution of the HST is sufficient to observe the cooling of particles within the optical-IR band. Moreover, since particle lifetimes $\tau \propto \nu^{-1/2}$, one should expect to see components that are smaller in the ultraviolet than in the infrared, since the particle lifetime at a wavelength of 2000 Å would be only about 30% of that for particles at 2 μm. Note that while the high-energy particles are traveling along the flow, they are also likely diffusing out from their sites of acceleration, thus the prediction is for components that are both shorter as well as narrower at shorter wavelengths.

While Sparks *et al.* [22] has made optical/radio comparisons only at a resolution of 0.8″ (Figure 6.2), it should be noted that the naive application of synchrotron theory (above) would predict that the narrowing of components across the optical-UV should be observable even at a resolution of 60 pc. The fact is, however, that we do not see components being 3× narrower in the UV than in the optical (e.g., [48]) in M 87. This indicates that the loci of particle acceleration are not fully resolved and/or are distributed, as also suggested by the Chandra data.

While the optical morphology has been studied in detail for several other jets, the data do not allow a detailed characterization of the structure. What can be said is that M 87 appears to be rather knottier than the norm for the FR I jets, as the three next nearest ones to be studied deeply (3C 264, 3C 78 and 3C 66B; all of which are 5–10 times more distant than M 87 [49, 50]) all have a less knotty, smoother appearance both in the radio and optical, with knot-like structures seen only in the highest-resolution UV images. Other, somewhat more distant FR Is are knottier (e.g., 3C 371 [49, 51]; PKS 0521-36 [52]; 3C 15 [53]; all of which have knotty morphology, similar to M 87) and thus more suitable for an analysis such as the above. However, due to their greater distance HST is unable to resolve the particle cooling length in these objects, because at $z = 0.05$, 0.1 arcsec corresponds to $\sim$ 100 pc.

Observations of FR II optical jets are hampered even more significantly by their greater distance, as the nearest well-studied FR II optical jet is that of 3C 273, at $z = 0.158$ (Table 6.1), which gives an image scale of 2.7 kpc arcsec^{-1}. For this object, the differences across the IR-UV band are much more subtle, with the only significant difference being that as photon energy increases, the downstream components get fainter, while those near the upstream end get brighter. A few other FR II jets are at similar distances, that is, 3C 303 at $z = 0.141$ and 3C 403 at $z = 0.201$; however, these jets have lower surface brightness (n.b., the surface brightness of the 3C 273 jet is about 1/100th that of the M 87 jet, at typically 21st mag per square arcsecond compared to 16th mag per square arcsecond for M 87) and have fewer knots. Most also lack multiwavelength data in the optical/IR. As a result, it is clear that this wavelength-dependent brightness gradient is not easily observable in other FR II jets with the HST.

One can do a similar analysis for optical hot spot emission. In particular, the hot spots are consistent with being the loci of particle acceleration, as also seen in optical jets. The extent of the emission is a few tenths of an arcsecond, which at the redshift of 3C 445 (image scale 1.07 kpc arcsec^{-1}), corresponds to around a thousand light-years. As the typical equipartition magnetic field for hot spots is around 100 μG, we are in a similar situation to the FR I jets, namely that the particle lifetimes are a few hundred years at the V band, and closer to 1000 years at the K band. Thus the observed component widths, shown in Figure 6.7, are roughly consistent with the lifetimes of the emitting particles. Similar morphologies are seen in other optically emitting hot spots. We do not observe any narrowing of the components with wavelength in the IR-optical; however, most have been observed only with ground-based telescopes, so that the resolution is only ~ 0.5 arcsec.

6.4 Optical and Broadband Spectra

The jets and hot spots we observe have a variety of spectral shapes, both within the IR-UV band, as well as across the electromagnetic spectrum. As already discussed, the general framework that they fit into is synchrotron emission, due to the fact that their emissions are highly polarized in the optical (see next section) and also because generally the optical emission is near the power-law extrapolation of the spectra observed in the radio-mm. However, as with so many other things, beyond this generalization are many more details, and there may well be exceptions! For the details, as well as the general and specific trends that are observed, it is once again useful to turn to the best-observed jets and hot spots as prototypes, and then discuss each class after that.

Figure 6.8 shows maps of the M 87 jet in optical spectral index α_o and radio-to-optical spectral index α_{ro}, compared to optical flux. Runs of the same spectral indices are plotted in Figure 6.9. Also plotted in Figure 6.8 are maps of injection index (related to the slope of the particle energy distribution p via $p = 2(\alpha_{in} + 1)$) as well as synchrotron break frequencies. The latter refers to the frequency at which the combined synchrotron plus inverse-Compton losses exceed the relativistic bremsstrahlung plus adiabatic losses (see for example [54]). As can be seen, while the two spectral indices are similar over much of the jet, there is more variation in α_o than in α_{ro}. The optical spectral index is strongly anticorrelated with

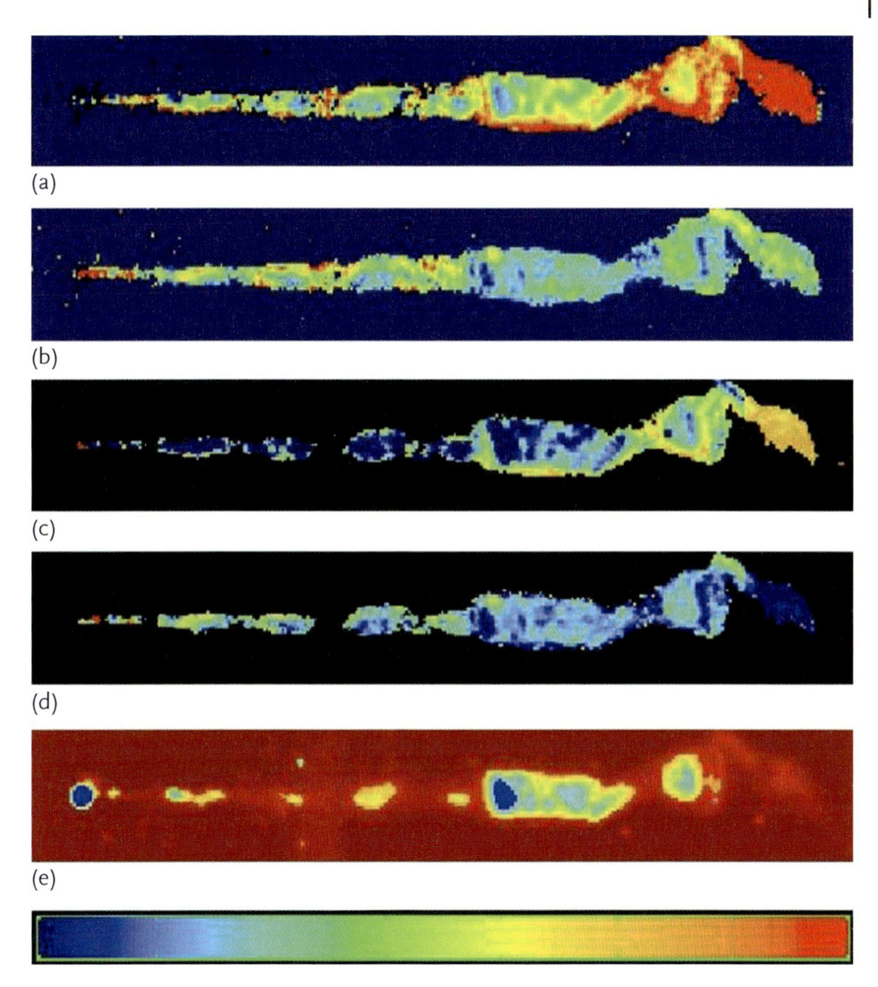

Figure 6.8 Spectral indices and information from spectral fits for the jet of M 87. (a) At the top, we show the optical spectral index image, while in (b) we show the radio-optical spectral index image. We also show (c) maps of the synchrotron break frequency and (d) injection index in the KP model. (e) For reference, the optical flux in the F814W band is shown (roughly Johnson I). Identical color tables were used for parts. The color scale in (a) runs from 1.2 (red) to 0.4 (blue), while that in (b) and (d) run from 0.85 (red) to 0.6 (blue), and that for the break frequency map (c) runs from 10^{15} (red) to 10^{16} Hz (blue). Reproduced from Perlman *et al.* [48] with permission from the AAS.

the optical flux – that is, smaller spectral indices (harder optical spectra) are seen within knots, and the hardest spectra of all are seen at knot maxima. The correlation of $\alpha_{\rm ro}$ with optical flux is much less strong. While several knot maxima do show decreases in both spectral indices, they are not always well correlated in $\alpha_{\rm ro}$, and furthermore the change in $\alpha_{\rm o}$ is usually much more drastic. The differences between the runs of these spectral indices can be seen on the $\alpha_{\rm o}-\alpha_{\rm ro}$ plot, which shows large swings near several flux peaks – usually in the sense that it goes posi-

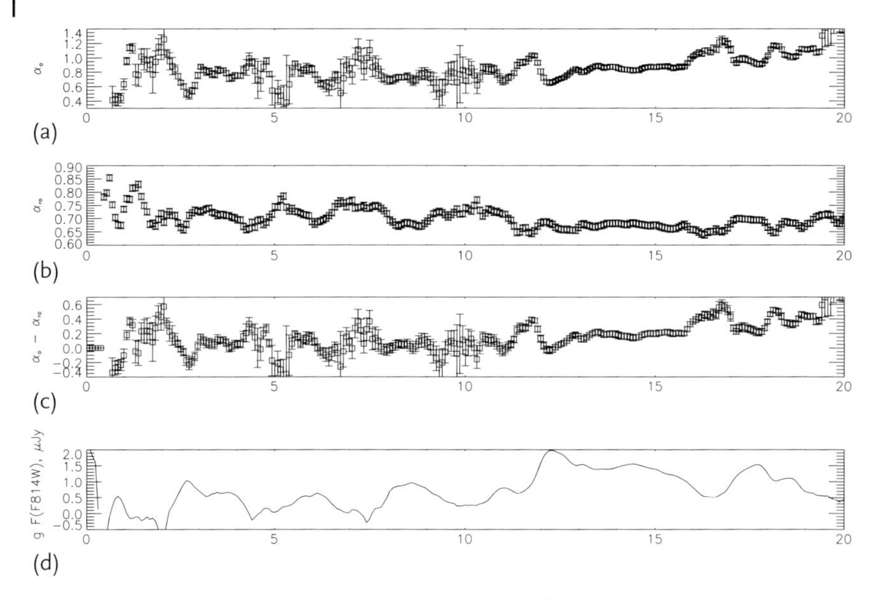

Figure 6.9 Runs of α_o (a), α_{ro} (b), $\alpha_o - \alpha_{ro}$ (c), and F814W flux (d), plotted as a function of distance along the jet axis. These values are flux weighted, that is, flux is summed across the jet at each pixel in the *x*-direction before spectral indices were computed. After [48].

tive (i.e., α_{ro} flatter or harder than α_o) upstream of the knot's flux maximum, and negative as one goes downstream from that. The optical spectral index also shows a gradual steepening in outer parts of the jet, beginning at about 13 arcsec from the nucleus, just after knot A, that is not seen in α_{ro}. All of these changes are reflected in the synchrotron break frequency map (Figure 6.9d), which varies significantly depending on what model for synchrotron emission is adopted. As detailed in Perlman and Wilson [24], the favored model is that of continuous injection of particles, but the data suggest that the fraction of the jet cross-section that is participating in the particle injection, decreases with increasing distance from the nucleus. In this case the turnover frequencies are $10^{15}-10^{16}$ Hz.

Figure 6.10 shows maps and runs of spectral indices in the 3C 273 jet from ultraviolet through radio, and the optical flux profile, integrated across the jet. As with the M 87 jet, the spectral index maps show a wealth of detail, much of which is correlated with the optical flux profile. Whereas the jet's spectrum in the radio is nearly constant at $\alpha_r = 0.9$, the $\alpha_{1.3}^{IR}$, α_{IR}^{opt} and α_{opt}^{UV} show considerably more variation. The jet spine seen in the optical shows a significantly harder $\alpha_{1.3}^{IR}$ than other regions, even though when the entire jet is looked at in Figure 6.10d, most of that variation is not seen. This accounts for the fact that the jet's outer "sheath", seen in the radio, does not appear in the optical. The inner knots' maxima show distinctly harder α_{IR}^{opt} and α_{opt}^{UV} than seen in interknot regions; however, little correlation is seen at distances $> 16''$ from the core.

There are also some interesting relationships between the spectral indices, with the α_{IR}^{opt} and α_{opt}^{UV} being steeper than α_{radio} through most of the jet and steepening

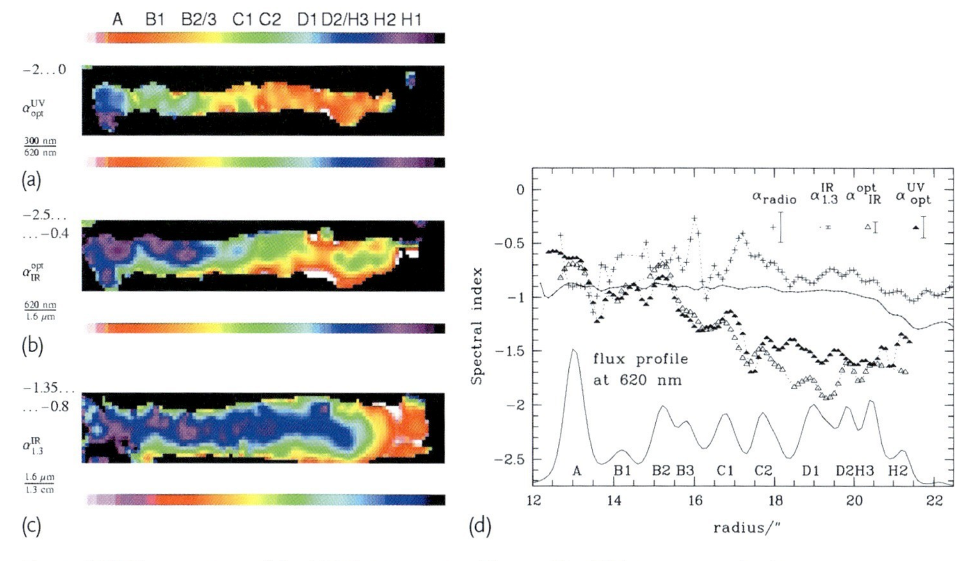

Figure 6.10 The spectrum of the 3C 273 jet in the optical. We show the spectral index from (a) optical to UV, (b) near-IR to optical, and (c) radio through near-IR. In each plot is a legend that includes the limit of the color scales. (d) We show runs of the spectral indices from radio through UV along the jet ridge line, all compared to one another as well as the optical flux profile. All data are convolved to a constant resolution of 0.3 arcsec. Parts (a), (b), and (c) parts are reproduced from Jester *et al.* [39] with permission from Astronomy and Astrophysics. Part (d) is reproduced from Jester *et al.* [38] with permission from the Royal Astronomical Society.

gradually with increasing radius, while $\alpha_{1.3}^{IR}$ is slightly flatter than α_{radio} through most of the jet. The former trend is consistent with the fact that within the optical band, one sees the general trend in the flux profile change from flux increasing with radius in the IR to decreasing with radius in the UV. Equally interesting is the fact that the IR-optical spectrum is very similar in shape to that seen in the optical-UV (seen between 12 and 16.5″ radius) or steeper than the optical-UV one (seen at larger radii), a trend which is not seen in the M 87 jet. This trend continues as one goes further into the ultraviolet, as shown in Figure 6.11, where every knot shows a pronounced hardening of the spectrum. This last fact is discussed below.

As with the jet morphologies, then, we notice both similarities and differences between the spectral evolution of M 87 and 3C 273 along the jet. In both objects, the broad outlines of the spectral shapes are consistent with the morphological trends noted in Section 6.3, and indicate that the dominant radiation mechanism is synchrotron emission throughout the radio and through optical band. For both jets, the radio through optical spectral data have been analyzed quite deeply, and synchrotron models have been used to produce estimates of magnetic fields. In the case of the M 87 jet, this process yields mean magnetic fields of $\sim$ 30–40 μG (assuming $\Gamma = 5$–10 and $\theta \sim 15°$ [57]), which is a factor 1.5–5 below the equipartition value, and a jet that is in rough pressure equilibrium with the surrounding interstellar medium. In the 3C 273 jet, this process yields lower magnetic fields

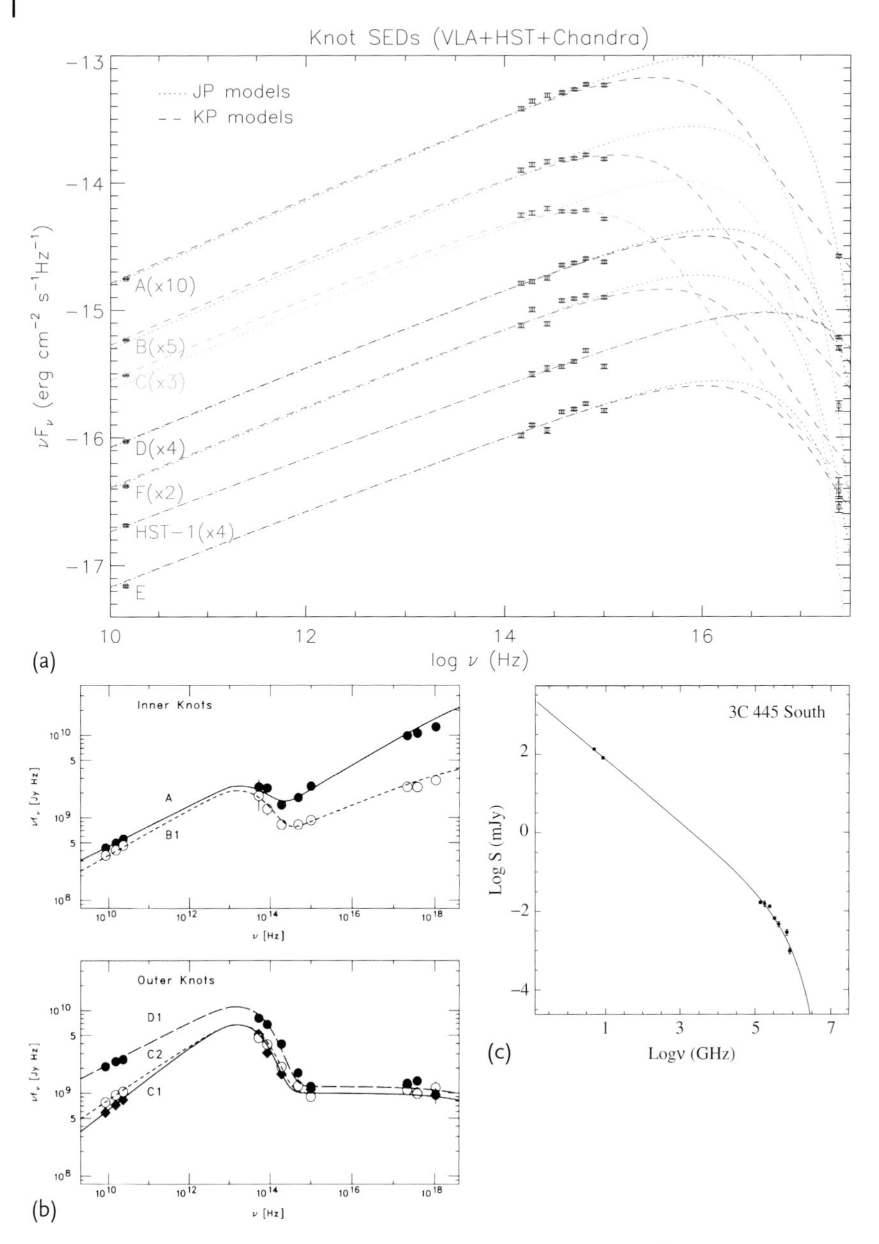

Figure 6.11 Synchrotron model fits for knots of the M 87 jet (a) and the 3C 273 jet (b), as well as the southeast hot spot of the 3C 445 jet (c). As can be seen, the broadband spectra of these two jets show very different overall shapes, with a single synchrotron component being able to fit the entire radio through X-ray SED of the M 87 jet's knots (albeit with particle acceleration taking place only in a fraction of the jet, as discussed in [24]), while the 3C 273 jet requires two clearly distinct emission components, as discussed in the text. The spectrum for the southeast hot spot of the 3C 445 jet steepens gradually towards the optical-ultraviolet. The plot for M 87 is adapted from [55], while the plot for 3C 273 is taken from [56]; reproduced with permission from the AAS. 3C 445 is taken from [46]; reproduced with permission from the Royal Astronomical Society.

$\sim$ 10 μG [58], and in addition, as Jester *et al.* [59] showed, even if one includes beaming, one cannot account for the presence of optical synchrotron emission without local particle acceleration. Also, in both jets, a gradual downstream steepening of the optical spectrum is seen over a large part of the jet, with flatter values of optical spectral index seen in and near knot maxima. While naively one could say that the overall trend of downstream steepening is consistent with increased particle aging at larger radii, it must be remembered that the mere existence of optical synchrotron emission in these jets requires *in situ* particle acceleration. It is therefore more likely that the efficiency of particle acceleration decreases in some way as distance from the nucleus increases, either because the shocks get weaker, or because the spectrum of the irregularities in the magnetic field changes. The smaller-scale trends, namely the flattening of the optical and UV-optical spectral indices at knot maxima, and steepening downstream of the maxima can be interpreted as enhanced particle acceleration at the maxima, and radiative losses immediately downstream. In the M 87 jet, one can see additional detail in the $\alpha_{ro}-\alpha_o$ plot as the spectral indices vary out of phase with one another, which, as discussed in Perlman *et al.* [48], is consistent with gradual cooling as the particles move away from where they are accelerated.

The spectral shapes of the two jets differ strongly towards the ultraviolet and X-rays (Figure 6.11). The spectrum of most regions of the M 87 jet does not change significantly from the optical into the ultraviolet [24, 48, 60]. Instead, the spectral curvature occurs between the optical and X-ray band, as indicated by the continuous injection models displayed in Figure 6.11a,b. The curvature is less in knot maximum regions than in interknot regions, consistent with the idea of particle acceleration in the knots. But interestingly, particle acceleration is also required in the interknot regions, and if one supposes that particle acceleration only occurs in a fraction of the jet's cross-section, one finds some interesting results [24]: in the inner jet, particle acceleration takes place over nearly the same cross-sections in the optical as in the soft ($\sim$ 0.3 keV) end of the Chandra spectral range, but the fraction of the jet cross-section accelerating particles to 10 keV is lower by a factor 3. The fraction of the jet that accelerates particles also decreases as one goes outwards along the jet, decreasing by a factor of $\sim$ 10 between 2.5 and 20 arcsec from the nucleus.

By comparison, every knot in the 3C 273 jet displays a hardening of the spectrum within the ultraviolet, with the second component pointing directly towards the observed X-ray spectrum [38, 59]. This can be interpreted in two different ways. If, as proposed by Sambruna *et al.* [61], the X-ray emission is inverse-Compton scattered CMB radiation, that component would need to start in the optical-UV for the entire jet, representing emission from very low energy ($\gamma \sim 10$) particles. Alternatively, the emission could be due to synchrotron radiation [38, 56, 62] from a second population of electrons, which must be accelerated *in situ*, and requires acceleration mechanisms both within knots as well as throughout the jet. This debate is currently unresolved, although optical polarimetry (Section 6.5) offers excellent prospects to discriminate uniquely between them.

Detailed spectral studies of other jets in the optical are few and far between. The only FR I jets with detailed optical spectral work at HST resolutions are those of 3C 15 [53], 3C 264 [50] and PKS 0521-36 [52]. In each of these, the optical emission is consistent with being synchrotron radiation, from a single component that steepens gradually in the IR-optical to connect to the observed X-ray emission, although in one knot in the 3C 15 jet an additional component is needed to connect to the X-ray emission. For FR IIs, detailed work exists on only two other jets. One of these, the jet of PKS 1136-135, displays hardening within the IR-optical band in most of its components that point towards the X-ray flux, similar to 3C 273 [41, 43]. However, the interpretation of its optical-to-X-ray spectral shape is subtly different than that of 3C 273, for reasons outlined in Section 6.5: namely that some knots require a second synchrotron-emitting population of electrons, but two others are more consistent with the optical and X-ray emission being the result of the inverse-Compton process. In this case the emitting electrons have very low energies, that is, $\gamma \sim 10$, and must carry the bulk of the jet's kinetic flux. The story is rather different for the optical spectrum of the jet of PKS 0637-752, which declines uniformly towards the ultraviolet in all knots, undershooting the observed X-ray flux by orders of magnitude [42, 63]. Here the interpretation of the optical emission is clear, being consistent with the tail of the synchrotron emission component.

An interesting possibility was discussed in Georganopoulos *et al.* [64], namely the idea of using IR emissions from the kiloparsec-scale jet as a possible test of the composition of the jet. If the jet contains a significant population of "cold" (i.e., $\gamma \sim$ few) particles, they will produce an inverse-Comptonized CMB component that is detectable with HST. Depending on the level and spectrum observed, it is then possible to determine whether the jet's matter composition is only leptonic, or contains a significant hadronic component, as well as better constrain the balance between the radiative and kinetic energy output of the jet at these distances. Unfortunately, deep, near-IR observations of the PKS 0637-752 jet failed to detect this emission component [42].

Multiband studies of the optically emitting hot spots in the optical-IR reveal that the optical emission is consistent with the high-energy tail of the synchrotron radiation seen in the radio [46, 65]. Generally, the spectrum is observed to steepen as one goes from the optical into the ultraviolet, as seen in the southern 3C 445 hot spot (Figure 6.11), and the western hot spot of Pic A [44]. However, there is no evidence in any hot spot for a single component that connects the optical emission to that seen at higher energies, which is generally believed to be due to the synchrotron self-Compton process. There is evidence for the hot spots as loci of particle acceleration [66], which in fact is required to explain the existence of particles with γ as high as 10^6 in these hot spots at distances up several hundred kiloparsec from the nucleus.

6.5 Polarimetry

The synchrotron nature of much of the optical emission allows us to exploit one additional tool: polarimetry. As mentioned in the previous section, the radio through optical spectral energy distributions of jets are consistent with synchrotron emission from a single population of electrons that can be extrapolated from radio through optical frequencies. However, the fact that the SED shape is consistent with that expected for synchrotron emission (see Chapter 3) is not in and of itself proof of the synchrotron nature of the emissions. Polarization provides that proof, and more. An electron population with a power-law index p gyrating in a uniform magnetic field produces a synchrotron spectrum with energy index $\alpha_{sy} = (p-1)/2$ (3.41). The fractional linear polarization P_S is defined as

$$P_S = \frac{P_\perp - P_\parallel}{P_\perp + P_\parallel} , \tag{6.1}$$

where $P_\perp$ and $P_\parallel$ are the radiative powers emitted with polarizations perpendicular and parallel to the projection of the magnetic field onto the plane of the sky. For a power-law distribution of electrons with index p, the maximum fractional polarization theoretically possible is

$$P_S^{max} = \frac{p+1}{p+7/3} \tag{6.2}$$

independent of the electron pitch angle distribution (the *pitch angle* of an electron is the angle between its momentum vector and the magnetic field vector) [67, 68]. In the frame of the emitting plasma, the polarization direction (the preferred electric field direction) is perpendicular to the projected direction of the magnetic field lines. For a typical electron spectral index $p = 2$ the photon index is $\alpha_{sy} = 0.5$ and the maximum polarization fraction for a perfectly uniform magnetic field is $P_S^{max} = 69.2\%$. Measured polarization fractions are expected to be smaller as a variation of the magnetic field direction leads to a smaller net polarization. The measurement of very high (up to $\sim 50\%$) radio and optical polarization fractions, close to the theoretical maximum, establishes the emission in both bands as synchrotron emission from electrons [49, 50] moving in regions with a nonisotropic magnetic field direction. Liang [69] pointed out that even a random magnetic field orientation can produce the maximum polarization degree as long as the random magnetic field component is compressed into a two-dimensional plane.

Polarimetry therefore provides a direct look at the magnetic field structure in the emission region. Thus it can often yield constraints that are unique as well as powerful. The combination of morphological, spectral and multiband polarimetric data can, in particular, allow one to disentangle the jet's energetic and magnetic field structure in three dimensions, as well as how that structure links into the dynamics and the acceleration of energetic particles. Unfortunately, the observations here are fewer, due to the difficulties inherent in polarimetric observations, which tend to be light-starved (due to the fact that polarizing filters must screen out cross-polarized light and thus cut the efficiency of any observations by a factor of about 2.5) and yet require very high signal-to-noise (1σ limits on position angle of 5° require a signal-to-noise of about 30 per pixel) to be useful. As with the previous sections, it is useful to discuss prototype sources. Here the best constraints can be obtained from the jets of M 87 [70] and PKS 1136-135 [41], which will be discussed in detail below.

In the M 87 jet, the optical polarization has a complex morphology, which is rather different both from that seen in the optical total flux as well as that seen in radio polarization (see Figure 6.12) [70]. Much of the jet is highly polarized, with fractional polarization averaging about 30%. The polarization fraction varies widely, however, with higher values seen near the northern and southern edges of some components, as well as near the flux maxima of outer jet knots (A, B and C), and much lower values near the flux maxima of inner jet knots (D, E, F and I), as well as along a "channel" that can be seen through the outer jet. The inner jet knots, in fact, display an anticorrelation of optical polarization with flux, with minima near zero at the flux maxima. The predominant direction of the magnetic field vectors is parallel to the jet flow direction. But just upstream of the knot maxima in all but one of the inner jet knots one sees rotations of 90° in the magnetic field direction. Perpendicular magnetic field vectors are also seen in the flux maximum regions of knots A, C and HST-1, with knots A, B and C revealing hints of a filamentary structure. These patterns do not all show up in the radio, where little if any decrease of polarization is seen at flux maxima in the inner jet, as well as less-prominent rotations in the magnetic field direction. However, the changes in magnetic field position in knots A and C are seen in the radio, as are the hints of filamentary structure in the A-B-C region. Interestingly, these changes in optical polarization are well correlated with the location of flux maxima in the X-rays [24], as seen in the Chandra images.

The high polarizations strongly confirm the synchrotron nature of the optical emissions. Similarly high optical polarizations are observed in other jets [49], thus confirming that the synchrotron mechanism dominates in the optical in all FR I jets. The near-90° rotations in the direction of the magnetic field are consistent with the knots being shocks (consistent with models of active regions in jets, for example [71–79]), possibly perpendicular to the jet direction, although an oblique shock morphology [80, 81] is also possible. Other jet regions, particularly knot E and the A/B region, show evidence for filamentary structures that could be the signature of Kelvin–Helmholtz instabilities and other normal modes [82, 83].

Strong differences are seen, however, between the radio and optical polarization characteristics of the inner jet knots (Figure 6.12). In particular, the flux maxima of several inner jet knots (in particular, HST-1, D-East, E and the upstream maximum of F) exhibit a polarization minimum consistent with zero polarization. These declines in polarization are not seen in the radio. Also, the magnetic field direction rotates strongly (up to 90°) upstream of the same flux maxima, which is not seen in the radio. Given these strong differences, we must conclude that the optical and

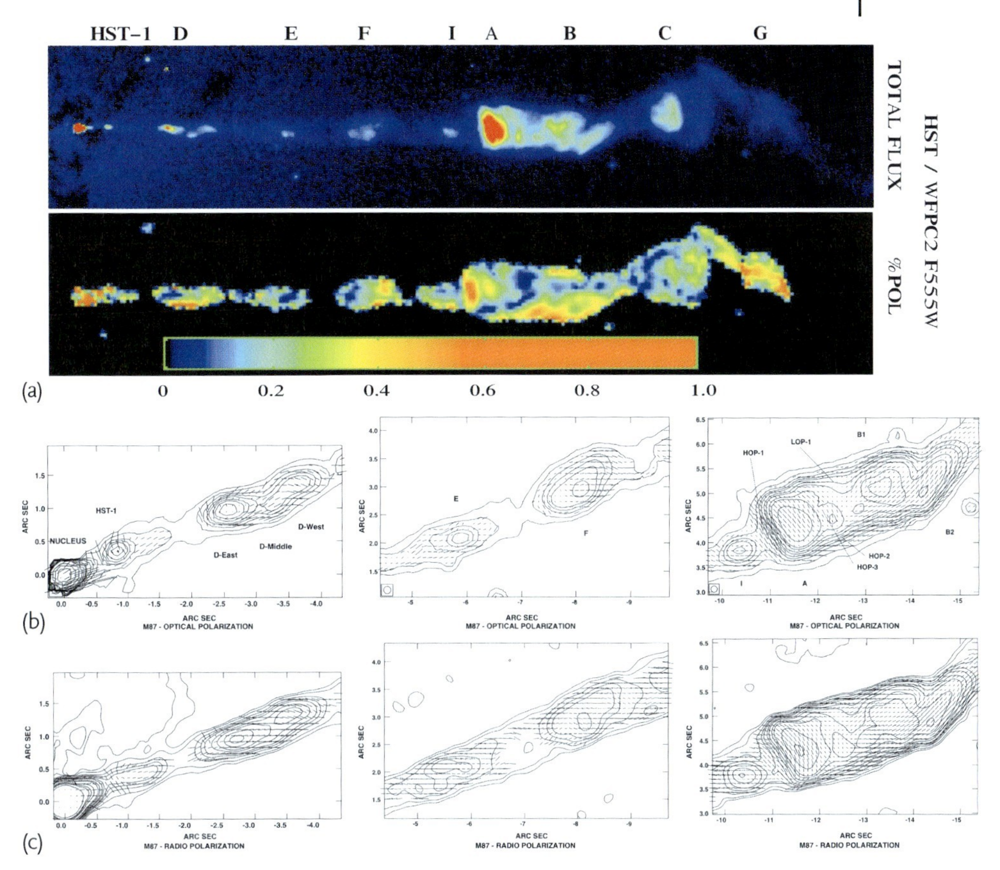

Figure 6.12 The jet of M 87, as seen in polarized optical light. (a) The jet in total optical flux and fractional polarization. The most polarized regions (~ 60%) are red, while the lowest polarization regions are deep blue. Three different regions of the M 87 jet are also shown, with optical polarization (b), and radio polarization (c); these show total flux density as contours. The length of the polarization vectors is proportional to the polarization (a vector 0.1 arcsec long corresponds to 30% polarization), while their direction indicates the direction of the magnetic field. Reproduced from Perlman *et al.* [70] with permission from the AAS.

radio emissions do not come from the same location – if they did, it would be impossible for us to observe such different polarization characteristics.

HST monitoring of the M 87 jet during 2002–2007 [84] reveals that in the flaring knot HST-1, the degree of polarization was seen to be strongly correlated with the optical flux. During that time, the direction of polarization stayed constant to within a few degrees, nearly perpendicular to the direction of the jet. This is consistent with compression due to a shock, with the maximum polarization coming at the time of maximum compression, when the flux was at its maximum. Also, within the few months immediately before and after the flux maximum, the optical spectrum of HST-1 exhibited "hard lags" (see Section 8.3.2.1), usually interpreted as nearly equal acceleration and cooling time scales [85], but difficult to reconcile with the X-ray variability characteristics (Section 7.3.2.3), which show similar variability time scales and which are also consistent with synchrotron emission.

Thus the combination of optical and radio polarimetry gives us a unique look into the energetic and field structure in the M 87 jet. Perlman *et al.* [70] interpreted the observed polarization characteristics as being the result of a stratified jet, with high-energy particles concentrated in the jet interior, where shocks compress the magnetic field and accelerate particles *in situ*, while lower-energy particles are located in the jet sheath. A cartoon of this model is shown in Figure 6.13a. The link between the knots and particle acceleration has already been drawn above, due to the spectral changes seen at these loci. The changes in the polarization morphology, seen in the optical, are attributed to shocks located in the jet interior. The shocks accelerate particles, which diffuse outward into the jet. The shocks are largely perpendicular to the jet direction, and their strong association with particle acceleration explains why the rotated magnetic field vectors are seen most strongly in the optical. Downstream of the initial shock, the flux has peaked in both the optical and radio, but we are observing two superposed regions, as some of the higher-energy, optically emitting particles have diffused outwards from their acceleration sites to the radio-emitting sheath, where the magnetic field is still parallel to the jet. This proposal is similar to the internal shock model put forth by Lazzati *et al.* [87] and Spada *et al.* [88]. As Stawarz *et al.* [89] pointed out for the more powerful quasar jets, if we consider the simplest situation, where two colliding portions of the jet flow differ only in bulk velocity, with $\Gamma_2 > \Gamma_1 \gg 1$, a double-shock structure will develop that is symmetric in the frame of the contact discontinuity. The contact discontinuity will have a Lorentz factor $\Gamma_{\rm ct} = \sqrt{\Gamma_1 \Gamma_2}$, and if one knows Γ_1 and Γ_2 one can then estimate the velocities of the forward and reverse shocks from the shock-jump conditions [90]. In that case the shocked region along the jet will have an extent of $\delta_l' \sim 2c|\beta_{\rm sh}'|\delta t'$ in the primed frame of reference of the contact discontinuity, where $\beta_{\rm sh}$ is the velocity of the forward and reverse shock fronts and $\delta t'$ is the time since the formation of the shock structure. While Stawarz *et al.* point out that this result produces a knot size similar to that seen in quasar jets, their point applies to FR I jets such as M 87 also, as for magnetic fields $\sim 100\,\mu$G and relatively modest Lorentz factors ~ 3–10, knot sizes of hundreds of parsecs can be produced – which are of the same order as those seen in M 87, when one accounts for a viewing angle $\sim 15°$.

More recently, Nakamura *et al.* [86] have carried this interpretation further, by modeling the knots downstream of HST-1 in terms of quad relativistic magnetohydrodynamic shock fronts (both forward/reverse fast and slow modes) in a narrow jet with a helically twisted magnetic field structure (see Figure 6.13b). Their motivation for this structure was the combination of the M 87 jet's broadband spectrum, where in knots one sees spectral indices that typify models for shock acceleration, combined with the above polarization maps. The shocks take on a quad structure as a result of the fact that the toroidal component of the magnetic field, B_ϕ plays a role analogous to the perpendicular component in the simpler planar oblique

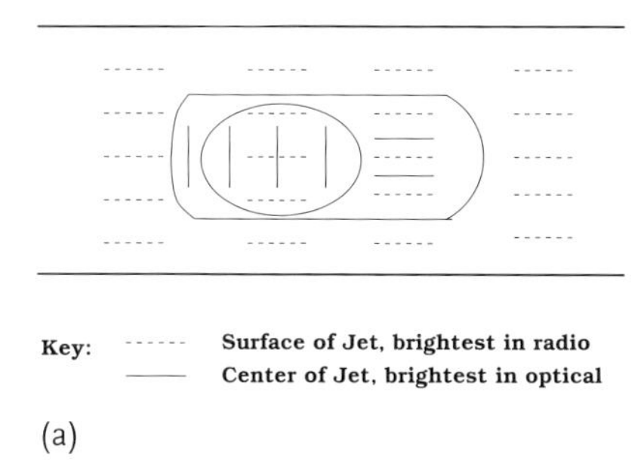

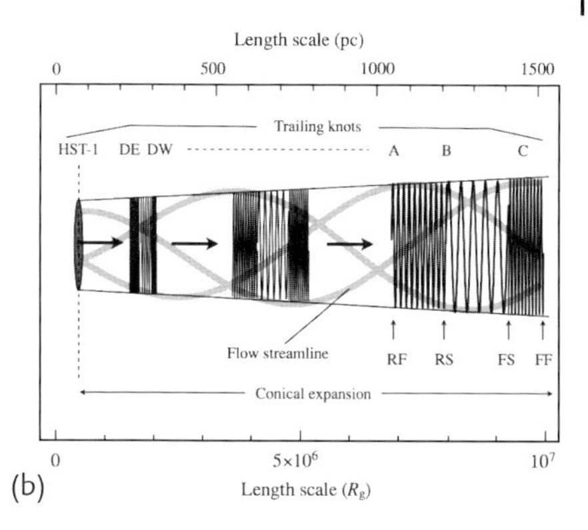

Figure 6.13 Structure of the M 87 jet and its knots. (a) The structure for knots proposed in Perlman *et al.* [70]. Shocks begin in the jet interior, with magnetic field vectors perpendicular to the jet flow direction. Magnetic field vectors in the jet interior are denoted by solid lines, while those in the jet's surface are denoted by dashed lines. Zero polarization would be seen at the maximum of knots, where perpendicular vectors are superposed. (b) An MHD model for the knots (from [86]). The knots are characterized by quasi-periodic MHD shock structures, showing the forward fast (FF), forward slow (FS), reverse slow (RS), and reverse fast (RF) shocks, as well as flow streamlines. Part (a) is reproduced from Perlman *et al.* [70], (b) is reproduced from Nakamura *et al.* [86] with permission from the AAS.

shock case. The dominance of the perpendicular magnetic field vectors in some knots is interpreted as longitudinal compression by a passing shock.

The strong link between the magnetic field structure and particle acceleration has recently been confirmed by polarimetric monitoring of the M 87 jet. During the flare in knot HST-1, a strong correlation was seen between the fractional polarization and the optical flux, revealing that the magnetic field in the flaring component was much more ordered than in the surrounding, quiescent region of the jet [84]. Interestingly, however, the polarization position angle remained constant during the flare at about 5–10° from perpendicular to the jet direction.

Observations of other FR I jets reveal a complicated morphological zoo, of which M 87 is just one exemplar, rather than the rule. Two classes of polarization characteristics are seen [49]. The first class is exemplified by 3C 15 [53], which resembles M 87: showing an anticorrelation between optical flux and polarization and flat optical spectra in regions of low optical polarization. 3C 15 also exhibits significant field rotations in knots that indicate a stratified flow. Another object in this class, 3C 346 [91], exhibits rotations in magnetic field at and before the location of a large bend in the jet, indicating a torsional "kink" and shock in the jet that is seen 0.8″ further upstream in the radio from where the bend occurs. The second class, exemplified by 3C 264 [50], has quite different properties, showing reduced polarization near the jet centerline, but only minor changes in fractional polarization, magnetic field direction, as well as optical spectral index, and overall similar radio and optical polarization characteristics. In all of the FR Is where the angular resolution allows, there are links between the optical polarization and X-ray emission properties. For example, in the 3C 15 and 3C 346 jets, the polarization properties (including differences between radio and optical polarized structure) were interpreted in the framework of oblique shocks that were also associated with particle acceleration at the X-ray-emitting knots [53, 91]. However, it should be pointed out that none of these jets reveal the 90° differences between optical and radio polarization direction, seen in some regions of the M 87 jet. This latter fact should, however, be treated with caution – while it could be the result of these jets having very different energetic structures than M 87, it is much more likely that the much lower physical resolution is to blame (note, in particular, that the well-resolved knots of M 87 would almost all be unresolved at the redshifts of most of the next nearest FR I jets).

In FR II jets, polarimetry has a potential second use, in that the X-ray emissions are often interpreted in the framework of inverse-Comptonized CMB radiation [92, 93], which requires a jet that remains highly relativistic out to distances of hundreds of kpc from the nucleus (see Chapter 7 for more detailed discussion of this model). As shown by Uchiyama and Coppi [94] as well as McNamara *et al.* [95], one would not expect significant polarization from inverse-Comptonized CMB emission because the seed photon population is unpolarized. By contrast, synchrotron emission will be highly polarized, as will be any synchrotron self-Compton radiation. Thus polarimetry has the potential to serve as a discriminant between competing models for the jet emission in the optical for FR II jets, and potentially into the X-rays as well, particularly in objects where a single component is seen to connect the X-ray and optical emission, such as 3C 273 and PKS 1136-135 [38, 39, 43, 56].

The confirmation of high-energy emission from either the synchrotron or the self-Compton mechanism in an FR II, which would be implied by the observation of high optical polarization, would have important implications for our knowledge of the physics in those jets. Optical or X-ray synchrotron self-Compton emission at kiloparsec scales, which would additionally predict optical polarization characteristics very similar to those observed in low-frequency radio observations, would require a jet that is massively out of equipartition if the emitting volume is actually as large as the upper limit determined from the observations. This would drastically increase the kinetic power of the jet and making it highly inefficient as a radiator. A second, high-energy synchrotron population extending from the optical into the X-rays, would require particle energies up to $\gamma = 10^8$ (higher than those seen in FR Is due to the lower magnetic fields presumably seen at these distances from the black hole) in a second, distinct population that could well be located in a different part of the jet from the radio emission (as in the case of M 87). Thus in this case the optical polarization characteristics could be very different from those seen in the radio, with the differences being intimately linked to the jet's energetic structure and the physics of the particle acceleration process, as in M 87.

Unfortunately, despite its promise, very few FR II jets have high-quality HST polarimetry. The jet of 3C 273 was observed fairly early in HST's mission [96]. Those data reveal high polarizations in several knots, as well as polarization vectors that largely appear to follow the "filamentary" structure noted in Section 6.3. They also show perpendicular magnetic fields in the hot spot H3 region. However, they have S/N $\sim$ 5–10 in many regions, far below what is required to perform any tests relating to emission mechanisms or jet structure. In addition, there are large conflicts between HST and ground-based optical polarimetry that cannot easily be explained by beam dilution in the ground-based data [97] but are more consistent with low S/N and background problems. As a result, 3C 273 badly needs to be reobserved both from the ground and with HST.

More promising results have been obtained very recently [41] for the jet of the quasar PKS 1136+135 ($z = 0.554$), where a single emission component is seen to link the optical and X-ray emission of multiple knots, similar to 3C 273. These results are shown in Figure 6.14. Knot A (the X-ray brightest) has a polarization of $36 \pm 6\%$ and apparent magnetic field vectors aligned with the jet, similar to that seen in high-frequency radio maps. This knot therefore cannot be emitting via the IC-CMB process either in optical or X-rays, and the characteristics we see in the optical polarization map as well as the broadband spectrum are most consistent with synchrotron radiation from high-energy particles. A similar statement can be made for the knot CDE region, which is even more highly polarized. But in two other knots (α and B) we do not detect significant polarization (2σ upper limits of 15 and 10%, respectively). These are more consistent with IC-CMB, where we

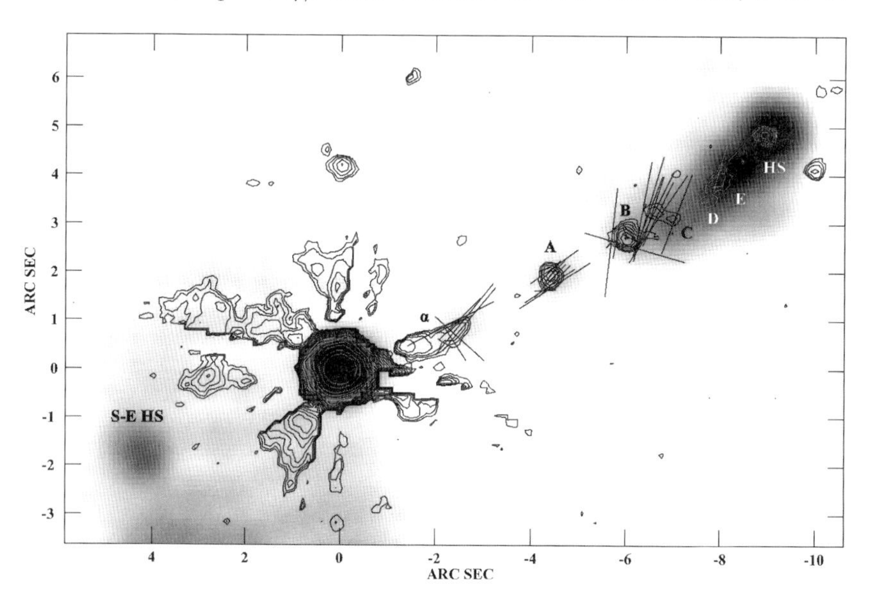

Figure 6.14 The jet of PKS 1136-135 as seen in optical polarized light (contours and vectors) with the HST and in the radio (grayscale) with the VLA. This image represents 21 orbits of observations with the HST/WFPC2, illustrating the demanding nature of the observations. Contours are shown at intervals of $\sqrt{2}$, and components are labeled according to the scheme in Sambruna *et al.* [40] and Uchiyama *et al.* [43]. The optical image has been galaxy-subtracted. Vectors are proportional to the fractional polarization (100% polarization corresponds to a vector 1″ long) and indicate the direction of the implied magnetic field. Reproduced from Cara *et al.* [41] with permission from the AAS.

are for the first time probing the lowest energy particles ($\gamma \sim 10$–100) – which contribute most of the jet's mass-energy budget. This notion is supported by their spectral energy distributions, which in the case of knot α shows a very hard optical spectrum with νF_ν below the radio-X-ray extrapolation and with the extrapolation of the optical spectrum overshooting the observed X-ray flux by orders of magnitude. Particularly in the case of knot α, particles with very low $\gamma \sim 10$ are required to explain the optical emission, as well as the very steep spectrum. More observations of FR II jets with HST are needed to elucidate the emission processes and jet structures.

While there is a similar controversy about the X-ray emission mechanisms of optical hot spots, as both the synchrotron self-Compton and synchrotron mechanisms currently appear viable [98], optical polarimetry does not have nearly the diagnostic power for hot spots as it does for FR II jets. There are two reasons for this. The first and most critical is the broadband SED shape observed in hot spots, where in every case so far the optical emission appears to come from the tail of the synchrotron emission, rather than from a component that can be extrapolated into the X-rays as with the jets of 3C 273 and PKS 1136-135. Also, synchrotron self-Compton emission, as detailed above, should be expected to be polarized, albeit with possibly different characteristics as it will be linked to the low-frequency radio emission rather than that seen at high frequencies.

Only one hot spot has been observed polarimetrically, namely the western hot spot of Pictor A [99]. These observations, shown in Figure 6.15, revealed very high polarizations – ranging from 30% to upwards of 50%, similar to the highest figures seen in jet components and indicative of a highly ordered magnetic field. The magnetic field vectors are roughly perpendicular to the jet direction, and are parallel to the western boundary. This agrees with the usual model of hot spots as being due to terminal shocks in the jet [100] and cements the connection of the optical emission with the high-energy tail of synchrotron emission. The direction

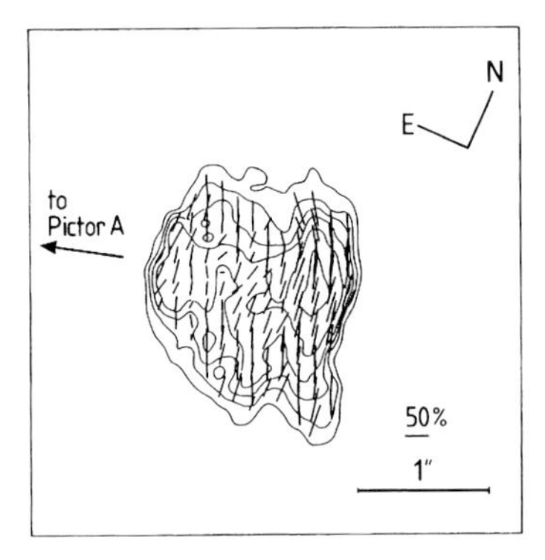

Figure 6.15 Polarization map of the optical emission from the western hot spot of Pictor A, superposed on a contour plot of the total intensity. The vector lengths are proportional to the fractional polarization in the optical and indicate the magnetic field direction. The direction to the host galaxy is shown. Reproduced from Thomson *et al.* [99] with permission from the AAS.

of the magnetic field vectors seen in the HST data for the western hot spot also agrees with that seen in the radio maps [101, 102]. Five other optical hot spots were observed polarimetrically by Lähteenmäki and Valtaoja [45], those being the only other polarimetric observations of this class of object. Those observations detected significant polarizations in one object (3C 111), with a nearly 3σ detection in one other case. 3C 111 also showed a polarization position angle perpendicular to the jet direction and similar to that seen in the radio. Optical polarimetry of additional hot spots needs to be done with HST and/or adaptive optics to further explore the characteristics of this class and to test jet models.

6.6 Conclusion

In this chapter, we have discussed optical emission from jets, and have shown that it is a powerful tool to explore jet physics. In most jet components and all hot spots, the optical emission appears to be due to the synchrotron process; however, there are at least a few exceptions among the optical jet components, where optical emission may be due to inverse-Comptonization of the CMB by the lowest-energy electrons. In either case, the optical emission tells us something unique about the physics of jet sources. In the case of the synchrotron-emitting sources, we are seeing radiation from particles that have radiative lifetimes of hundreds to thousands of years and thus must be accelerated *in situ*, as the emission regions are often many kiloparsecs from the core. What we have seen in these cases is that optical emission often has a morphology that differs somewhat from what is seen in the radio, with the optical emission coming more often from knot regions, and the trends being for increased knottiness and a thinner jet as one goes from the infrared towards the optical. The spectrum seen in the optical is often a continuation of that seen in the radio, but steepening is often seen, consistent with the optical emission being from the high-energy tail of the particle distribution. Harder optical spectra are seen near knot maxima in both FR Is and FR IIs. With these differences in both morphology and spectra between the optical and the radio, we should not expect the optical emission to come from the same regions of jets as that seen in the radio, and indeed the optical polarimetry of several FR I sources now appear to show strong evidence for structured jet flows, with optical emission restricted to regions near jet knots and also possibly within the jet interior. The optical polarimetry also contains powerful evidence of shocks as well as normal oscillation modes. Overall, optical jet emission has taught us important lessons about jet physics, and promises to teach us much more in the future. Much more work remains to be done in this fascinating field.

References

1 Martin, P., Roy, J.-R., Noreau, L., and Lo, K.Y. (1989) *Astrophys. J.*, **345**, 707.

2 Wolstencroft, R.D., Tully, R.B., and Perley, R.A. (1984) *Mon. Not. R. Astron. Soc.*, **207**, 889.

3 van Breugel, W., Miley, G., Heckman, T., Butcher, H., and Bridle, A. (1985) *Astrophys. J.*, **290**, 496.

4 Tremblay, G.R., Chiaberge, M., Sparks, W.B., Baum, S.A., Allen, M.G., Axon, D.J., Capetti, A., Floyd, D.J. E., Macchetto, F.D., Miley, G.K., Noel-Storr, J., O'Dea, C.P., Perlman, E.S., and Quillen, A.C. (2009) *Astrophys. J. Suppl. Ser.*, **183**, 278.

5 Russell, D.M. and Fender, R.P. (2008) *Mon. Not. R. Astron. Soc.*, **387**, 713.

6 Margon, B. (1984) *Annu. Rev. Astron. Astrophys.*, **22**, 507.

7 Hjellming, R.M. and Johnston, K.J. (1981a) *Astrophys. J.*, **246**, L141.

8 Hjellming, R.M. and Johnston, K.J. (1981b) *Nature*, **290**, 100.

9 Russell, D.M. and Fender, R.P. (2010) Powerful jets from accreting black holes: evidence from the optical and infrared, in *Black Holes and Galaxy Formation* (eds A.D. Wachter and R.J. Propst), Nova Science Publishers, pp. 295–320, arXiv:1001.1244.

10 Corbel, S. *et al.* (2002) *Science*, **298**, 196.

11 Russell, D.M., Fender, R.P., and Jonker, P.G. (2007) *Mon. Not. R. Astron. Soc.*, **379**, 1108.

12 Liu, F.K., Zhang and Y.H. (2002) *Astron. Astrophys.*, **381**, 757.

13 Sambruna, R.M., Gambill, J.K., Maraschi, L., Tavecchio, F., Cerutti, R., Cheung, C.C., Urry, C.M., and Chartas, G. (2004) *Astrophys. J.*, **608**, 698.

14 Marshall, H.L. *et al.* (2005) *Astrophys. J. Suppl. Ser.*, **156**, 13.

15 Marshall, H.L. *et al.* (2010) *Astrophys. J. Supp.*, **193**, 15.

16 Parma, P. *et al.* (2003) *Astron. Astrophys.*, **397**, 127.

17 Antonucci, R.R.J. and Ulvestad, J.S. (1985) *Astrophys. J.*, **294**, 158.

18 Hardcastle, M.J., Worrall, D.M., Kraft, R.P., Forman, W.R., Jones, C., and Murray, S.S. (2003) *Astrophys. J.*, **593**, 1659.

19 Scarpa, R. and Urry, C.M. (2002) *New Astron. Rev.*, **46**, 405.

20 Zirbel, E.L. and Baum, S.A. (1995) *Astrophys. J.*, **448**, 521.

21 Hardcastle, M.J. (2006) *Mon. Not. R. Astron. Soc.*, **366**, 1465.

22 Sparks, W.B., Biretta, J.A., and Macchetto, F.D. (1996) *Astrophys. J.*, **473**, 254.

23 Boksenberg, A. *et al.* (1992) *Astron. Astrophys.*, **261**, 393.

24 Perlman, E.S. and Wilson, A.S. (2005) *Astrophys. J.*, **627**, 140.

25 Biretta, J.A., Sparks, W.B., and Macchetto, F. (1999) *Astrophys. J.*, **520**, 621.

26 Harris, D.E., Cheung, C.C., Stawarz, L., Biretta, J.A., and Perlman, E.S. (2009) *Astrophys. J.*, **699**, 305.

27 Madrid, J. (2009) *Astron. J.*, **137**, 3864.

28 Perlman, E.S., Harris, D.E., Biretta, J.A., Sparks, W.B., and Macchetto, D.F. (2003) *Astrophys. J.*, **599**, L65.

29 Biretta, J.A., Zhou, F., and Owen, F.N. (1995) *Astrophys. J.*, **447**, 582.

30 Canvin, J.R., Laing, R.A., Bridle, A.H., and Cotton, W.D. (2005) *Mon. Not. R. Astron. Soc.*, **363**, 1223.

31 Laing, R.A. and Bridle, A.H. (2004) *Mon. Not. R. Astron. Soc.*, **348**, 1459.

32 Cheung, C.C., Harris, D.E., and Stawarz, L. (2007) *Astrophys. J.*, **663**, L65.

33 Ly, C., Walker, R.C., and Junor, W. (2007) *Astrophys. J.*, **660**, 200.

34 Kovalev, Y.Y., Lister, M.L., Homan, D.C., and Kellermann, K.I. (2007) *Astrophys. J.*, **668**, L27.

35 Ford, H.C. *et al.* (1994) *Astrophys. J.*, **435**, L27.

36 Lobanov, A., Hardee, P., and Eilek, J. (2003) *New Astron. Rev.*, **47**, 629.

37 Perlman, E.S., Marshall, H.L., and Biretta, J.A. (2002) in *Mass Outflow in Active Galaxies: New Perspectives* (eds D.M. Crenshaw, S.B. Kraemer, and I.M. George), ASP, San Francisco, p. 233.

38 Jester, S., Meisenheimer, K., Martel, A.R., Perlman, E.S., and Sparks, W.B. (2007) *Mon. Not. R. Astron. Soc.*, **380**, 828.

39 Jester, S., Röser, H.-J., Meisenheimer, K., and Perley, R. (2005) *Astron. Astrophys.*, **431**, 477.

40 Sambruna, R.M., Gliozzi, M., Donato, D., Maraschi, L., Tavecchio, F., Cheung, C.C., Urry, C.M., and Wardle, J.F.C. (2006) *Astrophys. J.*, **641**, 717.

41 Cara, M. and Perlman, E. (2011) *Optical Polarimetry of the Jet of PKS 1136-135: Emission Mechanisms and Jet Physics*, (in preparation).

42 Mehta, K.T., Georganopoulos, M., Perlman, E.S., Padgett, C.A., and Chartas, G. (2009) *Astrophys. J.*, **690**, 1706.

43 Uchiyama, Y. *et al.* (2007) *Astrophys. J.*, **661**, 719.

44 Meisenheimer, K., Yates, M.G., and Röser, H.-J. (1997) *Astron. Astrophys.*, **325**, 57.

45 Lähteenmäki, A. and Valtaoja, E. (1999) *Astron. J.*, **117**, 1168.

46 Mack, K.-H., Prieto, M.A., Brunetti, G., and Orienti, M. (2009) *Mon. Not. R. Astron. Soc.*, **392**, 705.

47 Orienti, M., Prieto, M.A., Brunetti, G., Mack, K.-H., Massaro, F., and Harris, D.E. (2011) *Mon. Not. R. Astron. Soc.*, in press, arXiv: 1109.4895.

48 Perlman, E.S., Biretta, J.A., Sparks, W.B., Macchetto, D.F., and Leahy, J.P. (2001) *Astrophys. J.*, **551**, 206.

49 Perlman, E.S. *et al.* (2006) *Astrophys. J.*, **651**, 735.

50 Perlman, E.S. *et al.* (2010) *Astrophys. J.*, **708**, 171.

51 Sambruna, R.M., Donato, D., Tavecchio, F., Maraschi, L., Cheung, C.C., and Urry, C.M. (2007) *Astrophys. J.*, **670**, 74.

52 Falomo, R. *et al.* (2009) *Astron. Astrophys.*, **501**, 907.

53 Dulwich, F., Worrall, D.M., Birkinshaw, M., Padgett, C.A., and Perlman, E.S. (2007) *Mon. Not. R. Astron. Soc.*, **374**, 1216.

54 Condon, J.J. (1992) *Annu. Rev. Astron. Astrophys.*, **30**, 575.

55 Marshall, H.L., Miller, B.P., Davis, D.S., Perlman, E.S., Wise, M., Canizares, C.R., and Harris, D.E. (2002) *Astrophys. J.*, **564**, 683.

56 Uchiyama, Y. *et al.* (2006) *Astrophys. J.*, **648**, 910.

57 Heinz, S. and Begelman, M.C. (1997) *Astrophys. J.*, **490**, 653.

58 Röser, H.-J., Meisenheimer, K., Neumann, M., Conway, R.G., and Perley, R.A. (2000) *Astron. Astrophys.*, **360**, 99.

59 Jester, S., Röser, H.-J., Meisenheimer, K., and Perley, R. (2002) *Astron. Astrophys.*, **385**, L27.

60 Waters, C.Z. and Zepf, S.E. (2005) *Astrophys. J.*, **624**, 656.

61 Sambruna, R.M., Urry, C.M., Tavecchio, F., Maraschi, L., Scarpa, R., Chartas, G., and Muxlow, T. (2001) *Astrophys. J.*, **549**, L161.

62 Jester, S., Harris, D.E., Marshall, H.L., and Meisenheimer, K. (2006) *Astrophys. J.*, **648**, 900.

63 Uchiyama, Y. *et al.* (2005) *Astrophys. J.*, **631**, L113.

64 Georganopoulos, M., Kazanas, D., Perlman, E., and Stecker, F.W. (2005) *Astrophys. J.*, **625**, 656.

65 Brunetti, G., Mack, K.-H., Prieto, M.A., and Varano, S. (2003) *Mon. Not. R. Astron. Soc.*, **345**, L40.

66 Gopal-Krishna, Subramanian, P., Wiita, P.J., and Becker, P.A. (2001) *Astron. Astrophys.*, **377**, 827.

67 Rybicki, G.B. and Lightman, A.P. (1986) *Radiative Processes in Astrophysics*, Wiley-VCH Verlag GmbH.

68 Longair, M.S. (1994) High Energy Astrophysics, in *Stars, the Galaxy and the Interstellar Medium*, vol. 2, 2nd edn., Cambridge University Press.

69 Liang, E.P.T. (1981) *Nature*, **292**, 319.

70 Perlman, E.S., Biretta, J.A., Zhou, F., Sparks, W.B., and Macchetto, F.D. (1999) *Astron. J.*, **117**, 2185.

71 Fraix-Burnet, D. and Pelletier, G. (1991) *Astrophys. J.*, **386**, 87.

72 Gomez, J.L., Alberdi, A., and Marcaide, J.M. (1994a) *Astron. Astrophys.*, **284**, 51.

73 Gomez, J.L., Alberdi, A., Marcaide, J.M., Marscher, A.P., and Travis, J.P. (1994b) *Astron. Astrophys.*, **292**, 33.

74 Hardee, P.E. and Clarke, D.A. (1995) *Astrophys. J.*, **449**, 119.

75 Hardee, P.E., Clarke, D.A., and Rosen, A. (1997) *Astrophys. J.*, **485**, 533.

76 Hughes, P.A., Aller, H.D., and Aller, M. (1985) *Astrophys. J.*, **298**, 301.

77 Hughes, P.A., Aller, H.D., and Aller, M. (1989) *Astrophys. J.*, **341**, 54.

78 Hughes, P.A., Aller, H.D., and Aller, M. (1991) *Astrophys. J.*, **374**, 57.

79 Mioduszewski, A.J., Hughes, P.A., and Duncan, G.C. (1997) *Astrophys. J.*, **476**, 649.

80 Cawthorne, T.V. and Cobb, W.K. (1990) *Astrophys. J.*, **350**, 536.

81 Bicknell, G.V. and Begelman, M.C. (1996) *Astrophys. J.*, **467**, 597.

82 Hardee, P.E. (1983) *Astrophys. J.*, **269**, 94.

83 Hardee, P.E. (2007) *Astrophys. J.*, **664**, 26.

84 Perlman, E.S., Cara, M., Harris, D.E., Adams, S.C., Bourque, M., Sparks, W.B., Biretta, J.A., and Cheung, C.C. (2011) *Astrophys. J.*, (in press) arXiv: 1109.6252.

85 Kirk, J.G., Rieger, F.M., and Mastichiadis, A. (1998) *Astron. Astrophys.*, **333**, 452.

86 Nakamura, M., Garofalo, D., and Meier, D.L. (2010) *Astrophys. J.*, **721**, 1783.

87 Lazzati, D., Ghisellini, G., and Celotti, A. (1999) *Mon. Not. R. Astron. Soc.*, **309**, L13.

88 Spada, M., Ghisellini, G., Lazzati, D., and Celotti, A. (2001) *Mon. Not. R. Astron. Soc.*, **325**, 1559.

89 Stawarz, L., Sikora, M., Ostrowski, M., and Begelman, M.C. (2004) *Astrophys. J.*, **608**, 95.

90 Blandford, R.D. and McKee, C.F. (1976) *Phys. Fluids*, **19**, 1130.

91 Dulwich, F., Worrall, D.M., Birkinshaw, M., Padgett, C.A., and Perlman, E.S. (2009) *Mon. Not. R. Astron. Soc.*, **398**, 1207.

92 Schwartz, D.A. *et al.* (2000) *Astrophys. J.*, **540**, L69.

93 Celotti, A., Ghisellini, G., and Chiaberge, M. (2001) *Mon. Not. R. Astron. Soc.*, **321**, L1.

94 Uchiyama, Y. and Coppi, P. (2010) *Astrophys. J.*, submitted.

95 McNamara, A.L., Kuncic, Z., and Wu, K. (2009) *Mon. Not. R. Astron. Soc.*, **395**, 1507.

96 Thomson, R.C., Mackay, C.D., and Wright, A.E. (1993) *Nature*, **365**, 133.

97 Röser, H.-J. and Meisenheimer, K. (1991) *Astron. Astrophys.*, **252**, 458.

98 Hardcastle, M.J., Croston, J.H., and Kraft, R.P. (2007) *Astrophys. J.*, **669**, 893.

99 Thomson, R.C., Crane, P., and Mackay, C.D. (1995) *Astrophys. J.*, **446**, L93.

100 Saxton, C.J., Bicknell, G.V., and Sutherland, R.S. (2002) *Astrophys. J.*, **579**, 167.

101 Perley, R.A., Röser, H.-J., and Meisenheimer, K. (1997) *Astron. Astrophys.*, **328**, 12.

102 Burke-Spolaor, S., Ekers, R.D., Massardi, M., Murphy, T., Partridge, B., Ricci, R., and Sadler, E.M. (2009) *Mon. Not. R. Astron. Soc.*, **395**, 504.

7
Observational Details: X-Rays

Rita Sambruna and Daniel E. Harris

7.1
Introduction

7.1.1
The Dawn

Entering the 1990s it was clear that extragalactic jets in radio galaxies were capable of accelerating particles to energies of $\gamma \sim 10^6$ as demonstrated by the detection of a few of the brightest jets at optical and X-ray wavelengths (see Chapter 6). The next obvious question, from an observational point of view, was whether X-ray detections could be achieved for fainter jets; the main expectation and prediction being that at these shorter wavelengths the jet X-ray emission would be weak because of synchrotron losses, producing high-energy cutoffs in the spectral distribution. Another possibility was that X-rays could be produced via thermal processes by the ISM heated by the jet itself through collisions. However, with the poor angular resolutions of the pre-Chandra X-ray observatories the chance of directly observing such thermal X-ray emission was low.

In the pre-Chandra era, X-ray observations of radio and optical jets were mainly performed with the *Einstein* and ROSAT satellites, the first two X-ray observatories designed for imaging. On both of these satellites, the gas proportional counters did not have sufficient angular resolution to distinguish jets from nuclei, but both missions had additional focal plane instruments, and in particular each had microchannel plates (dubbed "High-Resolution Imager", or HRI) with angular resolutions of the order of 6″ sufficient to resolve and detect the brighter jets in nearby sources. Already with *Einstein*, credible detections of the jets in Cen A, M 87, and 3C 273 were obtained. With somewhat better sensitivity, ROSAT observations increased the jet detections by a few, as well as detecting the hot spots of Cyg A and 3C 390.3.

The need for the angular resolution of the HRI detectors precluded X-ray spectral analyses, so the best one could do was to measure an X-ray flux and try to associate the X-ray morphology with known jet components. However, even without spectral

Relativistic Jets from Active Galactic Nuclei, First Edition. Edited by M. Böttcher, D.E. Harris, H. Krawczynski.

information, it was possible to rule out thermal processes for the genesis of the X-rays.

7.1.2
The Chandra X-Ray Observatory

The Chandra X-ray Observatory, one of NASA's Great Observatories, was launched in July 1999 and is still fully operational at this time (2011). Chandra operates in the energy range 0.2–10 keV, with imaging and spectroscopic capabilities. Its main advantage over the previous X-ray satellites is its improved angular resolution, and the ability to perform spatially resolved spectroscopy. The size of the CCD pixels on the ACIS camera, Chandra's primary imaging-spectroscopy device, is 0.5″, while the FWHM of the point spread function (PSF) on-axis is a bit less than 1″ at 1 keV. With this resolution, Chandra can easily separate two point-like X-ray knots at a physical distance $\sim$ 6 kpc in a jet at redshift $z = 0.7$. Together with its low instrumental background, the small PSF makes ACIS a milestone device for detecting and studying extragalactic jets.

A sense of the giant progress allowed by Chandra in the field of X-ray jets can be gleaned by comparing an image of the jet of M 87 as observed with the ROSAT HRI to the Chandra ACIS view of the same jet (Figure 7.1). It comes as no surprise, then, that the field of X-ray emission from AGN jets underwent a renaissance period after the launch of Chandra, with several new discoveries that have changed our perspective. A growing list of X-ray jets is maintained at http://hea-www.harvard.edu/XJET/ (accessed 8 September 2011) see also [1].

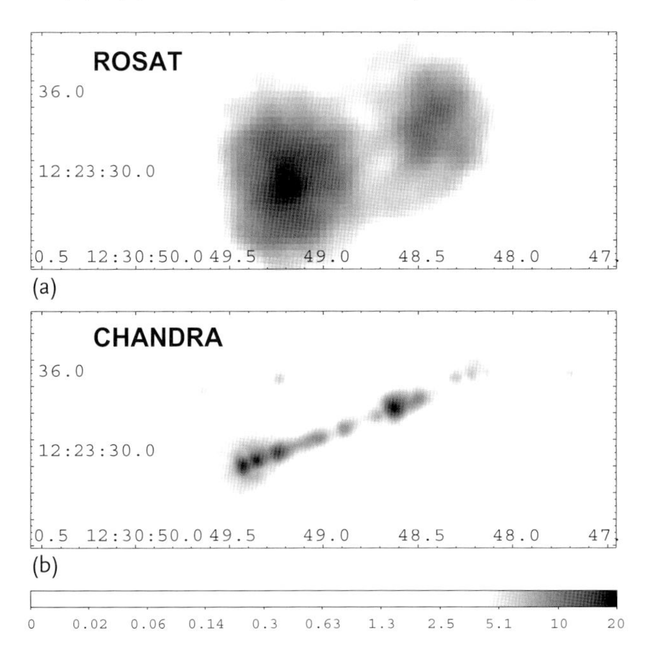

Figure 7.1 The M 87 jet as observed with the ROSAT HRI (a) and with Chandra ACIS (b).

In the rest of this chapter we describe the observations and theories that were spurred by the last 10 years of Chandra's operations. For the sake of discussion, we will divide the sources in three main groups based on apparent X-ray luminosities: high ($L_{X,\text{jet}} \gtrsim 10^{45}\,\text{erg}\,\text{s}^{-1}$), low ($L_{X,\text{jet}} \lesssim 10^{43}\,\text{erg}\,\text{s}^{-1}$), and intermediate ($10^{43} \lesssim L_{X,\text{jet}} \lesssim 10^{45}\,\text{erg}\,\text{s}^{-1}$). The grouping is made purely on the basis of convenience of discussion, and roughly corresponds to the division into FR IIs, FR Is, and "intermediate" sources, respectively. Interestingly, as we will see below, the properties of the high- and low-luminosity groups are quite distinct, while the intermediate sources are somewhat mixed. For a discussion see [2].

7.2 X-Ray Jets at Higher Luminosities

7.2.1 The First Chandra Jet

The first Chandra detection of an X-ray jet was a surprise. The very first celestial source observed with Chandra, the distant ($z = 0.651$) quasar PKS 0637–752, had been chosen as a point source for calibration and in-focus measurements of the telescope. However, after the first 2 ks exposure with ACIS-S, it became apparent that the quasar was *not* point-like – a fuzzy source of X-ray photons to the west of the core was present on the image, connected to the center by a tenuous bridge. PKS 0637 marked the first serendipitous discovery of an extragalactic X-ray jet with *Chandra* [3].

The properties of the quasar jet were striking. First, the jet is large. Figure 7.2 shows the Chandra ACIS image of PKS 0637 obtained by stacking several exposures, for a total observing time of 100 ks [4]. The X-ray jet is apparent, extending $11''$ to the west of the bright, unresolved core, and consisting in this deeper image of a narrow bridge ending in a large structure resolved into several knots (labeled in the figure; WK5.7, and so on).[1] On a physical scale the jet angular size corresponds to ~ 100 kpc, well outside the host galaxy.

Second, the X-ray emission matches closely the radio. Figure 7.2b shows the ATCA 8.6 GHz radio image of PKS 0637, smoothed to a resolution comparable to that of ACIS. There is an almost one-to-one correspondence between the X-ray and radio maps up to the $11''$ knot, where the jet bends in the radio and goes dark at X-rays. Assuming the properties of the emitting medium remain constant, the disappearance of the X-rays after the bend suggests that beaming may play a more important role at X-rays than at longer wavelengths.

Third, the jet is relatively weak in the optical compared to the radio and X-rays. An archival HST image showed only weak emission from the three main knots

1) The labeling convention used here is the following: WK = west knot, while the numbers represent the distance from the nucleus to the knot in arcseconds as measured in the X-rays.

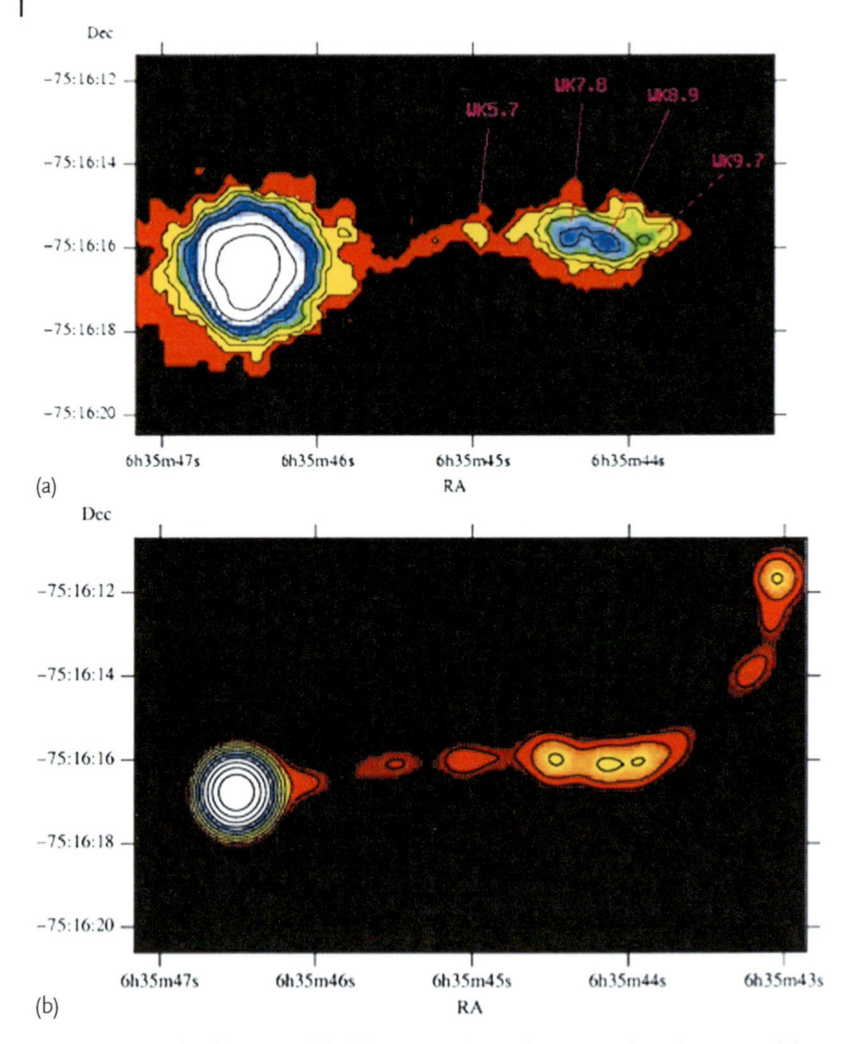

Figure 7.2 (a) Chandra image of the Mpc-scale X-ray jet in the $z = 0.651$ radio-loud quasar PKS 0637–752, obtained by stacking several short (2–3 ks each) ACIS-S exposures deconvolved using a maximum likelihood technique, for a total final exposure around 100 ks. The effective resolution for the stacked deconvolved image is 0.37″ (σ of the equivalent Gaussian). The labels indicate the knots detected at X-rays. (b) Radio image of the jet obtained with the Australian Telescope Compact Array (ATCA) at 8.6 GHz, restored with a circular beam of 0.84″ FWHM. Note the almost one-to-one correspondence between the X-ray and radio emission features. With the assumed cosmology, $1'' = 6.9$ kpc. Reproduced from Chartas *et al.* [4] with permission from the AAS.

at $\sim 10''$ [3]. With its larger than expected X-ray to optical flux ratio, the jet in PKS 0637 clearly represented a new phenomenon.

What is the origin of the prodigious X-ray emission from the jet of PKS 0637? A thermal origin from the galaxy ISM, that is, bremsstrahlung emission, seemed

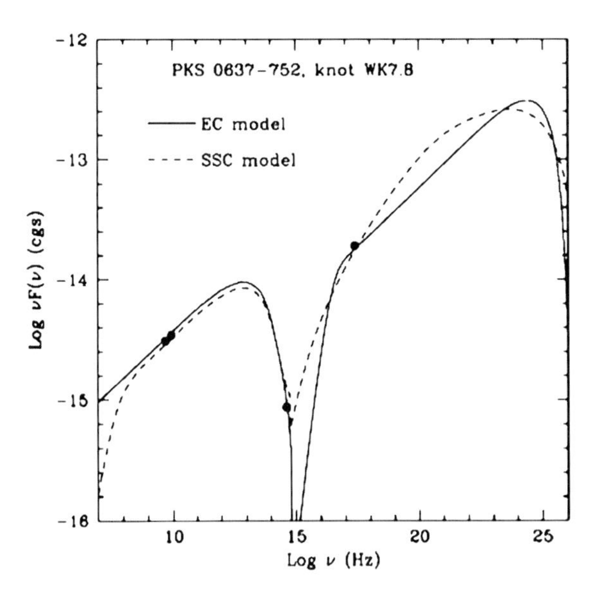

Figure 7.3 The spectral energy distribution (SED) of the knot WK7.8 in the jet of the quasar PKS 0637-752. The filled circles represent the observed radio (4.8 and 8.6 GHz from ATCA), optical (archival HST), and X-ray (Chandra) noncontemporaneous fluxes. A new feature of this SED compared to X-ray jets in the pre-Chandra era is its *upward-curved* optical-to-X-ray continuum. This indicates that the X-ray flux cannot originate from the same emission process as the radio-to-optical one. The dashed and continuous lines are fits with models incorporating an inverse-Compton origin for the X-rays, with seed photons either internal to the jet (SSC) or external (EIC). Reproduced from Tavecchio *et al.* [5] with permission from the AAS.

improbable because the measured X-ray flux would imply an unreasonably high electron density and a larger than observed rotation measure [3]. The only remaining possibility was nonthermal emission from the jet plasma itself.

The jet spectral energy distribution (SED) is shown in Figure 7.3 [5]. The observed fluxes are plotted with filled circles. A common zero-order diagnostic of emission processes, derived from the multiwavelength studies of blazars, is offered by the shape of the radio-to-optical and optical-to-X-ray continua: generally, if the X-rays lie on the extrapolation of the radio-to-optical continuum, synchrotron from the same electron population is accepted as a common emission process. If, however, the optical-to-X-ray continuum shows an upturn, other processes are invoked for the high-energy emission, such as inverse-Compton scattering of ambient photons off the jet electrons: either the same synchrotron photons internal to the jet (synchrotron self-Compton, SSC) or external photons from the disk, broad-line region, torus, and so on (external inverse-Compton, EIC).

As shown in Figure 7.3, the X-ray emission from the jet of PKS 0637 cannot originate via synchrotron from the same electron population producing the radio-to-optical wavelengths (solid and dashed lines), as the predicted X-ray flux would fall short of the observed one by several orders of magnitude. An inverse-Compton (IC) origin for the X-ray emission seemed inevitable.

A straightforward application of the theory showed that the standard SSC model was disfavored because it would imply a jet strongly out of equipartition, that is, strongly dominated by the energy of relativistic particles with an increase of the total (particle + magnetic) energy of a factor 1000 [3]. A more analytic approach that takes into account a simplified jet geometry and makes assumptions on the particle and magnetic field structures [5] shows that the SSC model fails also to reproduce the shape of the X-ray spectrum (dashed line in Figure 7.3).

Looking at external sources of photons, the large distance between the jet and the core excluded contributions from the stellar and nuclear radiation of the quasar [3, 6]. An alternative source of photons was needed.

7.2.2
A "New" Model: IC on the Cosmic Microwave Background Photons

Among the various external sources of seed photons, the cosmic microwave background (CMB) photons are particularly important. This is because, at the source's redshift z, their energy density is boosted by a factor $(1 + z)^4$; moreover, in the frame of the relativistic scattering particles in the jet, the photon energy density of the CMB is enhanced by a factor Γ^2, where Γ is the bulk Lorentz factor of the emitting plasma. For a discussion of the basic physics see Chapter 3; a detailed review and formulae derivation is given by Dermer [7]; see also [8].

Inverse-Compton scattering off the CMB photons as a source of cosmic X-rays in not a new model, having been used since the 1990s to explain X-ray emission from the giant lobes of radio galaxies detected at much lower spatial resolution, for example [9]. In this scenario, X-rays are produced by relativistic electrons with Lorentz factors $\gamma \approx 1000$. For the magnetic field strengths thought to characterize most radio lobes, electrons with these sort of energies emit synchrotron emission below 100 MHz, or even below the ionospheric cutoff, so the IC X-ray spectrum is sampling a segment of the electron spectrum not normally available to sensitive radio telescopes.

Figure 7.3 shows an application of the IC/CMB model to the SED of the PKS 0637 jet (solid line). The X-ray flux and spectrum are satisfactorily reproduced [5]. Overall, the synchrotron + IC/CMB model is a good representation of the SED of the jet and predicts conditions compatible with equipartition for a moderate value of beaming, $\delta \sim \Gamma_L \sim 10$, where Γ_L is the Lorentz factor of the plasma. The implied jet power is $\sim 10^{48}\,\mathrm{erg\,s^{-1}}$ [5], similar to the most powerful blazars, assuming a minimum Lorentz factor of the electrons $\gamma_{\mathrm{min}} \sim 10$. The latter value is set by the observed optical flux.

The discovery of the PKS 0637 jet with Chandra, and its unusual properties, raised the important question of whether this source was one of a kind, or a "prototype" of a new class of extragalactic X-ray jets. Additional jets started to be discovered serendipitously as AGN were observed as targets of programs with different science goals (e.g., Miller *et al.* [10], observations of a radio-loud broad absorption line quasar). Systematic surveys of samples of radio jets with Chandra, HST, and optical ground-based telescopes were initiated [11–13].

The surveys turned out to be very efficient in detecting PKS 0637-like jets even with short, snapshot-like exposures (2–3 ks), with a detection rate of X-ray counterparts to radio jets of roughly 50%. Important to note is that the samples targeted in these surveys were mostly bright, one-sided jets of quasars, thus introducing a strong bias for beaming. For example, all but one of the sources in the sample of Sambruna *et al.* [12] are core-dominated; in the only lobe-dominated source of the sample, PKS 0836+299, no X-ray counterpart to the radio jet was detected.

To study the jet properties in a quantitative way, the optical observations proved to be crucial. In all cases, the jet X-ray flux lay above the extrapolation from the radio-to-optical continuum, indicating a separate spectral component [12, 14]. Modeling of the high-energy emission in terms of IC/CMB implied plasma Lorentz factors $\Gamma \sim 3-15$ on kpc scales and relatively large total (magnetic + particle) jet powers, $L_{\rm jet} \sim 10^{48}\,{\rm erg\,s^{-1}}$, as in the case of PKS 0637, assuming equipartition and a proton-per-electron jet composition. The large bulk Lorentz factors derived from the SEDs were in contrast with those derived from radio observations, $\Gamma \sim 2$ [15, 16]. To reconcile the discrepancy, it was proposed that powerful quasar jets have a fast-moving spine responsible for the X-ray production, surrounded by a slower-moving sheath where the synchrotron radio photons are emitted [6]. Evidence for limb-brightening in jets is independently found at radio wavelengths [17].

The modest exposures of the survey observations did not permit the acquisition of X-ray spectra and jet radial profiles; thus limiting quantitative information on jet structure and physical processes. The best candidates were selected for deeper follow-up observations with longer exposures, typically $\sim$ 100 ks (e.g., [2, 14, 18–20]). The observed X-ray spectra were described by power laws with slopes well matching the predictions of the IC/CMB model, $\alpha_X \sim 0.5$. An interesting implication of these hard continua is that we are currently only sampling the rising portion of the high-energy component of their SED, and measuring a fraction of the total luminosity. Ideally, the Fermi Gamma-Ray Space Telescope should be able to set constraints on the high-energy cutoff (e.g., [21]), but its angular resolution is not sufficient to distinguish between the quasar nuclei and their jets.

The longer Chandra follow-up exposures permit the study of individual knots and the comparison of jet profiles at various wavelengths. In a few cases, for example, 3C 273 and 1136–135, it was found that while the radio flux increased going from the beginning of the jet to its end, the opposite was true at the higher energies – the X-ray flux was highest closer to the core and faded toward the end of the jet. In the optical, a somewhat intermediate profile was observed with similar emission levels from all the knots.

A simple explanation for the radio and X-ray profile behaviors is found in the context of the synchrotron+IC/CMB model if the plasma decelerates along the jet due to entrainment of ambient gas [12, 22, 23]. A similar idea had been previously discussed to account for the radio morphologies of FR I and FR II radio galaxies [24]. Indeed, if the bulk Lorentz factor decreases along the jet, the CMB radiation density will decrease in the reference frame comoving with the emitting plasma because of its dependency $\propto \Gamma^2$. This will yield a lower X-ray flux. At the same time, plasma compression will enhance the magnetic field and thus the synchrotron radio emissivity will increase by B^2 (see Chapter 3). For more details on jet deceleration, see [22, 23].

The synchrotron+IC/CMB model, although widely used, became the subject of scrutiny and criticism due to some observational shortcomings and other models were proposed (see below). In the next section we discuss some of the main challenges faced by the IC/CMB interpretation.

7.2.3 Challenges for the IC/CMB Model

It was immediately apparent that one of the challenges of the IC/CMB model was the difficulty to explain the relative sizes of the radio and X-ray knots. Already in the PKS 0637 jet, comparison of the width of knot W7.8 (Figure 7.4) at the two wavelengths showed that the X-ray intensity dropped faster than the radio. However, in the IC/CMB the X-rays are produced via scattering off a segment of the electron spectrum with $\gamma \approx 100$, whose synchrotron peak frequency is around 1–20 MHz for plausible values of their Lorentz factor and the magnetic field [8]. Recalling the equation of the synchrotron energy-loss time scale, $t_{\rm el} \sim \gamma^{-1}$, it can be seen that the lifetime of these electrons must be very long, in excess of 10^4 years. Observationally, we thus expect that the X-ray emitting knots, which trace the lower energy electrons, should be more extended than their radio (GHz) counterparts, opposite to what is observed (Figure 7.4).

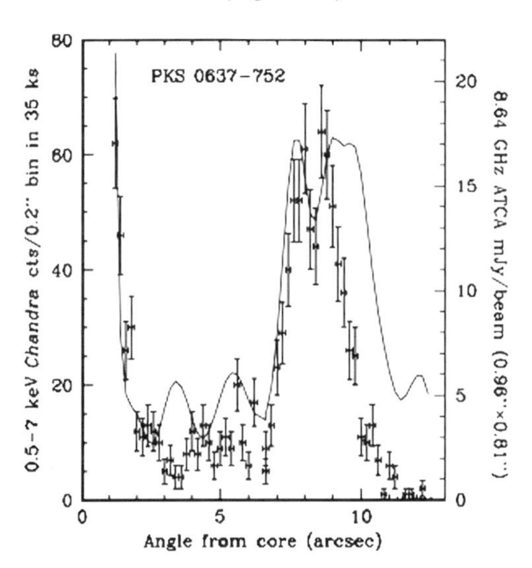

Figure 7.4 Comparison of the profiles at X-ray (data points) and radio (line) for the jet of PKS 0637-752 (from [8]). It illustrates how the X-ray flux drops faster than the radio at 10″ from the core. This is a problem for the IC/CMB model where the lifetime of the scattering electrons is expected to be very long, > 10^4 years. With kind permission from Springer Science + Business Media [74].

To circumvent this problem, strong subclumping of the emission region was proposed [25], with each smaller subregion expanding adiabatically. The crucial point of this model is to assume a more efficient cooling process than the radiative one for the emitting regions. This is achieved by allowing the region to be fragmented: each substructure (clump) then can move faster than the whole knot and dissipate energy more efficiently. The slower bulk motion implies a reduced X-ray yield via IC/CMB. However, as pointed out by Worrall [8], the need for IC/CMB would be unclear as the SSC process would be favored for X-ray production in such a slow jet.

A few of the largest quasar jets (PKS 0637, 3C 273, PKS 1136-135) were observed in the IR at moderate angular resolution with *Spitzer* [26–28]. The radio-to-IR continua of the brightest resolved knots are described by a single power law with a cutoff around 10^{13} Hz, while the optical emission appears to be consistent with the extension of the X-ray continuum provided by Chandra. This poses a problem for the IC/CMB model, where the electrons producing the X-rays should have $\gamma_{\rm min} \sim 10$ and thus the SED should show a cutoff between the optical and X-ray regions. Instead, a synchrotron origin from the same particle population, separate from the one producing the radio-to-IR, is advocated for the optical-UV-X-ray emission of these jets [26–28].

A point in favor of the synchrotron+IC/CMB model is the recent detection of several X-ray jets at cosmological redshifts, $z > 1$. Indeed their existence is a clear prediction of the model, as we will investigate later in Section 7.2.5. However, despite its success in accounting for several observed features of the jets, critical issues remain open for the IC/CMB interpretation. Alternative scenarios have been presented in the literature, as we discuss in the next section.

7.2.4 Alternative Scenarios to the IC/CMB

The first class of alternative scenarios include synchrotron emission from a separate population of electrons. Requiring a second particle population either in space or energy is necessary because the primary radio-to-optical synchrotron continuum in the SEDs cuts off around the UV energy range leaving the X-rays as a clearly separate spectral component.

Indeed, synchrotron X-ray emission is observed directly in some nearby FR IIs such as Pictor A [29], and others [30–32] . Moreover, synchrotron dominates in the inner knots of some of the IC/CMB jets of powerful FR II quasars [14]. There is a debate as to whether knot A in 3C 273 emits X-rays predominately via synchrotron [33] or IC/CMB [34], although Jester *et al.* [35] found an upturn in the UV favoring the former interpretation. Finally, FR I jets produce X-rays mainly via synchrotron (see below), although in these sources the X-ray flux is a smooth extension from the lower energies and does not require a separate particle population.

Another possibility is synchrotron emission from *protons* in the jet [36]. This model can successfully reproduce the X-ray flux of 3C 273 and offers a way to eliminate the need of multiple electron populations. However, it requires very large magnetic fields of order several Gauss, difficult to produce on scales of hundreds of kiloparsecs and a very large total energy budget.

An elegant alternative was proposed by Dermer and Atoyan [37]. In their scenario, a single population of electrons emits radio through optical via synchrotron in the jet; however, because of inefficient IC scattering in the Klein–Nishina regime, a hard tail develops at the higher energies in the electron distribution, which is responsible for the X-rays. In this model, the hard X-ray continuum observed in PKS 0637, 3C 273, and similar jets would be produced via synchrotron by the same electron population, but with a distorted spectrum. This model requires that IC losses are larger than synchrotron losses.

In summary, the origin of the copious X-ray emission from powerful FR II quasar jets remains open, and of central importance for the physics of these sources. Future observational tests will rely on theoretical progress on emission mechanisms incorporating realistic jet structures and more details from the observations. Specific tests will be enabled by the advent of technological advances. For instance, high-resolution, low-frequency radio telescopes such as LOFAR, SKA, and so on will allow us to evaluate synchrotron emission from the low-energy electrons responsible for up-scattering the CMB photons. In the optical, the structure of knots and possible clumpiness will be probed by JWST and the next generation of large, ground-based facilities.

7.2.5 Jets at High-z

A prediction of the IC/CMB model is that X-ray jets should be common at large redshifts. Assuming the physical properties (magnetic fields, Doppler factors, electron energy distributions) are similar at high- and low-z, jets should be bright at X-rays compared to the radio because the increase of the CMB density compensates for the distance effect, while the radio emission follows the usual decline due to distance [38]: $f_X/f_R \propto U_{CMB} \propto (1+z)^4$.

Indeed, X-ray counterparts of radio jets at $z > 3$ were discovered, mostly serendipitously (see [39, 40] and references therein). Their X-ray to radio flux ratios are shown in Figure 7.5 for the cases where this ratio could be clearly determined. Also shown are the values for the jet of PKS 1136-135 [14] at $z = 0.5$, chosen to be representative of the spread within the jet population at $z \lesssim 1$. The curved lines represent two values of the bulk Lorentz factor and magnetic field strength approximately encompassing the parameters at low redshift.

The dashed lines in Figure 7.5 represent the expected behavior of f_X/f_R with z according to the IC/CMB model for given values of the model parameters. Indeed, the highest-z (> 3) jets tend to have larger total f_X/f_R than nearer sources [41] despite the expected spread in intrinsic parameters. Thus, at least qualitatively, it appears that a basic prediction of the IC/CMB model – that quasar jets have increasing X-ray to radio flux ratios with increasing redshifts – is supported by the evidence.

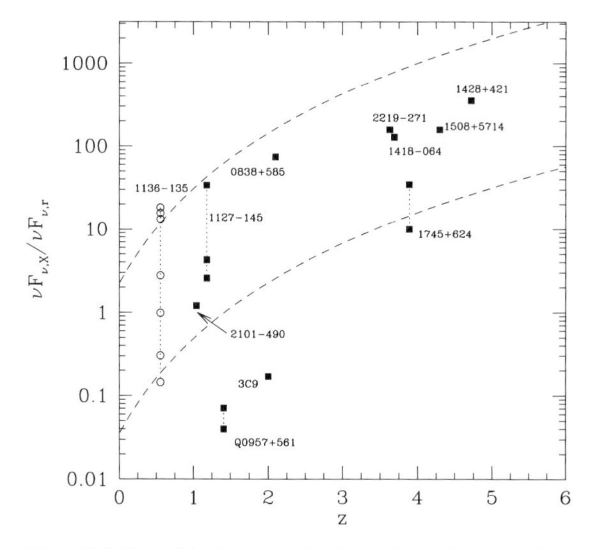

Figure 7.5 Plot of the X-ray to radio flux ratios of high-redshift ($z > 1$) jets serendipitously discovered with Chandra. The nearby jet of PKS 1136-135 at $z = 0.5$, chosen to be representative of the spread within the jet population at $z \lesssim 1$, is shown for comparison. The curved lines show the $(1 + z)^4$ behavior of the flux ratio expected for a jet feature of fixed parameters (i.e., constant B, $N(\gamma)$, and Γ) if it were to be seen at various distances. The upper and lower curves were chosen so as to accommodate the spread of intrinsic parameters at low redshift. A basic prediction of the IC/CMB model is that X-ray jets should be found at increasingly larger z; the data confirm this expectation, albeit showing a large scatter of X-ray to radio flux ratios.

A complicating issue, however, is introduced by the possible effects of the environment through deceleration. As mentioned in a previous section, deceleration is believed to be responsible for decreasing f_X/f_R along the same jet first observed in 1136-135 and 3C 273 [14, 22, 23, 34]. Deceleration is probably due to mass loading via entrainment of the ambient gas (e.g., [42]). Significant deceleration sets in when the entrained mass is of the order of m_{jet}/Γ, with m_{jet} being the jet mass. In these models, the jet power is dominated by the bulk kinetic energy of protons. This implies that high-power jets can travel long distances from the core and reach the hot spots unperturbed, consistent with the X-ray morphology observed in many powerful FR II jets. On the other hand, low-power jets are quenched closer to the core, in agreement with the X-ray morphologies of FR Is where the X-ray jet is shorter than at radio (see below). Due to the increasing IGM density, the effects of deceleration on large scales should be increasingly important with increasing distance from the central engine. A detailed discussion of deceleration can be found in [22] and references therein.

Owing to deceleration effects, which change the radio and X-ray morphologies, it is thus difficult to compare the jet structure at the two wavelengths in $z > 3$ jets as the radio jets are generally short and faint [43], due to the distance. An optimal experiment would be to find a few sufficiently long, bright, and knotty radio jets for observation with Chandra, in order to properly test the IC/CMB model. Such jets are rare.

A Chandra-HST program targeting intermediate redshift ($z \sim 2$–2.5) jets, optimized to taking into account possible deceleration effects, is under way (Cheung *et al.* 2012, in preparation). All jets in the survey were detected at X-rays, with the X-ray knots coinciding with the brightest radio ones.

7.3 X-Ray Jets at Lower Luminosities

7.3.1 Morphologies and Emission Process

At lower apparent luminosities, X-ray jets are usually associated with low-power radio galaxies of the FR I class. We know from the radio that the jets in these sources are often double-sided, poorly collimated, and brighter near the core. They start out weak and present a "flaring" point at which the radio emission brightens; a classical example is 3C 31 [44], also detected at X-rays [45].

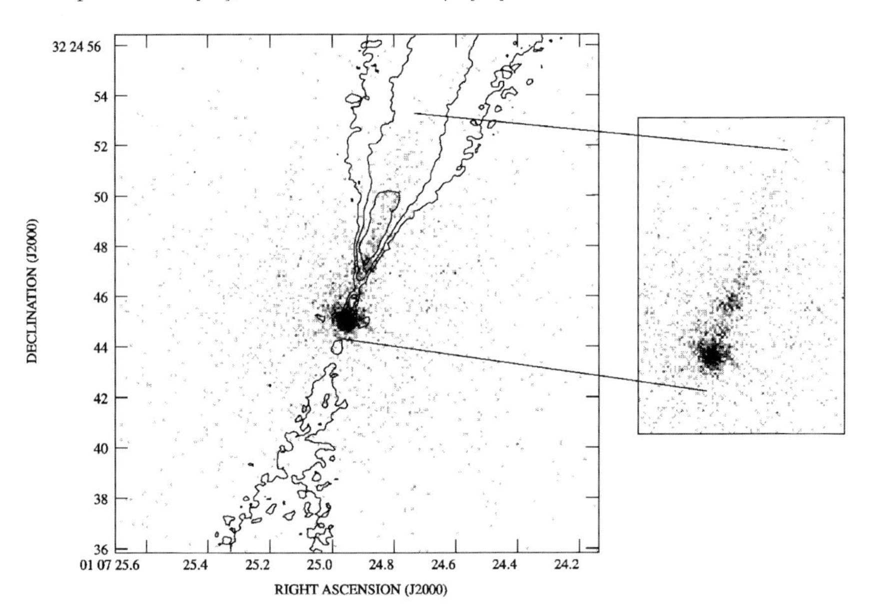

Figure 7.6 The X-ray jet of 3C 31, a prototypical FR I, from a Chandra observation (b), and with 8.4 GHz radio contours overlaid (a) [45]. As in many low-power radio galaxies, the X-ray morphology observed by Chandra correlates well with the radio, peaking at the so-called radio "flaring" point at a few kiloparsec from the core and fading downstream. The smooth SED strongly supports a synchrotron origin for the X-rays from high-energy electrons. Reproduced from Hardcastle *et al.* [45] with permission from the Royal Astronomical Society.

Early observations with Chandra found FR I jets to be X-ray emitters [46]. Their X-ray spectra are well described by a single power law, and in general the X-ray flux lies on the extrapolation of the optical continuum, supporting the idea that the same electron population emitting the radio and optical radiation via synchrotron is also responsible for the X-rays. The X-ray spectra are soft, $\alpha_x \gtrsim 1$ [46].

The jet X-ray morphology is illustrated in Figure 7.6, where the ACIS image of the archetypal FR I 3C 31 is shown with the radio contours overlaid (from XJET). The X-rays appear to be in good correlation with the radio, peaking at the radio flaring point and fading afterward. Optical counterparts are generally weak or absent, with typical upper limits to the flux densities from sensitive HST observations.

Evidence for diffuse gas around the cores is found, but generally the thermal emission is too weak to contribute significantly to the jet X-ray emission.

7.3.2 A Case Study: M 87

M 87 is a relatively uncomplicated giant elliptical at the center of the Virgo Cluster. It has a bright jet (radio/optical/X-rays) which terminates in the western (inner) radio lobe after about 2 kpc; that is, still within the optical galaxy, at least in projection. Because M 87 is relatively nearby, the spatial scale is 77 pc arcsec^{-1}, affording good spatial resolution of the 25″ long jet. Since the radio image shows inner lobes on both sides whereas only the western side has a jet, we can assume that there is an invisible (hidden by Doppler beaming) jet on the eastern side as well. Thus we deduce that the beaming factor, $\delta \geq$ a few, and the angle to the line-of-sight, $\theta \leq 50°$. The basic X-ray related properties of the jet were based on early Chandra observations [47–49].

There is also radio emission on larger scales and since the complex radio morphology does not lend itself to a simple interpretation of the relationship between the inner and outer radio structures, we suspect that the large-scale radio emission may lie mostly along our line-of-sight, producing strong projection effects.

7.3.2.1 Morphology

Before describing the sizes and shapes of the jet knots, it is important to remember that we are limited in angular resolution: roughly 1″ in the X-rays (Chandra); something a bit better than 0.1″ in the optical/UV (HST); and of order 0.01″ in the radio (VLBA). Thus care must be taken when comparing images from different bands, and also when attempting to deduce the sizes of emitting volumes. Figure 7.7 illustrates our dilemma: are the knots of M 87 actually comprised of many smaller emitting volumes within each knot? The answer is probably a qualified "yes" and "no". "Yes", as illustrated by Cen A, "no" as deduced by the failure of VLBA to detect very small, high brightness features within radio knots (although see Section 7.3.2.3).

Figure 7.8 illustrates the following aspects. At arcsec resolution we are presented with an inner string of quasi-regularly spaced knots, followed by the much wider knot A at 12″. Downstream of knot A are further complex emission regions, not

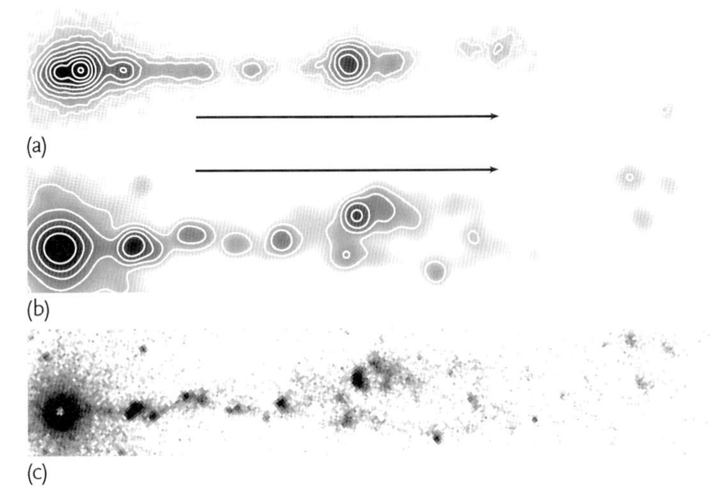

Figure 7.7 A comparison of the M 87 jet with that of Centaurus A if viewed by *Chandra* at the distance of M 87: (a) a grayscale Chandra image (contours increase by factors of two), (b) the Cen A jet, smoothed so as to have the same spatial resolution, and (c) the event file for Cen A. Notice how every jet knot of Cen A actually is comprised of several much smaller emitting regions. All images have been rotated so as to align the jets horizontally. The scale bars are 1 kpc long.

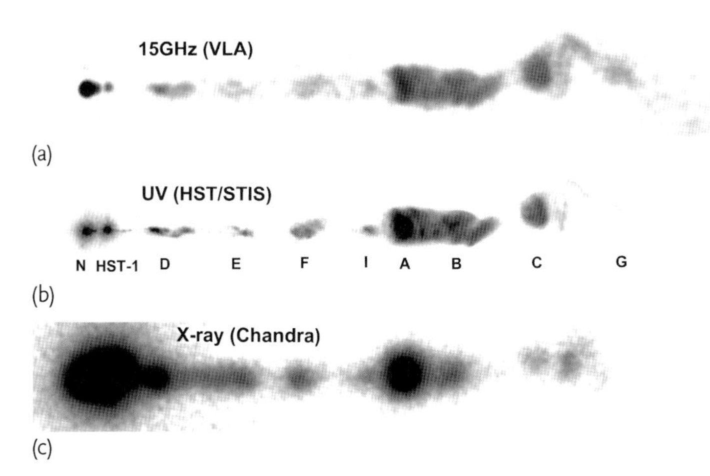

Figure 7.8 Comparative images of the M 87 jet: (a) a 15 GHz VLA observation, and (b) an image constructed by stacking many HST/STIS images (fits file kindly supplied by J. Madrid; see also [50]). The labels are historical. (c) A stacked image from over 60 ACIS-S observations made with 0.4 s frame times. Each observation was about 5 ks long. There is close agreement between the three bands except for the region just beyond knot C where there is substantial X-ray emission, but the radio and optical emissions detour around this area. The saturated region around HST-1 in (c) comes from the inclusion of many observations in 2005 when the giant flare of HST-1 peaked. The radio image was obtained in 2003 when HST-1 was still relatively faint. For other views of this jet see Figures 6.2 and 6.3.

clearly separated from each other, then a sudden excursion to the north before merging into the W lobe. At higher resolution we see that the inner knots are worm-like, wiggling back and forth within a cone of opening angle about 10° (this cone was defined by the VLBA inner jet). We also see that the upstream edge of knot A has a high gradient in brightness as does the downstream edge of knot C. If the M 87 jet were to be similar to some models of GRB jets, it might be tempting to identify knot C as a forward shock, coupled with knot A being a reverse shock.

7.3.2.2 **Spectral Properties**

The SEDs of the knots are generally consistent with one-zone synchrotron models: that is, concave downwards. Figure 7.9 shows the spectra of knots A, B, and D. Note the spectral steepening between the optical and X-ray which progressively increases moving out along the jet: D, A, B.

The value of α_x also increases progressively moving down the jet although at knot A, the spectrum hardens a bit. The general case of spectral steepening moving down the jet will be discussed in Section 7.6.1, but we note that M 87 displays both this trend and the associated behavior of "offsets" (the progressive downstream displacement of peak intensity of a knot with decreasing frequency). This morphology is shown in Figure 7.10.

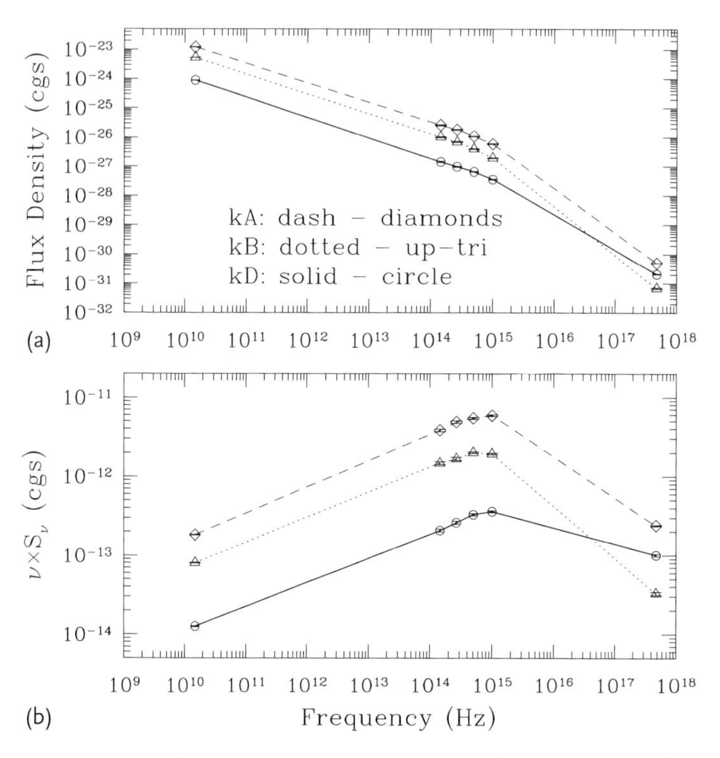

Figure 7.9 Spectra for knots D, A, and B in the M 87 jet: (a) the usual plot of log flux density vs. log frequency, and (b) shows the same data, but plotted as $\log(\nu * S_\nu)$ vs. $\log \nu$.

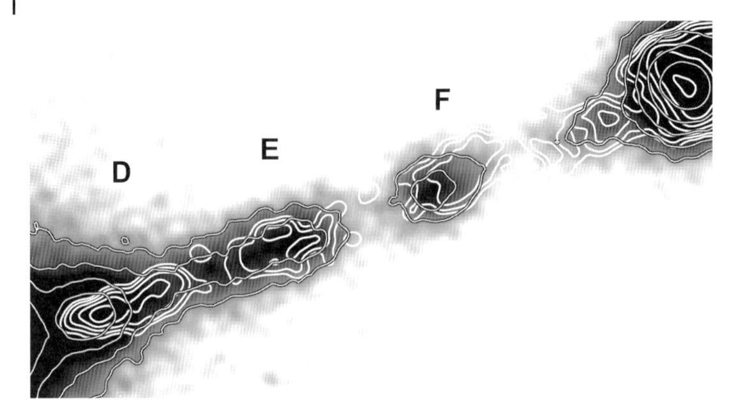

Figure 7.10 The central section of the M 87 Jet. The white contours are from a VLA image at 15 GHz (slightly better resolution than the X-ray image). The grayscale image and the contours of mixed black and white are from Chandra. Both sets of contours increase by factors of two. The offset between the peak brightness of X-ray and radio emissions is most evident in knot F. In knot D, the X-ray brightness peaks at the upstream end of the knot, and for the western half of the knot, drops to a brightness level roughly equivalent to that of the interknot region between D and E.

7.3.2.3 **Variability**

For radio-loud AGN, a major component (perhaps the dominant component) of the nuclear X-ray emission most likely comes from the inner (unresolved) jet rather than from an accretion disk or corona. Thus, the fact that most X-ray nuclei of AGN are variable, provides constraints on the sizes of emitting regions of the inner jet. The exquisite example of HST-1 in the M 87 jet is another case where we can derive reasonable estimates of physical parameters. Combining the variability time scale (which sets limits on the size of the emitting volume modified by the beaming factor) with the equipartition assumption and intensity information yields a consistent set of physical parameters [51]. Although we cannot be certain of the ultimate causes of the giant flare of HST-1 (which peaked in 2005 at over 50 times the intensity observed in 2000), the behavior of the lightcurves at radio, UV, and X-rays were consistent with a synchrotron broadband emitting region with size of order 20 light days or less, a beaming factor of 5 ± 2, a magnetic field strength of order 1 mG, and a power-law spectrum of electrons extending up to $\gamma \approx 10^7$ with a slope $p = 2\alpha + 1 \geq 3.4$. The decay of the lightcurve is thus consistent with synchrotron emission; IC/CMB X-ray emission would not explain the longer decay times at lower frequencies.

As part of the monitoring campaign on M 87, the VLBA was employed at 1.6 GHz [52]. Superluminal motions were found for components of HST-1: apparent velocities of $4c$, moving downstream were observed. It was also noticed that the upstream end of the complex was at essentially the same location as that found in the 1990s when apparent velocities of $6c$ were found for optical components [53]. Finally, the L band VLBA data showed the appearance and subsequent fading of a new component at the upstream end in 2006. The lightcurve of this component matched the secondary X-ray flare which occurred during the decay phase of the primary flare. Since HST-1 lies near the northern edge of the cone of the (here mostly invisible) jet and would intercept less than 1% of the cross-sectional area, it is speculated that the flare is a temporal example of the transfer of jet power to an emitting plasma.

7.4 X-Ray Jets at Intermediate Luminosities

7.4.1 Detection of X-Ray Jets in BL Lacs

High-energy emission was also detected from the jets of a few BL Lacertae sources [54, 55]. This may sound surprising, as according to the classical definition, in BL Lacs the jet is seen almost end-on and because of foreshortening effects these sources should thus appear point-like. Detection of extended structures in BL Lacs is then a rare occurrence because of the need for jet bending. According to unification scenarios, BL Lacs and FR Is are the same type of AGN seen at small and larger jet angles, respectively. It is then reasonable to expect that in the former, extended jets should exhibit morphologies and SEDs similar to those of their parent galaxies.

Indeed, in the first BL Lac jet observed with Chandra, PKS 0521-365, X-ray emission was detected with properties reminiscent of an FR I [56]. A synchrotron optical jet was already known from HST and ground observations by the time of the Chandra observation [57, 58], indicating the presence of a population of electrons with $\gamma \sim 10^6$ at kpc scales. The ACIS image (Figure 7 of [56]) shows that the X-ray morphology is reminiscent of an FR I jet, with the X-rays peaking at the radio flaring point and decaying thereafter. Overall, the radio, optical, and X-ray morphologies are close to each other as expected if the same population of electrons emits synchrotron at all three wavelengths.

The brighter knot X-ray spectrum is well described by single power-law model with $\alpha_X \sim 1.4$. The radio-to-X-ray SED has a break of $\Delta\alpha \sim 0.7$, larger than for continuous-injection models, for which $\Delta\alpha \sim 0.5$, and inconsistent with the exponential cutoff expected for single-injected, aged electron spectrum [56]. As in the case of FR Is (see above), local reacceleration is occurring, as required to sustain the production of X-ray photons.

X-ray counterparts to previously known synchrotron optical jets were also detected in 3C 371 and PKS 2201 [14, 59]. Figure 7.11 shows a gallery of the multiwavelength images of 3C 371 from radio (with MERLIN and VLA), IR, optical, and UV (with HST), and X-rays, illustrating the complex morphology of the jet at the higher resolutions. From radio to IR/optical and UV, the jet emission appears concentrated in a few bright knots at $0.9''$ (knot α) and $1.3''$ (knot β), then fades somewhat around knot B at $1.8''$, and picks up again in the lobe near knot A, which is itself fragmented into substructures.

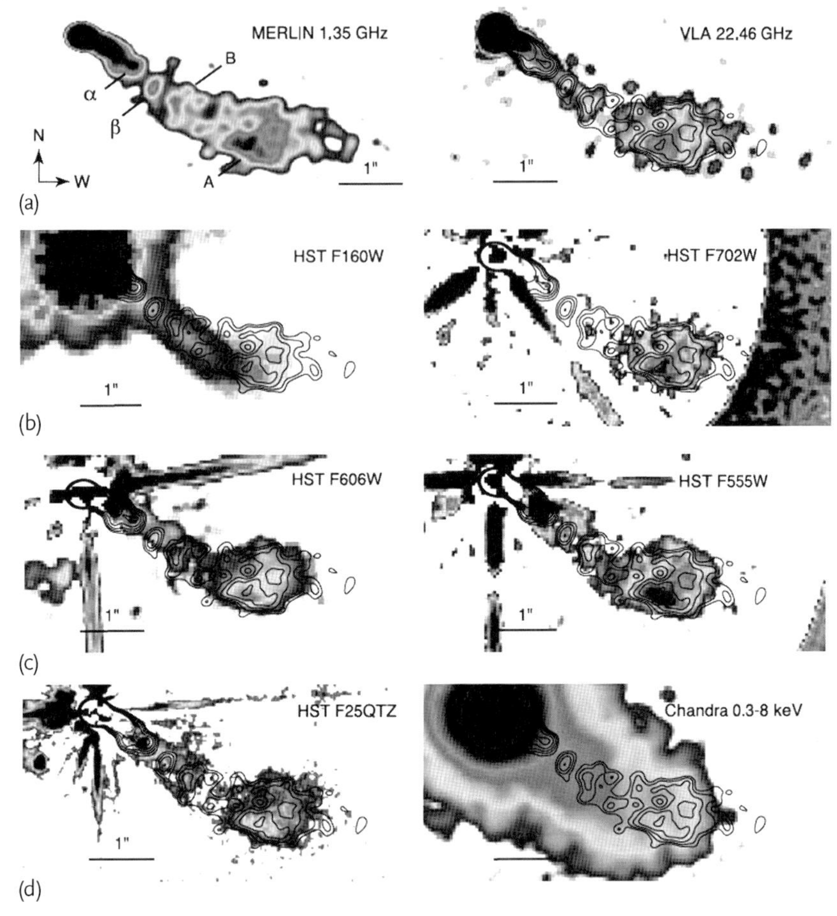

Figure 7.11 Images of the inner part of the jet of 3C 371 at various wavelengths: (a) MERLIN (1.35 GHz) and VLA (22.46 GHz); (b) NICMOS (F160W) and WFPC2 (F702W); (c) ACS (F606W) and WFPC2 (F555W); and (d) STIS (F25QTZ- NUV-MAMA) and Chandra 0.3–8 keV. The radio images have a resolution of $0.18'' \times 0.15''$ and $0.2''$, respectively. The optical images have been convolved using fgauss in FTOOLS with an elliptical Gaussian function of 1 pixel (2 pixels for STIS), with final resolution $0.15''$. In the NICMOS image, the inner part of the jet is buried under a diffraction spike. The X-ray image has been rebinned by a factor of 5 (final image pixel of $0.1''$) and then smoothed with the csmooth in CIAO with a circular Gaussian of $0.1''$, with final resolution $0.9''$ FWHM. In all cases, the 1.35 GHz radio contours are overlaid on the image. Reproduced from Sambruna *et al.* [18] with permission from the AAS.

At the lower Chandra resolution, the X-ray emission has a broad shoulder in correspondence to knots β and B, fading near the end of the jet. The X-ray spectra of the resolved structures are described by single power laws with $\alpha_X \sim 1.0$, steeper than in FR IIs but not as soft as in FR Is.

Overall, the properties of the jets in 3C 371 and PKS 0521 are reminiscent of FR I sources, while their one-sidedness is more typical of FR IIs. Interestingly, the core emission from these sources exhibit broad and narrow optical emission lines during states of low intensity of the nonthermal BL Lac continuum, indicating that these sources are possible "hybrids" between bona fide FR Is and FR IIs. Indeed, FR Is with radio morphologies and powers similar to FR IIs were already discovered by Kollgaard *et al.* [60] from a sample of radio sources.

While PKS 0521, 3C 371, and to an extent, PKS 2201 are "hybrid" cases with properties between FR Is and FR IIs, a different situation is presented by S5 2007+777. This BL Lac from the classical 1 Jy radio sample of Stickel *et al.* [61] shows a long (19″) radio jet which was *not* known to emit in the optical at the time of its selection as a Chandra target [18]. The main idea with this source was to test, at least qualitatively, the "spine-sheath" scenario: if BL Lacs are more aligned versions of FR Is, and if FR I jets have a spine-sheath structure, the X-ray emission from BL Lac jets should be highly beamed, and most likely due to IC-CMB. With its bright radio surface brightness and knotty structures, S5 2007+777 qualified as the primary candidate for this test.

A 100 ks Chandra exposure returned a spectacular image of the jet. X-ray emission was detected from all the resolved knots, with a close correspondence to the radio, similar to what is observed in FR II jets, and unusually hard continua, $\alpha_x \sim 0$. Optical upper limits from archival HST data implied concave radio-to-X-ray SEDs, similar to those of FR IIs. Attributing the X-ray emission to IC/CMB with equipartition requires strong beaming ($\delta = 13$) and a very large scale (Mpc) jet. Alternatively, synchrotron emission from a second population of very high-energy electrons is also viable. The X-ray jet of S5 2007+777 stands out as unusual among those observed so far in BL Lacertae sources, and promises to help us understand the connection between radio galaxies and quasars.

7.5 X-Ray Emission Processes

Since almost all X-ray jets are observed to be single-sided in spite of the almost ubiquitous double radio lobe structures, we infer that Γ, the bulk Lorentz factor of the emitting regions associated with the jet, is at least a few. Given that the emitting regions are moving close to the speed of light, it seems self-evident that the jet itself is also relativistic, possibly with a much larger Γ.

In this section, we will examine emission processes for the production of X-rays from kpc-scale jets since we are unable to achieve angular resolutions comparable to those obtainable with VLBI techniques. Moreover, we will not indulge in speculations on exotic emission processes (Chapter 3) but rather focus on electron synchrotron and electron IC processes. We dismiss thermal emission since the observed X-ray emission from jets is normally a part of a wide-band emission spectrum, often joining seamlessly with radio and optical data where linear polarization is detected.

7.5.1 Challenges for Synchrotron Models

With the usual assumptions for equipartition magnetic fields, typical parameters for knots and hot spots are $B \approx 100\,\mu\text{G}$; and for the highest-energy electrons responsible for X-ray synchrotron emission, $\gamma = 10^7$, gyroradius $\approx 0.01\,\text{pc}$, and half-life, $\tau, \approx 1-10\,\text{years}$. Thus we expect X-ray variability on human time scales.

In general, the most common process invoked to obtain these high-energy electrons is some sort of shock acceleration which will produce a rather steep power law of electrons. The high-energy cutoff is normally estimated by equating the acceleration and loss time scales. The loss time scale is estimated by combining E^2 losses (i.e., synchrotron and IC losses) with estimates of the time it takes for an electron to escape from the acceleration region.

There are, however, other scenarios such as the "proton-induced cascade" (PIC, Mannheim, Krulls, and Biermann [62]) where high-energy electrons are the result of cascades from higher energies, and the massive power-law distribution at lower energies is not present. Magnetic reconnection can generate electric fields along magnetic fields leading to efficient acceleration of charged particles. Depending on the geometry, it may be possible to generate electrons of a characteristic energy, again without a massive power law extending to low energies. Particle acceleration is discussed in more detail in Chapter 9; below we consider the ramifications since there have been so many successful models based on synchrotron emission with electrons distributions extending up to $\gamma = 10^7$.

The basic tenets of X-ray (single-zone) synchrotron models from a power-law distribution of electron energies $N(\gamma) = k_e \gamma^{-p}$ (where γ is the electron energy and k_e is the amplitude of the electron spectrum) include:

- A concave downwards spectrum; often with spectral breaks. We expect $\alpha_x > \alpha_r$ because of E^2 losses and/or proximity to a high-energy cutoff in $N(\gamma)$.
- Every X-ray-emitting region is an acceleration region. Since X-ray-emitting electrons lose their energies in short times, they are unable to travel any astrophysically interesting distances, and since emission is not limited to isolated "knots", this deduction requires more than the occasional isolated shock (e.g., shear acceleration, turbulent acceleration).
- Jet knots are emission regions produced by the conversion of jet energy to particles and fields. This follows from the requirement that whatever carries jet energy must not lose its energy before arriving at the far end of the jet, coupled with the fact that electrons with $\gamma \geq$ a few thousand will suffer inescapable IC losses to photons of the CMB [63].

7.5.2 Estimating Synchrotron Parameters

Because the intensity of synchrotron emission depends on both the amplitude of the electron spectrum (k_e) and the magnetic field, there has been a long standing practice of assuming equipartition between the energy densities in the relativistic particles and that of the magnetic field. This condition approximately minimizes the total nonthermal energy of any system. If alternate conditions are imposed (e.g., equal pressures in field and particles, minimum total energy, etc.), there is little change in the resulting equations: only a small change in the coefficient of the square brackets in (7.1):

$$B_{\text{eq}} = a\left[\frac{c_{12}(1+k)L}{\phi V}\right]^{2/7}, \tag{7.1}$$

where L is the bolometric synchrotron luminosity, ϕ is the filling factor (1.0 or less), k is the factor by which the energy density of relativistic protons exceeds that of electrons, c_{12} is a Pacholczyk variable which is a function of the spectral index and limits of integration over energy [64], V is the volume, and a is a constant of order two.

In spite of the favorably small value of the exponent, there are usually a number of uncertainties conspiring to render any estimates insecure. In particular, we cite the limited resolution and the effects that has on ϕ (see Figure 7.7) and V (which also has the uncertainty as to the dimension along the line-of-sight). Because we cannot measure intensities over all frequencies, we are unsure of the integration limits and of the luminosity. Finally, in general, we have no information about the proton component of the emitting plasma (the "$1+k$" term).

In spite of these caveats, it is common to calculate synchrotron parameters for features for which some estimate of (or limit to) the volume is obtainable. In addition to the total nonthermal energy, one can derive parameters such as which energy electrons are responsible for which observed frequency bands, corresponding half-lives for synchrotron losses, gyroradii, and so on

Although the literature contains countless examples of claims that equipartition does not hold, we suspect that in most astrophysical plasmas in more or less a steady state (i.e., continuous rather than explosive injection of particles and field), the equipartition assumption is valid to within a factor of a few.

7.5.3
Synchrotron Self-Compton Emission

Unless knots in jets consist of countless much smaller emitting volumes (a situation that should be investigated further – see Figure 7.7), SSC models fail to predict X-ray emission at the observed intensities. However, the compact, radio-bright features at the ends of FR II (and many Q) jets ("hot spots") often have sufficient photon energy density from the observed radio emission to construct consistent SSC models [65]. With the caveats involved in calculating B_{eq} (Section 7.5.2), the B fields required to produce the observed X-ray intensities are consistent with B_{eq}.

In a study of the then available observations of X-ray hot spot detections [66], it was found that for the lower-luminosity sources, the X-ray intensity was significantly larger than predicted by SSC models. The authors suggested that an additional emission component (e.g., synchrotron X-rays) might explain the observed excesses.

7.5.4 IC Emission from Photons Originating in Other Components

Another possibility for excess X-ray emission from hot spots was advanced by Georganopoulos and Kazanas [67]. They pointed out that as the jet approaches the terminal hot spot, the synchrotron radiation from the hot spot will produce an increasing photon energy density, $u(\nu)$, as the distance to the hot spot diminishes, and further, this energy density will be increased by Γ^2 in the frame of the jet. The resulting X-ray emission will be beamed into a cone around the jet, unlike the isotropic SSC emission. A plausible example of this scenario is the S hot spot of 3C 105 [68].

Another example of this sort of emission has been invoked for jet knots as a so-called two-zone model. In the particular case of a jet with a fast-moving spine and a slower sheath, the suggestion was advanced that in the frame of the fast-moving spine, $u(\nu)$ from the slower sheath will be augmented by the square of the difference in Γ values [6]. Generally speaking the introduction of a second zone, adds to the free parameters so that almost any observed conditions can be modeled successfully.

7.5.5 IC/CMB Emission from Jets with Large Γ

IC/CMB with beaming (i.e., boosting the CMB energy density by a factor of Γ^2 as seen in the jet frame) was advanced as a solution to the vexing problem that small intensities in the optical/UV for many quasar knots precluded a one-zone synchrotron model [5, 6].

The essential idea is to increase the ratio of IC to synchrotron emission by increasing $\mathrm{u}(\nu)$; in this case $\mathrm{u}(\nu)$ is that of the CMB. Conceptually one obtains the normalization of the electron spectrum from radio data coupled with an estimate of the average magnetic field strength, often the equipartition field. Then assuming that the exponent of the electron distribution is $p = 2\alpha_\mathrm{r} + 1$, one extrapolates $N(\gamma)$ down to low energies to find the number of low-energy electrons. At that point, the hypothesized bulk Lorentz factor of the jet, Γ, is increased until the observed X-ray intensity is produced.

This model requires the following conditions:

- $\Gamma \geq 10$ for many quasar jets.
- The electron distribution has no low-energy spectral breaks or cutoffs in the range $\gamma = 50$–5000 or more, that is, energies producing synchrotron emission below the frequencies observed with good resolution in the cm band. However, a cutoff in the range $\gamma = 10$–50 may be required in some knots so as not to overproduce the optical emission.

Since the introduction of the IC/CMB model, many questions have been raised. Are there really enough $N(\gamma \approx 100)$ electrons? Is the extrapolation using α_r correct? Is it true that there are no spectral breaks to flatter spectra at low energies? What about low-energy cutoffs to the power law? Many acceleration scenarios work only above $\gamma = 2000$ (e.g., Stawarz *et al.* [69]). If IC/CMB dominates the X-ray emission, why are jets knotty instead of smooth and why does the X-ray brightness fade faster moving downstream than does the optical and radio brightnesses? Most of these question remain open because there is little or no data available from other lines of investigation on the behavior of the low-energy end of $N(\gamma)$ or on the magnitude of Γ on kpc scales.

7.5.6 Estimating IC/CMB Parameters

There have been several presentations of how to derive the physical parameters of the jet under the IC/CMB beaming model (e.g., [8, 70, 71]); here we review a few of the relationships for particular parameters, mostly taken from [70].

The "extra beaming factor", engendered by the fact that head-on collisions between electrons and photons are favored compared to overtaking scattering events, is given by [72]

$$\xi = \left\{1 + \frac{\mu\Gamma - \sqrt{\Gamma^2 - 1}}{\Gamma - \mu\sqrt{\Gamma^2 - 1}}\right\}^2 , \tag{7.2}$$

where Γ is the bulk Lorentz factor of the emitting region and $\mu = \cos\theta$.

Thus while emissions which are isotropic in the jet frame (i.e., synchrotron emission) are beamed into the forward direction defined by the bulk jet motion, the IC/CMB is more beamed in this direction.

In solving for the equipartition magnetic field strength, one generally finds that the field required (B'), is equal to the field calculated in the case of no beaming, divided by the beaming factor, δ to the power 1 or some other value not too different from one.[2] From the appendix of [70], the relevant expressions are:

$$B'_{\text{eq}} = \left[\frac{18.85\, C_{12}(1 + k) L'_{\text{s}}}{\phi V'}\right]^{2/7} \text{G} , \tag{7.3}$$

where L'_{s} is the synchrotron luminosity in the jet frame, V' is the emitting volume, ϕ is the volume filling factor, and C_{12} is a Pacholczyk parameter which is a slowly varying function of the spectral index, α, and the limits of integration, ν'_1, and ν'_2 (e.g., $C_{12} = 5.7 \times 10^6$ for $\alpha = 0.8$, $\nu'_1 = 10^7$, $\nu'_2 = 10^{15}$). When the correct expression for C_{12} is used in (7.3), it introduces an extra factor of $\delta^{0.5}$ to the numerator within square brackets, so that together with the δ^4 (which appears in the denominator) required to relate the observed luminosity, $L(\text{obs})$, to L'; the final

2) There is some debate on how to relate observed sizes to the emitting volume in the jet frame. Stawarz *et al.* [73] finds $B' = B(1)/\delta^{5/7}$.

dependency within the square brackets goes as $\delta^{-7/2}$ and hence

$$B'_{\mathrm{eq}} = \frac{B(1)}{\delta} , \tag{7.4}$$

where $B(1)$ is the equipartition field calculated for no beaming ($\delta = 1$).

In order to understand the extent of the extrapolation necessary to move from electron energies responsible for the observed radio synchrotron emission to energies producing the IC/CMB emission, we find the following.

The electron energy responsible for a particular synchrotron frequency is:

$$\gamma^2_{\mathrm{sync}} = \frac{2.380 \times 10^{-7} \nu_{\mathrm{s}}(\mathrm{obs})(1+z)}{B'\delta} , \tag{7.5}$$

where B' is the magnetic field in the jet frame.

The electron energy responsible for an IC frequency is:

$$\gamma^2_{\mathrm{ic}} = \frac{6.25 \times 10^{-12} \nu_{\mathrm{ic}}(\mathrm{obs})}{(1+\mu'_{\mathrm{j}})\delta\Gamma} , \tag{7.6}$$

where μ'_{j} is the cosine of the angle between the jet direction and the emitted photons which the observer receives (i.e., the angle in the jet frame).

The half-life for E^2 losses (synchrotron and IC) in the jet frame is:

$$\tau' = \frac{10^{13}}{\gamma'_{\mathrm{e}}\left\{1.016B'^2 + 10.398(1+z)^4\left(\Gamma^2 - \frac{1}{4}\right)\left[\left(1+\beta'_{\mathrm{e}}\mu'_{\mathrm{j}}\right)^2 - \frac{(1+\beta'_{\mathrm{e}}\mu'_{\mathrm{j}})}{\gamma'^2_{\mathrm{e}}}\right]\right\}} , \tag{7.7}$$

where B' is in μG and τ' is in years.

Making the usual approximations for quantities close to one, and since time intervals in the jet frame are observed at the Earth as

$$\tau_{\mathrm{o}} = \tau'\frac{(1+z)}{\delta} , \tag{7.8}$$

an approximate expression for the observed half-life is:

$$\tau_{\mathrm{o}} = \frac{10^{13}(1+z)}{\gamma'_{\mathrm{e}}\delta\left\{B'^2 + 40(1+z)^4\Gamma^2\right\}}\ \mathrm{yr} , \tag{7.9}$$

where again, B' is in μG.

7.6
Summary, Conclusions, Future Work

We know that both synchrotron and IC are "mandatory" physical processes in any (magnetized, which of course they all are) relativistic plasma; the only questions are, do we have enough electrons of the required energies and enough $u(\nu)$ and B to produce the observed fluxes in the observed bands? So the real question should be phrased as *which process dominates in any particular band?* The fascinating aspect of nonthermal X-ray emission from jets and hot spots is that synchrotron X-rays will normally come from the high end of $N(\gamma)$ (i.e., γ of 10^7–10^8), whereas IC/CMB (with beaming) arises from the bottom end of $N(\gamma)$ which is difficult or impossible to access from any other type of observation. At the high end of $N(\gamma)$, we can investigate the effects of short half-lives, cutoffs, and E^2 losses. On the other hand, with IC/CMB we can learn about critical aspects of the acceleration process. Given the remarkable differences between these two processes, it is surprising that so many observables (e.g., brightness, intrinsic luminosity, ratio of X-ray flux to radio flux, morphology) are so similar.

In these closing sections we review a few of the ways that X-ray data constrain or inform us about the general nature of relativistic jets.

7.6.1 The Nature of Offsets and Spectral Progressions

In the study of jets with detections at both radio and X-rays of several individual knots we often (but not always) find both "progressions" and "offsets". By "progression" we mean that there is a steepening of spectra progressively moving down the jet and by "offsets" we mean that when knots are observed with comparable angular resolutions at all bands, we find individual knots for which the peak brightness shifts downstream as the wavelength increases. This behavior is seen at small and large physical scales, that is, in both FR I jets and quasar jets. It seems likely that these two effects are two sides of the same coin for if the jet in 3C 273, which displays a very large progression [47], were to be observed with a single resolution element, it would have a strong offset. Although a plausible scenario for synchrotron models could be an increasing field strength moving downstream (thereby limiting the maximum energy reached by acceleration processes and shortening the half-life of the higher-energy electrons), there is no agreed upon explanation for offsets under the IC/CMB model.

7.6.2 The Nature of Knots

The existence of knotty (as opposed to continuous) jets was never a serious problem for radio (and optical) synchrotron models. Knots were identified as the sites of acceleration and/or regions with stronger B fields (e.g., internal shocks). Interknot regions would be volumes of low emissivity because of low field, and/or the energy transferred from the jet to the emitting plasma has become exhausted; the next knot downstream would be a new site of energy transfer.

Knotty jets are more of a problem for the IC/CMB model because the low-energy electrons responsible for the X-rays would not lose their energy for 10^4 years or longer. Why are they not emitting strongly in low field interknot gaps? It is possible that low-energy electrons follow a helical path so that sometimes the beaming cone is away from our line-of-sight or the actual jet creation might be episodic so that there are segments of the jet devoid of energy transport. The latter option implies that kpc-scale knots are moving downstream, a conjecture not supported by HST-1 in the M 87 jet, which seems to be the site of a stationery shock.

Another possibility would be that each knot represents the creation of an emitting plasma which is initially moving close to the intrinsic jet speed, but decelerates, reducing u' (CMB) in the jet frame. In that way, the synchrotron component may persist for a longer time, producing the offsets observed. Then, the next knot would be a separate event (i.e., transfer of jet energy to emitting plasma). Although such a scenario is appealing, it is extremely *ad hoc*, and the deceleration should not be allowed to make the bulk motion subrelativistic because then the isotropic (synchrotron) radiation would be visible from all angles (i.e., we would see all counterjets at radio frequencies).

7.6.3
Future Possibilities

Among observational approaches are the new generation of low-frequency radio telescopes such as LOFAR with good angular resolution and sensitivity below 100 MHz. These hold the promise of providing independent estimates of the number of low-energy electrons. Higher-energy X-ray data with sufficient angular resolution and sensitivity could test cutoffs expected from the synchrotron model; however, none is likely to be available in the foreseeable future. Likewise X-ray polarimetry could demonstrate the dominance of synchrotron emission, but once again there is no prospect of achieving the required resolution and sensitivity any time soon.

More observational work is needed for two-zone models. We must find a way to observe each zone in more than a single band. It is too easy to say a spine produces the jet emission in one band and a sheath does the job for another band.

More work is needed to discover additional occurrences of variability. Given our limited angular resolution, the time scale of variability can provide much better constraints on the size of the emitting volume than those available from direct imaging.

The X-ray-TeV connection is just beginning to be exploited. IC (SSC or EIC) from $\gamma = 10^7$ electrons scattering photons in the range of 10^{11}–10^{14} Hz should produce TeV emission. Although it is currently difficult to ensure both instruments are observing the same emitting volume, angular resolution should improve for the next generation of Cherenkov telescopes.

In this chapter we have attempted to cover the basic concepts, and focus on the common properties. Thus while most jets display very good morphological correspondence from radio to X-rays, there are puzzling discrepancies. In the case of the M 87 jet, the radio and optical emission makes an excursion to the north beyond knot C, thus departing from the well-defined straight jet. The X-ray emission however has a component within the hoop formed by this excursion (Figure 7.8). In the case of 3C 273, just beyond knot A, the brighter radio emission deviates to the south whereas the optical and X-ray emission ridge lines continue on the straight path. Another example is the end of the jet in quasar 4C 19.44. In this case, the ratio of X-ray to radio brightness is almost the same all the way down the jet until the radio jet becomes very faint while the X-ray emission persists at its previous brightness for an additional resolution element ($\approx 1''$).

References

1 Harris, D.E. and Krawczynski, H. (2006) *Annu. Rev. Astron. Astrophys.*, **44**, 463.
2 Sambruna, R.M., Donato, D., Cheung, C.C., Tavecchio, F., and Maraschi, L. (2008) *Astrophys. J.*, **684**, 862.
3 Schwartz, D.A. *et al.* (2000) *Astrophys. J.*, **540**, L69.
4 Chartas, G. *et al.* (2000) *Astrophys. J.*, **542**, 655.
5 Tavecchio, F., Maraschi, L., Sambruna, R.M., and Urry, C.M. (2000) *Astrophys. J. Lett.*, **544**, L23.
6 Celotti, A., Ghisellini, G., and Chiaberge, M. (2001) *Mon. Not. R. Astron. Soc.*, **321**, L1–5.
7 Dermer, C.D., *Astrophys. J.*, **446**, L63.
8 Worrall, D.M. (2009) *Astron. Astrophys. Rev.*, **17**, 1.
9 Feigelson, E.D., Laurent-Muehleisen, S.A., Kollgaard, R.I., and Fomalont, E.B. (1995) *Astrophys. J.*, **449**, L149.
10 Miller, B.P. *et al.* (2006) *Astrophys. J.*, **652**, 163.
11 Sambruna, R.M., Maraschi, L., Tavecchio, F., Urry, C. Megan, Cheung, C.C., Chartas, G., Scarpa, R., and Gambill, J.K. (2002) *Astrophys. J.*, **571**, 206.
12 Sambruna, R.M., Gambill, J.K., Maraschi, L., Tavecchio, F., Cerutti, R., Cheung, C.C., Urry, C.M., and Chartas, G. (2004) *Astrophys. J.*, **608**, 698.
13 Marshall, H.L., Schwartz, D.A., Lovell, J.E.J., Murphy, D.W., Worrall, D.M., Birkinshaw, M., Gelbord, J.M., Perlman, E.S., and Jauncey, D.L. (2005) *Astrophys. J. Suppl. Ser.*, **156**, 13.
14 Sambruna, R.M., Gliozzi, M., Donato, D., Maraschi, L., Tavecchio, F., Cheung, C.C., Urry, C.M. and Wardle, J.F.C. (2006) *Astrophys. J.*, **641**, 717.
15 Wardle, J.F.C. and Aaron, S.E. (1997) *Mon. Not. R. Astron. Soc.*, **286**, 425.
16 Bridle, A.H., Hough, D.H., Lonsdale, C.J., Burns, J.O., and Laing, R.A. (1994) *Astron. J.*, **108**, 766.
17 Giovannini, G., Taylor, G.B., Arbizzani, E., Bondi, M., Cotton, W.D., Feretti, L., Lara, L., and Venturi, T. (1999) *Astrophys. J.*, **522**, 101.
18 Sambruna, R.M., Donato, D., Tavecchio, F., Maraschi, L., Cheung, C.C., and Urry, C.M. (2007) *Astrophys. J.*, **670**, 74.
19 Schwartz, D.A. *et al.* (2007) *Astrophys. Space Sci.*, **311**, 341.
20 Tavecchio, F. and Ghisellini, G. (2008) *Mon. Not. R. Astron. Soc.*, **385**, L98.
21 Georganopoulos, M. (2000) *Astrophys. J.*, **543**, L15.
22 Tavecchio, F., Maraschi, L., Sambruna, R.M., Gliozzi, M., Cheung, C.C., Wardle, J.F.C., and Urry, C.M. (2006) *Astrophys. J.*, **641**, 732.
23 Georganopoulos, M. and Kazanas, D. (2004) *Astrophys. J.*, **604**, L81.
24 Bicknell, G.V. (1995) *Astrophys. J. Suppl. Ser.*, **101**, 29.
25 Tavecchio, F., Ghisellini, G., and Celotti, A. (2003) *Astron. Astrophys.*, **403**, 83.
26 Uchiyama, Y., Urry, C.M., Van Duyne, J., Cheung, C.C., Sambruna, R.M., Takahashi, T., Tavecchio, F., and Maraschi, L. (2005) *Astrophys. J.*, **631**, L113.
27 Uchiyama, Y., Urry, C.M., Cheung, C.C., Jester, S., Van Duyne, J., Coppi, P., Sambruna, R.M., Takahashi, T., Tavecchio, F., and Maraschi, L. (2006) *Astrophys. J.*, **648**, 910.
28 Uchiyama, Y., Urry, C.M., Coppi, P., Van Duyne, J., Cheung, C.C., Sambruna, R.M., Takahashi, T., Tavecchio, F., and

Maraschi, L. (2007) *Astrophys. J.*, **661**, 719.

29 Wilson, A.S., Young, A.J., and Shopbell, P.L. (2001) *Astrophys. J.*, **547**, 740.

30 Kataoka, J., Stawarz, Ł., Harris, D.E., Siemiginowska, A., Ostrowski, M., Swain, M.R., Hardcastle, M.J., Goodger, J.L., Iwasawa, K., and Edwards, P.G. (2008) *Astrophys. J.*, **685**, 839.

31 Kraft, R.P., Hardcastle, M.J., Worrall, D.M., and Murray, S.S. (2005) *Astrophys. J.*, **622**, 149.

32 Worrall, D.M. and Birkinshaw, M. (2005) *Mon. Not. R. Astron. Soc.*, **360**, 926.

33 Marshall, H.L. *et al.* (2001) *Astrophys. J.*, **549**, L167.

34 Sambruna, R.M., Urry, C.M., Tavecchio, F., Maraschi, L., Scarpa, R., Chartas, G., and Muxlow, T. (2001) *Astrophys. J.*, **549**, L161.

35 Jester, S., Röser, H.-J., Meisenheimer, K. and Perley, R. (2002) *Astron. Astrophys.*, **385**, L27.

36 Aharonian, F.A. (2000) *New Astron.*, **5**, 377.

37 Dermer, C. and Atoyan, A.M. (2002) *Astrophys. J.*, **568**, L81.

38 Schwartz, D.A. (2002) *Astrophys. J.*, **569**, L23.

39 Cheung, C.C., Stawarz, Ł., and Siemiginowska, A. (2006) *Astrophys. J.*, **650**, 679.

40 Cheung, C.C., Healey, S.E., Landt, H., Verdoes Kleijn, G., and Jordán, A. (2009) *Astrophys. J. Suppl. Ser.*, **181**, 548.

41 Cheung, C.C. (2004) *Astrophys. J.*, **600**, L23.

42 Bicknell, G.V. (1994) *Aust. J. Phys.*, **47**, 669.

43 Cheung, C.C., Stawarz, Ł., Siegmiginowska, A., Harris, D.E., Schwartz, D.A., Wardle, J.F.C. Gobeille, D., and Lee, N.P. (2008) *The Highest Redshift Relativistic Jet. Extragalactic Jets: Theory and Observation from Radio to Gamma Ray*, ASP Conference Series, vol. 386, proceedings of the conference held 21–24 May 2007 in Girdwood, Alaska, USA (eds A. Rector and D.S. De Young), Astronomical Society of the Pacific, San Francisco, p. 462.

44 Laing, R.A. and Bridle, A.H. (2002) *Mon. Not. R. Astron. Soc.*, **336**, 328.

45 Hardcastle, M.J., Worrall, D.M., Birkinshaw, M., Laing, R.A., and Bridle, A.H. (2002) *Mon. Not. R. Astron. Soc.*, **334**, 182.

46 Worrall, D.M., Birkinshaw, M., and Hardcastle, M.J. (2001) *Mon. Not. R. Astron. Soc.*, **326**, L7.

47 Marshall, H.L., Miller, B.P., Davis, D.S., Perlman, E.S., Wise, M., Canizares, C.R., and Harris, D.E. (2002) *Astrophys. J.*, **564**, 683–687.

48 Wilson, A.S. and Yang, Y. (2002) *Astrophys. J.*, **568**, 133–140.

49 Perlman, E.S. and Wilson, A.S. (2005) *Astrophys. J.*, **627**, 140.

50 Madrid, J.P. (2009) *Astron. J.*, **137**, 3864.

51 Harris, D.E., Cheung, C.C., Stawarz, Ł, Biretta, J.A., and Perlman, E.S. (2009) *Astrophys. J.*, **699**, 305.

52 Cheung, C.C., Harris, D.E., and Stawarz, Ł. (2007) *Astrophys. J.*, **663**, L65.

53 Biretta, J.A., Sparks, W.B., and Macchetto, F. (1999) *Astrophys. J.*, **520**, 621.

54 Wardle, J.F.C., Moore, R.L., and Angel, J.R.P. (1984) *Astrophys. J.*, **279**, 93.

55 Kollgaard, R.I., Wardle, J.F.C., Roberts, D.H. and Gabuzda, D.C. (1992) *Astron. J.*, **104**, 1687.

56 Birkinshaw, M., Worrall, D.M., and Hardcastle, M.J. (2002) *Mon. Not. R. Astron. Soc.*, **335**, 142.

57 Macchetto, F. *et al.* (1991) *Astrophys. J.*, **369**, L55.

58 Falomo, R., Scarpa, R., and Bersanelli, M. (1994) *Astrophys. J. Suppl. Ser.*, **93**, 125.

59 Pesce, J.E., Sambruna, R.M., Tavecchio, F., Maraschi, L., Cheung, C.C., Urry, C.M., and Scarpa, R. (2001) *Astrophys. J.*, **556**, L79.

60 Kollgaard, R.I., Feigelson, E.D., Laurent-Muehleisen, S.A., Spinrad, H., Dey, A., and Brinkmann, W. (1995) *Astrophys. J.*, **449**, 61.

61 Stickel, M., Padovani, P., Urry, C.M., Fried, J.W., and Kuehr, H. (1991) *Astrophys. J.*, **374**, 431.

62 Mannheim, K., Krulls, W.M., and Biermann, P.L. (1991) *Astron. Astrophys.*, **251**, 723.

63 Harris, D.E. and Krawczynski, H. (2007) *Constraints on the Nature of Jets from kpc*

Scale X-Ray Data in "Triggering Relativistic Jets", Revista Mexicana de Astronomia y Astrofisica, Serie de Conferencias, 27, Contents of Supplementary CD, (eds W.H. Lee and E. Ramirez-Ruiz), p. 188.

64 Pacholczyk, A.G. (1970) *Radio Astrophysics*, W.H. Freeman, San Francisco.

65 Harris, D.E., Carilli, C.L., and Perley, R.A. (1994) *Nature*, **367**, 713.

66 Hardcastle, M.J., Harris, D.E., Worrall, D.M., and Birkinshaw, M. (2004) *Astrophys. J.*, **612**, 729.

67 Georganopoulos, M. and Kazanas, D. (2003) *Astrophys. J.*, **589**, L5.

68 Massaro, F. *et al.* (2010) *Astrophys. J.*, **714**, 589.

69 Stawarz, Ł., Cheung, C.C., Harris, D.E., and Ostrowski, M. (2007) *Astrophys. J.*, **662**, 213.

70 Harris, D.E. and Krawczynski, H. (2002) *Astrophys. J.*, **565**, 244.

71 Worrall, D.M. and Birkinshaw, M. (2006) Multiwavelength evidence of the physical processes in radio jets, in *Physics of Active Galactic Nuclei at All Scales* (eds S. Alloin, R. Johnson, and P. Lira), Lecture Notes in Physics, vol. 693, Springer, Berlin, Heidelberg, p. 39.

72 Massaro, F., Harris, D.E., Chiaberge, M., Grandi, P., Macchetto, F.D., Baum, S.A., O'Dea, C.P., and Capetti, A. (2009) *Astrophys. J.*, **696**, 980.

73 Stawarz, Ł., Sikora, M., Ostrowski, M. and Begelman, M.C. (2004) *Astrophys. J.*, **608**, 95.

74 Worral, D.M. (2009) *Astron. Astrophys. Rev.*, **17**, 1, Figure 9.

8
Unresolved Emission from the Core: Observations and Models

Henric Krawczynski, Markus Böttcher, and Anita Reimer

8.1
Introduction

Even though state-of-the-art telescopes achieve impressive angular resolutions, they are not sufficient to spatially resolve the immediate vicinities of the supermassive black holes at the cores of AGNs. Even for the black hole in the Milky Way and the supermassive black hole in the massive elliptical galaxy M 87, the VLBA radio interferometer (angular resolution: $\sim$ 0.5 mas) can only resolve features with a projected angular extent exceeding $\sim$ 50 Schwarzschild radii. Jets are thought to form at distances of between 10 and 10^3 Schwarzschild radii from the black hole. So, even for the supermassive black holes with the largest aspect ratios, current technology only starts to be able to resolve the jet formation region. However, even in the absence of imaging information, we can study the central engine and the jet formation process with the help of spectrally and temporally resolved observational data. Multiwavelength observations can give information about the mass and spin of the black hole, the properties of the accretion disk, and the composition and structure of jets.

In this chapter, the spatially unresolved emission from the cores of AGNs will be reviewed. In Section 8.2, we will describe observations of radiation coming from the accretion disk and from various components orbiting the central black hole, that is, emission line clouds, various types of X-ray absorbing material, an obscuring region of molecular material, and fast and slow outflows. We discuss these components with an emphasis on how they constrain the properties of the central engine, that is, the mass and spin of the supermassive black hole, and the accretion flow. Subsequently, we describe observations of the spatially unresolved emission from the jets themselves and models for their interpretation in Section 8.3. We conclude with a summary of the chapter and discuss avenues for further progress in Section 8.4.

Relativistic Jets from Active Galactic Nuclei, First Edition. Edited by M. Böttcher, D.E. Harris, H. Krawczynski.

8.2
Emission from Various Nonjet Components

AGN exhibit complex spectra in the IR/optical, UV and X-ray bands. Some of the emission comes from the inner accretion disk and gives information about the mass and spin of the supermassive black hole. Other emission components are believed to originate from atomic and molecular material at different distances from the black hole. Figure 8.1 shows a sketch of the geometrical configuration of the emission components. The reader should refer to this figure as the individual emission components are discussed in turn.

In some AGNs we believe that we are able to detect thermal emission from the accretion disk itself. It emerges as a feature called the big blue bump extending from the (rest frame) optical/UV band to (sometimes) the soft X-ray band (e.g., [2–4]). The accretion disk is believed to be partially covered by a corona of hot – yet still thermal – material. Emission from the accretion disk can gain energy by being scattered by the hot coronal plasma. This Comptonization of the disk emission gives rise to a high-energy tail extending into the hard X-ray regime. Alternative explanations of the hard X-ray emission include a hot inner accretion flow (e.g., [5, 6]), the "lamp-post model" of an X-ray source illuminating the accretion disk from above (e.g., [7, 8]), and multiple magnetic field reconnection flares heating tenuous plasma above the accretion disk [9]. Extensive discussions of the emission from accretion disks can be found in [10–13].

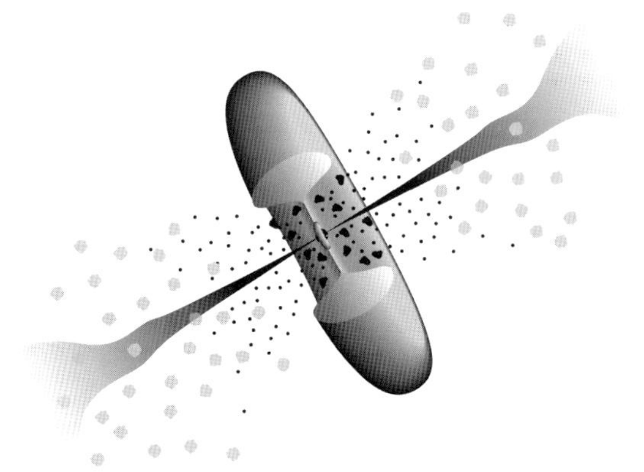

Figure 8.1 Schematic drawing of the central engine of an AGN including several components which can explain most of the observed properties of AGNs (not to scale; example lengths for an $M = 10^8\,M_\odot$ black hole in parentheses). The central black hole (Schwarzschild radius $R_S = 2GM/c^2 \approx 3 \times 10^{13}$ cm) is surrounded by an accretion disk (~ 1–30×10^{14} cm). The broad emission lines originate in clouds orbiting above the disk (at ~ 2–20×10^{16} cm). Molecular material, traditionally envisioned as dusty torus (inner radius $\sim 10^{17}$ cm) or a warped accretion disk, obscures the BLR when the AGN is seen from the side. However, the torus geometry as shown in the figure should not be taken literally, as the obscuring material may be intertwined with outflowing accretion disk winds and BLR/NLR material and does almost certainly not assume a toroidal shape. The graph is from Urry and Padovani [1] and reproduced with permission from the ASP.

In the X-ray spectra of some AGN one detects a broad X-ray emission line due to K_α fluorescence of partially ionized iron. The line is believed to be emitted by the inner accretion disk material orbiting the black hole with velocities that are a sizable fraction of the speed of light. The line encodes unique information about the mass and spin of the black hole. The relativistic orbital motion of the emitting atoms leads to the Doppler broadening of the line. The line is further widened by the fact that some of the emission originates so close to the black hole that the photons lose a substantial fraction of their energy when they escape the gravitational potential of the black hole. This effect is called the gravitational redshift. X-ray observations of the broad iron K_α line count as one of the best observational proofs for the existence of astrophysical black holes (e.g., [14–18]). The Seyfert galaxy MCG 6-30-15 exhibits one of the best-studied iron K_α lines. The analysis of the line shape can be used to constrain the black hole spin. This is based on the result from general relativity that the innermost stable circular orbit of disk material is at $3R_S$ for a nonrotating black hole, but can be as close as $(1/2)R_S$ for a maximally rotating black hole. Indeed, it seems likely that the dimensionless spin of the black hole in MCG 6-30-15 has a value $a > 0.987$ close to its theoretical maximum of $a = 1$ [19]. Until now, only a handful of AGNs have been established to unambiguously emit relativistically broadened iron K_α fluorescence lines.

An important diagnostic tool for AGNs are optical emission lines. One distinguishes broad and narrow emission lines, originating from different regions, called the broad-line region (BLR) and the narrow-line region (NLR), respectively. These lines are presumably emitted by cold clouds of interstellar material orbiting the black hole [20] and result from photoexcitation and photoionization of gas by the UV and X-ray emission from the accretion disk, and the subsequent recombination and deexcitation. The BLR clouds are believed to be located at $\sim$ 10 light days from the supermassive black hole, while the NLR clouds orbit the black hole at pc to kpc distances. The line centroids can be used to measure the redshifts and thus the distances of AGNs. Intensive observation campaigns of the BLR emission and the continuum emission from the accretion disk have been carried out to measure the mass of the central black hole based on a technique called reverberation mapping [21, 22]. The technique combines the widths of the BLR lines (which constrain the orbital velocities of the emitting clouds) with measured time lags between variations of the continuum flux from the accretion disk and the BLR flux. As the BLR emission stems from reprocessing the continuum flux, the time lag can be used to estimate the distance of the BLR clouds from the central engine. The information about the velocity of the BLR clouds and the distance of the BLR clouds from the supermassive black hole constrain the orbital parameters of the BLR clouds sufficiently to make it possible to estimate the mass of the supermassive black hole. This technique can be used for AGNs which are so far away that stellar orbits close to the black hole cannot be resolved.

Measurements of the luminosities of the emission lines can be used to estimate the accretion rate. Two classical papers used the line luminosities as estimators of the accretion rate and studied the correlation of the line luminosities with the kinematic power of the jet. Rawlings and Saunders [23] estimated the jet power based on the energetics of the radio lobes and correlated the estimated jet power with the [O III] narrow-line luminosities. The authors find a significant correlation between these two quantities for a sample of radio galaxies. Celotti, Padovani and Ghisellini [24] performed a similar study: they estimated the jet luminosities from VLBI observations of radio-loud AGNs, and used broad-line luminosities as a measure of the accretion rate. The study also showed evidence for a correlation (see also [25, 26]).

Some of the differences between observational AGN classes can be explained with the presence of a dusty absorber, often referred to as "torus", which can obscure accretion disk and BLR emission from view and emits reprocessed emission from the central engine in the infrared (see [27], and references therein). Observations constrain the dust torus to be clumpy and compact. Modeling of the data indicates that the torus extends from a distance of $R_\mathrm{d} = 0.4 \times \sqrt{L_{45}}$ pc, where L_{45} is the accretion disk luminosity in units of 10^{45} erg s^{-1}, to an outer distance of $> 5 R_\mathrm{d}$ [28–30]. Inside R_d, the continuum AGN emission destroys the dust, ionizes the atoms and creates the material making up the BLR and X-ray obscuring clouds [27].

The presence of blueshifted broad (BAL) and narrow absorption lines (NAL) in the optical, UV, and X-ray regime show evidence for fast AGN outflows or winds (see [31], and references therein). While the NAL outflows are believed to be launched at $\sim$ pc distances from the central engine and are largely radiatively driven, the location and the driving mechanism of the fast ($0.2c$) BAL outflows are still highly uncertain. Quite possibly, jets and winds are two related byproducts of black hole accretion.

8.3 Emission from the Inner Jet

8.3.1 Blazars

Blazars are a subclass of AGN with a set of characteristic properties including a strong continuum emission extending from the radio all the way to the γ-ray regimes, rapid variability (at all frequencies and on all time scales probed so far), high polarization (at both optical and radio frequencies) in most objects, and, in some objects, a lack of any strong optical emission lines. Blazars are thought to be AGNs with jets which are closely aligned with the line-of-sight. Owing to the relativistic motion of the jet material with a velocity $v_\mathrm{jet} = \beta_\Gamma c$ with $\beta_\Gamma \approx 1$, the νF_ν flux of the continuum emission from the jet is amplified by the fourth power of the relativistic Doppler factor δ introduced in Chapter 2 (2.30). Further below, we will describe the observational evidence for Doppler factors between 10 and 50. The relativistic amplification (boosting) by factors of δ^4 explains why blazars are among the brightest objects in the Universe. Blazars allow us to gain unique insights into relativistic components of AGN jets.

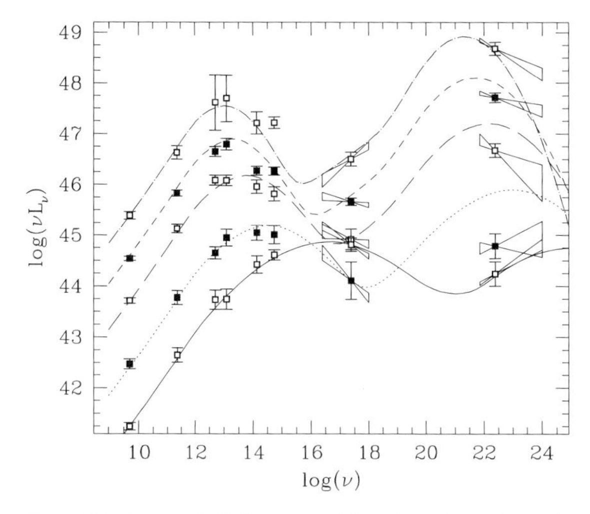

Figure 8.2 Average SED for sources binned in different classes according to their radio fluxes. The luminosities are comoving luminosities calculated under the assumption that the emission of the blazars is isotropic. The lines show phenomenological curves which connect the data points. Reproduced from Fossati *et al.* [32] with permission from the Royal Astronomical Society.

Blazars emit a characteristic spectrum of continuum emission with at least two broad emission components. Figure 8.2 shows average blazar spectral energy distributions (SEDs) for several luminosity classes. In an SED, one shows the photon frequency ν times the differential energy flux F_ν as a function of $\log \nu$. The merit of an SED plot is that it clearly shows which part of the spectrum carries most of the energy. This can be understood by noting that $F_\nu = dF/d\nu$ is the energy flux per linear frequency interval, and $\nu F_\nu = dF/d \ln \nu$ is the energy flux per logarithmic frequency interval. If one plots νF_ν against $\log \nu$, the area below the curve is proportional to the energy emitted by photons in a certain $\log \nu$ interval.

The general trend in Figure 8.2 is that more luminous blazars are "redder" (meaning that their SEDs peak at lower frequencies) than the less luminous blazars. However, there are substantial variations from source to source and individual sources can deviate substantially from this trend. BL Lacs designate a class of blazars called after the prototypical object BL Lacertae. The defining feature of BL Lacs is the absence of strong, broad emission lines in their optical spectra. Specifically, this is defined as the equivalent width of their emission lines being $EW < 5\,\text{Å}$. Depending on the location of the SED peaks one distinguishes low-frequency peaked BL Lacs (LBLs), intermediate-frequency peaked BL Lacs (IBLs), and high-frequency peaked BL Lacs (HBLs). Large samples of blazar SEDs have been compiled by Giommi *et al.* [33] and Abdo *et al.* [34]. In the case of LBLs, the low-energy component peaks at infrared frequencies and the high-energy component peaks in the MeV energy range [34]. HBLs have been detected with the low-energy component peaking at $> 100\,\text{keV}$, and the high-energy component extending all the way to > 10 TeV energies [35, 36]. The blazar class includes sources with similar SEDs as BL Lac objects but with quasar-like emission lines. Depending on the properties used for classification (optical variability, optical polarization or radio spectral index), the quasar type blazars are called OVVs (Optically Violent Variables), HPQs (High-Polarization Quasars) or FSRQs (Flat-Spectrum Radio Quasars), respectively. One of the main aims of current blazar research is to explain the differences between the various blazar classes in terms of the properties of the accreting supermassive black holes and their environments. Furthermore, one would like to identify the physical processes which determine the broadband spectral properties of blazars.

8.3.2
Blazar Models

Constraining the SEDs of blazars requires observations with a multitude of observatories to cover the entire electromagnetic spectrum. Observations should ideally be simultaneous, as blazar SEDs evolve continually. Snapshots of small portions of the SED taken at different times are of little value to constrain blazar models.

Figure 8.3 shows well-sampled SEDs of the objects 3C 279 (a famous FSRQ) and Mrk 421 (a prototypical HBL). As mentioned above, the SEDs are characterized by two broad humps, a low-energy emission component and a high-energy emission component. The current paradigm is that a single population of electrons emits the low-energy component as synchrotron emission. Evidence for the synchrotron origin of the low-energy component comes from measurements of very high ($\sim$ 50%) polarization degrees (e.g., [37, 38]). Whereas synchrotron emission can explain such high polarization degrees naturally (see Section 6.5) most other emission processes are expected to lead to much lower polarization degrees. Note that high polarization fractions imply an anisotropic magnetic field distribution, but not necessarily a completely uniform magnetic field. Laing [39] emphasized that a magnetic field isotropic in two dimensions can lead to as high polarization degrees as a uniform magnetic field (see also [40]).

There are two fundamentally different models for the origin of the high-energy (X-ray–γ-ray) component of blazar SEDs. In *leptonic models*, it is assumed that the high-energy emission is dominated by Compton scattering of low-energy photons off the same ultrarelativistic electrons that produce the synchrotron emission. If protons are also present in the emission region, they are assumed not to have sufficiently high energies to contribute substantially to the high-energy emission. However, if ultrarelativistic protons are also present in the emission region, they may produce a substantial level of proton synchrotron emission. Furthermore, if they exceed the threshold for photo-pion production on the synchrotron or external radiation fields, they will initiate electromagnetic cascades which may dominate the high-energy radiation output. Models based on the latter scenario are called *hadronic models*.

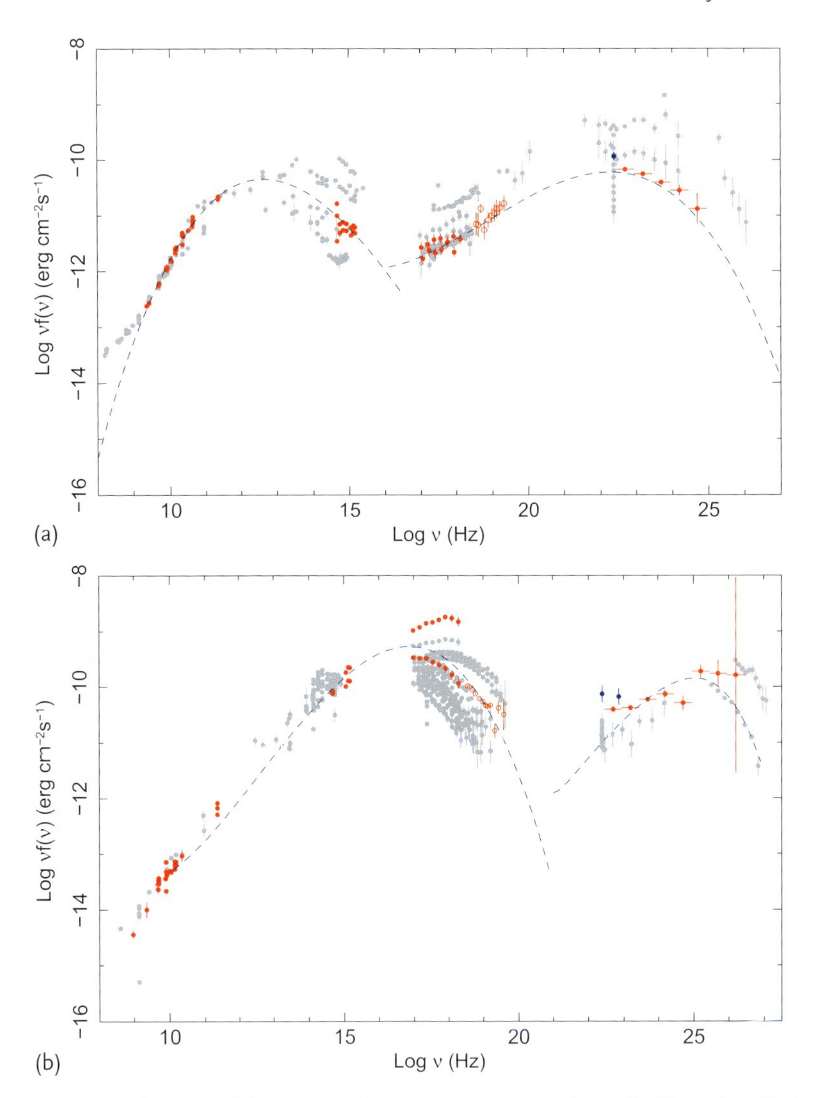

Figure 8.3 The SEDs of the FSRQ 3C 279 (a) and the HBL Mrk 421 (b) (from [34]). Quasi-simultaneous data are shown as large filled red symbols, and nonsimultaneous archival measurements are shown as small open gray points. The dashed lines show fits to the quasi-simultaneous data (see [34], for details). Reproduced from Abdo *et al.* [34] with permission from the AAS.

8.3.2.1 Leptonic Models: Synchrotron Self-Compton and External Compton Models

In leptonic models, the high-energy component originates as inverse-Compton emission of high-energy electrons scattering off lower-energy target photons. In the typical environments of blazars, there are several radiation fields that can potentially serve as target photons for Compton scattering off relativistic electrons (see Figure 8.4). These include (i) cospatially produced synchrotron emission, (ii)

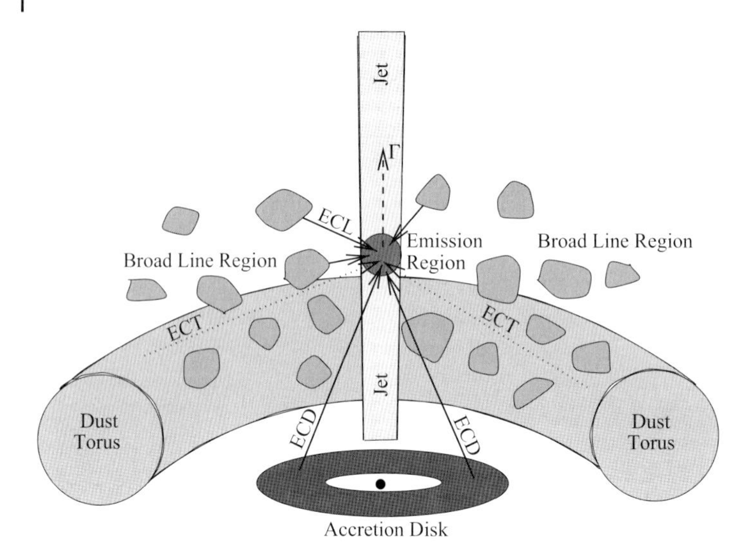

Figure 8.4 Sketch of the central region of an active galaxy, illustrating the various external radiation fields that may be Compton scattered to form the high-energy emission observed from AGNs.

optical/UV/X-ray emission from the central accretion disk, (iii) line emission from the broad- and narrow-line region, (iv) infrared emission from warm dust, and (v) the cosmic microwave background (CMB) radiation, to name just the most commonly considered contributions.

It is often assumed that the magnetic field does not have a preferred orientation (i.e., it is "tangled") in the co-moving frame of the emission region. Since we usually assume that the electron distribution is isotropic in the comoving frame of the emission region, so is the synchrotron radiation and, hence, also the SSC emission, which corresponds to (i) above. Therefore, the simple flux boosting laws of (2.41) and (2.44) apply.

However, the situation is somewhat more complicated when we consider external radiation fields. Due to relativistic aberration, external radiation fields will generally *not* be isotropic in the frame of the emission region. The angle-dependent external photon density then needs to be evaluated according to (2.54) for the transformation of the radiation energy density $u_\nu(\Omega)$. In case (ii) above, this transformation can become quite cumbersome, in particular close to the accretion disk, where there is already a strong intrinsic angular dependence of the spectral shape and intensity of the disk radiation in the rest frame of the AGN (e.g., [41–43]).

In cases of more extended sources of external radiation fields, for example, the broad-line region, the dust torus, or, in particular, the CMB, the intrinsic angular dependence of the field is typically rather weak. Therefore, the angular characteristics of the external radiation field in the comoving frame of the emission region will be dominated by relativistic aberration due to the relativistic bulk motion with Lorentz factor Γ. In that case, (2.57) for the transformation of an isotropic exter-

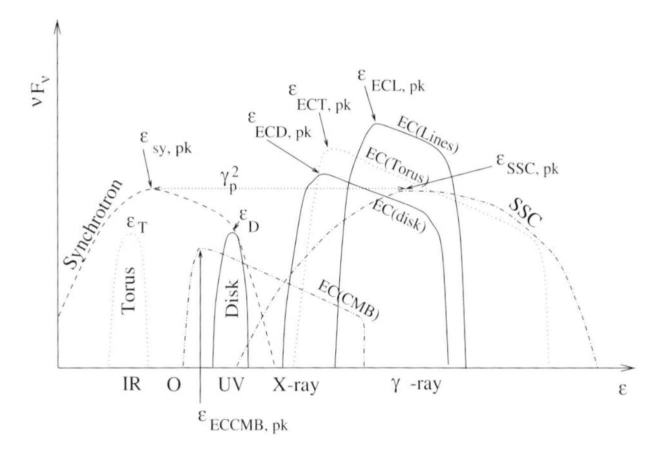

Figure 8.5 Sketch of synchrotron and Compton spectra from various synchrotron and external photon sources.

nal radiation field can be considered an appropriate approximation. Hence, the energy density of the external radiation field will be boosted by approximately a factor $(4/3)\Gamma^2$ into the emission-region frame. Now, in the Thomson regime, the luminosity due to Compton scattering in the comoving frame will be proportional to u^{em} (see (3.63)). Therefore, the received external Compton flux will not only depend on the δ^3 boosting of intrinsically isotropic emission, but will have an additional Γ^2 dependence due to the boosting of the external radiation field into the emission-region frame [44].

It is beyond the scope of this book to present the details of the transformations of the various radiation fields into the emission-region rest frame, and of the calculation of the resulting Compton spectra. The interested reader can find several examples of detailed calculations of the transformations of radiation fields and the resulting Compton spectra relevant for relativistic jets in AGN in [43]. We will restrict this exposition to a general discussion of the salient features of spectra arising from the Comptonization of the radiation fields listed above. We will assume that the electrons in the emission region are characterized by a broad distribution with a low-energy cutoff at γ_1, a peak (where most of the energy in the electron distribution is concentrated) at $\gamma_{\text{p}} \equiv 10^3 \gamma_{\text{p},3}$, and a high-energy cutoff $\gamma_2 \equiv 10^5 \gamma_{2,5}$. Above the peak, the distribution is characterized by a power law with index p. The emission region moves with a bulk Lorentz factor $\Gamma \equiv 10\Gamma_1$, and the viewing angle is such that the Doppler factor is $\delta \equiv 10\delta_1$. The characteristics of the Compton spectra can then be estimated as discussed below, and sample spectra are sketched in Figure 8.5.

1. *Synchrotron self-Compton* (SSC) [45–47]: SSC, the Compton scattering of the synchrotron emission produced by the same electron population that produces the high-energy emission, typically results in a very smooth, featureless broadband spectrum. This is because the synchrotron seed photon field is already a broad

distribution, so that for any scattered photon energy ϵ_s, there is typically a wide range of combinations of seed photon energy and electron energy that can result in the production of a photon of ϵ_s. Furthermore, towards the highest energies, Klein–Nishina effects are gradually beginning to suppress Compton scattering. Therefore, any spectral features in the synchrotron and electron spectra are smeared out. In the Thomson regime, the peak of the SSC spectrum can be directly related to the peak of the synchrotron spectrum through

$$\epsilon^{\text{obs}}_{\text{SSC, pk}} \sim \epsilon^{\text{obs}}_{\text{sy, pk}} \gamma_{\text{p}}^2 \sim 10^6 \gamma_{\text{p},3}^2 \epsilon^{\text{obs}}_{\text{sy, pk}} . \tag{8.1}$$

Because the synchrotron spectrum typically extends down to low energies ($\epsilon \ll 10^{-6}$), there are usually still photons available to Compton scatter in the Thomson regime ($\epsilon\gamma_2 \ll 1$) even for ultrarelativistic electrons ($\gamma_2 \gtrsim 10^6$). In the Klein–Nishina regime, the high-energy end of the SSC spectrum is typically related to γ_2 through

$$\epsilon^{\text{obs}}_{\text{SSC, max}} \sim \gamma_2 \delta \sim 10^6 \delta_1 \gamma_{2,5} . \tag{8.2}$$

This corresponds to an observed photon energy of $500\delta_1\gamma_{2,5}$ GeV. Thus, if the electron distribution extends out to $\gamma_2 \sim 10^6$, the SSC process can easily produce photons up to TeV energies.

2. *Direct accretion-disk emission* [41, 48, 49]: As indicated above, the transformation of the accretion disk radiation field into the emission-region rest frame is complicated by the radial structure of the accretion disk and the strong angular dependence of the disk emission. In particular, keep in mind that, if the jet is highly relativistic ($\Gamma \gg 1$), photons entering the emission region under an angle $\theta \gtrsim 1/\Gamma$ with respect to the jet axis, will still appear to enter the emission region from the front due to relativistic aberration. The situation becomes simpler if the emission region is at a sufficiently large distance from the accretion disk ($d \gg \Gamma^2 R_{\text{d}}$, where R_{d} is the characteristic radius of the disk). In that case, most accretion disk photons enter the emission region essentially from behind and will be deboosted by a factor $\sim 1/\Gamma$. Let us characterize the accretion disk emission by a peak photon energy $\epsilon_{\text{D}} \sim 10^{-5}$. This is in the ultraviolet region of the spectrum, typical for accretion disks around supermassive black holes. The peak of the Compton scattered accretion disk spectrum (ECD = external Compton of disk emission) will then be observed at

$$\epsilon^{\text{obs}}_{\text{ECD, pk}} \sim \epsilon_{\text{D}} \frac{\delta}{\Gamma} \gamma_{\text{p}}^2 \sim 10 \frac{\delta}{\Gamma} \gamma_{\text{p},3}^2 , \tag{8.3}$$

which corresponds to $\sim 5(\delta/\Gamma)\gamma_{\text{p},3}^2$ MeV, that is, soft- to medium-energy γ-rays. The high-energy end of the spectrum will be determined either by the onset of the Klein–Nishina suppression at $\epsilon_{\text{D}}\gamma/\Gamma \sim 1$, or by the maximum electron energy at γ_2. For typical values of $\gamma_2 \lesssim 10^6$, it is usually the latter constraint that determines the high-energy end of the ECD spectrum at

$$\epsilon^{\text{obs}}_{\text{ECD, max}} \sim 10^5 \frac{\delta}{\Gamma} \gamma_{2,5}^2 , \tag{8.4}$$

which corresponds to $\sim 50(\delta/\Gamma)\gamma_{2,5}^2$ GeV and is in the high-energy to very-high-energy γ-ray domain. The spectrum between $\epsilon_{\text{ECD, peak}}$ and $\epsilon_{\text{ECD, max}}$ will be represented by a power law with index $\alpha = (p-1)/2$ and should consequently resemble the high-frequency portion of the synchrotron spectrum.

3. *Line emission from the BLR* [50–52]: The line emission from the broad-line region (BLR) of an AGN is often dominated by optical and near-UV emission lines (e.g., Ly_α) with $\epsilon_\text{L} \sim 10^{-5}$. If the emission region is located within the BLR, the line radiation field will only have a weak angular dependence and is often approximated as isotropic in the AGN frame. In the emission-region rest frame, it will then appear blueshifted by a factor $\sim \Gamma$. Therefore, the peak of the Compton spectrum (ECL = external Compton of line emission) is expected to appear at

$$\epsilon_{\text{ECL, pk}}^{\text{obs}} \sim 10^3 \delta_1 \Gamma_1 \gamma_{\text{p},3}^2 , \tag{8.5}$$

which is at ~ 100 MeV–a few GeV, that is, in the range of high-energy γ-rays observable by the Fermi Gamma-Ray Space Telescope. The high-energy end of the ECL spectrum is likely dominated by the Klein–Nishina cutoff ($\gamma\Gamma\epsilon_\text{L} \sim 1$) and will be at

$$\epsilon_{\text{ECL, max}}^{\text{obs}} \sim 10^5 \frac{\delta}{\Gamma} , \tag{8.6}$$

which is still in the multi-GeV regime of high-energy γ-rays. The spectrum between $\epsilon_{\text{ECL, pk}}$ and $\epsilon_{\text{ECL, max}}$ is expected to represent a power law with index $\alpha = (p-1)/2$, as for synchrotron and ECD discussed above.

4. *IR emission from dust* [53, 54]: The infrared emission from warm dust around the central engine of the AGN typically peaks around $\epsilon_\text{T} \sim 10^{-7}$. With these low seed photon energies, Klein–Nishina effects are expected to play a negligible role. Also, due to the expected large extent of the warm dust, the IR emission from the warm dust will enter the emission region preferentially from the side in the AGN frame. Due to relativistic aberration, it will then appear to enter the emission region in the comoving frame at an angle $\cos\theta \sim -\beta_\Gamma$, that is, almost from the front. It will therefore appear blueshifted by a factor $\sim \Gamma$. Hence, we can estimate the characteristic frequencies of the spectrum from external Comptonization of torus photons (ECT) as

$$\epsilon_{\text{ECT, pk}}^{\text{obs}} \sim 10\delta_1 \Gamma_1 \gamma_{\text{p},3}^2 \quad \text{and} \quad \epsilon_{\text{ECT, max}}^{\text{obs}} \sim 10^5 \delta_1 \Gamma_1 \gamma_{2,5}^2 , \tag{8.7}$$

with the usual $\alpha = (p-1)/2$ spectrum between those energies. Thus, the ECT spectrum may span several orders of magnitude in frequency in the γ-ray domain. For bulk Lorentz factors Γ and Doppler factors δ exceeding 10, and values of $\gamma_2 \sim 10^6$, the ECT component can also plausibly produce emission in the very-high-energy range > 100 GeV.

5. *The CMB* [55]: The cosmic microwave background peaks at a photon energy around $\epsilon_\text{CMB} \sim 10^{-9}$. Therefore, Klein–Nishina effects are not expected to play a role when it is Compton scattered. Due to its intrinsic isotropy, the emission

region will "see" these photons blueshifted by a factor $\sim \Gamma$. The characteristic photon energies of the external Compton of CMB (ECCMB) radiation spectrum are therefore

$$\epsilon^{\text{obs}}_{\text{ECCMB, pk}} \sim 0.1 \delta_1 \Gamma_1 \gamma^2_{\text{p,3}} \quad \text{and} \quad \epsilon^{\text{obs}}_{\text{ECCMB, max}} \sim 10^3 \delta_1 \Gamma_1 \gamma^2_{2,5} , \tag{8.8}$$

spanning the soft X-ray to medium-energy γ-ray regime. The ECCMB process has been invoked to explain the X-ray emission from spatially resolved knots in the jets of several radio galaxies, and has been discussed in more detail in Chapter 7.

In addition to the target photon sources discussed above, more elaborate multi-component jet scenarios have been considered by various authors. For example, Georganopoulos and Kazanas [56] and Ghisellini, Tavecchio and Chiaberge [57] consider that synchrotron emission from slowed down jet components or from a slow outer layer of the jet illuminates the population of highly relativistic electrons. Ghisellini and Tavecchio [58] calculated blazar SEDs with inverse-Compton components resulting from the superposition of various seed photon populations.

As discussed in Chapter 3 (3.65), a comparison of the levels of high-energy (Compton), L_{C}, and low-energy (synchrotron), L_{sy}, emissions allows one to derive the ratio of the energy densities in the magnetic field and the target radiation field for Compton scattering,

$$\frac{L_{\text{sy}}}{L_{\text{C}}} = \frac{u_B}{u_{\text{rad}}} \tag{8.9}$$

as long as the scattering of the target photons occurs in the Thomson regime. The equation demonstrates that every synchrotron radiation emitting electron population also emits SSC photons. For a given magnetic field strength and number of emitting electrons, the SSC component is strong if the emission region is compact. The larger the emission region, the more diluted are the synchrotron photons, the lower is the synchrotron contribution to the target photon energy density u_{rad}, and the lower is the self-Compton luminosity.

Various authors developed codes to model the SEDs of blazars. The simplest codes assume a certain electron energy distribution and compute the emitted synchrotron and inverse-Compton emission (e.g., [59–61]). Usually one assumes that diffusive acceleration at strong shocks produces an electron energy spectrum with a spectral index $p = 2$ in the case of nonrelativistic shocks and $p = 2.2$–2.3 for relativistic (parallel) shocks. At higher energies one expects a steepening of the electron energy spectrum due to radiative cooling. The radiative energy-loss rates per electron per unit time in the comoving frame can be expressed as [62]

$$\frac{d\gamma}{dt} = -\frac{4}{3}\sigma_{\text{T}} c \gamma^2 \frac{u}{m_{\text{e}} c^2} , \tag{8.10}$$

where $u = u_B + u_{\text{rad}}$ is the sum of the energy densities of the magnetic field and all radiation fields (cf. (3.27) and (3.63)). The time scale on which an electron loses a substantial fraction of its energy is called the *cooling time* and is given by the expression

$$\tau = \frac{\gamma}{|d\gamma/dt|} . \tag{8.11}$$

From (8.10) we infer that the radiative cooling time scales proportional to γ^{-1}. Thus, the highest-energy electrons are the first to radiate away their energy leading to a steepening of the electron energy spectrum by $\Delta p = 1$ above a characteristic electron Lorentz factor γ_b (see [63], for an analytical treatment). γ_b is the electron Lorentz factor at which the radiative cooling time equals a characteristic dynamical time scale, $\tau(\gamma_b) = t_{dyn}(\gamma)$, e.g., the time scale for escape of the electrons from the emission region, or the time elapsed since the onset of the injection/acceleration process. An electron energy spectrum with $dN_e/dE_e \propto E_e^{-p_1}$ for $\gamma \leq \gamma_b$ and $dN_e/dE_e \propto E_e^{-p_2}$ for $\gamma > \gamma_b$ leads to a synchrotron spectrum with a low-energy photon index of $\alpha_1 = (p_1 - 1)/2$ and a high-energy photon index of $\alpha_2 = (p_2 - 1)/2$. If the break is exclusively due to radiative cooling through synchrotron radiation and Compton scattering in the Thomson regime, its magnitude is expected to be $\Delta\alpha = 0.5$.

A simple SSC code has the following fitting parameters: the radius of the emission region R (assumed to be spherical), the Doppler factor δ of the emission volume, the minimum and maximum Lorentz factors γ_{min} and γ_{max} of accelerated electrons, the Lorentz factor γ_b of the break of the electron energy spectrum, the spectral indices p_1 and p_2 before and after the break, and a normalization factor relating to the total number of emitting electrons. The values of some of these parameters can be directly inferred from the observed shape of the SED [64]. However, typically models are degenerate, and, in addition, there are only rough lower or upper bounds on some of the model parameters. An example of two degenerate model parameters is the jet magnetic field and the jet Doppler factor. The SED constrains only combinations of these two parameters. External Compton models have additional model parameters describing the properties of the external radiation field in the jet frame, that is, its spectral properties and the intensity.

In more sophisticated codes, the electron energy spectrum is derived self-consistently by modeling the acceleration and cooling of electrons (e.g., [42, 65–69]). Even more sophisticated treatments involve multizone radiation transfer models. In particular, internal shocks resulting either from the collision of isolated shells of relativistic materials ejected with different bulk Lorentz factors, or from standing shocks (Mach disks) plausibly resulting from the recollimation of jets [70–73] have met with substantial success in reproducing SEDs and light curve features of blazars.

In the framework of leptonic modeling of blazars, some authors explain the variation of the emission properties along the blazar sequence FSRQ → LBL → IBL → HBL through a decreasing contribution of external radiation fields to the radiative cooling of electrons [74]. Indeed, the emission of HBLs has often been fitted with pure SSC models, while FSRQs often require a substantial EC component. This interpretation is consistent with the observed strong emission lines in FSRQs, which are absent in BL Lac objects. At the same time, the denser circumnuclear environment in quasars might also lead to a higher accretion rate and hence a more powerful jet, consistent with the overall trend of bolometric luminosities along the blazar sequence. As as source evolves through cosmic time, it may transition from exhibiting FSRQ type emission properties early on to BL Lac type emission properties [75]. The AGN activity may initially be triggered by galaxy mergers, resulting in dense circumnuclear environments and quasar-like characteristics. With time, the circumnuclear fuel is depleted as it is either accreted onto the black hole, or ejected in winds and jets, leading to a gradual transition to BL Lac-like characteristics.

In a few cases, it is possible to measure time lags between the variability patterns at different frequencies. This has been particularly successful with X-ray observations of HBLs, where the X-ray emission is dominated by synchrotron emission from the highest-energy electrons. Assuming that the time lags between the emission at two frequencies can be interpreted as the difference in the radiative cooling time scale of the electrons emitting at those frequencies, (8.10) and (8.11) can be used to derive an estimate of a combination of the magnetic field and the Doppler factor. Let us recall from Section 3.2 (3.35) that the characteristic synchrotron frequency for an electron with Lorentz factor γ corresponds to

$$E_{sy} = 1.74 \times 10^{-11} B_G \gamma^2 \frac{\delta}{(1+z)} \text{ keV} , \tag{8.12}$$

where z is the redshift of the source and B_G is the magnetic field in units of gauss. We can invert (8.12) to find the Lorentz factor γ corresponding to an observed photon energy $E = E_{keV}$ keV. Let us furthermore parameterize the importance of Compton cooling by a Compton parameter k such that

$$u = u_B + u_{rad} \equiv u_B(1+k) . \tag{8.13}$$

The value of k can usually be estimated from the ratio L_C/L_{sy} (see (8.9)). In the case of HBLs, the two broad emission peaks carry approximately equal luminosity, so that $k \sim 1$. A time scale τ in the comoving frame of the emission region will be observed as $\tau_{obs} = \tau(1+z)/\delta$. Now, plugging (8.13) and (8.12) into (8.11), and identifying an observed time lag τ_{obs} with the difference in radiative cooling time scales, we find

$$\tau_{obs} = 3.22 \times 10^3 (1+k)^{-1} \sqrt{\frac{1+z}{\delta}} B_G^{-3/2} \left(E_{1,\,keV}^{-1/2} - E_{2,\,keV}^{-1/2} \right) \text{ s} , \tag{8.14}$$

which can easily be inverted to find an estimate of the magnetic field from the observed time lags between the variations at two photon energies E_1 and E_2, if an independent estimate of the Doppler factor δ is available [76].

8.3.2.2 **Hadronic Models**

Hadronic blazar models are characterized by the existence of a relativistic hadron component (in addition to relativistic pairs) that contributes to the overall photon spectral energy distribution as typically observed from blazar jets. Such considerations are deeply connected to the still unknown composition of the relativistic particle population in jets: protons and electrons, or pairs. There is meanwhile mounting evidence that jets of powerful AGN may be energetically and dynamically dominated by protons and/or ions, albeit with an uncertain electron-to-proton number ratio, e/p. This is implied by the limits from the jet's pair content through the observed lack of soft X-ray excesses and/or X-ray precursors from Comptonization on cold pairs [77, 78], through modeling knots and features in radio galaxies considering the Comptonization of the cosmic microwave background radiation [79], and through broadband modeling of a large collection of blazars using leptonic emission processes [80]. In order to support the observed dissipated luminosity of blazars and radio lobes of quasars protons are required as the main jet energy carriers. Furthermore, on kpc-scales the $p\,dV$ work of observed X-ray cavities (which are interpreted as expanding bubbles from radio sources) in galaxy clusters requires a significant matter component beyond what pairs can provide in order to ensure pressure equilibrium (e.g., Cyg A [81]). Hydrodynamic simulations show that the shape of the observed cavities can be reproduced by low-density jets that are dominated by relativistic cosmic rays [82]. Further motivation to consider the role of relativistic protons in AGN jets comes from the detection of ultra-high-energy cosmic rays (UHECRs), whose hadronic component extends up to $\sim 10^{20}$ eV. An argument invoked by Hillas provides an upper limit of the maximum attainable energy from the requirement that the particles are confined to the acceleration region: $E_{\rm max} < 3 \times 10^{19} (B/10\,{\rm G})(R/10^{16}\,{\rm cm})\,{\rm eV}$ [83] in the case of protons, with R the size of the acceleration site, and B the magnetic field strength. Thus ultrahigh energies may be possible for cosmic rays in highly magnetized environments ($\geq$ several tens of gauss) for typical "blob" sizes in jets of AGN. Another requirement concerns the overall jet power: For magnetized relativistic outflows a minimum power of $P_{\rm jet} > 10^{46} \Gamma^2 \beta^{-1} (E_{\rm p}/10^{20}\,{\rm eV})^2$ erg/s is necessary [84]. Powerful high-energy AGN generally fulfill this requirements and are therefore among the prime candidate populations for the origin of the UHECRs.

Relativistic protons may interact with target photons or material to produce secondary particles, if above the corresponding particle production threshold: $\simeq 0.145$ MeV for photo-meson production and $\simeq 1.22$ GeV for inelastic proton–proton interactions in the nuclear rest frame. The latter requires jets with sufficient dense ambient matter to produce observable fluxes. In addition, if the π^0-decay photons from the proton–proton (pp) interaction represent part of the dominant photon production mechanism, then high densities of the target material, $n'_{\rm T}$ (jet frame), are needed to explain flux variability. By requiring the interaction time scale for pp-collisions to be smaller than the variability time scale, one finds (e.g., [85]) $n'_{\rm T} \gg 10^9/(t_{\rm var,day}\delta_{10})\,{\rm cm}^{-3}$ with $t_{\rm var,day}$ being the observed variability time scale in days, $\delta = 10\delta_{10}$ the bulk Doppler factor. Here the pp mean interaction time has been estimated from $t'_{\rm pp} = (n'_{\rm T} c \kappa_{\rm pp} \sigma_{\rm pp})^{-1}$ with $\kappa_{\rm pp} \approx 0.5$ being the mean inelasticity, and $\sigma_{\rm pp} \approx 30$ mb the nuclear interaction cross-section. The reader interested in heavy mass-loaded blazar jet models may consult, for example [86, 87]. Useful expressions for secondary particles from proton–proton interactions may be found, for example, in [88].

In the following we will disregard heavily mass-loaded jets, and focus instead on radiation-dominated jets where particle-photon interaction rates in general dominate over particle-particle interaction rates above the respective threshold.

In order to accelerate protons to ultrarelativistic energies high magnetic field strengths of at least several tens of Gauss are required. In such fields charged particles lose an appreciable amount of their initial energy by synchrotron radiation. There is overall consensus that synchrotron emission of the primary electrons dominates most portions of the low energy SED of AGN. If the field strengths are sufficiently high ($\geq$ several tens of Gauss) then the relativistic protons start to radiate significantly due to interactions with the magnetic field [89, 90], with important implications for the overall predicted neutrino fluxes from AGN jets.

Particle–photon interaction processes in hadronic models include photo-meson production,[1] Bethe–Heitler pair production[2] for protons, and inverse-Compton scattering of pairs. As noted in the previous paragraphs AGN jets offer a selection of target photon fields for such processes: internal jet synchrotron photon fields (e.g., [91–93]), and fields external to the jet such as direct accretion disk radiation [94, 95], jet or accretion disk radiation reprocessed in the BLR [85], or the radiation fields provided by the dusty torus and the CMB. For those processes that produce particles (here photo-meson production and Bethe–Heitler pair production) to take place, the injected protons must possess sufficiently high energies E_p to overcome the respective threshold conditions (see Chapter 3). These are $\gamma_p > 7\times10^4\,\Gamma_1$, $\gamma_p > 7\times10^2/\Gamma_1$, $\gamma_p > 7\times10^4/\Gamma_1$ and $\gamma_p > 7\times10^6/\Gamma_1$ for photo-meson production in direct accretion disk target photon fields ($\epsilon_D \sim 10^{-5}$), emission from the BLR ($\epsilon_L \sim 10^{-5}$), dust torus ($\epsilon_T \sim 10^{-7}$) and the CMB radiation field ($\epsilon_{CMB} \sim 10^{-9}$), respectively. Observations of the low-energy hump of blazar SEDs reveal a broad range of peak energies: while the low-energy component of LBL/IBL-like blazars and FSRQs typically peak in the optical – IR energy range ($\epsilon_{LSP} \sim 10^{-7...-5}$), HBLs can reach very hard X-rays with the peak around $\epsilon_{HSP} \sim 10^{-4...-3}$. Photo-meson production in these internal target radiation fields thus requires proton energies of at least $\gamma_p > 7\times10^{4...6}\delta_1$ in LBL/IBL- and FSRQ-like AGN, and $\gamma_p > 7\times10^{2...3}\delta_1$ in HBL-like radiation fields.

The corresponding threshold energies in case of Bethe–Heitler pair production are $\gamma_p > 4\times10^2\,\Gamma_1$, $\gamma_p > 4\Gamma_1$, $\gamma_p > 4\times10^2/\Gamma_1$ and $\gamma_p > 4\times10^4/\Gamma_1$ (for the external photon fields; same order as above), and $\gamma_p > 4\times10^{2...4}\delta_1$ and $\gamma_p > 4\times10^{0...1}\delta_1$ (for the internal photon fields; same order as above). Obviously, to produce neutrinos in target photon fields external to the jet, the accelerated/injected proton component must reach $\sim 10^3$ GeV. In this respect, photo-meson production in internal HBL-like photon fields seem favorable. However, as discussed below in flaring HBLs the gamma-rays are more likely due to mostly proton synchrotron radiation owing to

1) Above its threshold of $\sqrt{s} = m_p c^2 + m_{\pi^0} c^2$ for pγ.
2) Above its threshold of $\sqrt{s} = m_p c^2 + 2m_e c^2$ for pγ.

the low (jet frame) density of their internal jet radiation field. In this case, neutrino production is significantly suppressed in favor of photon production.

In case of sufficiently high magnetic field strengths, the produced (through photo-pion production) charged mesons (mostly pions) and leptons (mostly muons) may also lose part of their energy by synchrotron radiation before they decay [92, 93, 96]. The condition for this situation is obviously that the synchrotron-loss time scale shall be shorter than the decay time scale. At the particle energy where these time scales are equal, a (cooling) break in the emitting particle spectrum is expected, which impacts both the neutrino and photon spectrum (see Section 3.2.4). The so produced $\mu^{\pm}$ and $\pi^{\pm}$ synchrotron radiation may constitute a nonnegligible contribution at gamma-ray energies in the overall SED. In particular, because of the mass dependence of the synchrotron radiation ($\nu_c \propto Z E^2/m^3$ with E the particle energy; see (3.35)) one finds the muon/pion synchrotron component extending to higher energies than the proton synchrotron component, albeit often on a much smaller flux level.

The mesons eventually decay into γ-rays, pairs and neutrinos (ν) (see Chapter 3 for the respective source functions and overall distribution of the initial energy into the various channels). Because the initial proton energy is very high (required to overcome the threshold conditions!), the produced secondary particles are also very energetic, and consequently the emission region optically thick to π^0 decay γ-rays and synchrotron and Compton photons produced by the secondaries: a pair cascade develops which redistributes the power from very high to lower energies as explained in Section 3.3. Depending on the overall magnetic field to photon energy density in the emission region, synchrotron- [85, 91–93] or Compton-supported pair cascades [85] may dominate. For example, in the synchrotron-proton blazar (SPB) model [92, 93] where charged muons and pions, if produced, often contribute to a μ-, π- (and to a lesser amount also *kaon*)-synchrotron photon component, those pair cascades are clearly synchrotron-supported. Furthermore, as discussed in Section 3.3, these cascades are unsaturated, and therefore terminate usually already after a few generations. Such electromagnetic cascades in AGN jets can be initiated by photons from the following: π^0-decay ("π^0 cascade"), electrons from the $\pi^{\pm} \rightarrow \mu^{\pm} \rightarrow e^{\pm}$ decay ("$\pi^{\pm}$ cascade"), p-synchrotron photons ("p-synchrotron cascade"), and μ-, π- and K-synchrotron photons ("$\mu^{\pm}$-synchrotron cascade"). The work in [92] and [93] has shown that the "π^0 cascades" and "$\pi^{\pm}$ cascades" generate featureless γ-ray spectra, in contrast to "p-synchrotron cascades" and "$\mu^{\pm}$-synchrotron cascades" that produce a two-component spectrum. Note that the radiating particles in the various cascades are electrons and positrons of very large energies with their corresponding (short) loss time scales.

Hadronic interactions have also the property to convert part of the initial protons into neutrons (and vice versa) by means of an isospin flip.[3] As a consequence collimated neutron beams may form [85, 98] which can be used to transport a significant portion of the initial energy to large distances from the black hole.

3) For example, in hadronic pγ-interactions 30–70% of the initial protons are converted into neutrons [97].

As in the case of leptonic models, and also hadronic models, the evolution of the injected particle distributions (electrons and protons), and the secondary particle populations produced through the various processes is followed according to the (coupled) kinetic equations outlined in Section 3.3. This means that one-zone hadronic models possess in general more free fitting parameters than one-zone leptonic blazar models, where the emitting electron distribution is derived self-consistently: the magnetic field strength (assumed homogenous in the emission region), the size of the emission region, its bulk Doppler factor, the injected primary electron spectrum (spectral index and minimum/maximum Lorentz factor), the injected proton spectrum (spectral index, mostly assumed to be identical to the electron injection spectral index, and maximum Lorentz factor) and its energy density, and the electron-to-proton number ratio. Unlike in the case of leptonic models, multizone hadronic models have not been developed yet.

The resulting photon SED consists of several contributions. If high magnetic field strengths (several tens of gauss) are present in the emission region (e.g., in the SPB model [92, 93]), proton and $\mu^{\pm}$ synchrotron radiation and their reprocessed components are mainly responsible for the high-energy bump in blazars, whereas the low-energy bump is dominated by synchrotron radiation from the primary electrons, with some contribution from secondary pairs. The higher the field strength for a given target photon density, the higher the importance of proton synchrotron radiation ([89] for a corresponding hadronic model) relative to processes initiated by photo-meson production. If charged meson and proton synchrotron losses are not considered (e.g., in the "proton blazar" of [99]) the gamma-rays are solely the result of cascades initiated by the secondary pairs and π^0-decay photons (π cascades). In both models, the synchrotron emission from the primary electrons is the main target photon field for photo-meson production and the cascading for most model applications. When lowering the magnetic field strength both Compton- and synchrotron-supported pair cascades must be considered (e.g., [85] for a corresponding hadronic model), and charged meson and proton synchrotron emission becomes less important. Models where the external BLR radiation field constitutes the dominating target photon field (e.g., [85]) have the advantage that the pion production threshold can be reached by somewhat lower proton energies because the BLR photon energies are blueshifted by a factor Γ in the comoving frame of the jet. On the other side, if the emission region is very close to the central engine, the accretion disk emission may represent the dominating target radiation field (e.g., [94, 95]). In situations where the emission region is sufficiently far from the black hole, compared to the extent of the accretion disck, $d \gg \Gamma^2 R_{\mathrm{D}}$, most photons will enter the emission region from behind, and appear therefore deboosted by a factor Γ^{-1} in the comoving frame of the jet. This leads to a factor Γ^2 higher proton energy requirement with respect to the case of a dominating BLR radiation field (provided the emission region is inside the BLR boundary) to meet the threshold condition for meson production. Generally, hadronic AGN emission models consider also all leptonic processes owing to the presence of ultrarelativistic electrons and positrons in the emission region. For high field strengths, any inverse-Compton component is, however, strongly suppressed, leaving the proton-

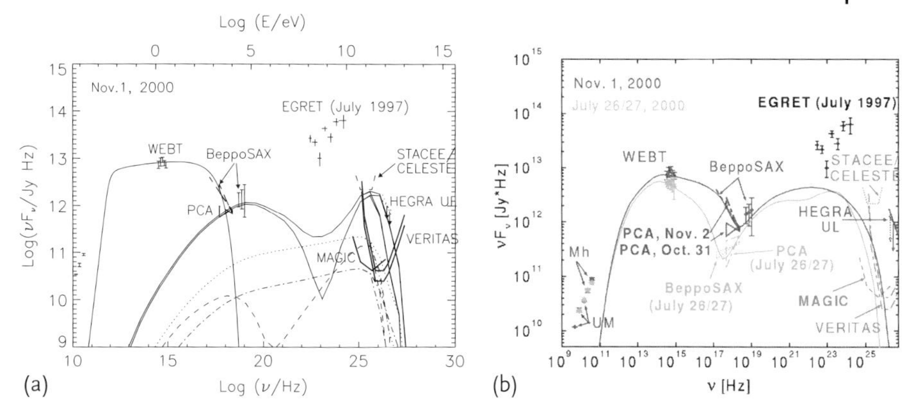

Figure 8.6 (a) The quasi-simultaneous broadband SED from 1 November 2000 of the LBL BL Lacertae and fits using the hadronic SPB model [92, 93]. The solid left line shows the synchrotron spectrum of the primary electrons, the right solid line represents the total cascade spectrum, which is the sum of proton synchrotron cascade (dashed line), μ/π synchrotron cascade (triple-dot-dashed line), π^0 cascade (dotted line), and $\pi^{\pm}$ cascade (dot-dashed line). The model fluxes are corrected for absorption in the Extragalactic Background Light radiation background. (b) Leptonic model fit to the same data using the time-dependent SSC + ERC model of [69]. The comparison shows that the hadronic models are more naturally predicting VHE γ-ray emission as later detected by MAGIC (adapted from [101]). Reproduced from Böttcher and Reimer [101] with permission from the AAS.

initiated radiation components as the dominant high-energy emission. Can we put constraints on the jet composition from multifrequency modeling of blazar SEDs? This avenue has been followed by a number of publications (e.g., [100–104]) where both a hadronic and a leptonic model has been applied to the same blazar SED. An example is depicted in Figure 8.6. So far both types of models were able to meet all requirements deduced from observations for the respective blazars. Note also that often the model fits are ambiguous, that is, several sets of the parameters can reproduce the observed model SED. The main difference between the resulting model parameters from hadronic and leptonic models is a general higher magnetic field in the emission region in the case of the SPB model, and often the requirement of a larger overall jet power.

An inevitable byproduct of hadronic interactions, and possibly the clearest signature of the existence of relativistic protons and/or ions in blazar jets, is the production of *neutrinos* through the decay of the photo-produced mesons (see Section 3.2.4). Because of the potentially denser target photon field provided by powerful FSRQs as compared to BL Lacs, members of the former source class are considered more promising neutrino emitters than the nearby BL Lacs. Within the BL Lac class HBL type AGN are predicted to produce the lowest neutrino fluxes (e.g., [93]). In the extreme case of the gamma-rays produced solely through proton synchrotron radiation no neutrinos are produced at all. Consequently, the AGN contribution to the diffuse neutrino background is probably dominated by objects with intrinsically dense target photon fields. Because of the tight connection between the production of secondary gamma-rays and neutrinos in photo-pion production (see Section 3.2.4), attempts have been made to estimate a limit to the diffuse neutrino flux from the sole scaling of the extragalactic diffuse gamma-ray background. The results of such estimates may, however, be considered with caution as upper limits in view of the assumptions involved.

If UHE hadrons exist in AGN jets those protons can be converted into neutrons, which then can escape from any magnetic confinement in the source due to their charge neutrality. After their decay into protons during propagation, they may then contribute to the observed cosmic rays. Consequently, a relation between observed cosmic ray and expected neutrino flux has been proposed by [105] where simplifying assumptions were used to derive an energy-independent diffuse neutrino flux limit. More realistic and detailed propagation calculations have subsequently revealed the full energy-dependent bound [106].

Hadronic models which predict high neutrino fluxes (e.g., [94, 106–108]) are currently challenged by observational flux limits from various neutrino telescopes which reach meanwhile cubic kilometer sizes (e.g., AMANDA-II, IceCube, Km3Net, and so on). Note, however, that the uncertainties and ambiguities in the model parameters, blazar luminosity functions, and so on, translate into corresponding uncertainties of the predicted neutrino fluxes which can extend up to several orders of magnitude (e.g., [93]).

At the time of writing observational muon neutrino flux limits on the search from potentially (preselected) extraterrestrial point sources are still compatible with instrumental background fluctuations, and 90% C.L. upper limits (assuming an E^{-2} spectrum) on the muon neutrino fluxes are on the order of a few tens of $10^{-12}\,\mathrm{TeV\,cm^{-2}\,s^{-1}}$ [109, 110]. Current observational limits on the diffuse muon neutrino flux at TeV-PeV energies has reached the order of $\sim 10^{-7}\,\mathrm{GeV\,cm^{-2}\,s^{-1}\,sr^{-1}}$ (for an E^{-2} spectrum and assuming a $1:1:1$ flavor ratio) [110, 111].

8.3.3
Blazar Multiwavelength Observations

Intense observation campaigns have been performed to unambiguously identify the emission mechanisms and to constrain the parameters describing the emitting plasmas in blazar jets. A well-known campaign during the EGRET era lead by Wehrle *et al.* [112] targeted the FSRQ 3C 279 (see also [113]). The gamma-ray flux increased by a a factor of 10 within 10 days (Figure 8.7a). Inspecting the SED shows that the entire high-energy component brightened – possibly because of an increase of the EC target photon energy density (Figure 8.7b). The low-energy component showed less-dramatic flux variability.

In the simplest versions of one-zone leptonic models, one would expect the emission at the various wavelength ranges to be closely correlated, since the entire SED is produced by essentially the same electron population. Hartman *et al.* [114] performed a detailed analysis of cross-correlations between the optical, X-ray, and γ-ray emissions during the EGRET era. Surprisingly, no significant persistent cor-

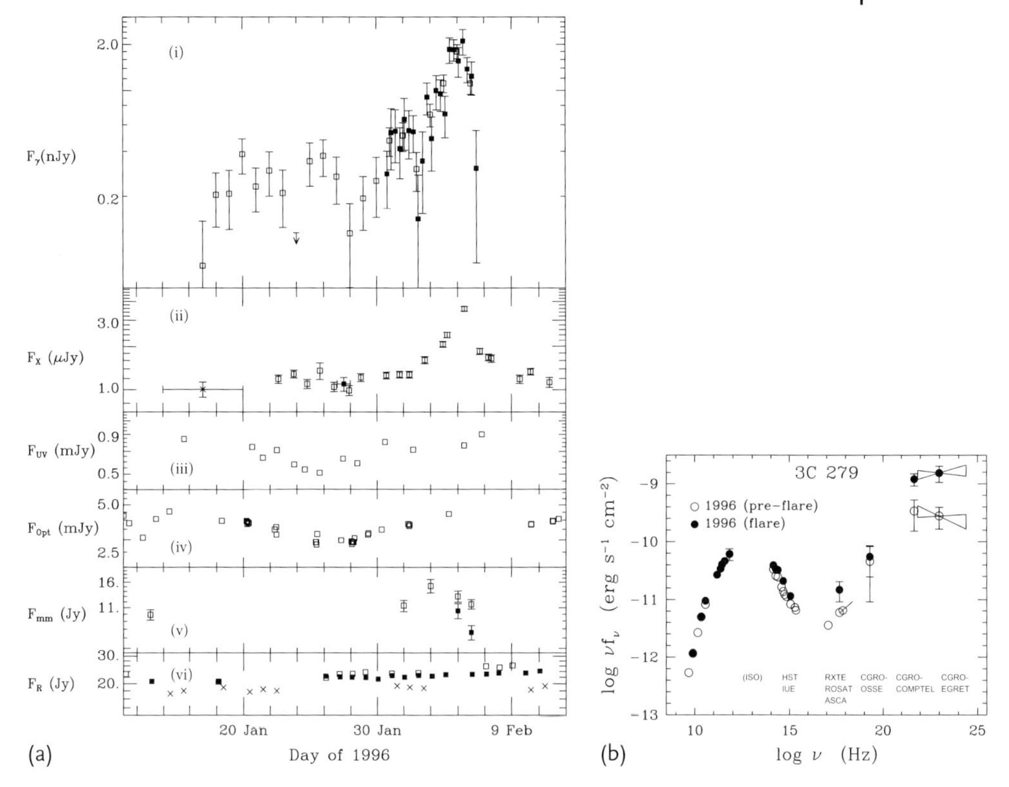

Figure 8.7 (a) Light curves from a multiwavelength campaign targeting the FSRQ 3C 279 during 16 January–6 February 1996 [112]. The fluxes in the gamma-ray band (i), X-ray fluxes (ii), UV fluxes (iii), optical R band fluxes (iv), 0.8 mm (open squares) and 0.45 mm (filled squares) fluxes (v), and radio data (vi) at 37 GHz (open squares), 22 GHz (filled squares), and 14.5 GHz (crosses). (b) SED for a "preflare time interval" (24–28 January) and a "flare time interval" (4–6 February). See [112] for more information. Reproduced from Wehrle *et al.* [112] with permission from the AAS.

relations were found. While quite often optical and γ-ray flares accompany each other, there are other instances where this is not the case. The X-ray and radio emissions often do not show any significant correlation with the optical and γ-ray emissions at all. A similarly inconclusive trend was also found, e.g., for the quasar type blazar PKS 1510-089 (see Figure 8.8 [115]). The lack of a clear correlation with the radio regime is most likely explained by the fact that within the "blazar zone", the emission region is still optically thick to radio emission, so that the observed radio flux originates further out along the jet. The X-ray emission in FSRQs like 3C 279 and PKS 1510-089 is most likely produced by SSC emission from relatively low-energy electrons (below the break energy in the electron spectrum), which are cooling on longer time scales than the high-energy electrons producing the γ-ray (Compton) and optical (synchrotron) emission. This may explain the lack of a close temporal correlation of the X-rays with γ-rays and optical. In a more recent observation campaign on 3C 279 [116] the gamma-ray flare seems to be accompanied

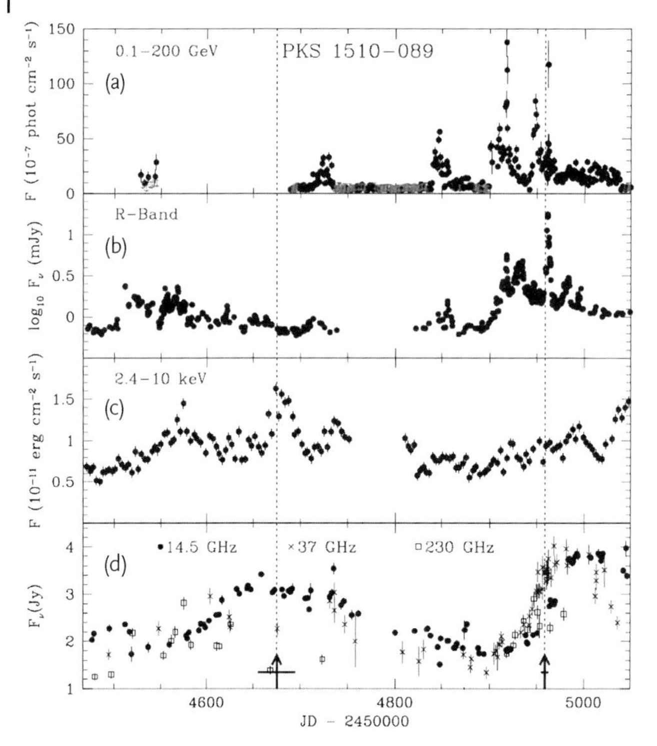

Figure 8.8 Multiwavelength light curves of the FSRQ PKS 1510-089 [115]: (a) Fermi γ-rays, (b) optical R band, (c) X-rays, and (d) radio. While some optical and γ-ray variability patterns seem well correlated, such a correlation is not always present. The X-ray light curve does not show any obvious correlation with the optical or γ-ray light curves. Reproduced from Marscher *et al.* [115] with permission from the AAS.

by a swing of the optical polarization. This may provide evidence for blazar flares originating in strong shocks.

A large number of observation campaigns targeted the HBLs Mrk 421, Mrk 501, PKS 2155-304, and 1ES 1959+650. In the case of these HBLs, X-ray satellites can be used to observe the low-energy component of the SED close to its peak and ground-based TeV gamma-ray telescopes can be used to observe the high-energy component close to its respective peak. The observations are attractive as both components exhibit fast flux variability on time scales of minutes, and observation campaigns lasting a few days can be used to study the correlation of the X-ray (presumably synchrotron) and TeV gamma-ray (presumably inverse-Compton) fluxes in great detail. Interestingly, the X-ray and gamma-ray light curves of HBLs show extremely fast flares with the shortest flux doubling times Δt of a few minutes [117–119]. The rapid gamma-ray flux variations have been used to argue that blazar jets move highly relativistically with bulk Lorentz factors and relativistic Doppler factors > 10 or even ~ 50 [24, 120]. If the emission volumes did not move relativistically, the gamma-ray emission could not escape without being absorbed in pair absorption processes (see Section 3.2.3) through pair production on cospatially emitted synchrotron photons observed in the IR regime. The relativistic motion ameliorates this problem in several ways: the emission volume can be larger by a factor of δ than inferred from the time scale of the flares; the gamma-rays have lower energies in the moving reference frame, so that they can pair produce only with the less-abundant higher-frequency photons; last but not least, the photon energy densities are lower in the jet frame, as the high brightness of the observed flares results from the relativistic boosting of the emission. Importantly, independent evidence for large bulk Lorentz factors and relativistic Doppler factors on the order of 50 came from modeling the SEDs of HBLs with leptonic models [121].

One of the main results of the observation campaigns targeting HBLs has been that the X-ray and TeV gamma-ray fluxes are correlated, see Figure 8.9. Although some flares show a tight correlation (e.g., [35, 122]) even during multiple flares (e.g., [123]), others do not (e.g., [61, 122, 124]). The variations in the X-ray to TeV

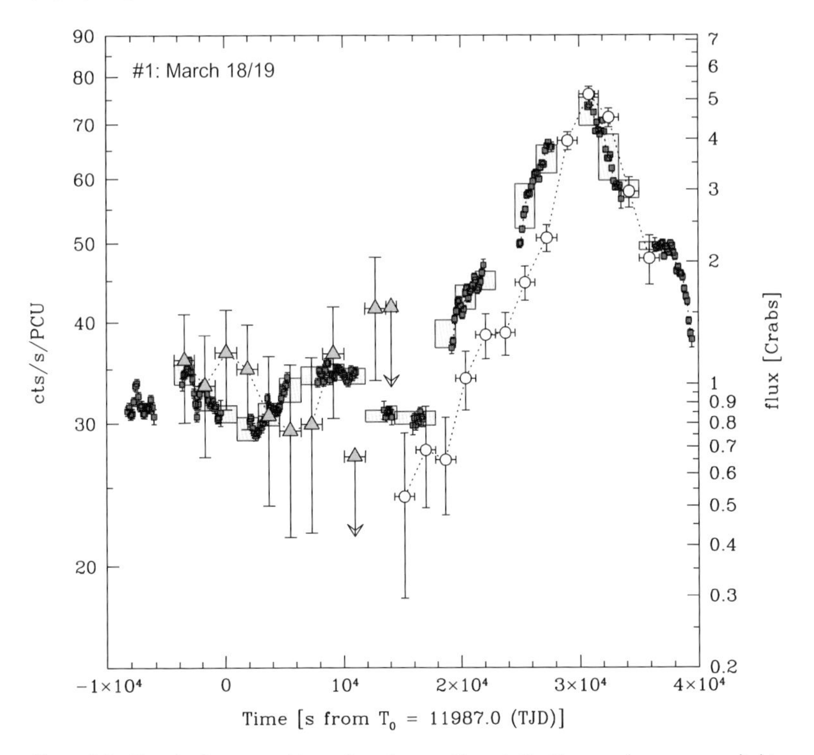

Figure 8.9 Results from a multiwavelength observation of the HBL Mrk 421 (from [122]). The light gray triangles and white circles show TeV gamma-ray fluxes, and the dense dark gray points show 2–10 keV X-ray fluxes. The shaded boxes represent the average and variance of the X-ray data for each (longer duration) TeV bin. The left (right) axis shows the units used for the X-ray (gamma-ray) data. Although the X-ray and gamma-ray light curves were measured at frequencies approximately eight orders of magnitude apart, the fluxes are well correlated bolstering the hypothesis that the X-ray and TeV gamma-ray emission is synchrotron and inverse-Compton emission from one population of relativistic electrons. See [122] for details. Reproduced from Fossati *et al.* [122] with permission from the AAS.

gamma-ray flux correlations can be explained with SSC models but require variations of more than one model parameter.

Modeling HBL (and more recently IBL) multiwavelength data has so far not been successful in distinguishing between SSC and EC models. The following conclusions can be drawn: SSC and EC models describe the data rather satisfactorily, even though some observations require the presence of several emission zones (e.g., [61, 125]). Whereas SSC models require electron (and positron) energy densities of

$$u_\mathrm{e} = \int_{\gamma_\mathrm{min}}^{\gamma_\mathrm{max}} d\gamma \frac{d N_\mathrm{e}}{d\gamma} \gamma m_\mathrm{e} c^2 \tag{8.15}$$

exceeding the energy density of the corresponding magnetic field u_B by up to several orders of magnitude, EC models can fit the data with u_e and u_B in rough equipartition: $u_\mathrm{e} \sim u_B$ [56, 57, 126, 127].

From a theoretical standpoint, an equipartition between the particle and magnetic field energy densities is preferred: one of the favored jet formation scenario starts with magnetically dominated jets (e.g., [128–135]). Magnetic reconnection may lead to a conversion of magnetic field energy into particle energy (e.g., [78, 136, 137]). However, this process is expected to stop at the very latest when the particle energy density (electrons and protons) equals the magnetic field energy density and is thus unlikely to produce a plasma which is strongly dominated by the energy of electrons (and positrons). An electron energy density greatly exceeding the magnetic field energy density would also imply that the pressure of the electron gas exceeds the pressure of the magnetic field. This would make magnetic jet collimation very problematic and seems to contradict the observational fact that the jets of AGN tend to remain well collimated out to > kpc scales. Some authors tried to constrain the jet composition. Celotti and Fabian [138] and Sikora and Madejski [77] both argue that electrons and pairs cannot be the main carriers of energy and momentum of blazar jets – at least not all the way from the formation of the jet to kpc distances. Celotti *et al.* [138] use an SSC model of the blazar emission to constrain the electron/positron density in the jet, and VLBI observations of blazar radio cores to estimate the diameter of the jets. The authors find that the pairs cannot transport a sufficient energy to Mpc scales. Sikora and Madejski [77] show that if the jet power of OVVs was carried exclusively by pairs, the pairs would Compton scatter UV photons from the accretion disk and would produce X-rays through inverse-Compton emission in excess of observed levels. Both groups argue that protons are the dominant carriers of the jet energy and momentum. The conclusions should be taken with a grain of salt as there might be ways to circumvent them, for example, the magnetic field may carry a substantial fraction of the jet energy and momentum.

Recently, several observation campaigns have combined broadband spectral monitoring with imaging observations with the VLBI. Marscher *et al.* [139] reported an optical, an X-ray, and a gamma-ray flare in temporal coincidence with a swing of the optical polarization direction in BL Lacertae, followed by a delayed radio flare. The authors interpreted the observations with the help of a model in which a helical magnetic field threads the jet, and accelerates and confines it. Analytically, the model was studied by [133, 134, 140] and others. General relativistic magnetohydrodynamic models are starting to validate the analytical results (e.g., [135, 141–143]).

Acciari *et al.* [144] reported on an exceptionally strong episode of TeV gamma-ray flares from the radio galaxy M 87 accompanied by one of the strongest X-ray flares from the core of M 87 ever observed, and a $\sim$ 2 week long increase of the radio activity from the central VLBA resolution element – at a projected distance of less than 50 Schwarzschild radii away from the supermassive black hole. The authors interpret the observations with the ejection of shells of plasma with non-thermal particles, followed by the slow expansion of these shells. While the radio synchrotron emission is self-absorbed at the time of the ejection, the shells slowly become transparent as they expand adiabatically. Although the statistical significance of the association of the flares in different bands is limited for the two campaigns, the observations show the diagnostic power of combining radio interferometry with polarimetry and with broadband spectral monitoring.

8.4 Conclusions and Outlook

Observations of the continuum emission from blazars allows us to obtain information about a highly relativistic core of the jet, the "spine". The synthesis of observations and modeling can contribute to our understanding of how accretion works, which role highly relativistic jets play for the accretion process, and how jets are structured. Although magnetic models of jet formation are able to explain highly relativistic jets, they predict magnetically dominated jets – in contrast to the results from modeling the SEDs of blazars – which show that in the blazar emission zone the particle energy density is higher than the magnetic field energy density. There are a number of avenues which have promise to lead to further progress:

- The spaceborne Fermi MeV-GeV gamma-ray observatory and the ground-based H.E.S.S., MAGIC, and VERITAS TeV gamma-ray observatories are currently sampling the gamma-ray emission from blazars over 6 orders of magnitude in energy. The long lever arm of the observations might make it possible to detect unambiguously the imprints of particle acceleration and particle cooling processes. The aim of the observations is the identification of the emission process and the measurement of the parameters describing the emitting plasma.
- A worldwide consortium of members of the H.E.S.S., MAGIC, and VERITAS collaborations is planning the construction of a next-generation Cherenkov Telescope Array (CTA). The current plan is to construct two arrays, one in the northern and one in the southern hemisphere, of $\sim$ 100 Cherenkov telescopes each. This will significantly improve the sensitivity of current-generation Cherenkov telescope facilities, and lower the energy threshold to below 100 GeV. This will open up the avenue to extend the searches for VHE γ-ray blazars at larger red-

shift $z \gtrsim 0.2$, and increase the sample of blazars that do not belong to the HBL class. This will substantially improve our understanding of the origin of the γ-ray emission in blazars and the effects of $\gamma\gamma$ absorption local to the source as well as in intergalactic space.

- NASA plans to launch the Gravity and Extreme Magnetism SMEX (GEMS) mission in 2014 [145]. GEMS will be able to measure the polarization of the X-ray emission of many blazars. In the previous section we discussed optical polarimetry results. In the case of blazars which emit X-rays as synchrotron emission, the X-rays may exhibit a higher degree of polarization – because X-ray-emitting electrons lose their energy faster as they propagate away from shocks. The X-ray polarization signal is thus potentially cleaner than the optical polarization signal. If the jet magnetic field is indeed helical, X-ray polarization swings may be ubiquitous. GEMS has thus the potential to validate the helical structure of the jet magnetic field and to establish the role of the magnetic field in AGN jets.

References

1. Urry, C.M. and Padovani, P. (1995) *Publ. Astron. Soc. Pac.*, **107**, 803.
2. Czerny, B. and Elvis, M. (1987) *Astrophys. J.*, **321**, 305.
3. Koratkar, A. and Blaes, O. (1999) *Publ. Astron. Soc. Pac.*, **111**, 1.
4. Collin, S., Boisson, C., Mouchet, M. *et al.* (2002) *Astron. Astrophys.*, **388**, 771.
5. Ichimaru, S. (1977) *Astrophys. J.*, **214**, 840.
6. Narayan, R., Quataert, E., Igumenshchev, I.V., and Abramowicz, M.A. (2002) *Astrophys. J.*, **577**, 295.
7. Henri, G. and Pelletier, G. (1991) *Astrophys. J. Lett.*, **383**, L7.
8. Malzac, J., Jourdain, E., Petrucci, P.O., and Henri, G. (1998) *Astron. Astrophys.*, **336**, 807.
9. Galeev, A.A., Rosner, R., and Vaiana, G.S. (1979) *Astrophys. J.*, **229**, 318.
10. Poutanen, J. (1998) in *The Theory of Black Hole Accretion Disks* (eds M. Abramowicz, G. Bjornson, and J. Pringle), Cambridge University Press, Cambridge, p. 100.
11. Leighly, K.M. (1999) *Astrophys. J. Suppl. Ser.*, **125**, 317.
12. Done, C. (2002) *Philos. Trans. R. Soc. Lond. A*, **360**, 1967.
13. Blaes, O. (2007) The Central Engine of Active Galactic Nuclei, ASP Conference Series, vol. 373, Astronomical Society of the Pacific, p. 75, [astro-ph/0703589].
14. Fabian, A.C. *et al.* (1989) *Mon. Not. R. Astron. Soc.*, **238**, 729.
15. Laor A. (1991) *Astrophys. J.*, **376**, 90.
16. Reynolds, C.S. and Nowak, M.A. (2003) *Phys. Rep.*, **377**, 389.
17. Yaqoob, T. (2007) The Central Engine of Active Galactic Nuclei, ASP Conference Series, vol. 373, Astronomical Society of the Pacific, p. 109, [arXiv:astro-ph/0612527].
18. Reynolds, C.S. and Fabian, A.C. (2008) *Astrophys. J.*, **675**, 1048.
19. Brenneman, L.W. and Reynolds, C.S. (2006) *Astrophys. J.*, **652**, 1028.
20. Véron-Cetty, M.P. and Véron, P. (2000) *Annu. Rev. Astron. Astrophys.*, **10**, 81.
21. Peterson, B.M. (2007) The Central Engine of Active Galactic Nuclei, ASP Conference Series, vol. 373, Astronomical Society of the Pacific, p. 3, [astro-ph/0703197].
22. Kaspi, S. (2007) The Central Engine of Active Galactic Nuclei, ASP Conference Series, vol. 373, Astronomical Society of the Pacific, p. 13, [arXiv:0705.1722].

23 Rawlings, S. and Saunders, R. (1991) *Nature*, **349**, 138.

24 Celotti, A., Padovani, P., and Ghisellini, G. (1997) *Mon. Not. R. Astron. Soc.*, **286**, 415.

25 Maraschi, L., and Tavecchio, F. (2003) *Astrophys. J.*, **593**, 667.

26 Kawakatu, N., Nagao, T., and Woo, J.-H. (2009) *Astrophys. J.*, **693**, 1686.

27 Elitzur, M. (2007) The Central Engine of Active Galactic Nuclei, ASP Conference Series, vol. 373, Astronomical Society of the Pacific, p. 415, [astro-ph/0612458].

28 Nenkova, M., Ivezić, Ž., and Elitzur, M. (2002) *Astrophys. J.*, **570**, L9.

29 Nenkova, M., Sirocky, M.M., Nikutta, R., Ivezić, Ž., and Elitzur, M. (2008a) *Astrophys. J.*, **685**, 160.

30 Nenkova, M., Sirocky, M.M., Ivezić, Ž., and Elitzur, M. (2008b) *Astrophys. J.*, **685**, 147.

31 Chelouche, D. (2007) The Central Engine of Active Galactic Nuclei, ASP Conference Series, vol. 373, Astronomical Society of the Pacific, p. 277. http://adsabs.harvard.edu/abs/2007ASPC..373..277C (accessed 9 August 2011).

32 Fossati, G., Maraschi, L., Celotti, A., Comastri, A., and Ghisellini, G. (1998) *Mon. Not. R. Astron. Soc.*, **299**, 433.

33 Giommi, P., Capalbi, M., Fiocchi, M., Memola, E., Perri, M., Piranomonte, S., Rebecchi, S., and Massaro, E. (2002) A Catalog of 157 X-ray Spectra and 84 Spectral Energy Distributions of Blazars observed with BeppoSAX, Proc. Blazar Astrophysics with BeppoSAX and other Observatories (eds P. Giommi, E. Massaro, and G. Palumbo), Frascati, 10–11 December 2001, [arXiv:astro-ph/0209596], http://www.asdc.asi.it/blazars/ (accessed 9 August 2011).

34 Abdo, A.A. *et al.* (2010a) *Astrophys. J.*, **716**, 30.

35 Pian, E. *et al.* (1998) *Astrophys. J. Lett.*, **492**, L17.

36 Aharonian, F.A. *et al.* (1999) *Astron. Astrophys.*, **349**, 11.

37 Angel, J.R.P. and Stockman, H.S. (1980) *Annu. Rev. Astron. Astrophys.*, **8**, 321.

38 Scarpa, R. and Falomo, R. (1997) *Astron. Astrophys.*, **325**, 109.

39 Laing, R.A. (1980) *Mon. Not. R. Astron. Soc.*, **193**, 439.

40 Begelman, M.C. (1993) *Jets in Extragalactic Radio Sources* (eds H.J. Roser and K. Meisenheimer), Lecture Notes in Physics, vol. 421, Springer Verlag, p. 14.

41 Dermer, C.D. and Schlickeiser, R. (1994) *Astrophys. J. Suppl. Ser.*, **90**, 945.

42 Böttcher, M., Mause, H., and Schlickeiser, R. (1997) *Astron. Astrophys.*, **324**, 395.

43 Dermer, C.D. and Menon, G. (2009) *High Energy Radiation from Black Holes*, Princeton University Press.

44 Dermer, C.D. (1995) *Astrophys. J.*, **446**, L63.

45 Marscher, A.P. and Gear, W.K. (1985) *Astrophys. J.*, **298**, 114.

46 Maraschi, L., Celotti, A., and Ghisellini, G. (1992) *Astrophys. J.*, **397**, L5.

47 Bloom, S.D. and Marscher, A.P. (1996) *Astrophys. J.*, **461**, 657.

48 Dermer, C.D., Schlickeiser, R., and Mastichiadis, A. (1992) *Astron. Astrophys.*, **256**, L27.

49 Dermer, C.D., Sturner, S.J., and Schlickeiser, R. (1997) *Astrophys. J. Suppl. Ser.*, **109**, 103.

50 Sikora, M., Begelman, M.C., and Rees M.J. (1994) *Astrophys. J.*, **421**, 153.

51 Blandford, R.D. and Levinson, A. (1995) *Astrophys. J.*, **441**, 79.

52 Ghisellini, G. and Madau, P. (1996) *Mon. Not. R. Astron. Soc.*, **280**, 67.

53 Błażejowski, M. *et al.* (2000) *Astrophys. J.*, **545**, 107.

54 Arbeiter, C., Pohl, M., and Schlickeiser, R. (2002) *Astron. Astrophys.*, **386**, 415.

55 Harris, D.E. and Krawczynski, H. (2002) *Astrophys. J.*, **565**, 244.

56 Georganopoulos, M. and Kazanas, D. (2004) *Astrophys. J. Lett.*, **604**, L81.

57 Ghisellini, G., Tavecchio, F., and Chiaberge, M. (2005) *Astron. Astrophys.*, **432**, 401.

58 Ghisellini, G. and Tavecchio, F. (2009) *Mon. Not. R. Astron. Soc.*, **397**, 985.

59 Inoue, S. and Takahara, F. (1996) *Astrophys. J.*, **463**, 555.

60 Tavecchio, F. *et al.* (2001) *Astrophys. J.*, **554**, 725.

61 Krawczynski, H. *et al.* (2004) *Astrophys. J.*, **601**, 151.

62 Rybicki, G.B. and Lightman, A.P. (1986) *Radiative Processes in Astrophysics*, Wiley-VCH Verlag GmbH.

63 Nikishov A.I. (1962) *Sov. JETP*, **14**, 393.

64 Tavecchio, F., Maraschi, L., and Ghisellini, G. (1998) *Astrophys. J.*, **509**, 608.

65 Coppi, P.S. (1992) *Mon. Not. R. Astron. Soc.*, **258**, 657.

66 Mastichiadis, A. and Kirk, J.G. (1997) *Astron. Astrophys.*, **320**, 19.

67 Coppi, P.S. and Aharonian, F.A. (1999) *Astrophys. J. Lett.*, **521**, L33.

68 Li, H. and Kusunose, M. (2000) *Astrophys. J.*, **536**, 729.

69 Böttcher, M. and Chiang, J. (2002) *Astrophys. J.*, **581**, 127.

70 Sokolov, A., Marscher, A.P., and McHardy, I.A. (2004) *Astrophys. J.*, **613**, 725.

71 Mimica, P., Aloy, M.A., Müller, E., and Brinkmann, W. (2004) *Astron. Astrophys.*, **418**, 947.

72 Sokolov, A. and Marscher, A.P. (2005) *Astrophys. J.*, **629**, 52.

73 Graff, P.B., Georganopoulos, M., Perlman, E.S., and Kazanas, D. (2008) *Astrophys. J.*, **689**, 68.

74 Ghisellini, G. *et al.* (1998) *Mon. Not. R. Astron. Soc.*, **301**, 451.

75 Böttcher, M. and Dermer, C.D. (2002) *Astrophys. J.*, **564**, 86.

76 Takahashi, T. *et al.* (1996) *Astrophys. J.*, **470**, L89.

77 Sikora, M. and Madejski, G. (2000) *Astrophys. J.*, **534**, 109.

78 Sikora, M., Begelman, M.C., Madejski, G.M., and Lasota, J.-P. (2005) *Astrophys. J.*, **625**, 72.

79 Mehta, K.T., Georganopoulos, M., Perlman, E.S. *et al.* (2009) *Astrophys. J.*, **690**, 1706.

80 Celotti, A. and Ghisellini, G. (2008) *Mon. Not. R. Astron. Soc.*, **385**, 283.

81 Wilson, A.S., Smith, D.A., and Young, A.J. (2006) *Astrophys. J.*, **644**, 9.

82 Guo, F. and Mathews, W.G. (2011) *Astrophys. J.*, **728**, 121, arXiv:1009.1388.

83 Hillas, A.M. (1984) *Annu. Rev. Astron. Astrophys.*, **22**, 425.

84 Waxman, E. (2004) *Pramana*, **62**, 483, arXiv:0310079.

85 Atoyan, A.M. and Dermer, C.D. (2003) *Astrophys. J.*, **586**, 79.

86 Beall, J.H. and Bednarek, W. (1999) *Astrophys. J.*, **510**, 188.

87 Pohl, M. and Schlickeiser, R. (2000) *Astron. Astrophys.*, **354**, 395, and Erratum: (2000) *Astron. Astrophys.*, **355**, 829.

88 Kelner, S.R., Aharonian, F.A., and Bugayov, V.V. (2006) *Phys. Rev. D*, **74**(3), 034018.

89 Aharonian, F.A. (2000) *New Astron.*, **5**, 377.

90 Mücke, A. and Protheroe, R.J. (2000) *AIP Conf. Proc.*, **515**, 149.

91 Mannheim, K. and Biermann, P.L. (1992) *Astron. Astrophys.*, **253**, 21.

92 Mücke, A. and Protheroe, R.J. (2001) *Astropart. Phys.*, **15**, 121.

93 Mücke, A., Protheroe, R.J., Engel, R., Rachen, J.P., and Stanev, T. (2003) *Astropart. Phys.*, **18**, 593.

94 Protheroe, R.J. (1997) *Accretion Phenomena and Related Outflows*, IAU Colloquium 163, ASP Conference Series, vol. 121 (eds D.T. Wickramasinghe, G.V. Bicknell, and L. Ferrario), Astronomical Society of the Pacific, p. 585.

95 Bednarek, W. and Protheroe, R.J. (1999) *Mon. Not. R. Astron. Soc.*, **302**, 373.

96 Rachen, J.P. and Mészáros, P. (1998) *Phys. Rev. D*, **58**, 123005.

97 Mücke, A., Engel, R., Rachen, J.P., Protheroe, R.J., and Stanev, T. (2000) *Comput. Phys. Commun.*, **124**, 290.

98 Eichler, D. and Wiita, P.J. (1978) *Nature*, **274**, 38.

99 Mannheim, K. (1993a) *Astron. Astrophys.*, **269**, 67.

100 Böttcher, M., Mukherjee, R., and Reimer, A. (2002) *Astrophys. J.*, **581**, 143.

101 Böttcher, M. and Reimer, A. (2004) *Astrophys. J.*, **609**, 576.

102 Aharonian, F.A. *et al.* (2005) *Astron. Astrophys.*, **430**, 865.

103 Reimer, A., Joshi, M., and Böttcher, M. (2008) *AIP Conference Proc.*, **1085**, 502.

104 Böttcher, M., Reimer, A., and Marscher, A.P. (2009) *Astrophys. J.*, **703**, 1168.

105 Waxman, E. and Bahcall, J. (1999) *Phys. Rev. D*, **59**, 023002.

106 Mannheim, K., Protheroe, R.J., and Rachen, J.P. (2000) *Phys. Rev. D*, **63**, 023003.

107 Stecker, F.W., Done, C., Salamon, M.H., and Sommers, P. (1991) *Phys. Rev. Lett.*, **66**, 2697.

108 Halzen, F. and Zas, E. (1997) *Astrophys. J.*, **488**, 669.

109 Abbasi, R. *et al.* (2009) *Phys. Rev. Lett.*, **103**(22), 221102.

110 DeYoung, T. *et al.* (2009) 2009 Meeting of the Division of Particles and Fields of the American Physical Society, 26–31 July 2009, Detroit, Michigan, SLAC Electronic Proceedings Repository.

111 Ackermann, M. *et al.* (The IceCube collaboration) (2008) *Astrophys. J.*, **675**, 1014.

112 Wehrle, A.E. *et al.* (1998) *Astrophys. J.*, **497**, 178.

113 Hartman, R.C. *et al.* (2001a) *Astrophys. J.*, **553**, 683.

114 Hartman, R.C. *et al.* (2001b) *Astrophys. J.*, **558**, 583.

115 Marscher, A.P. *et al.* (2010) *Astrophys. J.*, **710**, L126.

116 Abdo, A.A. *et al.* (2010b) *Nature*, **463**, 919.

117 Gaidos, J.A. *et al.* (1996) *Nature*, **383**, 319.

118 Albert, J. *et al.* (2007) *Astrophys. J.*, **669**, 862.

119 Aharonian, F. *et al.* (2007) *Astrophys. J. Lett.*, **664**, L71.

120 Begelman, M.C., Fabian, A.C., and Rees, M.J. (2008) *Mon. Not. R. Astron. Soc.*, **384**, L19.

121 Krawczynski, H. *et al.* (2001) *Astrophys. J.*, **559**, 187.

122 Fossati, G. *et al.* (2008) *Astrophys. J.*, **677**, 906.

123 Krawczynski, H., Coppi, P.S., Maccarone, T., and Aharonian, F.A. (2000) *Astron. Astrophys.*, **353**, 97.

124 Rebillot, P.F. *et al.* (2006) *Astrophys. J.*, **641**, 740.

125 Błażejowski, M. *et al.* (2005) *Astrophys. J.*, **630**, 130.

126 Acciari, V.A. *et al.* (2008) *Astrophys. J. Lett.*, **684**, L73.

127 Abdo, A.A. *et al.* (2011) *Astrophys. J.*, **726**, 43.
Erratum: (2011) *Astrophys. J.*, **731**, 77.

128 Weber, E.J. and Davis, L., Jr. (1967) *Astrophys. J.*, **148**, 217.

129 Blandford, R.D. and Znajek, R.L. (1977) *Mon. Not. R. Astron. Soc.*, **179**, 433.

130 Phinney, E.S. (1983) A theory of radio sources, PhD Thesis. University Cambridge.

131 Camenzind, M. (1986) *Astron. Astrophys.*, **162**, 32.

132 Lovelace, R.V.E., Wang, J.C.L., and Sulkanen, M.E. (1987) *Astrophys. J.*, **315**, 504.

133 Li, Z.-Y., Chiueh, T., and Begelman, M.C. (1992) *Astrophys. J.*, **394**, 459.

134 Vlahakis, N. and Königl, A. (2004) *Astrophys. J.*, **605**, 656.

135 McKinney, J.C. (2006) *Mon. Not. R. Astron. Soc.*, **368**, 1561.

136 Giannios, D., Uzdensky, D.A., and Begelman, M.C. (2009) *Mon. Not. R. Astron. Soc.*, **395**, L29.

137 Sikora, M., Stawarz, L., Moderski, R., Nalewajko, K., and Madejski, G.M. (2009) *Astrophys. J.*, **704**, 38.

138 Celotti, A. and Fabian, A.C. (1993) *Mon. Not. R. Astron. Soc.*, **264**, 228.

139 Marscher, A.P., Jorstad, S.G., D'Arcangelo, F.D. *et al.* (2008) *Nature*, **452**, 966.

140 Blandford, R.D. and Payne, D.G. (1982) *Mon. Not. R. Astron. Soc.*, **199**, 883.

141 Komissarov, S.S., Barkov, M.V., Vlahakis, N., and Königl, A. (2007) *Mon. Not. R. Astron. Soc.*, **380**, 51.

142 Krolik, J.H. and Hawley, J.F. (2010) *The Jet Paradigm* (ed. T. Belloni), Lecture Notes in Physics, vol. 794, Springer Verlag, Berlin, p. 265.

143 Spruit, H.C. (2010) *The Jet Paradigm* (ed. T. Belloni), Lecture Notes in Physics, vol. 794, Springer Verlag, Berlin, p. 233.

144 Acciari, V.A. *et al.* (2009) *Science*, **325**, 444.

145 Swank, J. *et al.* (2010) GEMS Gravity and Extreme Magnetism SMEX, http://gems.gsfc.nasa.gov (accessed 9 August 2011).

Part Four Crucial Aspects under Investigation

9
Particle Acceleration in Turbulent Magnetohydrodynamic Shocks

Matthew G. Baring

9.1
Introduction

The existence of astrophysical shocks in the Universe is ubiquitous because of the presence of drivers for supersonic flows. Supernovae are classic examples of powerhouses that propel fast, nonrelativistic outflows that subsequently impact their environment, and have provided the core paradigm for the generation of galactic cosmic rays. Black holes are much more powerful drivers, and so can be expected to precipitate flows of much higher speeds. This is observed to be the case for both galactic superluminal jet sources, and the extragalactic jets associated with supermassive black holes in AGNs that are the focus of this book. The speeds of the outflows are clearly relativistic in AGN jets, signaled by the observation in radio of superluminal motions (e.g., [1]; see [2], for a review, and Chapter 3), and lower bounds to their bulk Lorentz factors from pair transparency inferences in the gamma-ray band ([3]; see also Chapter 3). The dramatic deceleration of such fast kinetic flows at outer boundaries, namely in the Mpc-scale radio lobes at the extremities of powerful radio galaxies, and outer shells in supernova remnants (SNRs), naturally spawns rampant dissipation of the kinetic energy of the flow at one or more external shocks. This is also expected for shocks that arise within jets during their transit from the central black holes out to the outer regions of active galaxies. Such shocks may be precipitated by turbulent entrainment of boundary layer material at the jet/ISM interface (see Chapter 11 for jet structure considerations), or may reflect inhomogeneities in the driving stimulus at the base of the magnetohydrodynamic (MHD) jet, that is, at its formation zone near the black hole. In either case, the collision of fast, relativistic fluids with slower ones forms shock discontinuities.

At these shocks, the dissipation precipitated by the rapid deceleration of hot, ionized plasma entails the creation of electrodynamic turbulence, confined cross-shock potentials, spatially diffusive transport of charges, and particle acceleration. All of these phenomena are not just anticipated from global theoretical perspectives, they are observed or inferred either using astrophysical data or in computer

Relativistic Jets from Active Galactic Nuclei, First Edition. Edited by M. Böttcher, D.E. Harris, H. Krawczynski.

simulations. The core focus here is on particle acceleration, and as will become evident, this process is intimately connected to other phenomena. Nonthermal radiation is present in different varieties of AGN, spanning the radio, optical, X-ray and gamma-ray bands, as discussed in Chapters 5–8. Nonthermal particle populations naturally must generate these photon signatures, and the leading paradigm for AGN jet studies is that acceleration of charges at plasma shocks within relativistic jets is the natural seed for such emission. The theory pertaining to the acceleration of these nonthermal particles is the focus of this chapter, providing an underpinning for the radiation models addressed in Chapter 3. The exposition starts with a brief overview of the nature of electromagnetic turbulence pertinent to jet shocks. Then it moves on to the global conditions governing flow and field spatial profiles in shock environs, before developing a major focus on the character and key properties of diffusive acceleration at relativistic shocks. Alternative energization mechanisms are then addressed, before finally summarizing some key questions for shock acceleration theory that are pertinent to studies of jets in active galaxies.

9.2 Electromagnetic Turbulence in Jet Shocks

In the absence of a continual driver, the incoming plasma far upstream of a shock is approximately of uniform number density since, on long time scales, most density fluctuations are smoothed by space-time convolutions, even in the presence of MHD turbulence. Yet, it must be noted that stirring is present in our best-known outflow/shock system, the solar wind, and this wind possesses moderate density variations even on hour-long and day-long time scales. The situation is expected to be similar for blazar/AGN jets. The probable presence of a multitude of shocks along the extent of a jet may offer the possibility that one shock may impact a neighboring one, feeding it turbulence and accelerated particles. There are two classes of turbulence that will be discussed briefly here, as preparatory material for the shock structure and acceleration presentations. These are Alfvén wave turbulence, and Weibel instability driven fluctuations.

The linearized electrodynamic response of a magnetized plasma is asymmetric in character, and so leads to a dielectric response tensor ϵ_{ij} as the appropriate descriptor. This expresses the coupling between the field and plasma fluctuations, as a perturbative solution to Maxwell's equations and the space charge continuity equation in a locally uniform $\boldsymbol{B}$ field. By choosing appropriate basis states for the wave perturbations, the tensor can be diagonalized and a determinant formed to obtain a dispersion relation, which characterizes a long-lived electromagnetic response to a charge/current perturbation. Details of this analysis are given in the books by [4] and [5]. For a magnetized, nonrelativistic hydrogenic plasma, the eigenmodes are naturally of circular polarization. They then generate a dispersion relation at frequencies considerably below the proton cyclotron frequency, $\Omega_{\rm p} = eB/m_{\rm p}c$, of the form

$$\frac{\omega}{k_{\|}} = v_{\mathrm{A}} \,, \quad v_{\mathrm{A}} = \frac{B}{\sqrt{4\pi\rho}} \,, \tag{9.1}$$

for waves of angular frequency ω and wavenumber $\boldsymbol{k}$. The component of the wavenumber along $\boldsymbol{B}$ is $k_{\|} = \boldsymbol{k} \cdot \hat{\boldsymbol{B}}$, so that if $\boldsymbol{B}$ is directed along the x-axis, the waves possess sinusoidal character $\exp\{i(k_{\|}x - \omega t)\}$. These torsional (shear) waves are known as *Alfvén waves,* having been predicted by [6], and propagate at the *Alfvén speed* v_{A}, *predominantly along the magnetic field.* Since the mass of the protons has not explicitly appeared in the dispersion relation, other than implicitly in the mass density ρ, such wave modes exist in electron–positron pair plasmas also, but only below the electron cyclotron frequency $\Omega_{\mathrm{e}} = eB/m_{\mathrm{e}}c$. They also exist in relativistic plasmas, where the Alfvén speed is of the form in (9.35), modified to take into account the rest mass of the plasma.

Such small amplitude waves abound in space plasmas in disparate environs, and naturally provide a means of transmitting energy and momentum between one test charge and another (i.e., accelerated particles) through wave-particle interactions. They are particularly germane to tenuous plasmas found in the cosmos because the wave-particle collisional time scale is typically on the scale of the inverse gyrofrequency of a charge, and is often much smaller than the Coulomb collisional *Spitzer time scale.* They are central to the diffusive transport that underpins the conventional shock acceleration mechanism that is the emphasis of this chapter. Observe that at higher frequencies, the dispersion relation also admits solutions of a different character. These include ion-cyclotron waves at $\omega \sim \Omega_{\mathrm{i}}$, whistler modes at $\Omega_{\mathrm{i}} \ll \omega \ll \Omega_{\mathrm{e}}$, and electron cyclotron waves at $\omega \sim \Omega_{\mathrm{e}}$, also discussed at length in [4, 5]. All of these can contribute to turbulent diffusion of charges if they are generated with sufficient power spectral densities. Yet it is a generic property of turbulence that lower frequencies possess greater power, particularly near the stirring scale, so that the Alfvén mode branch of the dispersion is naturally favored in space plasmas. Evidence to support this is readily found in the MHD turbulence spectra for the solar wind.

A completely different type of turbulence is that commonly elicited in relativistic shocks. Consider a pair shock. At the flow and field discontinuity, assume that the field is not totally laminar, and possesses a perturbation that is oscillatory in character, $\delta\boldsymbol{B} = \hat{y}B_y \cos kz$. For now, we assume that the shock is in the (y, z)-plane, though this specialization can be relaxed as needed. Add to this electrons that stream back and forth in the x-direction with speeds $\boldsymbol{v} = \pm\hat{x}v_x$. The Lorentz force on these electrons, due to the field perturbation, is

$$\delta\boldsymbol{F} = -e\left(\frac{\boldsymbol{v}}{c}\right) \times \delta\boldsymbol{B} \,. \tag{9.2}$$

Given the oscillatory nature of the field perturbation, the direction of this force alternates twice each spatial period $2\pi/k$. Therefore electrons will be deflected from their trajectories in a manner that forces sequential bunching and "debunching" in the z-direction. This drives a fluctuating electron density perturbation $\delta\rho$ and therefore a fluctuating electron current density $\delta \boldsymbol{j}$ in the x-direction, approximately coherently tied to the field fluctuations. Accordingly, $\delta \boldsymbol{j}$ reinforces $\delta \boldsymbol{B}$ and the field perturbation grows, as does the current one. This is known as the *Weibel instability*, first being predicted by [7] and subsequently interpreted in a simple physical manner by [8]; a nice description of its nature and pertinent literature can be found in [9]. It applies to hydrogenic plasmas too, and does not require idealized sinusoidal perturbations to initiate it, just spatially varying signs of $\delta \boldsymbol{B}$. To develop, it requires an anisotropic distribution, and its growth is terminated roughly when a sizable faction of the kinetic energy of the incoming plasma is converted into turbulent field energy. This renders this instability a powerful dissipative influence in shock layers. The Weibel growth rate is essentially the plasma frequency $\omega_{\mathrm{p}} = \sqrt{4\pi e^2 n/m}$, but if the ambient field is high, say becoming comparable to the bulk kinetic energy of the upstream flow, the growth gets stunted. The Weibel instability serves as the foundation for many particle-in-cell plasma simulations of relativistic shocks (e.g., [10–13]).

These are not the only two types of MHD/plasma turbulence exhibited by ionized gases, but they are the most pertinent to the focus of this chapter. They are complementary in their principal impact on a shock environs. Alfvén modes communicate information along field lines, and therefore can potentially do so to large distances from a shock, depending on how prolifically they are damped. They also can persist for large ambient MHD fields. In contrast, the Weibel instability is most powerful when the ambient field is low, and requires significant plasma anisotropy in order to be most active. Therefore, Weibel-driven turbulence is more likely to be confined to a shock layer; it is certainly created there, sampling the locally strong anisotropies. It is an open and important question as to whether and how information from the Weibel instability in relativistic shocks can be propagated to, or originate at, large distances upstream and downstream. This connects to the issue of whether or not the Weibel mechanism is directly involved in the diffusion of energetic charges on large spatial scales, as opposed to three-dimensional transport mediated by Alfvén turbulence tied to the field lines.

9.3
Structure of Relativistic Shocks

A formal definition of a shock is a dissipative interface between fast, supersonic (*upstream*) and slower, subsonic (*downstream*) fluid flows. This interface is not infinitesimally thin. The faster upstream flow cannot be effectively stopped instantaneously, and its rapid deceleration precipitates a confined zone of chaos where bulk kinetic energy is transformed into heating the plasma and electrodynamic turbulence. There are two key scale lengths that control the spatial scale of this zone. Before identifying these, a few definitions need to be posited. Presuming that on large scales the shock layer defines an approximately planar geometry, the component of the fast upstream flow velocity $\boldsymbol{u}_1$ normal to this plane is u_{1x}, and its downstream counterpart is u_{2x}. The traditional convention of using subscripts 1 and 2 to denote upstream and downstream MHD quantities will be adopted throughout this chapter. In addition, the shock geometry is ascribed a Cartesian coordinate system with the x-direction being assigned to the shock normal; further details of these choices will be expounded in Section 9.3.3. The shock Lorentz factor is that for the relative motion of the *total* upstream flow to the shock, namely $\Gamma_1 = [1 - \beta_1^2]^{-1/2}$, where $\beta_1 = |\boldsymbol{u}_1|/c$ is the dimensionless upstream speed.

The first scale defining dissipation in the shock layer is a gyrational length or *Larmor radius.* Since this length is dependent on the energy of the charge, the appropriate measure is obtained by adopting a speed of $\beta_1 c$, that is, that of an incident upstream beam as viewed in the reference frame where the shock is stationary. The result is then readily derived from the electromagnetic Newton–Lorentz force equation $d\boldsymbol{p}/dt = q(\boldsymbol{E} + \boldsymbol{v} \times \boldsymbol{B}/c)$ in the case of zero electric field:

$$r_{g1} = \Gamma_1 \beta_1 \frac{mc^2}{eB_2} . \tag{9.3}$$

This is the maximal radius of circular axial projections of the helical orbits in a uniform downstream $\boldsymbol{B}$ field of strength B_2, corresponding to particle pitch angles of $\pi/2$. Again, the subscript 1 is employed here to signify the specific choice of $|\boldsymbol{v}| = u_1$. In a shock where the field lies in or almost parallel to the plane of the discontinuity, in the absence of electrodynamic adjustments (discussed just below), the gyroscale represents a typical smoothing scale of the shock. Information from the upstream region penetrates into the downstream zone on this scale. For the case where the upstream beam is aligned along the shock normal, this gyrational scale is oriented in the x-direction. For more general field inclinations, the downstream gyrations are tilted and the depth of upstream penetration can be reduced. In principle, when the field lies along the shock normal, the scale could collapse to zero. However, diffusion in the shock layer turbulence introduces its own spatial scale. In general, if the turbulence is of, or similar in nature to, the Alfvén variety discussed in Section 9.2, the dominant contribution to stochastic diffusion is inherently gyroresonant. Then, for a component $v_\parallel$ of the particle's speed parallel to $\boldsymbol{B}$, charges of Lorentz factor γ interact primarily with waves that satisfy the *Doppler resonance condition*

$$\omega - n\frac{\Omega}{\gamma} = k_\parallel v_\parallel , \quad n = 0, \pm 1, \pm 2, \ldots \tag{9.4}$$

As above, the waves possess an angular frequency ω and a parallel wavenumber component $k_\parallel = \boldsymbol{k} \cdot \boldsymbol{B}$. Here $\Omega = eB/mc$ is the nonrelativistic gyrofrequency, and $n = 0$ is termed the *Landau resonance.* For Alfvén modes with $\omega \ll \Omega_p \ll \Omega_e$, this resonant restriction (for $n \neq 0$) implies that the diffusive mean free path $\lambda \sim 1/k_\parallel \sim \gamma v_\parallel/\Omega = \gamma \beta_\parallel mc^2/(eB)$ is of the order of, or greater than, a charge's gyroradius r_g, and automatically imposes a scale $\gtrsim r_{g1}$ for the diffusive transmission of information from the upstream beam into the downstream region. Hence, a diffusive thickness to the shock layer either is comparable to or exceeds that in (9.3), and is only moderately sensitive to the directional orientation of $\boldsymbol{B}$. Observe that this gyroscale is mass-dependent, underpinning important consequences for hydrogenic shocks to be discussed at the end of Section 9.4.3. Also, thermal contributions to the upstream speeds of the charges only modify this estimate by a factor on the order of unity.

The second core dissipation scale is electrostatic in origin. In the absence of a driver, the far upstream plasma possesses a pervasive localized charge neutrality, since in the fluid, transient electric fields cannot be sustained for long, being shorted out by the flow of charges. Sudden disruption of the upstream convective flow at the shock can, and does, permit localized departures from charge neutrality. Particle density gradients and spatially dependent coupling of the charges to the fields permit strong electrodynamic fluctuations that are short-lived and spatially confined. Yet they are sufficient to offer the prospect of significant energization to charges. Again, the Lorentz force equation $d\boldsymbol{p}/dt = q(\boldsymbol{E} + \boldsymbol{v} \times \boldsymbol{B}/c)$ can be employed, but this time focusing on the role of the electric field. If λ_e represents the typical *electrostatic scale length* along the shock normal in the shock rest frame (in which all quantities are represented in this discussion), then charges of speed u_{1x} transit this length in a time $dt \sim \lambda_e/u_{1x}$, if a rectilinear trajectory is presumed. A circular path in the presence of a $\boldsymbol{B}$ field lengthens this time estimate to $dt \sim 2\pi\lambda_e/u_{1x}$ when averaging over all gyrophases; the $\boldsymbol{v} \times \boldsymbol{B}$ drift motion contributes only parallel to the plane of the shock. The electrostatic deceleration in the x-direction is substantial if the momentum increment is $dp \sim \Gamma_1 m u_{1x}$. These combine in the Lorentz force to yield $|\boldsymbol{E}| \sim \Gamma_1 m u_{1x}^2/(2\pi q \lambda_e)$ when the influence of $\boldsymbol{B}$ can be neglected. This can be equated to the electric field that is estimated using Gauss' law for electrostatics, $\nabla \cdot \boldsymbol{E} = \rho$. Since $\rho = qn$ for a number density n of the charges q, if the gradient is set equal to $1/\lambda_e$, then one arrives at $|\boldsymbol{E}| \sim en\lambda_e$. Eliminating the electric field strength, the result of this line of reasoning is

$$\lambda_e \sim \frac{u_{1x}}{\omega_p}\sqrt{2\Gamma_1}\,, \quad \omega_p = \sqrt{\frac{4\pi q^2 n}{m}}\,. \tag{9.5}$$

Here ω_p is the *plasma frequency* of the charge, the natural oscillation frequency for perturbative electrostatic fluctuations. Its appearance in this calculation is no surprise since the physical environment essentially mimics a complex of incoherently (or coherently) oscillating electric dipoles. This order of magnitude estimate should not be interpreted beyond just that: the chaos of the shock layer and the interference between species with different oscillation scales (e.g., electrons and ions) will complicate the picture.

In an electrodynamic situation, the electric response to charge perturbations is governed by the same physical considerations, and so λ_e is still the natural scale for electric field fluctuations, such as for the Weibel instability (e.g., see [9]) discussed in Section 9.2. Accordingly, λ_e defines the scale for the shock layer physics in particle-in-cell (PIC) full plasma simulations (e.g., [10, 11, 13]), where it is often termed the collisionless *skin depth*. Turbulent dissipative activity therein seeds substantial plasma heating. This electrostatic scale is also particularly germane to plasmas with species of disparate masses, notably hydrogenic plasmas. In such systems, when the magnetic field is inclined to the shock normal, there arises an intrinsic charge separation due to the different gyrational scales of components with distinct masses. The concomitant *cross-shock potentials* are confined to regions within distances $\sim \lambda_e$ of the flow discontinuity, with this scale length being mostly controlled by the smallest (and therefore most mobile) mass. Such potentials can have a profound influence on the thermal energy balance in the shock layer; conceivable consequences of this electrostatic adjustment are mentioned briefly in Section 9.4.3.

Inextricably intertwined with the dissipative zone that defines the shock is the onset of diffusive particle acceleration, the injection domain. The resulting energetic particles that are the central focus of this chapter form only a small minority of the plasma population, and therefore are an insignificant influence on dissipative smoothing on scales of the order of r_{g1} and λ_e. However, it is noteworthy that they can be influential on larger scales, for example via a hydrodynamic feedback phenomenon that is briefly discussed in Section 9.4.6. Beyond the fuzzy dissipative interface, the shock environment is more organized in the sense that the partitioning of energy and momentum between the gas and the fields is more slowly varying. On these larger spatial scales, one can conveniently express the global character of the fluid and field flows through relations for the conservation of the energy, momentum and number fluxes of the charged particles and the electromagnetic fields. These are the so-called *Rankine–Hugoniot* jump conditions, discussed in many texts (e.g., see Chapter 6 of [14], for a nonrelativistic MHD exposition, and Chapter 15 of [15] for a fully relativistic hydrodynamical presentation) and many papers in the research literature. These conservation laws are at the foundation of shock structure, and indeed underpin various astronomical inferences of physical quantities in shock environs, and so form the focus of the remainder of this section, after the thermodynamic preliminaries.

9.3.1
Relativistic Thermal Gases

Anywhere in the shock neighborhood, upstream or downstream, we define a local fluid frame to be that in which the net momentum of the ensemble of particles is zero. In this local fluid frame, which moves with a velocity $\boldsymbol{u}$ with respect to the shock, the plasma is predominantly thermalized. In principle, it does not have to be isotropic in this frame, though in practice it often is, due to dissipative influences. Anisotropizing agents include the magnetic field, which may bias the angular distribution of charges if the turbulence that enables diffusive transport is highly anisotropic. Hereafter, it is assumed that the fluid frame gas is isotropic, to leading order. Each species in its particle population therefore can be approximated by a relativistic Maxwell–Boltzmann distribution:

$$n_p dp = \frac{n}{\Lambda_T} e^{-\gamma/T} p^2 dp \; , \quad \gamma = \sqrt{1 + p^2} \; , \tag{9.6}$$

with

$$\Lambda_{\mathcal{T}} = \mathcal{T} K_2 \left(\frac{1}{\mathcal{T}} \right) , \tag{9.7}$$

where $K_\nu(z)$ is a modified Bessel function of the second kind [16] of order ν, often named a Macdonald function. Here,

$$\mathcal{T} = \frac{kT}{mc^2} \tag{9.8}$$

is the dimensionless gas temperature, which is conveniently deployed throughout this exposition. This distribution is expressed here in momentum space, the natural choice for shock acceleration considerations. It uses a variable $p \equiv \gamma\beta$ for the dimensionless particle momentum, that is, that divided by mc, with γ being the particle Lorentz factor. Integrating (9.6) over momentum yields the normalization condition

$$n = \int_0^\infty n_\mathrm{p} dp \tag{9.9}$$

for the particle number density. Using γ as the integration variable, this result can be derived using identities 3.547.9 and 8.486.13 of [16], the latter of which is the recurrence relation $d/dz\{K_1(z)/z\} = -K_2(z)/z$. These identities are also employed in deriving the results in (9.10) and (9.11) just below. Other moments of the Maxwell–Boltzmann distribution can similarly be obtained. Weighting by the particle, the Lorentz factor γ gives the energy density $\mathcal{E}$, scaled by mc^2:

$$\mathcal{E} = n \left\{ \frac{K_3(1/\mathcal{T})}{K_2(1/\mathcal{T})} - \mathcal{T} \right\} . \tag{9.10}$$

The extreme nonrelativistic ($\mathcal{T} \ll 1$) and ultrarelativistic ($\mathcal{T} \gg 1$) limits of this energy density, when divided by the number density (i.e., corresponding to the specific energy or energy per particle), yield $\mathcal{E}/n - 1 \approx 3\mathcal{T}/2$ and $\mathcal{E}/n \approx 3\mathcal{T}$, respectively. These results can be deduced using two asymptotic forms for the modified Bessel function, namely $K_\nu(z) \approx \sqrt{\pi/(2z)}[1 + (4n^2 - 1)/(8z)] \exp(-z)$ for $z \gg 1$ and $K_\nu(z) \approx 2^{\nu-1}\Gamma(\nu) z^{-\nu}$ for $z \ll 1$. Such limits are very familiar, with the rest-mass energy contribution being subtracted from the nonrelativistic limit to generate the three-dimensional ideal gas equation form.

The pressure $\mathcal{P}$ can be obtained similarly, but requires more subtle considerations. It too is manifestly an energy density. The force exerted by a particle on a surface area element $\boldsymbol{dA}$ is proportional to $|-\Delta \boldsymbol{p}| = 2p\mu$, where μ is the angle cosine between the particle momentum vector and the normal to the element. This presumes that the particle specularly reflects off the element. The probability that the particle actually encounters $\boldsymbol{dA}$ is proportional to $v\mu$, since this scales with the flux of crossings of the element; particles that move along the element normal ($\mu = 1$) encounter it much more frequently. Therefore, the pressure per particle is $p v \mu^2 n_{\rm p}$, where a factor of 1/2 has been introduced because constructing the pressure in this way demands that only one side of $\boldsymbol{dA}$ is impacted, that is, half the isotropic distribution in (9.6) is sampled. The angle average of μ^2 must then be formed, which for isotropy reduces to a factor of 1/3. In this way, the appropriate weighting of the Maxwell–Boltzmann distribution to form a pressure is via introduction of a factor of $p v/3 = p^2/(3\gamma)$. Then,

$$\mathcal{P} = n\mathcal{T} \tag{9.11}$$

represents the pressure per mc^2, and this clearly is a scaled form of the ideal gas equation. These core results for Maxwell–Boltzmann statistics define the so-called *Jüttner–Synge equation of state* [17–19] for a relativistic Maxwellian; an extensive discussion can be found in the books by Synge [19], Chapter 4 and by de Groot *et al.* [20], Chapter 4.

The three thermodynamic quantities $\mathcal{E}$, $\mathcal{P}$ and n can be combined in a single relation to express how pressure changes in adiabatic (i.e., isentropic) response to adjustments in density. Yet the temperature cannot be eliminated due to the transcendental nature of the Bessel functions. The standard practice is to introduce the *adiabatic index* $\gamma_{\rm F}$, which is the ratio of specific heats for the fluid under system changes at constant volume and constant pressure. Defining $c_{\rm P}$ to be the heat capacity per particle at constant pressure, and $c_{\rm V}$ to be the capacity at constant volume, the form for $\gamma_{\rm F}$ is established via

$$\gamma_{\rm F} \equiv \frac{c_{\rm P}}{c_{\rm V}}\,, \quad c_{\rm P} = \left(\frac{\partial}{\partial \mathcal{T}}\left[\frac{\mathcal{E}+\mathcal{P}}{n}\right]\right)_{\rm P} \quad \text{and} \quad c_{\rm V} = \left(\frac{\partial}{\partial \mathcal{T}}\left[\frac{\mathcal{E}}{n}\right]\right)_{\rm V} \,. \tag{9.12}$$

Both heat capacities are scaled by $k_{\rm B}/(mc^2)$, where $k_{\rm B}$ is Boltzmann's constant, factors that are immaterial to the evaluation of $\gamma_{\rm F}$. Using (9.10) and (9.11), it is quickly determined that

$$\gamma_{\rm F} = \frac{w'(1/\mathcal{T})}{w'(1/\mathcal{T}) + \mathcal{T}^2}\,, \quad w(z) = \frac{K_3(z)}{K_2(z)}\,. \tag{9.13}$$

The $w(z)$ function represents the *enthalpy* $\mathcal{W} = \mathcal{E} + \mathcal{P}$ divided by n, that is, it constitutes a *specific enthalpy*. The pertinent asymptotic limits of its derivative are $w'(z) \approx -5/(2z^2)$ for $z \gg 1$ and $w'(z) \approx -4/z^2$ for $z \ll 1$. These readily yield $\gamma_{\rm F} \approx 5/3$ in the nonrelativistic, $\mathcal{T} \ll 1$ limit, and $\gamma_{\rm F} \approx 4/3$ for the ultrarelativistic ($\mathcal{T} \gg 1$), pseudophoton gas. Using standard manipulations of thermodynamic differentials ($d\mathcal{P}$, dn, $d\mathcal{T}$) under the constraint of constant entropy, the adiabatic form of the equation of state (EOS) then can be stated in polytropic form via

$$\mathcal{P} \propto n^{\gamma_{\rm F}}\,. \tag{9.14}$$

Accordingly, the monotonic decline of $\gamma_{\rm F}$ with increasing temperature renders relativistic Maxwell–Boltzmann fluids more compressible than their nonrelativistic counterparts; common terminology is that relativistic gases possess *a softer equation of state*. Since such compressibility directly couples to the global hydrodynamic structure of a shock, this property has implications for particle distributions obtained in diffusive acceleration theory, as addressed in Section 9.4.

The Jüttner–Synge EOS leads to a compact expression for the *sound speed* c_s for a relativistic, Maxwell–Boltzmann gas (e.g., see Chapter 7 of [19])

$$\frac{c_s}{c} \equiv \sqrt{\frac{\gamma_F \mathcal{P}}{\mathcal{E} + \mathcal{P}}} = \sqrt{\frac{\gamma_F \mathcal{T}}{w(1/\mathcal{T})}} . \tag{9.15}$$

Note that the square of this ratio is just $\partial\mathcal{P}/\partial\mathcal{E}$ subject to the isentropic restriction. It is simply deduced that $c_s \approx c/\sqrt{3}$ when $\mathcal{T} \gg 1$, and that $c_s \approx c\sqrt{5\mathcal{T}/3}$ if $\mathcal{T} \ll 1$ when the first formula in (9.15) collapses to the familiar form $(c_s/c)^2 = \gamma_F \mathcal{P}/n$ in the nonrelativistic limit. Finally, we note that frequently in the astrophysics literature (e.g., [21]), an alternative form for the equation of state is employed, namely

$$\frac{\mathcal{P}}{\gamma_g - 1} = \mathcal{E} - n , \quad \gamma_g = \frac{w(1/\mathcal{T}) - 1}{w(1/\mathcal{T}) - \mathcal{T} - 1} \approx \frac{5 + 14\mathcal{T} + 8\mathcal{T}^2}{3 + 10\mathcal{T} + 6\mathcal{T}^2} . \tag{9.16}$$

This choice can be an advantage in that the term on the left hand side expresses the internal or kinetic energy content of the gas. The *gas index* γ_g reduces to 5/3 when $\mathcal{T} \ll 1$ and 4/3 when $\mathcal{T} \gg 1$, that is, coincides with γ_F in these domains. However, it is not isentropic in character, and in general differs ($\gamma_g > \gamma_F$) from the value of γ_F, by as much as $\approx 5.2\%$ when $\mathcal{T} \approx 2.43$. For many applications, such a modest difference is not costly relative to the comparative mathematical simplicity of γ_g. In addition, the Padé approximation to γ_g in (9.16), obtained in [22], is accurate to better than 0.2% for all $\mathcal{T}$, an impressive result that renders it attractive for applications. Combining this with a $\gamma_F \to \gamma_g$ protocol yields

$$\left(\frac{c_s}{c}\right)^2 \approx \frac{2\mathcal{T}(1+\mathcal{T})^2}{2 + 9\mathcal{T} + 16\mathcal{T}^2 + 8\mathcal{T}^3} \frac{5 + 14\mathcal{T} + 8\mathcal{T}^2}{3 + 10\mathcal{T} + 6\mathcal{T}^2} , \tag{9.17}$$

which is a Padé approximation (accurate to 4.8%) useful for employing in hydrodynamic and MHD shock jump condition considerations. Numerical evaluations of γ_F, γ_g and c/c_s are plotted in Figure 9.1 as functions of $\mathcal{T}$. These illustrate the modest difference incurred by substituting γ_g for γ_F.

9.3.2 Hydrodynamic Jump Conditions

These thermodynamic preliminaries establish the background for exploring the hydrodynamic structure of a relativistic shock. On large spatial scales, which constitute being outside the dissipative shock layer, the fluid particle numbers, momenta and energies must be conserved. Since the fluid is convecting through space, then it is the fluxes of these quantities (e.g., normal to the shock) that are actually conserved. In other words, whatever flux is measured far upstream of the shock is precisely that far downstream. Since the subsonic downstream flow in the shock frame is necessarily slower than the upstream one, these conservation laws automatically

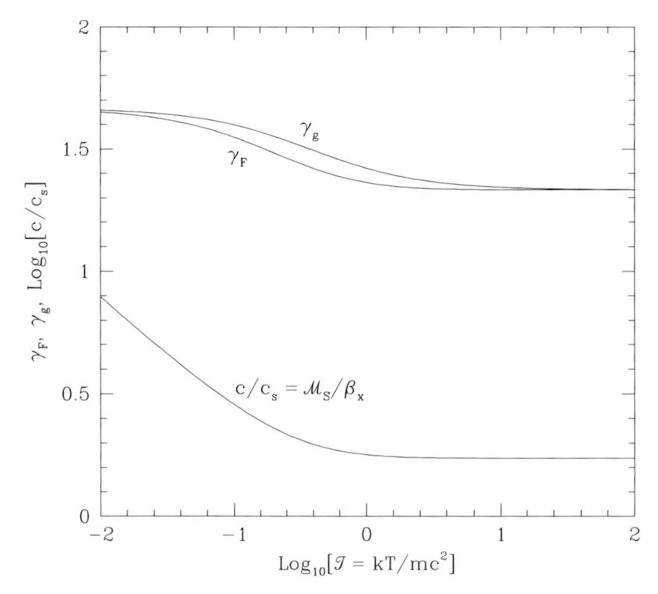

Figure 9.1 Two thermodynamic quantities, the adiabatic index γ_F in (9.13) and the gas index γ_g in (9.16), as functions of the dimensionless temperature $\mathcal{T} = kT/mc^2$. The difference between these two indices is modest. Also displayed is the scaled inverse of the sound speed, c/c_s (logarithmic scale) in (9.15), which couples to the sonic Mach number $\mathcal{M}_S$ of flows of speed $\beta_x c$, as defined in (9.25) below. For ultrarelativistic temperatures, $\mathcal{T} \gg 1$, the familiar result $c/c_s \to \sqrt{3}$ emerges.

guarantee that *the downstream gas is hotter than that upstream.* This is an essential "bottom line" to the nature of a shock: it converts bulk kinetic energy into heat. The details of this conversion in the dissipative shock layer are less fundamental than the actual mandate for dissipation, though the studies of the dissipation processes are of interest in their own right. For example, the relative apportionment of energy deposition in the particle populations and the electromagnetic turbulence has implications for both shock acceleration expectations and astrophysical diagnostics of shock environs.

The conservation of energy, momentum and the number of particles is most simply expressed when the large-scale magnetic fields embedded in the shocked plasma are dynamical uninfluential. In this simplification, the shock possesses a purely hydrodynamic structure, and the direction of the local $\boldsymbol{B}$ vector is immaterial: it equivalently can be presumed to lie along the shock normal, whereby its tension cannot compress the plasma in any way. It is instructive to consider this more elementary case first, which was explored at length in the context of relativistic outflows in spherical geometries by [21].

The energy and momentum content of a shocked flowing fluid is most conveniently encapsulated in the *stress-energy tensor,* $T_F^{\mu\nu}$, following the exposition of [23]. Here, the protocol of using dimensionless energies and pressures will continue to be adopted. Then, in the rest frame of the fluid, $T_F^{\mu\nu}$ is a diagonal tensor with $\mathcal{E}$ as its (00) time component and $\mathcal{P}$ for each of its diagonal space (11, 22, 33) components:

$$T_{\rm F}^{\mu\nu} = \begin{pmatrix} \mathcal{E} & 0 & 0 & 0 \\ 0 & \mathcal{P} & 0 & 0 \\ 0 & 0 & \mathcal{P} & 0 \\ 0 & 0 & 0 & \mathcal{P} \end{pmatrix} . \tag{9.18}$$

This summarizes the rest mass and thermal energy content of an isotropic gas in its rest (i.e., zero net momentum) frame. To define flux conservation relations, this must be boosted to the rest frame of the shock. Without loss of generality, it will be assumed that the flow is along the shock normal, since "transverse" boosts in the plane of the shock can always generate such a specialization. Flux conservation conditions established using planes parallel to the shock interface are not altered by transverse boosts. As there are no large-scale magnetic fields present in this hydrodynamic system, the compression at the shock is not accompanied by deflection of the fluid, a direct consequence of momentum conservation in directions within the plane of the shock. If the component of the flow speed in this normal incidence shock rest frame (NIF) along the shock normal is $\beta_x c$ at any x-position, and the corresponding flow Lorentz factor is $\Gamma = 1/\sqrt{1-\beta_x^2}$, then the transformation matrix for a Lorentz boost from the local fluid frame to the NIF shock rest frame is

$$\Lambda_{\alpha}^{\mu} = \begin{pmatrix} \Gamma & \Gamma\beta_x & 0 & 0 \\ \Gamma\beta_x & \Gamma & 0 & 0 \\ 0 & 0 & 1 & 0 \\ 0 & 0 & 0 & 1 \end{pmatrix} . \tag{9.19}$$

This is deployed to transform the stress energy tensor to the NIF according to the relation

$$T_{\rm S}^{\mu\nu} = \Lambda_{\alpha}^{\mu} T_{\rm F}^{\alpha\beta} \Lambda_{\beta}^{\nu} , \tag{9.20}$$

where the subscript F(S) refers to the plasma (shock) frame with the x-axis oriented normal to the shock in each case. The Einstein summation convention is adopted here and elsewhere in this chapter. The evaluation of the matrix products embodied in (9.20) is routine, and the result can be represented in the compact and familiar tensor form

$$T_{\rm S}^{\mu\nu} = (\mathcal{E} + \mathcal{P})\beta^{\mu}\beta^{\nu} + \mathcal{P}\eta^{\mu\nu} , \tag{9.21}$$

where $\eta^{\mu\nu}$ is the diagonal Minkowski metric tensor (00 component is unity, 11, 22, 33 components are -1), and $\beta^{\mu} = \Gamma(1, \beta_x, 0, 0)$ is the dimensionless four-velocity vector of the flow. In this form, the components of the stress energy tensor encapsulate the fluxes in different spatial directions (i.e., for $\nu = 1, 2, 3$ or $\mu = 1, 2, 3$) of the energy (0) and momentum components (1,2,3) of the fluid in the NIF. It is often also referred to as the *energy-momentum tensor*, and its components can be cast as integrations over the ensemble of contributing fluid particles, weighted by various four-momentum components (e.g., see Section 5.3 of [24]). This interpretation naturally leads to a straightforward extension when incorporating Poynting fluxes and global electromagnetic fields.

The flux conservation relations can now be assembled. The particle number flux $\mathcal{F}_n$ in the NIF shock frame along the x-direction is simply $\Gamma\beta_x n$, where the β_x factor represents the velocity flux in the Galilean sense, and the Γ factor appears because of length contraction in special relativity that influences the volume and therefore density. Continuity of energy and momentum fluxes across the shock is established by setting the appropriate divergence or covariant derivative $T^{\mu\nu}_{S;\nu}$ of the stress-energy tensor equal to zero, for directions locally normal to the shock. This divergence is integrated over a thin volume containing the shock plane, yielding conservation conditions across the plane of the shock by using $T^{\mu\nu}_S n_\nu = \text{const.}$, for $n_\nu = (0, 1, 0, 0)$ being the unit four-vector along the shock normal in the NIF. Accordingly, $T^{0\nu} n_\nu$ yields the energy flux $\mathcal{F}_e$, and $T^{1\nu} n_\nu$ yields the x-component of momentum flux $\mathcal{F}_{px}$, in the x-direction. The three conserved fluxes for a purely thermodynamic fluid are therefore

$$\begin{aligned} \mathcal{F}_n &= \Gamma\beta_x n \,, \\ \mathcal{F}_{px} &= \Gamma^2\beta_x^2(\mathcal{E}+\mathcal{P})+\mathcal{P} \,, \\ \mathcal{F}_e &= \Gamma^2\beta_x(\mathcal{E}+\mathcal{P}) \,. \end{aligned} \tag{9.22}$$

Observe that the energy flux can be understood as the total thermal energy content or enthalpy "Doppler boosted" or blueshifted by a Γ factor, weighted by the β_x velocity flux factor, again with an extra Γ factor emerging due to volume compression in the Lorentz boost from the fluid to the shock frame.

In a shock, the subscripts 1(2) are introduced to represent far upstream (downstream) quantities. The three flux conservation conditions $\mathcal{F}_{1\sigma} = \mathcal{F}_{2\sigma}$, for $\sigma = n, px, e$, then lead to three relations expressing the downstream number density n_2, energy density $\mathcal{E}_2$ and pressure $\mathcal{P}_2$ and flow speed β_{2x} in terms of their upstream counterparts. These can be manipulated as described in [23] and Chapter 15 of [15], with the algebra being simplified by dividing each conserved flux by the local number density n, so that the specific enthalpies $w_i = (\mathcal{E}_i + \mathcal{P}_i)/n_i$ and specific pressures $\mathcal{P}_i/n_i$ appear. The pressure differential across the shock is determined from the momentum flux equation, which is squared, and then the flow velocities are eliminated using the conservation of $\mathcal{F}_n$ and $\mathcal{F}_e$ equations. The result is

$$\frac{\mathcal{P}_2-\mathcal{P}_1}{n_1 n_2} = \frac{w_2^2-w_1^2}{w_1 n_2 + w_2 n_1} \,, \qquad w_i = \frac{\mathcal{E}_i+\mathcal{P}_i}{n_i} \,, \tag{9.23}$$

which is dubbed the *Taub adiabat* for the relativistic shock, thereby expressing the extent of heating by the shock. The upstream and downstream flow speeds can also be expressed compactly in terms of the thermodynamic quantities:

$$\beta_{1x}^2 = \frac{(\mathcal{P}_2-\mathcal{P}_1)(\mathcal{E}_2+\mathcal{P}_1)}{(\mathcal{E}_2-\mathcal{E}_1)(\mathcal{E}_1+\mathcal{P}_2)} \,, \qquad r \equiv \frac{\beta_{1x}}{\beta_{2x}} = \frac{\mathcal{E}_2+\mathcal{P}_1}{\mathcal{E}_1+\mathcal{P}_2} \,. \tag{9.24}$$

The mathematical expression for β_{2x}^2 can be obtained directly from that for β_{1x}^2 by the subscript substitution $1 \leftrightarrow 2$, but is here inferred indirectly instead using the *shock compression ratio* r of velocities, which is the standard measure of the compressive and dissipative strength of a shock. Equations (9.23) and (9.24) are generally applicable to any relativistic hydrodynamic shock. Observe that (9.24) can be cast in the form $\beta_{1x}\beta_{2x} = (\mathcal{P}_2 - \mathcal{P}_1)/(\mathcal{E}_2 - \mathcal{E}_1)$ so that if the gas downstream is ultrarelativistically hot so that $\gamma_{g2} = 4/3$ and $\mathcal{P}_2/\mathcal{E}_2 = 1/3$, then $\beta_{1x}\beta_{2x} = 1/3$, a familiar specialization of the jump condition. This result holds whether or not the upstream gas satisfies $\gamma_{g1} = 4/3$.

In order to provide closure, $\mathcal{P}_2$ must be eliminated using a form for the equation of state. The most convenient choice is (9.16), because of its algebraic simplicity; there is no physical reason why the pseudoadiabatic index γ_g cannot be used in lieu of its isentropic relation γ_F. Of course, the temperature $\mathcal{T}_2$ could also be employed as a parameter. The equation of state can then be suitably expressed using the adiabatic sound speed c_s in (9.15), or rather its $\gamma_F \to \gamma_g$ approximant, via the *sonic Mach number*

$$\mathcal{M}_S \equiv \frac{\beta_x c}{c_s} \approx \beta_x \sqrt{\frac{\mathcal{E} + \mathcal{P}}{\gamma_g \mathcal{P}}} \ . \tag{9.25}$$

This too will be subscripted 1(2) to denote upstream (downstream) evaluations. Then, the identity of upstream and downstream number fluxes $\mathcal{F}_n$ generates equalities for the density compression ratio, which can then be inserted in the momentum and flux equations. There are various ways to express the final result of these eliminations. Blandford and McKee [21] were the first to explicitly state a compact result for the hydrodynamic *Rankine–Hugoniot* shock jump conditions for a relativistic shock. The form stated here is

$$\begin{aligned} &\frac{\Gamma_1^2 \beta_{1x}^2}{(\gamma_{g1} - 1) r^2} - \frac{\gamma_{g1}\beta_{1x}^2}{(\gamma_{g1} - 1) r}\left(\Gamma_1^2 + \frac{1}{\mathcal{M}_{S1}^2 \gamma_{g1}}\right) + \Gamma_1^2 \\ &= \left(1 - \frac{\beta_{1x}^2}{\mathcal{M}_{S1}^2(\gamma_{g1} - 1)}\right)\sqrt{1 + \Gamma_1^2\beta_{1x}^2\left(1 - \frac{1}{r^2}\right)} \ , \end{aligned} \tag{9.26}$$

and is due to [25], applying only to the specific case of $\gamma_{g2} = \gamma_{g1}$. Here, $\mathcal{M}_{S1}$ is the upstream sonic Mach number. Completing the squares on (9.26) spawns a quartic in r, for which the trivial root $r = 1$ can be extracted, leaving a cubic. In general, the solution equation is algebraically more complicated when $\gamma_{g2} \neq \gamma_{g1}$. Yet, if the degree of dissipation heating is known, then γ_{g2} is determined, and the resulting equation can be solved for r given all upstream fluid parameters. Often, in relativistic shock applications, $\gamma_{g2} \approx 4/3$ and r is again routinely obtained. In any of these cases, other downstream quantities are then derived using substitutions in the various flux equations.

9.3.3
MHD Rankine–Hugoniot Conditions

The preceding exploration of hydrodynamic jump conditions lays the pedagogical foundation for addressing the fully magnetohydrodynamic case, when the ambient plasma magnetic field plays a substantial role. The MHD case is an essential one for the consideration of extragalactic jets, since the field is putatively a necessary contributor to the confinement and propagation of jets. In terms of the shock jump conditions, the magnetic field must be incorporated when it is a sizable fraction of the incoming kinetic energy of the upstream flow, that is, whenever $B_1^2/(8\pi) \gtrsim \epsilon \Gamma_1 \beta_{1x} n_1 mc^2$ for some small parameter ϵ. Its principal roles include the modification of the deceleration of the flow and the level of heating in the shock layer, and the deflection of the flow when crossing the shock. Maxwell's equations serve to constrain the field adjustments in transition across the shock.

The introduction of a magnetic field of arbitrary direction complicates the geometrical picture, so this needs to be detailed first. Let the global field be laminar on either side of the shock, and oriented in any point in the (x, z)-plane. This specialization is possible as the deflection of $\boldsymbol{u}$ and $\boldsymbol{B}$ at the shock is necessarily coplanar. Since $\boldsymbol{B}$ is time-independent on large spatial scales, $\nabla \times \boldsymbol{E} = \boldsymbol{0}$ everywhere and $\Gamma \boldsymbol{\beta} \times \boldsymbol{B}$ possesses the same direction upstream and downstream (see (9.34)). Like the fluid flow, the field is uniform upstream and downstream, but kinked at the shock because its magnitude must change due to the deceleration of the flow. The tension of the field and its nonuniformity at the shock necessarily forces the fluid flow to deflect there. With this construction, in the NIF shock frame the upstream flow yields a normalized four-velocity $\beta_1^\nu = \Gamma_1(1, \beta_{1x}, 0, 0)$, being oriented in the x-direction, and the downstream normalized four-velocity is $\beta_2^\nu = \Gamma_2(1, \beta_{2x}, 0, \beta_{2z})$. The deflection of the flow in the z-direction is generally small, that is, $\beta_{2z} \ll \beta_{2x}$.

In general, the magnetic field in the rest frame of the upstream fluid is oriented at an oblique angle Θ_{Bf1} to the shock normal in the upstream fluid frame. This is then also true in the downstream fluid rest frame, which moves relative to the upstream fluid. Accordingly, the obliquity angle Θ_{Bf2} of the downstream fluid frame field to the shock normal differs from Θ_{Bf1} unless $\Theta_{\mathrm{Bf1}} = 0°$. Such *oblique shocks* are termed *quasi-parallel* when $\Theta_{\mathrm{Bf1}} \lesssim 20°$, and *quasi-perpendicular* when $\Theta_{\mathrm{Bf1}} \gtrsim 70°$. In the NIF frame, the upstream field makes an angle Θ_{Bs1} to the x-direction, an angle that differs in general from Θ_{Bf1} due to relativistic transformation of the field; the two angles are simply related via the aberration formula $\tan \Theta_{\mathrm{Bs1}} = \Gamma_1 \tan \Theta_{\mathrm{Bf1}}$. Such a boost, if in the (x, y)-plane, normally introduces a finite $\boldsymbol{u} \times \boldsymbol{B}$ drift electric field in the shock frame, in the y-direction. There is often a unique boost that selects a shock rest frame where such an electric field is identically zero everywhere. This Lorentz transformation was isolated by de Hoffmann and Teller [26], and the resulting shock rest frame is known as the *de Hoffmann–Teller (HT) frame*. The HT frame is the shock rest frame where the flow is everywhere parallel to the local magnetic field; in it the upstream fluid flows with a dimensionless speed $\beta_{1\mathrm{HT}} = \beta_{1x}/\cos \Theta_{\mathrm{Bf1}}$. This HT flow speed $\beta_{1\mathrm{HT}}$ corresponds to a physical speed when it is less than unity, that is, the upstream field obliquity satisfies $\cos \Theta_{\mathrm{Bf1}} > \beta_{1x}$; the shock is then called *subluminal*. When mathematically $\beta_{1\mathrm{HT}} > 1$, no Lorentz boost from the local fluid frame can render the flow parallel to $\boldsymbol{B}$ and the de Hoffmann–Teller frame does not exist: the shock is said to be *superluminal*. The flow and field geometry in

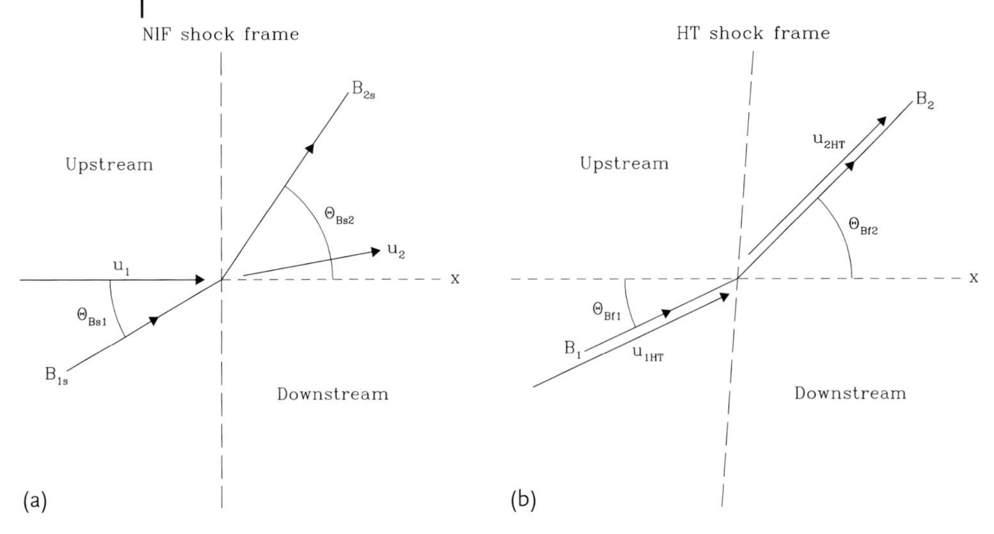

Figure 9.2 Plots of (a) the geometry in the normal incidence (NIF), and (b) de-Hoffmann Teller (HT) shock rest frames. Upstream flow speeds in the two reference frames are related by $u_{1\mathrm{HT}} \equiv \beta_{1\mathrm{HT}} c = u_{1x}/\cos\Theta_{\mathrm{Bf1}}$. Upstream and downstream quantities are denoted by subscripts 1 and 2, respectively. In general, the NIF field angle Θ_{Bs1} differs from the fluid frame/HT frame value Θ_{Bf1}. Also, the shock plane in the HT frame is rotated from that in the NIF due to relativistic aberration effects.

the NIF and HT shock frames are depicted in Figure 9.2. Observe that since the velocity boosts from the local fluid frame to these respective shock frames are oblique to each other, the plane of the shock is tilted between the NIF and HT frames (e.g., see [27]). Note also that in superluminal cases, it is possible to boost to a shock rest frame in which the $\boldsymbol{B}$ field is everywhere parallel to the shock plane.

The electromagnetic component adds a term to the overall stress-energy tensor that is imported into the Lorentz transformation in (9.20), which assumes the more general form

$$T_{\mathrm{S}}^{\mu\nu} = \Lambda_{\alpha}^{\mu} \left\{ T_{\mathrm{F}}^{\alpha\beta} + T_{\mathrm{EM}}^{\alpha\beta} \right\} \Lambda_{\beta}^{\nu} , \tag{9.27}$$

for a magnetohydrodynamic flow. The form for the electromagnetic stress-energy tensor $T_{\mathrm{EM}}^{\mu\nu}$ can be found in standard texts (e.g., [28, 29]) and many papers (e.g., [27, 30]). In the local fluid frame away from the shock, only a magnetic field exists, since transient electric fields are quenched on short temporal and spatial scales in the absence of a large gradient in the structure. This assumption simplifies fluid frame $T_{\mathrm{EM}}^{\mu\nu}$ to the form

$$T_{\mathrm{EM}}^{\mu\nu} = \frac{1}{8\pi} \begin{pmatrix} B_x^2 + B_z^2 & 0 & 0 & 0 \\ 0 & B_z^2 - B_x^2 & 0 & -2B_x B_z \\ 0 & 0 & B_x^2 + B_z^2 & 0 \\ 0 & -2B_z B_x & 0 & B_x^2 - B_z^2 \end{pmatrix} \tag{9.28}$$

for the specific shock geometry considered here. The nonidentity of the diagonal space components highlights the anisotropic nature of the field "pressure", and the presence of negative off-diagonal elements is a consequence of the tension possessed by the field. The shock frame electromagnetic stress energy tensor is then routinely developed, as are the various fluxes, but this time employing the transformation matrix for a Lorentz boost from the local fluid frame including nonzero velocity components in the z-direction:

$$\Lambda^{\mu}_{\alpha} = \begin{pmatrix} \Gamma & \Gamma\beta_x & 0 & \Gamma\beta_z \\ \Gamma\beta_x & 1+\frac{\Gamma\beta_x^2}{\Gamma+1} & 0 & \frac{\Gamma\beta_x\beta_z}{\Gamma+1} \\ 0 & 0 & 1 & 0 \\ \Gamma\beta_z & \frac{\Gamma\beta_z\beta_x}{\Gamma+1} & 0 & 1+\frac{\Gamma\beta_z^2}{\Gamma+1} \end{pmatrix} . \tag{9.29}$$

Observe that the particle number flux $\mathcal{F}_n$ remains unaltered by the presence of the field. A profound change introduced by the presence of the $\boldsymbol{B}$ field is that it forces the addition of a new flux equation involving $T^{3\nu}n_\nu$, that is, the x-component of momentum flux $\mathcal{F}_{pz}$, in the z-direction. This complication is a result of the stress imposed by a field bent at the shock. There are now four conserved fluxes for a combined thermodynamic/electromagnetic fluid:

$$\begin{aligned} \mathcal{F}_n &= \Gamma\beta_x n , \\ \mathcal{F}_{px} &= \Gamma^2\beta_x^2(\mathcal{E}+\mathcal{P}) + \mathcal{P} + \mathcal{F}_{px}^{\mathrm{EM}} , \\ \mathcal{F}_{pz} &= \Gamma^2\beta_x\beta_z(\mathcal{E}+\mathcal{P}) + \mathcal{F}_{pz}^{\mathrm{EM}} , \\ \mathcal{F}_e &= \Gamma^2\beta_x(\mathcal{E}+\mathcal{P}) + \mathcal{F}_e^{\mathrm{EM}} . \end{aligned} \tag{9.30}$$

This introduces three new electromagnetic flux contributions, whose evaluation is outlined in [30]:

$$\begin{aligned} \mathcal{F}_{px}^{\mathrm{EM}} &= \left[1+\frac{\Gamma^2\beta_x}{\Gamma+1}(\beta_x-\beta_z)\right]\left[1+\frac{\Gamma^2\beta_x}{\Gamma+1}(\beta_x+\beta_z)\right]\frac{B_z^2-B_x^2}{8\pi} \\ &\quad + \Gamma^2\beta_x^2\frac{B_x^2+B_z^2}{8\pi} - \frac{\Gamma^2\beta_x\beta_z}{\Gamma+1}\left[1+\frac{\Gamma^2\beta_x^2}{\Gamma+1}\right]\frac{B_x B_z}{2\pi} \\ \mathcal{F}_{pz}^{\mathrm{EM}} &= -\Gamma\left\{1+\frac{2\Gamma^3\beta_x^2\beta_z^2}{(\Gamma+1)^2}\right\}\frac{B_x B_z}{4\pi} + \frac{\Gamma^2\beta_x\beta_z}{8\pi} \\ &\quad \times\left\{B^2+\frac{\Gamma^4(\beta_x^2-\beta_z^2)}{(\Gamma+1)^2}\left[B_z^2-B_x^2\right]\right\} \\ \mathcal{F}_e^{\mathrm{EM}} &= \frac{\Gamma}{4\pi}\left\{\frac{\Gamma^2\beta_x\beta_z^2}{\Gamma+1}B_x^2 - \left[\frac{2\Gamma^2\beta_x^2}{\Gamma+1}+1\right]\beta_z B_x B_z \right. \\ &\quad \left. + \left[\frac{\Gamma^2\beta_x^2}{\Gamma+1}+1\right]\beta_x B_z^2\right\} . \end{aligned} \tag{9.31}$$

Conservation of the fluxes in (9.30) defines the first four of the six MHD jump conditions for a relativistic shock in the form

$$\mathcal{F}_{1\sigma} = \mathcal{F}_{2\sigma} , \quad \sigma = n, px, pz, e . \tag{9.32}$$

The remaining two emerge from Maxwell's equations, namely that the magnetic field is steady state, with $\nabla \cdot \boldsymbol{B} = 0$:

$$B_{1x} = B_{2x} , \tag{9.33}$$

an equation that derives its simple form from the independence of magnetic field components along the direction of a boost. Also, $\nabla \times \boldsymbol{E} = \boldsymbol{0}$ yields

$$\Gamma_1 (\beta_{1z} B_{1x} - \beta_{1x} B_{1z}) = \Gamma_2 (\beta_{2z} B_{2x} - \beta_{2x} B_{2z}) , \tag{9.34}$$

which completes the set of six jump conditions. In the limit of $(\Gamma_1 - 1) \ll 1$, these six constituent equations reproduce the standard continuity conditions at nonrelativistic shocks (e.g., see p. 117 of [14]).

This system of six equations relates seven unknown downstream quantities (β_{2x}, β_{2z}, B_{2x}, B_{2z}, $\mathcal{P}_2$, $\mathcal{E}_2$, and n_2) to their upstream counterparts. As with the hydrodynamical treatment, to obtain a closed set of equations for isotropic pressure, an assumed equation of state such as (9.16) is added to the analysis. Successive elimination of variables then generally leads to a 7th-order equation in the velocity compression ratio $r \equiv \beta_{1x}/\beta_{2x}$, with lengthy algebraic expressions for its coefficients (e.g., [31, 32]). The forms of the coefficients depend on the chosen equation of state and do not simplify easily; it is not particularly instructive to state them explicitly here. The reader is referred to [30] for a lengthy exposition on solutions to these relativistic MHD Rankine–Hugoniot conditions, some highlights of which are now outlined. It should be noted that in the de Hoffmann–Teller frame, the specialization $\beta_z B_x = \beta_x B_z$ everywhere yields dramatic simplification of the electromagnetic flux contributions in (9.31), reducing them to the level of complexity appearing in nonrelativistic MHD treatments. This expedient approach was adopted by [27] for subluminal shocks. Extragalactic jets possess shocks that may be superluminal, or near the luminal boundary, and so require flux conservation considerations in the NIF shock frame.

The solutions of the relativistic MHD jump conditions exhibit the following core features. For a single-component gas with particles of a single mass, the velocity compression lies in the range $1 \le r \le 4$, and this parameter defines the so-called *strength of the shock*. Strong shocks are those with larger r that normally exhibit more pronounced particle acceleration. Such larger compression ratios are realized only when both the sonic and Alfvénic Mach numbers ($\mathcal{M}_S$ and $\mathcal{M}_A$, respectively) are high, that is, the upstream gas is fairly cold, and the magnetic field is low. The definition of the sonic Mach number is given in (9.25). Here we employ a standard definition of the *Alfvénic Mach number* derivative of nonrelativistic expositions on MHD shocks:

$$\mathcal{M}_A \equiv \frac{\beta}{\beta_A} , \quad \Gamma_A \beta_A = \frac{B}{\sqrt{4\pi \rho c^2}} , \tag{9.35}$$

for $\Gamma_A = 1/\sqrt{1 - \beta_A^2}$. This fully relativistic form incorporates the mass loading of the plasma response to circularly polarized Alfvén wave MHD modes, discussed at length in the texts by [4, 5]. The RHS of the second identity in (9.35) just represents the ratio of the magnetic to particle energy densities, and is evaluated *in the local fluid frame*. The Alfvén speed $\beta_A c$ takes on all speeds between zero and c, and when nonrelativistic, it reduces to the familiar form $\beta_A c = B/\sqrt{4\pi\rho}$.

For $\Gamma_1 \approx 1$, strong shocks have $r \approx 4$, while the softer equation of state in the ultrarelativistic, $\Gamma_1 \gg 1$ shocks reduces the compressibility of the discontinuity to $r \approx 3$ so that $\beta_{2x} \approx 1/3$. Reducing $\mathcal{M}_S$ and $\mathcal{M}_A$ down close to unity weakens the shock to effectively eliminate it when r approaches unity. Observe that multicomponent fluids, such as explored by Heavens and Drury [33] (hydrodynamic case) and Ballard and Heavens [27] (full MHD) modify the solutions because of the introduction of different mass scales in the fluid contributions to the stress-energy tensor. Next, the magnetic field is amplified in strength ($B_2 > B_1$) and aligned more towards the plane of the shock ($\Theta_{Bf2} > \Theta_{Bf1}$). The deflection is significant in non- and mildly relativistic shocks, and is dramatic when $\Gamma_1 \gg 1$, where the downstream field generally becomes quasi-perpendicular. The magnitude of the downstream field is given by

$$B_2 = \sqrt{B_{1x}^2 + \Gamma_1^2 (r^2 - 1) B_{1z}^2} \,, \tag{9.36}$$

a result that is simply derived from the field jump conditions in (9.33) and (9.34), assuming that $\beta_{1z} = 0$ (NIF) and that $\beta_{2z} \ll \beta_{2x}$. This latter condition, which is tantamount to a small angle deflection of the fluid during shock passage, is a general property of relativistic shocks when the Mach numbers are not very low. Large deflections of the fluid require significant push from the magnetic field, and therefore are only realized in $\mathcal{M}_A \lesssim 10$, $\Gamma_1 \lesssim 2$ shocks. Representative illustrations of these MHD shock properties are given in Figure 2 and 3 of [30], with their quasi-perpendicular case $\Theta_{Bf1} = 85°$ also displayed in Figure 9.3. Double *et al.* [30] used a kinematic EOS of $\gamma_{g2} - 1 = (\Gamma_{rel} + 1)/(3\Gamma_{rel})$ in lieu of the more thermodynamic one in (9.16). Here $\Gamma_{rel} = 1/\sqrt{1 - \beta_{rel}^2}$ for a dimensionless relative speed $\beta_{rel} = (\beta_1 - \beta_2)/(1 - \beta_1\beta_2)$ between the upstream and downstream flows. Figure 9.3 also depicts solutions that are derived in [22] using the Jüttner–Synge equation of state. Comparison of the two sets of results clearly demonstrates that the choice of the EOS matters for mildly relativistic shocks, but not in the nonrelativistic ($\gamma_g \approx 5/3$) and ultrarelativistic ($\gamma_g \approx 4/3$) domains. Anticipating the discussion at the end of Section 9.4.3, the distribution index for mildly relativistic shocks is accordingly dependent on the choice of the equation of state. The influence of the EOS on the downstream field and flow obliquity angles is much more modest when $\Gamma_1\beta_1 \sim 1$ [22].

It is instructive to highlight the ultrarelativistic shock case, since this may reflect much of the character pertaining to the relativistic shocks in extragalactic jets. In the limit that Γ_1 approaches infinity, the details of the downstream equation of state become largely irrelevant because the shocked gas is hot, with $\gamma_{g2} \approx 4/3$. Then, a simple analytic solution can be found that establishes the departure of solutions from the well-known hydrodynamic solution in this domain, $\beta_{2x} = 1/3$. Defining

$$q = \frac{4\pi w_1}{B_{1z}^2} \equiv \frac{(\gamma_{g1} - 1)\mathcal{M}_{S1}^2}{(\gamma_{g1} - 1)\mathcal{M}_{S1}^2 - \beta_{1x}^2} \frac{\mathcal{M}_{A1}^2 - \beta_{1x}^2}{\beta_{1x}^2 \sin^2 \Theta_{Bf1}} \,, \tag{9.37}$$

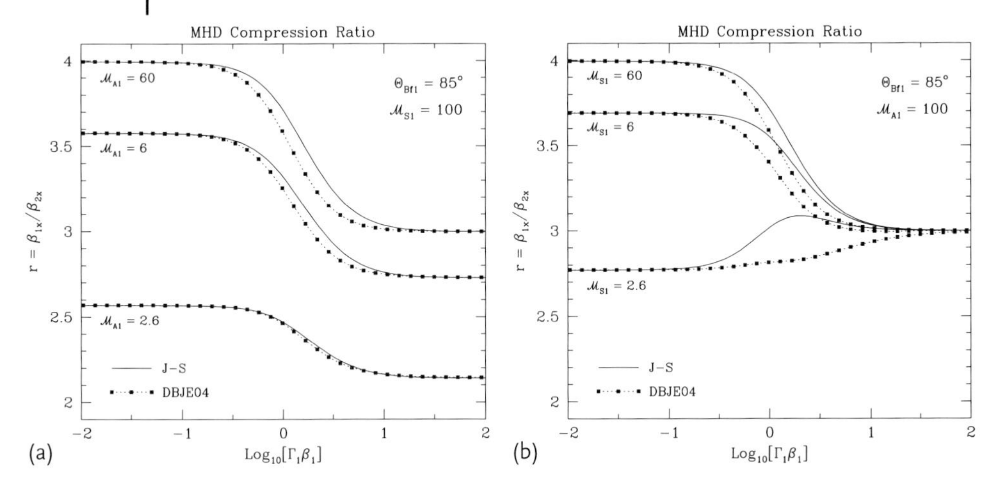

Figure 9.3 The compression ratio r of relativistic shocks, obtained numerically as a solution of the system of equations (9.32)–(9.34), as functions of the plasma's upstream dimensionless flow momentum $\Gamma_1\beta_1 \equiv \Gamma_1\beta_{1x}$. The shock obliquity was set at $\Theta_{Bf1} = 85°$, defining a quasi-perpendicular shock. Solutions are displayed for different sonic ($\mathcal{M}_{S1}$) and Alfvénic ($\mathcal{M}_{A1}$) Mach numbers, as marked: (a) keeps $\mathcal{M}_{S1} = 100$ fixed, while (b) holds $\mathcal{M}_{A1}$ constant at a value of 100 while varying $\mathcal{M}_{S1}$ is marked as a dotted vertical line for reference. The solid curves are for a Jüttner–Synge equation of state, as computed in [22], while the dotted curves labeled by DBJE04 correspond to numerical results presented in [30], which use a different equation of state. The dynamical influence of the magnetic field in reducing r below 3 in the $\Gamma_1 \gg 1$ domain is evident in (a).

for subscripts 1 on the Mach numbers denoting their upstream evaluation, one can write the asymptotic compression ratio as (see (47) of [30]):

$$r \approx \frac{1}{\beta_{2x}} \approx \sqrt{4 + 4q + \frac{q^2}{4}} - 1 - \frac{q}{2}\,, \quad \Gamma_1 \gg 1\,. \tag{9.38}$$

This approximation only applies in the $\beta_{1x} \to 1$ limit. The compression ratio clearly approaches $r \approx 3$ when $q \to \infty$, that is, when the Alfvénic Mach number is high, or the sonic Mach number approaches its minimum possible value (for $\beta_{1x} \to 1$) of $1/\sqrt{\gamma_{g1} - 1}$, corresponding to infinite specific enthalpy as the upstream gas temperature $\mathcal{T}_1 \to \infty$. The dependence of the compression ratio on temperature can be approximately determined by inserting (9.16) and (9.17) for the gas index and the sound speed/sonic Mach number into (9.37). The result is

$$q \approx \frac{2 + 9\mathcal{T}_1 + 16\mathcal{T}_1^2 + 8\mathcal{T}_1^3}{2(1 + \mathcal{T}_1)^2}\,\xi\,, \quad \xi \equiv \frac{\mathcal{M}_{A1}^2 - \beta_{1x}^2}{\beta_{1x}^2 \sin^2 \Theta_{Bf1}}\,, \tag{9.39}$$

and is illustrated in Figure 9.4, clearly displaying a monotonic increase of the compression ratio r with both the parameter ξ, that is, Alfvénic Mach number, and the upstream temperature $\mathcal{T}_1$. The smallest value of q is zero, realized when $\mathcal{M}_{A1} \to \beta_{1x} \approx 1$; this then yields $r \approx 1$ and the shock disappears, as expected when the magnetic field dominates the energy density; this dynamical influence of $\boldsymbol{B}$ is evident in Figure 9.3. Weaker fields generate stronger shocks with larger r. The

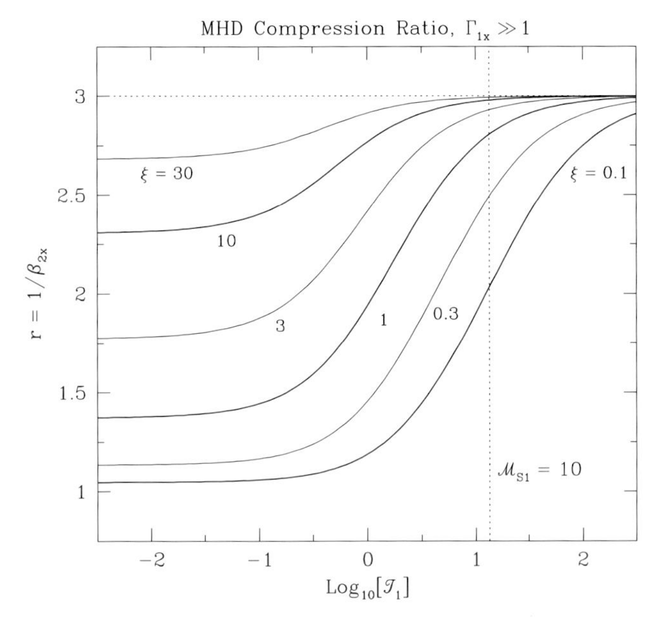

Figure 9.4 The compression ratio r of ultra-relativistic, $\Gamma_1 \gg 1$ shocks, given in (9.38), as functions of the plasma's upstream dimensionless temperature $\mathcal{T}_1 = k_B T_1/(mc^2)$. A sonic Mach number of $\mathcal{M}_{S1} = 10$ is marked as a dotted vertical line for reference. The curves are for different parameters ξ identified in (9.39); they encapsulate the Alfvénic Mach number $\mathcal{M}_{A1}$ and the shock field obliquity Θ_{Bf1}. When $\mathcal{M}_{A1} \gg 1$, the familiar $r \approx 3$ hydrodynamic result (horizontal dotted line) is generated.

dependence of r on the $\mathcal{T}_1$ is somewhat surprising, running counter to that experienced in nonrelativistic shocks where lower sonic Mach numbers reduce r. Heating the upstream gas enhances the enthalpy, driving up q, leading to an increase in the compression ratio. Physically, this corresponds to extremely large "ballistic" contributions to the upstream fluxes in (9.30) that both heat the downstream gas but also drive up its bulk motion. This character only persists up to temperatures $\mathcal{T}_1 \lesssim \Gamma_1$, beyond which the solution embodied in (9.37) breaks down and upstream heating has a profound impact, weakening the shock.

The downstream deflection angle θ_{us2} of the fluid flow at the shock, in the NIF, is encapsulated in the approximate equality (for $\Gamma_1 \gg 1$)

$$\tan\theta_{us2} \equiv \frac{\beta_{2z}}{\beta_{2x}} \approx \frac{3-r}{2\Gamma_1 \tan\Theta_{Bf1}} . \tag{9.40}$$

Clearly, this deflection is very small for ultrarelativistic flows, the hallmark of the extreme inertia of the upstream fluid. This $\beta_{2z} \ll \beta_{2x}$ property can then be combined with the field jump conditions in (9.33) and (9.34) to yield the approximation governing the bending of the field angle at the shock:

$$\tan\Theta_{Bf2} \equiv \frac{B_{2z}}{B_{2x}} \approx \Gamma_1 \tan\Theta_{Bf1}\sqrt{r^2-1} , \tag{9.41}$$

a result that uses the fact that $r/\Gamma_2 \approx \sqrt{1 - r^2}$ in this $\Gamma_1 \gg 1$ limit, and is consistent with (9.36) in this domain. Evidently, even modest obliquities upstream yield quasi-perpendicular field orientations downstream. Yet, it should be observed that due to the relativistic transformation of $\boldsymbol{B}$, upstream field obliquities in shocks in extragalactic jets are likely to be large if the convecting, fluid frame field contains even a small component in the plane parallel to the shock face.

9.4
The Character of Diffusive Acceleration in Relativistic Shocks

The global structure of relativistic shocks having been outlined, the emphasis turns to the acceleration of charges in their environment. The concept of diffusive acceleration of particles in space plasmas has been around for over six decades. Fermi [34] first postulated that cosmic rays could be produced via diffusion between collisions between interstellar clouds. If such clouds have predominantly random directions of motion, then the frequency of head-on collisions between the cosmic rays and the clouds would exceed the rate of tail-on encounters, leading to a net acceleration. The elegance of Fermi's idea was founded on the fact that when particles are confined in such a diffusive process, power-law particle distributions in momentum space are naturally produced, thereby modeling the cosmic ray observations well. It was subsequently realized that shocks in space plasmas could also provide such diffusive acceleration in an efficient manner: they tap the dissipative potential of the flow discontinuity by transferring the upstream fluid's kinetic energy to nonthermal populations both upstream and downstream of the shock, at the same time as heating the downstream gas. Thus the notion of *diffusive shock acceleration* or the *first-order Fermi mechanism* came about.

The modern era of shock acceleration theory began with a collection of seminal papers by [35–38] that outlined the basic properties of the process. Since then the field has grown substantially, with numerous approaches being developed, each with their different attributes. In this introductory material, the basic properties of shock acceleration theory that are most relevant to the modeling of extragalactic jets are summarized, starting with a brief outline of the principle of the Fermi mechanism. The main result of the linear or test particle theory of diffusive acceleration, namely the emergence of *canonical power-law distributions*, will be derived, specifically for nonrelativistic shocks. This will set the scene for the subsequent discussion of the predictions of acceleration theory at relativistic shocks, which is necessarily more complicated because of the inherent anisotropies in the energetic populations. This exposition will therefore borrow heavily from simulation results. There are many discussions of shock acceleration in the literature, however three principal reviews with complementary focuses and approaches prove informative for the reader who wishes to delve further. Drury [39] provides an in-depth analytic description of linear and nonlinear shocks, tailored for the specialist, while Blandford and Eichler [40] adopt a slightly tempered analytic approach and connect closely with astrophysical observations. Jones and Ellison [41] emphasize model developments and data/theory comparisons afforded by computer simulations.

9.4.1 The Principle of the Fermi Mechanism

The Fermi acceleration mechanism is now discussed in the context of collisionless shocks, that is, those that have energy and momentum transfer between particles mediated purely by plasma processes, with Coulomb scattering being negligible on relevant astrophysical time scales. It is instructive to review the principle of how the Fermi mechanism operates. Consider a parallel shock ($\Theta_{\mathrm{Bf1}} = 0°$) with upstream and downstream flow speeds u_{1x} and u_{2x}, respectively. Suppose particles of speed v in the rest frame of the shock originate on the upstream side of the shock, and diffuse around via collisions with magnetic plasma turbulence until they eventually cross the shock and move around downstream. Such diffusion tends to isotropize the angular distribution of the particles in the frame in which the upstream plasma is at rest. After a period in which the particles have experienced a plasma of speed u_{1x} the particles now collide with magnetic turbulence that is associated with the downstream plasma. If this turbulence in turn tends to isotropize the particles elastically, then this test particle population effectively senses a plasma moving towards it with speed $|u_{1x} - u_{2x}|$ upon arrival downstream. Elementary kinematics leads to the result that the process of quasi-isotropization then yields a net increase in the average particle speed in the rest frame of the shock. This kinematic guarantee of an increase in energy is akin to the gain that a photon experiences in inverse-Compton scattering collisions with relativistic electrons, an effect that relies on photon quasi-isotropization in the electron rest frame.

Eventually some of the particles will return to the upstream side of the shock, and these will see the upstream plasma moving, on average, towards them. Hence the process is repeated, and diffusion that leads to significant isotropization in the local plasma frame will yield particle speeds that are on average higher than those previously obtained on the downstream side of the shock. Sequential transport back and forth across the shock always leads to increases in particle speed, so that many shock crossing cycles afford significant acceleration. This is the principle of the first-order Fermi mechanism, where energy gains are cumulatively positive, and spatial diffusion allows the shock to dissipate its energy via its associated turbulence, imparting it to a nonthermal particle population. This situation is depicted in Figure 9.5, where a particle trajectory from an actual Monte Carlo simulation (e.g., see [42]) was used to demonstrate the coupling between shock crossings and energy increase, the latter being signified by the obvious increase in particle gyroradius. The velocity increase for nonrelativistic shocks is proportional to $|u_{1x} - u_{2x}|$, so that many cycles are required for particles to achieve high energies; this situation changes for relativistic discontinuities. For a simple visualization, the Fermi mechanism can be likened to specular reflection between two converging mirrors, for example, a ball bouncing elastically back and forth between two walls that approach each other.

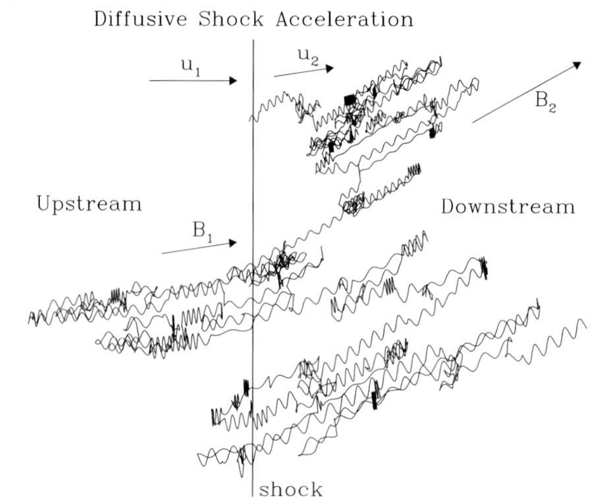

Figure 9.5 The trajectory of a simulated particle diffusing in a slightly oblique shock, with sporadic particle-turbulence interactions that are elastic in the local fluid frame. The fluid vectors $u_{1,2}$ and magnetic field vectors $B_{1,2}$ are indicated for both upstream (1) and downstream (2) regions. The particle path is projected onto a plane perpendicular to the shock face and computed in the rest frame of the shock. Helical gyrations are interrupted by large angle scattering encounters that, on average, isotropize the charge in local fluid frame. As the particle crosses the shock, it gains energy in the fluid frame and its gyroradius increases. The succession of such transits energizes the charge substantially before it eventually leaves the shock environs, due to convection, with a much larger gyroradius than it started.

The ions and electrons accelerated by the Fermi mechanism are called test particles when they do not have sufficient pressure (i.e., energy density) to influence the hydrodynamics of the shocked plasma, and hence are passive dynamical contributors. The analysis of the shock acceleration mechanism is then linear and therefore comparatively simple. When the test particle speeds far exceed that of the shock, u_{1x}, the system possesses no momentum scale; this is not true of nonlinear acceleration (see Section 9.4.6). Then, the natural solution for the distribution function of the nonthermal cosmic ray ions and electrons is a power-law distribution. This is the foremost property of the test-particle acceleration regime, and was a significant motivational factor in the application of Fermi's idea to supernova remnant shocks becoming so popular in explaining cosmic ray generation. It is instructive to derive the "canonical" power-law index for nonrelativistic shocks. Let the distribution function for particles of speed v and momentum $p = mv$ in, say, the shock frame of reference be dN/dp, and the cumulative distribution between p and infinity be $\mathcal{N}(p)$. Assume that $v \gg u_{1x}$, and that the particles are isotropic in both the upstream and downstream plasma frames. Under these conditions, the charges are isotropic also in the shock frame. This isotropy specialization is designated the *diffusion approximation*. Then it is possible to write down a differential equation governing $\mathcal{N}(p)$ that expresses the conservation of particle "fluxes" in momentum

(p) space. In a steady-state scenario, the system is described by

$$0 = t_{\text{cyc}} \frac{d\mathcal{N}}{dt} \equiv -\langle \Delta p \rangle \frac{d\mathcal{N}}{dp} - P_{\text{esc}} \mathcal{N} \,, \quad \mathcal{N}(p) = \int_{p}^{\infty} \frac{dN}{dp_1} dp_1 \,. \tag{9.42}$$

In this conservation equation, t_{cyc} is the mean time for a cycle of two transmissions through the shock, e.g., upstream to downstream and then back upstream again, and $\langle \Delta p \rangle$ is the mean (positive) net momentum gain in a cycle. Hence the $-\langle \Delta p \rangle \partial \mathcal{N} / \partial p$ term represents the supply of particles to higher energies via the acceleration process; it involves a derivative with respect to p because the momentum gains are approximately differential: $\Delta p \sim m|u_{1x} - u_{2x}|$. The negative sign appears in this term due to the use of a cumulative distribution. The remaining term represents a loss of particles from each cycle, with probability P_{esc}, beyond some downstream reference boundary that is on the order of a diffusive mean free path or less away from the shock. This convective loss escape probability is generally small for high-energy particles: when $v \gg u_1$, the spatial diffusion of the particles back and forth across this boundary dominates convection from the flow and $P_{\text{esc}} \ll 1$. The escape term is proportional to $\mathcal{N}(p)$ since particles that are lost at a particular momentum cannot contribute further to the acceleration process.

The form of (9.42) corresponds to the *first-order Fermi acceleration process* so named because the acceleration is first-order in momentum (or velocity) differentials, with the equation having only friction terms in momentum. Second-order (stochastic) Fermi acceleration for which there is diffusion in p in a Fokker–Planck type formalism can also occur, yielding power-law distributions too (e.g., see [43]). It naturally arises because scattering turbulence moves in the local fluid frame. Such stochastic contributions are generally minor, unless the shock speed u_{1x} is almost as slow as the Alfvén speed $\beta_A c$, which defines the representative speed of the magnetic disturbances in the upstream and downstream plasmas that effect particle scattering and diffusion. Specifically, inspection of Eqs. (8) and (10) in [44] of quasi-linear stochastic acceleration formalism, appropriate for small fluctuations $\delta \boldsymbol{B}/\boldsymbol{B}$, clearly indicates that the stochastic contribution to the spatial diffusion coefficients is smaller than the first-order Fermi one by an order of $1/\mathcal{M}_A^2$. For low Alfvénic Mach numbers, corresponding to weak shocks that are comparatively inefficient accelerators, the stochastic diffusion influences can be routinely incorporated via a momentum diffusion term, and solutions to a modified form of the conservation equation in (9.42) are readily obtained. For the strong shocks that are the focus here, the first-order acceleration mechanism is the dominant one.

For isotropy of the energetic particles, the probability of downstream escape of particles with $p \gg u_1 c$ from a cycle is $P_{\text{esc}} \approx 4u_{1x}/v$ for parallel shocks with $\Theta_{\text{Bf1}} = 0°$. The average momentum gain in a cycle for isotropic populations is $\langle \Delta p \rangle \approx 4u_{1x}(u_{1x} - u_{2x})p/(3u_{2x}v)$ (e.g., see [37, 43]; this is a result that can be deduced from the nonrelativistic interpretation of (9.45) below). Hence, for a shock compression ratio $r = u_{1x}/u_{2x}$, (9.42) admits power-law solutions

$$\frac{dN}{dp} \propto p^{-\sigma} \,, \quad \sigma = 1 + \frac{p\,P_{\text{esc}}}{\langle \Delta p \rangle} = \frac{r+2}{r-1} \,, \quad r = \frac{u_{1x}}{u_{2x}} \tag{9.43}$$

for $u_1 \ll c$. This is the canonical or universal power-law distribution that is frequently invoked in shock acceleration applications to astrophysics. It applies to both nonrelativistic and relativistic particles. The power-law index σ depends only on the compression ratio, $r = u_{1x}/u_{2x}$, an elegant feature of the first-order Fermi mechanism in nonrelativistic shocks that has underpinned its popularity. For high Alfvénic Mach numbers, the hydrodynamic, nonrelativistic shock solution for adiabatic index $\gamma_g \approx 5/3$ is

$$r = \frac{\gamma_g + 1}{\gamma_g - 1 + 2/\mathcal{M}_S^2} , \tag{9.44}$$

and so large sonic Mach numbers generate so-called strong shocks with $r = 4$ and $\sigma = 2$. The result in (9.43) extends to oblique shocks (e.g., [39]) with $\Theta_{Bf1} > 0°$. Furthermore, this canonical power law is applicable regardless of the relative contributions of particle diffusion along and orthogonal to $\boldsymbol{B}$ [43]. It is also robust in the sense that the type of scattering is immaterial: it applies for a wide variety of turbulence regimes. Alternative derivations of the universal power law are presented by [37, 39–41]. Note that while second-order Fermi acceleration also yields power laws, there is no coupling between σ and the compression ratio r, primarily because the index σ depends on two independent parameters, the speed of the scattering magnetic turbulence in the flow frame, and the residence time in the acceleration region. Hence there is no universality of the power law for second-order acceleration, underlining the beauty of the first-order process.

9.4.2
Diffusive Acceleration in Parallel, Relativistic Shocks

Diffusive acceleration in relativistic shocks is far less studied than that for nonrelativistic flows. Early work on relativistic shocks was mostly analytical in the test-particle approximation (e.g., [33, 45, 46]) in some cases using eigenfunction expansion techniques to solve the diffusion-convection equation, a simplification of the full Boltzmann transport equation. Complementary Monte Carlo techniques have been
employed for relativistic shocks by a number of authors, including test-particle analyses by Ellison *et al.* [47] for parallel, steady-state shocks, and extensions to include oblique magnetic fields by [48, 49]. A key characteristic that distinguishes relativistic shocks from their nonrelativistic counterparts is their *inherent anisotropy* due to rapid convection of particles through and away downstream of the shock, since particle speeds v are never much greater than the downstream flow speed $u_2 \sim c/3$. Accordingly, the diffusion approximation, the starting point for virtually all analytic descriptions of shock acceleration when $u_1 \ll c$, cannot be invoked since it requires nearly isotropic distribution functions. Hence analytic approaches prove more difficult when $\Gamma_1 \gg 1$, though advances in special cases such as the limit of extremely small angle scattering (*pitch angle diffusion*) are possible [46, 50]. This section outlines some of the distinctive properties of particle acceleration at relativistic shocks.

The existence of a canonical power-law solution embodied in (9.43) is an elegant feature of nonrelativistic shock acceleration theory that does not carry over to relativistic shocks because of their strong plasma anisotropy. As a consequence, while power laws are created when $u_{1x} \sim c$, the index σ becomes a function of the flow speed u_{1x}, the field obliquity $\Theta_{\rm Bf1}$, and the nature of the scattering, all of which intimately control the degree of particle anisotropy. In the specific case of parallel, ultrarelativistic shocks, the analytic work of Kirk *et al.* [50] demonstrated that as $\Gamma_1 \to \infty$, the spectral index σ asymptotically approached a constant, $\sigma \to 2.23$, a value realized once $\Gamma_1 \gtrsim 10$. This enticing result, which has been confirmed by Monte Carlo simulations in the papers by [49, 51, 52], is an important, but rather special case corresponding to compression ratios of $r = 3$ and the particular assumption of small angle scattering (SAS, that is, pitch angle diffusion, PAD). It applies specifically for incremental changes $\theta_{\rm scatt}$ in a particle's momentum with angle $\theta_{\rm scatt} \ll 1/\Gamma_1$ that would generally be realized in $\delta B/B \ll 1$ turbulence.

Here, we explore how this index σ can depend on several shock environmental parameters, displaying a myriad of possibilities, the contemplation of which is important for interpreting nonthermal blazar spectra in both the X-ray and gamma-ray windows. For parallel, $\Theta_{\rm Bf1} = 0°$ shocks, the spectral index of the power-law distribution is a declining function of the Lorentz factor for a fixed compression ratio, a characteristic evident in [27, 46, 50] for the case of pitch angle diffusion (PAD) where $\theta_{\rm scatt} \ll 1/\Gamma_1$. This is a property that extends to large angle scattering (LAS; see [47], and [51]), where stronger turbulence spawns deflections $\theta_{\rm scatt} \gtrsim 1/\Gamma_1$. Clearly, for ultrarelativistic shocks, LAS can be realized for quite modest deflections of the charges in the local fluid frame. Faster shocks generate flatter distributions if r is held constant, a consequence of the increased kinematic energization occurring at relativistic shocks (e.g., see [48]). For parallel shocks, the energy gain experienced by a particle in transiting from upstream to downstream, or vice versa, can be simply expressed in terms of the Lorentz transformation formulae

$$\gamma_2 = \Gamma_{\rm rel}\gamma_1\left(1 - \beta_{\rm rel}\beta_{1\parallel}\right) , \quad \gamma_1 = \Gamma_{\rm rel}\gamma_2\left(1 + \beta_{\rm rel}\beta_{2\parallel}\right) , \tag{9.45}$$

for a relative dimensionless speed $\beta_{\rm rel} = (\beta_1 - \beta_2)/(1 - \beta_1\beta_2)$ between the upstream (1) and downstream (2) flows. Here $\Gamma_{\rm rel} = 1/\sqrt{1-\beta_{\rm rel}^2}$, and the subscripts mark the *fluid frame determinations* of the particle Lorentz factor and velocity components. The $\beta_{i\parallel}$ denote components of the fluid frame speed along the shock normal, directed downstream. Accordingly, $-1 \le \beta_{1\parallel}, \beta_{2\parallel} \le 1$. For the anisotropic conditions realized in relativistic shocks, the *flux-weighted* angle averages, $\langle\beta_{1\parallel}\rangle$ and $\langle\beta_{2\parallel}\rangle$ of $\beta_{1\parallel}$ and $\beta_{2\parallel}$, generally differ from the isotropic values of $2\beta_1^2/3$ and $2\beta_2^2/3$ that arise in nonrelativistic shocks. Sequential application of the two identities in (9.45) can yield a net average energy gain described by $\langle\gamma\rangle$ in shock crossing cycles $1 \to 2 \to 1$, and this spans the range $1 < \langle\gamma\rangle < 4\Gamma_{\rm rel}^2$. The upper end of this range is sampled by LAS scattering scenarios, and clearly an increase in Γ_1, that is, $\Gamma_{\rm rel}$, should enhance the energization per cycle. Using the Monte Carlo simulation technique of EJR90 [53] illustrates the flattening of the accelerated particle distribution as Γ_1 increases, specifically in the limit of pitch angle diffusion, where $\langle\gamma\rangle \ll \Gamma_{\rm rel}^2$. Note that imposing a specific equation of state such as the Jüttner–Synge one renders r a declining function of Γ_1 so that this monotonicity property can disappear, as evinced in Figure 2 of [50].

It is instructive to define more precisely what is meant by a scattering angle $\theta_{\rm scatt}$. In analytic methods or some Monte Carlo simulations of diffusive transport, in turbulent magnetic fields, the departure of particle trajectories from pure helices is captured via a sequence of discrete scattering events. At every scattering, the direction of the particle's momentum vector $\boldsymbol{p}$ is deflected in the local fluid frame to a new value $\boldsymbol{p}'$. The scattering angle $\delta\theta = \cos^{-1}(\boldsymbol{p}\cdot\boldsymbol{p}'/|\mathrm{p}|\,|\mathrm{p}'|)$ is uniformly sampled within a solid angle $\delta\Omega$ up to a maximum deflection angle $\theta_{\rm scatt}$; this is a standard approach, but other constructs can be adopted. Because of the strong beaming anisotropy in relativistic shocks, the bifurcation of diffusive character occurs around the division point $\theta_{\rm scatt} \sim 1/\Gamma_1$, leading to the formal definitions:

$$\begin{aligned}\theta_{\rm scatt} &\lesssim \frac{1}{\Gamma_1} \Rightarrow \text{small angle scattering (SAS or PAD)}\,,\\ \theta_{\rm scatt} &\gtrsim \frac{1}{\Gamma_1} \Rightarrow \text{large angle scattering (LAS)}\,. \end{aligned} \tag{9.46}$$

While diffusion-convection differential equation approaches (e.g., [27, 46, 50]) are usually restricted to small angle scattering phase space that would be applicable to particle transport in quasi-linear field turbulence regimes, recently they have been generalized by [54, 55] to incorporate large angle deflections in MHD turbulence of larger amplitudes $\langle\delta B/B\rangle$. In contrast, Monte Carlo simulation techniques (e.g., EJR90 [42]) can routinely transition smoothly between SAS and LAS domains. Note that while the general principle of the diffusive acceleration process is identical for these two scattering regimes, the frequent deflections in the SAS domain destroy the episodes of coherent, helical trajectories that are illustrated for an LAS case in Figure 9.5.

The time, $\delta t_{\rm scatt}$, between scatterings in this frame is coupled (EJR90) to the mean free path, λ, and $\theta_{\rm scatt}$, via $\delta t_{\rm scatt} \approx \lambda\theta_{\rm scatt}^2/(6v)$ for particles of speed v. The resulting effect is that the gyrocenter of a particle with gyroradius r_g is shifted randomly by a distance of the order of $\theta_{\rm scatt} r_g$ in the plane orthogonal to the local field. Accordingly, *cross-field diffusion* (i.e., transport perpendicular to the mean local $\boldsymbol{B}$) emerges naturally from this discretization of the diffusion. For large angle scattering, the spatial transport is governed by kinetic theory [56, 57]. Then, the ratio of the spatial diffusion coefficients parallel ($\kappa_\parallel = \lambda v/3$) and perpendicular ($\kappa_\perp$) to the mean magnetic field is given by $\kappa_\perp/\kappa_\parallel = 1/(1+\eta^2)$. Here, the parameter $\eta = \lambda/r_g \gtrsim 1$ is the ratio of a particle's mean free path λ to its gyroradius r_g. Clearly, η controls the amount of cross-field diffusion, and is a measure of the level of turbulence present in the system, that is, it is an indicator of $\langle\delta B/B\rangle$. The Bohm limit of quasi-isotropic diffusion is realized when $\eta \sim 1$ and $\langle\delta B/B\rangle \sim 1$. This phenomenological discrete description of diffusion, usually implemented in Monte Carlo techniques, is most appropriate at high energies, and omits the details of microphysics present in plasma simulations such as PIC codes. In the injection

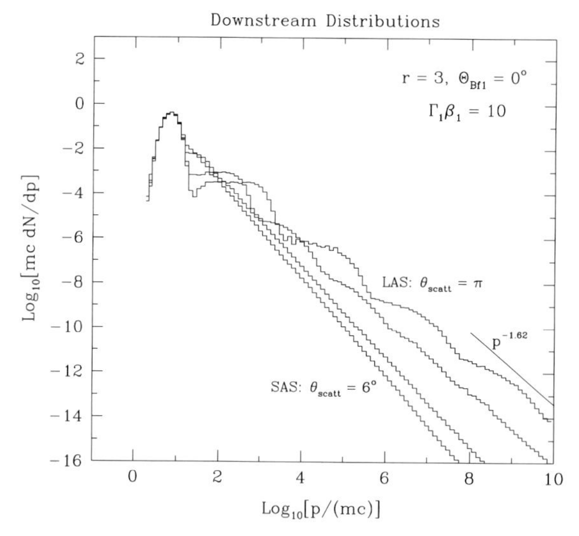

Figure 9.6 Particle distribution functions dN/dp from parallel ($\Theta_{\mathrm{Bf1}} = 0°$), $\Gamma_1\beta_1 = 10$ ($\beta_1 = 0.995$) relativistic shocks of upstream-to-downstream velocity compression ratio $r = u_{1x}/u_{2x} = 3$, as obtained from a Monte Carlo simulation of particle diffusion and gyrational transport (see [59]). Scattering off hydromagnetic turbulence is modeled by randomly deflecting particle momenta by an angle within a cone, of half-angle θ_{scatt}. The uppermost histogram is for a large angle scattering case $\theta_{\mathrm{scatt}} = \pi \gg 1/\Gamma_1$; at high p (not shown), this asymptotically approaches the power law $dN/dp \propto p^{-\sigma}$ indicated by lightweight lines, with $\sigma = 1.62$. Also exhibited are three smaller angle scattering cases, in order from top to bottom: $\theta_{\mathrm{scatt}} = 60°$, $\theta_{\mathrm{scatt}} = 20°$, and $\theta_{\mathrm{scatt}} = 6°$, corresponding to pitch angle diffusion.

domain at slightly suprathermal energies, the influences of complex turbulent and coherent electrodynamic effects become important, and will substantially modify the picture from pure diffusion.

To illustrate the dependence of σ on the size of θ_{scatt}, we employ here results from Monte Carlo simulations that have been documented in several recent papers [53, 58, 59]. Representative distributions, dN/dp, for $\Theta_{\mathrm{Bf1}} = 0°$ shocks, are depicted in Figure 9.6. Each distribution displays a prominent thermal peak at the lowest energies that constitutes the majority of the particles that are just compressively heated in their encounter with the shock. This is the general picture for all shocks, relativistic and nonrelativistic: most of the incident upstream plasma "beam" is just heated at the shock, dissipatively, and only a minority of the population is *injected* into the diffusive acceleration process to acquire an opportunity to be energized to extremely high energies. Note that these distributions are equally applicable to electrons or ions, and so the mass scale is not specified. The dN/dp in Figure 9.6 highlight the transition from SAS to LAS with $1/\Gamma_1 \lesssim \theta_{\mathrm{scatt}} \lesssim \pi$, which corresponds to diffusion in the shock layer that samples large field fluctuations $\langle \delta B/B \rangle \sim 1$; such turbulence is seen, for example, in PIC simulations of relativistic shocks driven by the Weibel instability [11, 13, 60].

This regime of LAS was first explored for $\Gamma_1 \lesssim 5$ by Ellison *et al.* [47]. Such large deflections produce huge gains in particle energy, of the order of Γ_1^2 as can be deduced from (9.45), in successive shock crossings. These gains are kinematic in origin, and are akin to those experienced by photons in inverse-Compton scattering, or to those obtained in the jet shear-layer acceleration scenario discussed by [61]. The result is an acceleration distribution dN/dp that is highly structured and much flatter on average than p^{-2} for strong, $\Theta_{\rm Bf1} = 0°$ shocks. The bumpy structure is kinematic in origin, corresponding to sequential shock transits [51], and becomes more pronounced [53, 58, 59] for large Γ_1. For ultrarelativistic shocks, information of the injection momentum scale becomes insignificant when $p \gg mc$, and the bumps asymptotically relax to form a power-law distribution $dN/dp \propto p^{-\sigma}$, with an index in the range of $\sigma \sim 1.6$. As $\theta_{\rm scatt}$ is lowered, Figure 9.6 demonstrates that the bumps disappear as the kinematic energy boosts decline (eventually down to factors on the order of two [51, 62], where upstream beaming along the shock normal sets $1 - \beta_{\rm rel}\langle\beta_{1\|}\rangle \sim 1/\Gamma_{\rm rel}^2$ in (9.45)) and the convective loss rates from the shock layer increase. The combination of these two ingredients leads to a steepening of the spectrum, yet only the SAS case of $\Theta_{\rm Bf1} = 6°$ yields σ as large as ~ 2.23. Astrophysically, there is no reason to exclude such intermediate $\theta_{\rm scatt}$ cases. Yet there is no spectral evidence in extragalactic jet sources that might suggest that the extreme case of $\theta_{\rm scatt} \sim \pi$ might be realized in nature. From the plasma physics perspective, magnetic turbulence in relativistic shocks could easily be sufficient to effect scatterings on intermediate to large angular scales $\theta_{\rm scatt} \gtrsim 1/\Gamma_1$, a proposition that is more enticing for ultrarelativistic shocks. Furthermore, it should be noted that generating distributions with $\sigma < 2$ may be required for the interpretation of inverse-Compton spectra in the TeV gamma-ray band for certain blazars (see [59] for a discussion).

As background for the reader, a few contextual words on the simulation technique used to generate these distributions are appropriate. It is a kinematic Monte Carlo method that has been employed extensively in supernova remnant and heliospheric contexts, and is described in detail in numerous papers (e.g., [41, 42, 47, 58]). It is conceptually similar to Bell's [37] test particle approach to diffusive shock acceleration, and essentially solves a Boltzmann transport equation for arbitrary orientations of the large scale MHD field $\boldsymbol{B}$. The background fields and fluid flow velocities on either side of the shock are uniform, and the transition at the shock is defined by the relativistic MHD Rankine–Hugoniot conservation relations discussed in Section 9.3.3. The diffusion characteristics are handled via the scattering algorithm described earlier in this section, with the general invocation that the mean free path λ is proportional to some power of a particle's gyroradius $r_{\rm g}$. The ability to span large ranges of particle momenta permits this technique to be validated against analytic techniques in suitable parameter domains. There are other Monte Carlo simulations in the market, most notably those of [49, 63] that inject prescribed magnetic turbulence and follow the chaotic particle trajectories in microphysical detail. It is notable that the general character of the simulated distributions from these two Monte Carlo approaches is fairly similar.

As a complementary tool, a recently popular approach to modeling particle acceleration at relativistic shocks employs full plasma or particle-in-cell (PIC) simulations (e.g., [11–13, 60]). PIC codes compute fields generated by mobile charges, and the response of the charges to the dynamic electromagnetic fields. Accordingly they are rich in their information on shock-layer electrodynamics and turbulence, but are computationally expensive. This presently limits them to small mass ratios when applied to hydrogenic shocks, and also to the exploration of thermal and suprathermal energies, even in pure pair shocks, so that full plasma simulations generally exhibit largely Maxwellian distributions. However, we note they are beginning to suggest [13, 64, 65] nonthermal tails spanning a relatively limited range of energies, generated by diffusive transport in PIC simulations, with the thermal population still dominating the high-energy tail by number. To interface with astrophysical spectral data, such as in blazars, a very broad dynamic range in momenta is desirable. This is the natural niche of Monte Carlo simulation techniques, since they are computationally expedient, and a principal reason why we focus on them in the following discussion.

9.4.3
Diffusive Acceleration in Oblique, Relativistic Shocks

The focus now turns to oblique shocks, adding an extra dimension to the shock parameter space. The key property that is highlighted now is that introducing field obliquity increases the de Hoffmann–Teller frame speed in the downstream frame, and so drives up convection power. If the scattering mean free path is not too long, cross-field diffusion can be significant, and aid the ability of particles to return to the shock from downstream and continue in the Fermi process. Yet, as Θ_{Bf1} increases, unless $\eta = \lambda / r_{\mathrm{g}}$ is very close to unity, the cross-field diffusion is insufficient to override strong convection. The consequence is that losses are significant and the index σ grows. Representative particle differential distributions to illustrate this property are depicted in Figure 9.7. These distributions are obtained just downstream of the shock from a Monte Carlo simulation (e.g., see [42]) and are measured in the normal incidence (NIF) shock rest frame (see Figure 9.2). Again, they are equally applicable to electrons or ions. They constitute results in the SAS regime where the particle momentum is stochastically deflected in the local fluid frame on arbitrarily small angular scales. In such cases, particles diffuse in the region upstream of the shock only until their velocity's angle to the shock normal exceeds around $1/\Gamma_1$, after which they are rapidly swept downstream of the shock. Figure 9.7 indicates clearly that when the field obliquity Θ_{Bf1} increases, so also does the index σ, with values greater than $\sigma \sim 3$ arising for $\Theta_{\mathrm{Bf1}} \gtrsim 50°$ for this mildly relativistic scenario. This is the signature of more prolific convection downstream and away from the shock.

The shock obliquity phase space can be explored more extensively, and the result is a more complicated picture: the distribution index is not necessarily a monotonic function of Θ_{Bf1}. In particular, it will soon become apparent that the MHD phase space of relativistic shocks bifurcates neatly into two regimes, the first be-

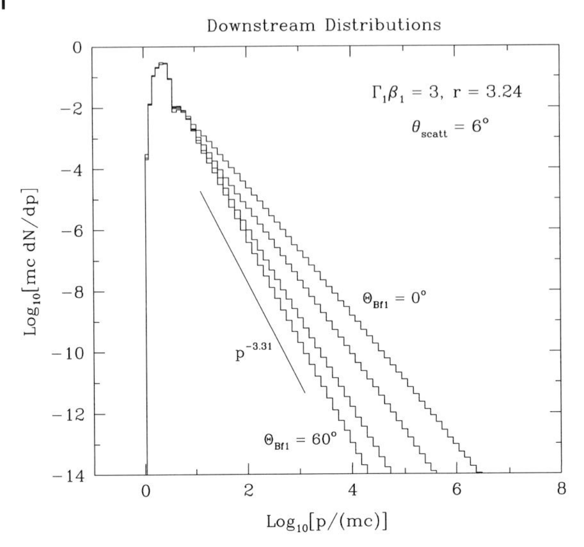

Figure 9.7 Particle distribution functions dN/dp from mildly relativistic shocks ($\Gamma_1\beta_1 = 3$, that is, $\beta_1 = u_1/c = 0.949$) of upstream-to-downstream velocity compression ratio $r = u_{1x}/u_{2x} \approx 3.24$. The upper distribution constitutes a parallel shock case ($\Theta_{Bf1} = 0°$); the remaining three histograms are oblique, superluminal shock cases $\Theta_{Bf1} = 20, 40$, and $60°$, in order from top to bottom. Diffusion due to magnetic turbulence was modeled as small angle scattering, $\theta_{scatt} = 6° \ll 1/\Gamma_1$. The ratio of the diffusive mean free path λ to the gyroradius r_g was fixed at $\eta = \lambda/r_g = 5$. The asymptotic power law had an index $\sigma = 2.21$ for $\Theta_{Bf1} = 0°$, and $\sigma = 3.31$ for $\Theta_{Bf1} = 60°$.

ing *subluminal* shocks with de Hoffmann–Teller frame flow speeds $\beta_{1HT} < 1$ (see Section 9.3.3). The upstream field obliquity then satisfies $\cos\Theta_{Bf1} > \beta_{1x} \equiv u_{1x}/c$. For larger obliquities Θ_{Bf1}, when $\beta_{1HT} = \beta_{1x}/\cos\Theta_{Bf1} > 1$, the de Hoffmann–Teller frame does not exist, and the shock is *superluminal*; this constitutes the second regime. Such a division naturally demarcates a dichotomy for the gyrational characteristics of charges orbiting in the shock layer. Subluminal shocks permit many gyrational encounters of charges with the shock interface, and therefore also reflection of them into the upstream region. This implies efficient trapping (e.g., see [42]), and effective continuation of acceleration. In contrast, for superluminal shocks, in the absence of deflections of particles by magnetic turbulence, the convective power of the flow compels particles to rapidly escape downstream (e.g., [66]), thereby suppressing acceleration. In such cases, particles sliding along the magnetic field lines would have to move faster than the speed of light in order to return to the upstream side of the shock. As mentioned above, such dramatic losses from the acceleration mechanism can only be circumvented by strong cross-field diffusion precipitated by large amplitude field turbulence fields (e.g., [57, 67]), that is, essentially close to the Bohm limit, $\eta = \lambda/r_g \sim 1$. Due to the current limitations on spatial scales in full plasma simulations, it is presently unclear whether Alfvén turbulence can be sustained at the Bohm limit in the environs of relativistic shocks.

The influence of the parameter λ/r_g that defines the effective frequency of scatterings on the particle distributions is profound, as is that of the upstream field obliquity Θ_{Bf1}. Representative particle (electrons or ion) differential distributions dN/dp that result from the simulation of diffusive acceleration at a mildly relativistic shock of speed $\beta_{1x} = 0.5$ are depicted in Figure 9.8. Again, these NIF frame distributions were generated for $\theta_{scatt} \lesssim 10°$, that is, in the SAS regime, and for low magnetic fields corresponding to Alfvénic Mach numbers $\mathcal{M}_A \gg 1$. Results are displayed for an obliquity $\Theta_{Bf1} = 59.1°$, corresponding to a de Hoffmann–Teller frame dimensionless speed of $\beta_{1HT} = \beta_{1x}/\cos\Theta_{Bf1} = 0.975$. This implies a marginally subluminal case. The distributions clearly exhibit an array of indices σ, including very flat power laws for $\eta = \lambda/r_g = 10^2, 10^3$. Observe that they are not monotonic functions of the key diffusion parameter η, which was held constant over the entire range of particle momenta. If λ/r_g is chosen to be momentum-dependent, curvature or breaks in the distributions would appear, highlighting departures from pure power-law forms. This broadband curvature may prove necessary for viable multiwavelength modeling of radio-to-X-ray synchrotron emission in jet sources. Figure 9.8 clearly emphasizes the fact that the normalization of the power laws relative to the low momentum thermal populations is a strongly declining function of λ/r_g. This also is a consequence of a more prolific convection of suprathermal particles downstream of the shock that suppresses diffusive injec-

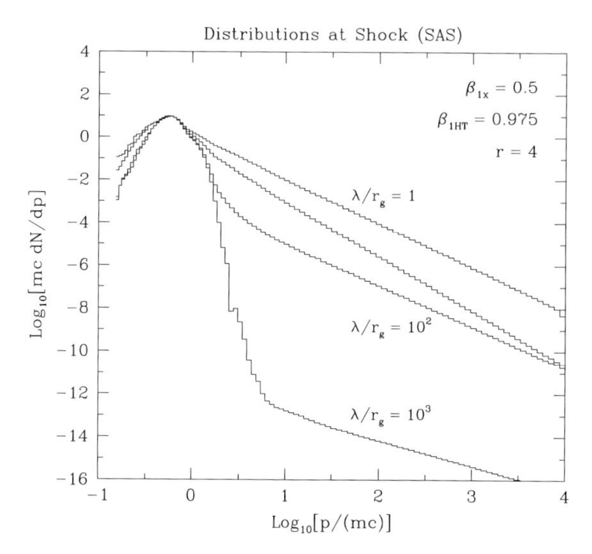

Figure 9.8 Particle distribution functions dN/dp from mildly relativistic subluminal shocks ($\Gamma_{1x}\beta_{1x} = 0.577$, that is, $\beta_{1x} = u_{1x}/c = 0.5$) of velocity compression ratio $r = u_{1x}/u_{2x} \approx 4$. Monte Carlo simulation results are depicted for upstream fluid frame magnetic field obliquity $\Theta_{Bf1} = 59.1°$, corresponding to de Hoffmann–Teller frame upstream flow speeds $\beta_{1HT} = \beta_{1x}/\cos\Theta_{Bf1} = 0.975$. All results were for small angle scattering (SAS). Distributions, from top to bottom, are for $\eta = \lambda/r_g = 1, 10, 10^2, 10^3$. A low sonic Mach number $\mathcal{M}_S \sim 4$ was chosen so as to effectively maximize the efficiency of injection from thermal energies. Adapted from [42].

tion from thermal energies into the acceleration process. Such losses are even more pronounced when $\lambda / r_g \geq 10^4$, to the point that acceleration is not statistically discernible for $\beta_{1HT} > 0.98$ cases [42]. Obviously, such low injection efficiencies mark turbulence regimes that are of far less interest for astrophysical jet environments.

To provide a synopsis of the array of indices realized in Fermi acceleration at relativistic discontinuities, a parameter survey for diffusive acceleration at a typical mildly relativistic shock is exhibited in Figure 9.9, where only the pitch angle diffusion limit was employed. The power-law index σ is plotted as a function of the de Hoffmann–Teller frame dimensionless speed $\beta_{1HT} = \beta_{1x} / \cos \Theta_{Bf1}$. It is clear that there is a considerable range of indices possible for nonthermal particles accelerated in mildly relativistic shocks. A feature of this plot is that the dependence of σ on field obliquity is nonmonotonic. When $\lambda / r_g \gg 1$, the value of σ at first declines as Θ_{Bf1} increases above zero, leading to very flat spectra. As β_{1HT} approaches and eventually exceeds unity, this trend reverses, and σ then rapidly increases with increasing shock obliquity. This is the character of near-luminal and superluminal shocks: it is caused by inexorable convection of particles away downstream of the shock, steepening the distribution dramatically. The only way to ameliorate this rapid decline in the acceleration efficiency is to reduce λ / r_g to values below around 10. Physically, this corresponds to increasing the hydromagnetic turbulence to high levels that force the particle diffusion to approach isotropy, that is, near the Bohm limit. This renders the field direction immaterial, and the shock behaves much like a parallel, subluminal shock in terms of its diffusive character. Charges can then be retained near the shock for sufficient times to accelerate and generate suitably flat distribution functions. This defines a second core property illustrated in Figure 9.9: σ is only weakly dependent on Θ_{Bf1} when $\lambda / r_g < 10$, even into the superluminal regime if the Bohm limit is realized.

In summary, it is noted that while these index results were obtained with one class of Monte Carlo simulation, they closely reproduce results from analytic techniques [68] in the $\lambda \gg r_g$ domain (see [42]), and concur in general character with the injected-turbulence simulations of [63]. Detailed confirmation using analytic techniques for low η domains is not yet possible, nor is comparison with PIC simulation indices, since they cannot yet achieve sufficient dynamic range to explore power-law tails.

At this juncture, it is desirable to briefly interpret these distribution index results in the context of extragalactic jet sources, specifically blazars and their gamma-ray spectra. Blazars have been one of the two brightest classes of sources in the gamma-ray sky (gamma-ray bursts being the other) ever since their discovery by the EGRET experiment aboard the Compton Gamma-Ray Observatory [69], and subsequent observation by ground-based Cherenkov telescopes at TeV energies (e.g., [70]). The TeV band signals typically exhibit steep photon spectra (e.g., Mrk 421 [71, 72]) that include the absorption due to pair producing interactions $\gamma\gamma \rightarrow e^+e^-$ with infrared and optical light generated by the intergalactic medium along the line-of-sight to the observer. The correction for this attenuation leads to the inference of extremely flat particle distributions in energetic gamma-ray blazars (see, e.g., [59]), with indices as low as $\sigma \sim 1.0$–1.5 in high-redshift sources. To accommodate these

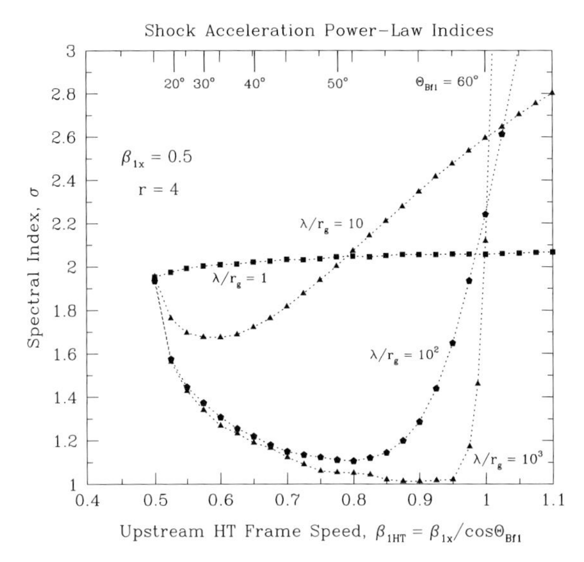

Figure 9.9 Power-law indices σ for distributions generated by Monte Carlo simulations in the limit of small angle scattering, for mildly relativistic shocks of upstream flow speed $\beta_{1x} \equiv u_{1x}/c = 0.5$, and an MHD velocity compression ratio $r = 4$. The indices are displayed as functions of the effective de Hoffmann–Teller frame upstream flow speed $\beta_{1HT} = \beta_{1x}/\cos\Theta_{Bf1}$, with select values of the fluid frame field obliquity Θ_{Bf1} marked on upper axis. The displayed simulation index results were obtained for different diffusive mean free paths λ parallel to the mean field direction, namely $\lambda/r_g = 1$ (squares), $\lambda/r_g = 10$ (triangles), $\lambda/r_g = 10^2$ (pentagons), and $\lambda/r_g = 10^3$ (triangles), as labeled. Observe that the indices for the $\beta_{1HT} = 0.975$ cases correspond to those of the distributions exhibited in Figure 9.8. Adapted from [42].

observational constraints would require subluminal shocks with very modest or low turbulence levels. Yet, such inferences are based upon measurements in a limited waveband subject to profound absorption. There is also the possibility that the TeV photons are emitted by particles close to the maximum energy of acceleration, that is, proximate to a distribution turnover or cutoff.

More improved diagnostics are now enabled by Fermi-LAT detections of blazars, which extend the observational window over a much larger energy range, nominally from 100 MeV to over 1 TeV, and most crucially, below the attenuation window. Accordingly, Fermi observations can more directly probe the underlying radiating particle population. A prime example of this is the multiwavelength campaign on the PKS 2155-304 blazar [73] in a nonflare state, whose combined Fermi-HESS spectrum from 300 MeV to 3 TeV indicates an unattenuated photon spectral index of $\alpha \sim 1.6$, steepening to $\alpha \sim 2.0$ above 1 GeV (for $dn_\gamma/d\varepsilon_\gamma \propto \varepsilon_\gamma^{-\alpha}$). For a standard inverse-Compton scattering interpretation with insignificant radiative cooling, this translates into an electron power-law index in the range $\sigma \sim 2.2$–3.0, so that acceleration at mildly superluminal oblique shocks should provide the best description. A similar conclusion is arrived at for other blazars with Fermi-LAT detections that are summarized in [74].

Returning to the injection issue, the presentation here has focused on results from single species simulations, that is, it can be regarded as applying to pure pair shocks, or pure ion systems. It is natural to ask whether they might differ if the shocks contain a mix of species, such as electrons/pairs plus ions. This question must be entertained because there is still an ongoing discussion of leptonic versus hadronic models for blazars (see Chapter 8 for characteristics of these contrasting scenarios). A complete answer must be deferred to future explorations of diffusive acceleration in hydrogenic plasma shocks, which are at present burdened with various complexities. Yet, it is expected that the index results presented here should be the same unless the turbulence generation is dramatically different when species of different masses are present. At high energies, while it is conceivable that electrons could be in a large angle scattering regime while protons satisfy the SAS requirements (or vice versa), for example, there is no compelling reason to presume such without simulation data to support such a contention. The same could be said for λ/r_g being a possible function of the mass of a charge. Currently, PIC simulations cannot access the momentum regimes to probe such an issue, being impeded by costly computational speed constraints when disparate mass scales are included in the plasma. Therefore, for the purpose of modeling shock acceleration implementations in extragalactic jet sources, it suffices to treat σ for hydrogenic shocks as being similar to that inferred here for single species explorations.

At energies below 1 GeV, the gyrational scales of protons and electrons of a given energy differ because the protons are at most only mildly relativistic. This naturally leads to significantly different gyroresonant interactions for e^- and p. Furthermore, at these energies, charge separation cross-shock potentials are anticipated to play a profound role in energy exchange between the two species (e.g., [75]). The larger inertial (gyrational) scale of protons relative to electrons generates a spatial separation of charges just downstream of the shock, if it is oblique. Resulting "capacitance" electric fields dissipatively heat the electrons and cool ions as the shock layer adjusts to largely quench the electric field. Details of this phenomenon and its underpinning physics are outlined in Section 9.3. The scale of the dissipation is electrostatic in origin, and therefore is $\propto u_{1x}/\omega_p$ as given in (9.5). Consequently, electrons should acquire a sizable fraction of the incoming proton beam kinetic energy in the form of thermal heat. This offers the possibility that injection of electrons into the acceleration process for $\lambda/r_g \gtrsim 30$ regimes will be enhanced by cross-shock potentials in hydrogenic shocks, well above the injection efficiencies indicated by Figure 9.8. Therefore, these two contributions should provide substantial differences in injection efficiency between pair shocks and hydrogenic or electron–ion ones. Future directions in research on relativistic shock acceleration are strongly motivated to tackle these important injection problems.

9.4.4 Shock Drift Acceleration

The focus of the previous section has been on the diffusive elements of the acceleration process. Yet, inextricably embedded in the results is another phenomenon that is primarily responsible for the origin of these flat indices in subluminal, $\lambda / r_g \gg 1$ cases. Clearly then, once a rare injection is achieved, energization is efficient, but largely in the absence of diffusion. Hence, the primary origin of the acceleration is then connected to coherence in the shock layer. To provide further insight [42] (see also [22]) tracked individual particle trajectories in their Monte Carlo simulation runs for $\lambda \gg r_g$ regimes, with those exhibiting profound energy gains being isolated. These exceptional "events" displayed episodes of continued gyration in the shock layer with concomitant increases in fluid frame and shock frame momenta. Diffusion had minimal impact on the gyrational motion, and the pitch angle evolved almost adiabatically in the particle-shock interaction to preserve gyrational coherence. Following these episodes there were ephemeral upstream excursions that yielded no energization. The cycle was repeated, and [42] noted that it was clear that the acceleration is directly coupled to periods when the particle's gyration straddles the shock. Moreover, inertial motion in the $\boldsymbol{u} \times \boldsymbol{B}$ direction accompanied these epochs of energization. All these characteristics are the hallmarks of *shock drift acceleration* (SDA), and they quickly disappear if λ / r_g drops below 10^2 and cross-field diffusion starts to reappear. Diffusive transport of particles perpendicular to the mean field was omitted in the diffusion-convection equation analysis of [68], who first identified low σ realizations in subluminal, relativistic shocks, and that are approximately reproduced by the $\lambda / r_g = 10^3$ example in Figure 9.9. Baring and Summerlin [42] therefore forged a logical connection that the combined absence of cross-field diffusion and action of shock drift acceleration was primarily responsible for the $\sigma \lesssim 1.5$ phase space being realized in subluminal cases when $\lambda / r_g \gtrsim 10^2$, that is, for laminar fields and almost pure gyrational motion.

The acceleration of particles in the drift electric fields associated with oblique shock discontinuities has been extensively studied in nonrelativistic contexts (e.g., [76–79]). The origin of the effect is the net work done on a charge by the Lorentz force in a zone of nonuniform magnetic field. The principal equation governing this is

$$\boldsymbol{p} \cdot \frac{d\boldsymbol{p}}{dt} = q\boldsymbol{p} \cdot \left\{ \boldsymbol{E} + \frac{\boldsymbol{v}}{c} \times \boldsymbol{B} \right\} \equiv q\boldsymbol{p} \cdot \boldsymbol{E} , \tag{9.47}$$

where $\boldsymbol{E}$ is the $\boldsymbol{u} \times \boldsymbol{B}$ drift field that exists in any oblique shock rest frame other than the de Hoffmann–Teller frame. In uniform $\boldsymbol{B}$ fields the energy gains and losses acquired during a gyroperiod exactly cancel, so that no net work is done, $dW = 0$. In contrast, when a charge's gyromotion straddles the shock discontinuity, the sharp field gradient induces an asymmetry in the time spent by the charge on either side of the shock, so that energy gains and losses do not negate each other. The compressive nature of the field discontinuity biases the net work done to positive increments in shock encounters between upstream excursions, and it is simply shown (e.g., [77]) that $dW = qE_y dy$, that is, this energy gain scales linearly with displacement along the drift coordinate y that marks the $\boldsymbol{u} \times \boldsymbol{B}$ direction.

The upstream excursions that [42] observed to be interspersed between these acceleration periods, where infrequent scattering slowly tries to isotropize the pitch angles, was a feature clearly identified for the SDA phenomenon at nonrelativistic shocks [79]. Particles that participate in the SDA initially have HT frame pitch angle cosines $\mu_{HT} = \boldsymbol{p} \cdot \boldsymbol{B}/|\boldsymbol{p}|\,|\boldsymbol{B}|$ considerably less than unity, so as to satisfy the reflection criterion (see, e.g., [39] for extensive discussions on adiabatic reflection at oblique shocks). Such large pitch angles are rare in near-luminal shocks, for which the incident upstream angular distribution is highly beamed around $\mu_{HT} \sim 1$. During upstream excursions that follow reflection, the particles tend to get swept back to the shock as soon as they are deflected and acquire momenta outside their Lorentz cone, that is, before reaching isotropy in the upstream fluid frame. Therefore, in these $\lambda/r_g \gg 1$ cases, μ_{HT} continues to satisfy the reflection criterion during these brief upstream epochs. The repetition of SDA and upstream excursions, that is, trapping in and near the shock, is therefore virtually guaranteed once the initial reflection is realized for a select particle. But, occasionally it is not, and this occurrence then terminates the SDA process as the charge is swept far downstream. In superluminal shocks, the convection is overwhelming and the reflection criterion is virtually never satisfied.

The identification of the significance of shock drift acceleration in controlling σ for relativistic shocks is an interesting academic issue. The reason it couples to unusually flat distributions ($\sigma < 1.5$) revolves around extremely efficient trapping in the shock layer, which permits repeated episodes of SDA in select particles with appropriately tuned gyrophases at the onset of shock-orbit interactions. The combination of the rapid energization rate, upstream hiatuses in SDA, and a slow leakage rate downstream leads naturally to an approximate $dN/dp \propto p^{-1}$ distribution, when $\lambda/r_g \gtrsim 10^3$. The introduction of turbulence easily disrupts the coherence and precipitates efficient convection downstream, quenching the effectiveness of SDA. The result then is a profound dominance of shock drift acceleration contributions by first-order Fermi diffusive ones when $\lambda/r_g \lesssim 30$. Coupled with the fact that injection from thermal energies is generally highly suppressed for the high λ/r_g conditions that underpin prolific shock drift energization, it appears that SDA-dominated cases may be of less practical interest than diffusion-dominated ones. Moreover, when $\lambda/r_g \gtrsim 10$, an inexorable sweeping of charges downstream in superluminal shocks overpowers both acceleration contributions and steepens the spectrum dramatically, as is evident in Figure 9.9. Accordingly, it is not yet clear how relevant this SDA regime is for extragalactic jet sources; further source spectral modeling will be necessary to elucidate this.

9.4.5 Acceleration Time Scales

In interpreting models of energetic emission in extragalactic jet sources, the spectral characteristics of diffusive shock acceleration at relativistic shocks must be supplemented by considerations of the acceleration time scale τ_{acc}. This connects directly to the issue of the maximum energies of particles realized in the shocked jet, since τ_{acc} is an increasing function of energy. Given a mean free path λ between diffusive collisions, say for example in turning an energetic particle around against the convective flow, diffusion theory determines the time scale for successive reversals, and therefore the temporal scale for successive shock transits. This accumulates to yield a net acceleration time, and the mathematical form for it is well known (e.g., see [39]) in the case of nonrelativistic shocks, $\tau_{\text{acc}} \to \tau_{\text{NR}}$:

$$\tau_{\text{NR}}(p) = \frac{3}{u_{1x} - u_{2x}} \int_{p_i}^{p} \left\{ \frac{\kappa_1(p')}{u_{1x}} + \frac{\kappa_2(p')}{u_{2x}} \right\} \frac{dp'}{p'} . \tag{9.48}$$

Here κ is termed the *spatial diffusion coefficient*, and in parallel shocks $\kappa = \kappa_{\parallel} \sim \lambda v/3$ for a particle of speed v and diffusive mean free path $\lambda(p)$. It is subscripted $i = 1, 2$ to identify the respective upstream and downstream contributions, which differ because the turbulent collisional rate is not the same on either side of the shock. Observe that $\kappa_{\parallel}$ is the component of the spatial diffusion coefficient along the magnetic field; $\kappa_{\perp}$ is its counterpart for diffusion across the field. If ν_{c} is the momentum-dependent collision frequency for a particle, then $\lambda = v/\nu_{\text{c}}$ so that $\kappa_{\parallel} = v^2/(3\nu_{\text{c}})$ and $\kappa_{\perp} = r_{\text{g}}^2 \nu_{\text{c}}/3$ (e.g., [43]), where r_{g} is the particle's gyroradius. The diffusion is generally anisotropic, so that $\kappa_{\perp} < \kappa_{\parallel}$, with equality only when approaching the Bohm limit, $\lambda = r_{\text{g}}$. Technically, the κ appearing in (9.48) is κ_{xx}, the component of a spatial diffusion tensor that describes transport of p_x in the x-direction. Accordingly, it becomes a geometrical mix of $\kappa_{\parallel}$ and $\kappa_{\perp}$, that is, of the diffusive mean free λ and the gyroradius r_{g} of the particle, and the magnetic field obliquity (e.g., see [57, 67]):

$$\kappa_{xx} = \kappa_{\parallel} \cos^2 \Theta_{\text{Bn}} + \kappa_{\perp} \sin^2 \Theta_{\text{Bn}} . \tag{9.49}$$

The appearance of flow speeds in the denominators of the integrand of (9.48) is a result of convection enhancing the *net* collision frequency in diffusive transport back and forth normal to the shock. Observe also that if the collisional process is *gyroresonant* in character, as is the case for particle interactions with MHD turbulence like Alfvén waves, then the collision frequency scales as the gyrofrequency $\nu_{\text{c}} \sim (\lambda/r_{\text{g}})\nu_{\text{g}}$ for $\nu_{\text{g}} = eB/(\gamma mc)$, where γ is the charge's Lorentz factor. The interested reader can refer to papers by [80–84] for quasi-linear field fluctuation theory and the determination of ν_{c} and the spatial diffusion tensors in terms of magnetic power spectra in nonrelativistic systems. Weaker turbulence naturally spawns larger λ/r_{g} ratios. For such gyroresonant collisions involving ultrarelativistic charges, (9.48) reduces to the simple proportionality

$$\tau_{\text{NR}}(p) \sim \frac{1}{\beta_{1x}^2 \nu_{\text{g}}} \frac{\lambda}{r_{\text{g}}} , \qquad p \gg mc . \tag{9.50}$$

Since $\nu_{\text{g}} \propto p^{-1}$, if $\lambda \propto p$ this time scales linearly with the particle momentum, or energy, with the approximate identity $1/\nu_{\text{g}} \approx 0.1 E_{\text{TeV}}/B_{\text{Gauss}}$ seconds *for any relativistic species* (i.e., protons or electrons). Details of how the coefficient of proportionality connects to upstream and downstream shock properties are discussed in [85] in the context of supernova remnant shocks. The essential conclusion is that the acceleration rate scales as the particle's gyrofrequency and is proportional to the square of the upstream flow speed. This behavior also maps over to relativistic shock regimes.

The result in (9.48) applies to nonrelativistic shocks of arbitrary field obliquity $\Theta_{\rm Bf1}$. In highly oblique shocks, [67] determined that κ_{xx} shrinks dramatically (see (9.49)) if cross-field transport is suppressed, thereby seeding extremely rapid acceleration when $\Theta_{\rm Bf1} \sim \pi/2$ and $\kappa_{\perp} \ll \kappa_{\parallel}$. This increase can be ascribed to the presence of electric fields in the shock layer that seed shock drift acceleration. Yet, there is a price to pay for this enticing result: [57] demonstrated that injection of the particles from thermal energies into the acceleration process in oblique shocks is profoundly inhibited when cross-field diffusion is small, that is, $\lambda \gg r_{\rm g}$. Essentially, the ballistic speed of the incident upstream plasma beam barely exceeds the downstream fluid speed in the de Hoffman–Teller frame, so that diffusive return to the shock to instigate acceleration is only significant when cross-field diffusion is strong. Such a trade-off between the enhanced rapidity of acceleration and the diminished injection probability renders this fast energization regime of acceleration phase space less relevant to sources containing shocks that are efficient radiators. The inefficiency of acceleration at $\Theta_{\rm Bf1} > 30°$, $\lambda \gg r_{\rm g}$ shocks was also demonstrated in relativistic discontinuities by [42], and is depicted in Figure 9.8. If $\lambda \sim r_{\rm g}$ and the diffusion is Bohm-limited (corresponding to strong field turbulence $\delta B/B \sim 1$), then cross-field diffusion is prolific and the injection is satisfactorily high, but $\tau_{\rm NR}$ is reduced to levels commensurate with those in quasi-parallel, $\Theta_{\rm Bf1} \sim 0°$ shocks [67]. Current models of field amplification in nonrelativistic shocks (e.g., [86]) and the Wiebel instability in relativistic systems (discussed in Section 9.2) suggest strong MHD turbulence seeding near-Bohm diffusion.

The reason acceleration time scales can be written down in closed form for $\beta_{1x} \ll 1$ shocks is that the *diffusion approximation*, that is, local isotropy of the gas in any pertinent shock frame, can be invoked. This reduces the governing transport equation to a Fokker–Planck form in terms of the collisional integrals, generating the so-called diffusion-convection equation. The temporal dependence can then be directly extracted as in classic mathematical solutions of the diffusion ordinary differential equation (e.g., see [39], for a lengthy exposition). This expedient development is not afforded by relativistic shocks. The power of convection, when flow speeds approach c, forces the energetic population to become highly anisotropic, in which case compact analytic formal solutions for $\tau_{\rm acc}$ are not easily obtained. This complexity essentially derives from the property that the anisotropy and transport information at one point in the flow is influenced by anisotropies and diffusion coefficients at other locales in a large, causally connected volume. Therefore determinations of acceleration times in relativistic shocks are usually more numerical in nature.

Various authors have researched this subject for relativistic shocks ([47], and [87–91]). In particular, the seminal work of EJR90 found that for large angle scattering (LAS) of particles, the acceleration time for a $\Gamma_1 \lesssim 5$ shock was only marginally shorter than that expected from classical nonrelativistic shock theory. Their approach was a numerical Monte Carlo simulation of convection and diffusion. The

Monte Carlo simulations of Bednarz and Ostrowski [87] (see also [88]) revealed similar modest reductions for both LAS and pitch angle diffusion (PAD). Observe that in spite of substantial energy gains per shock crossing in LAS cases, typically on the order of Γ_1^2, the particles then spend considerable time diffusing downstream, a time coupled to their inverse gyrofrequency. This lengthy downstream residency is the controlling factor in determining τ_{acc}.

To amplify the focus on relativistic shocks, Baring [90] computed acceleration times in the limit of pitch angle diffusion using the simulation of EJR90, specifically for application to jets in blazars. It was found that extrapolation of simulations into the relativistic regime revealed a hard lower bound on the total acceleration time τ_{acc} *as measured in the shock rest frame*. The time τ_{acc} monotonically decreases (for ultrarelativistic particles) to this limit as Γ_1 increases to infinity, yet proximity is achieved for $\Gamma_1 \gtrsim 10$. These results were obtained for $\Theta_{Bf1} = 0°$ shocks. Using ν_g to represent the energy-dependent gyrofrequency of an ultrarelativistic electron or ion, [90] obtained an empirical approximate fit to the velocity dependence of the acceleration times in plane-parallel shocks:

$$\tau_{acc}(p) = \left(\frac{1}{4} - \frac{0.18}{\Gamma_1 \beta_1} + \frac{1}{\Gamma_1^2 \beta_1^2} + \frac{0.22}{1 + \Gamma_1 \beta_1} \right) \frac{f}{\nu_g} \frac{\lambda}{r_g} . \qquad (9.51)$$

This form is accurate to around 1–3% when compared with the simulation numerics. The times are for a velocity compression ratio of $u_1/u_2 = 3$ (i.e., an ultrarelativistic gaseous equation of state), and the coefficient f describes details of the differences in diffusion between the upstream and downstream regions, and is on the order of unity and independent of Γ_1. Note that when $\beta_{1x} \ll 1$, the familiar nonrelativistic result emerges: $\tau_{acc} \propto \lambda/(r_g \nu_g \beta_{1x}^2)$. Introducing shock obliquity can speed up the acceleration, as in nonrelativistic cases, but at the price of dramatically steepening the distribution, a property discussed at length above. It is important to emphasize that this empirical result serves as a general guide, but is formally applicable only to strong, plane-parallel shocks. Exploring the parameter space of sonic and Alfvénic Mach numbers and magnetic field orientations will introduce changes, but mostly to the proportionality constant f in (9.51).

As $\Gamma_1 \to \infty$, (9.51) evinces a limiting form $\tau_{acc} \approx f/(4\nu_g)$, that is, one quarter of the extrapolation of the familiar nonrelativistic time scale to $\beta_{1x} \approx 1$ regimes. This bound arises due to the insensitivity of the downstream flow speed and diffusion in the downstream region to the upstream Γ_1. Downstream diffusion yields the dominant contribution to τ_{acc}, with upstream particles requiring only small deflections (accomplished in short times, e.g., see [88]) from the shock normal in order to return downstream. This automatically implies a hard lower bound to τ_{acc} as $\Gamma_1 \to \infty$, since the downstream speed saturates at $c/3$. Effectively, particles can never be accelerated at rates much faster than their gyrofrequency, if the wave-particle interaction is gyroresonant in character. The limit translates to a *comparable mathematical form for the limit in the upstream fluid frame*, which is often the observer's reference perspective, for example the interstellar medium surrounding a jet. This property follows from the connection between Lorentz transformations of times and energies, with the proper time of the particle being a Lorentz invariant. While there is a time-dilation Γ_1 factor appearing in computing τ_{acc} in the upstream fluid frame, the same factor modifies the detected particle energy. Accordingly, models of acceleration at relativistic shocks do not incur any increases to the energization rate other than through the enhancement of the magnetic field by a single Lorentz boost from an observer's frame.

The rate of acceleration is important to the interpretation of blazar X-ray observations. The traditional picture that the electron synchrotron component turns over in the 0.1–100 keV X-ray band provides significant constraints on the jet acceleration environment. If one envisages a cospatial competition between acceleration and synchrotron cooling, the turnover energy is defined when the rates of diffusive acceleration and synchrotron emission are equal. The former is $d\gamma_e/dt \propto (u_{1x}/c)^2 eB/(\eta m_e c)$, where $\eta = \lambda/r_g \geq 1$. This can be equated to the synchrotron loss rate $|d\gamma_e/dt| \propto \gamma_e^2 B^2$ in the comoving frame of the jet. The resulting maximum electron Lorentz factor $\gamma_e \equiv \gamma_c \propto u_1(\eta B)^{-1/2}$ for cooling-limited acceleration then yields a synchrotron peak/cutoff energy that is independent of the field strength:

$$E_{syn} \sim \frac{\delta}{\eta}\left(\frac{u_1}{c}\right)^2 \frac{m_e c^2}{\alpha_f} . \tag{9.52}$$

Here $\alpha_f = e^2/(\hbar c)$, and a blueshift factor δ describing Doppler boosting in the jet has been included. This is a well-known result, which for $\delta = 1$, $u_{1x} \approx c$ and $\eta = 1$ lies around 50 MeV; it was highlighted by De Jager *et al.* [92] for considerations of gamma-ray synchrotron emission at the relativistic pulsar wind termination shock in the Crab nebula.

To move E_{syn} into the classic X-ray band, it is possible to adjust η to fix $u_1 \sim c$, which quickly leads to fitting values $\eta \sim 10^5$, thereby dramatically reducing the rapidity of the acceleration process. This was the approach of [93] when exploring multiwavelength modeling of Mrk 421 spectra. However, as mentioned above, requiring such large values of η leads to extraordinarily inefficient injection of particles into the acceleration process (e.g., [42]), imposing uncomfortable constraints on blazar energetics. Moreover, such large values of η define essentially laminar fields that are not expected in shocks, which are inescapably turbulent. Therefore, it may be more likely that the acceleration and cooling regions are spatially distinct in extragalactic jets, a standpoint advocated by Garson *et al.* [94] for Mrk 421; this is a preferred paradigm in many blazar models.

9.4.6
Nonlinear Acceleration Effects

The extensive presentation on diffusive acceleration thus far has presumed that there is no interplay between the accelerated population and the global MHD structure of the shock that dictates several of the key properties of the acceleration. In principle, the accelerated particles can largely serve as test charges in the bigger picture of the accelerated flow, yet, as exemplified in PIC simulations, all charges are contributing to the overall turbulent electromagnetic picture, and ultimately to their own transport properties. When acceleration is extremely efficient, a sizable fraction of the total energy budget emerges as high-energy cosmic ray ions and electrons, an inevitable occurrence in an $r \approx 4$ nonrelativistic shock if the $dN/dp \propto p^{-\sigma}$ power-law extends to high enough energies, that is, when $\sigma \sim 2$. The pressure of these particles decelerates the upstream flow, i.e., provides significant additional contributions to the energy flux in (9.22) and (9.30). This flux contribution is biased toward the downstream side because of the net convective loss of particles downstream. Accordingly, the accelerated population compresses the upstream flow and increases the compression ratio above the *unmodified* test-particle result. This, in turn, provides feedback to the distribution of accelerated ions and electrons, since in general σ is a declining function of r. The direct consequence is that the fraction of energy going into these particles is enhanced by the hydrodynamic feedback, and the nonlinearity saturates.

This nonlinearity has been thoroughly explored in the literature (e.g., [39, 95–101]) and is a critical characteristic of efficient $u_{1x} \ll c$ shocks. The quintessential example is the Earth's bow shock immersed in the solar wind; it affords a nice data comparison between experiment and acceleration theory [47]. The flattest index in a nonrelativistic, strongly nonlinear shock is $\sigma = 1.5$ (e.g., [99]). As a result of the energy conservation that regulates the acceleration and the energy apportionment between thermal ions, electrons and high-energy cosmic rays, there are two distinctive features of such nonlinear shocks: (i) a distribution that deviates from pure power-law nature, exhibiting a characteristic upward concavity (i.e., $\sigma(p)$ is a declining function of momentum p); this is due to higher-energy particles sampling larger effective compression ratios, since their diffusive mean free paths are longer (i.e., $d\lambda/dp > 0$), and (ii) the thermal particles are somewhat cooler [102] than for test-particle shocks, since the shock is weakened on the smallest diffusive scales (i.e., the so-called subshock) and energy removed from the thermal ions and electrons. These phenomena can be probed to a certain extent by examining radiation from isolated supernova remnants, and at present, there are at best modest indications to support these theoretical predictions.

Turning now to relativistic shocks in extragalactic jets, the key ingredient that must underpin the importance of nonlinear acceleration effects is that the test particle distribution index should be flatter than $dN/dp \propto p^{-2}$. From the exposition in Sections 9.4.2 and 9.4.3, it is clear that this corresponds to one of the two following circumstances: (i) the shocks are subluminal, that is, of modest obliquity, and shock drift acceleration is active, if the scattering is in the pitch angle diffusion regime, or (ii) large angle scattering is operative and the shocks are either subluminal or at most mildly superluminal. It is clear that rampant convective action in highly superluminal, oblique shocks suppresses return of particles to the shock interface and thereby steepens the spectral index σ to a point that renders the contribution of the accelerated population to the particle energy flux insignificant. Contrasting the case for nonrelativistic shocks, there is no extensive literature on the investigation of nonlinear effects in $\Gamma_1 \gtrsim 2$ shocks. The reader is directed to [52] for a sample exploration that demonstrated that nonlinear modifications to σ are minimal in parallel, $\Gamma_1 = 10$ shocks where $\sigma \approx 2.23$ is the canonical test particle result obtained in Kirk *et al.* [50]. Obviously, investment in the study of nonlinear acceleration effects at relativistic shocks is needed, since these can significantly impact the interpretation of X-ray synchrotron spectra in HBL (blue) blazars and inverse-Compton signals seen by the Fermi Gamma-Ray Space Telescope from energetic blazars such as Mrk 421 and PKS 2155-304.

9.5 Acceleration by Magnetic Reconnection

Another potential acceleration mechanism in jets is *magnetic reconnection*, the astrophysical and laboratory aspects of which were recently reviewed at length by [103]. When neighboring field lines of opposite "polarity" or vector direction become very close, the environment is dynamically unstable. The spatially proximate currents required to sustain the field structure can intermingle, leading to the annihilation of adjacent field loops. This action of reconnection is energetically favored, driving transient electric fields as charges relocate, with the liberated energy being deposited as heat in the plasma. Essentially, transient fields in such current sheets can be extremely effective in energizing particles, both as thermal dissipation, but also beyond the thermal pool into an acceleration process. Magnetic reconnection is widely cited as a principal mechanism for the generation of relativistic electrons in solar coronal field loops that are inferred by the nonthermal bremsstrahlung emission in the hard X-ray and soft gamma-ray bands (e.g., see [104], for an extensive review): twisting of the loops in connection with a mobility of their footpoints naturally affords an opportunity for field line merging. Reconnection is also suggested as a possible means of lepton acceleration in and near the quasi-equatorial current sheet of the striped wind emanating from pulsars (e.g., [105, 106]). Both these astrophysical settings offer proximity of oppositely directed field lines, a circumstance that may or may not extend to extragalactic jets.

In the context of active galaxies, Nalawejko *et al.* [107] recently discussed the possibility of field reconnection seeding acceleration and radiative emission. Their focus was at the base of the jet, near the accretion disk, a quasi-equatorial environment where current sheets and striped-field outflows naturally occur. Yet, in the polar regions of a rotating MHD outflow, in principle it is much more difficult to sample proximate field reversals. In particular, for quasi-dipolar stellar rotators such as the sun and pulsars, sampling of lines sequentially connected to opposite poles is inhibited at high latitudes above the magnetic poles. In the case of supermassive black holes, field lines cannot thread the event horizon: at more remote locales at higher altitudes, the MHD configuration may somewhat resemble those of rotating stellar objects. Accordingly, it is unclear whether or not large-scale field reversals persist out to large radii in extragalactic jets. It is possible that they do, and an unambiguous indication could be obtained through high resolution radio/optical polarimetric observations of synchrotron emission. This defines a task for the future, going beyond time-dependent polarization angle swings measured in blazars such as 3C 279 [108].

The prospect that leptons in a jet may be energized via magnetic reconnection needs deeper exploration from a theoretical standpoint. Reconnection theory must make robust predictions of distributions and injection efficiencies (from thermal gas) of accelerated populations in order to connect effectively with blazar models and observations. Some progress has been made toward this goal, for example in the analyses of Larabee *et al.* [109], Jaroschek *et al.* [110] and Zenitani and Hoshino [111], suggesting the generation of very flat tail distributions of accelerated leptons. Yet the understanding of diffusive shock acceleration is more developed in this regard, and accordingly is the focus of this chapter. It is also quite possible that both reconnection acceleration and shock-related acceleration can be simultaneously active in jets, perhaps with reconnection generating some preacceleration for the diffusive acceleration mechanism. As a concluding remark, one anticipates that the solar corona may prove a powerful testing ground for honing models of reconnection in the same way that the solar wind has demonstrated the general viability of diffusive acceleration at nonrelativistic shocks.

9.6 Outstanding Questions

To conclude this chapter, it is clear that the state of knowledge of particle acceleration at relativistic shocks is sufficient for insightful deployment of its predictions in radiation models of extragalactic jets in different wavebands. Yet, the understanding is still limited, and energy in the theoretical community needs to be expended to validate different approaches to the acceleration problem. In an ideal world, agreement between particle-in-cell plasma simulations, Monte Carlo techniques and analytic approaches like solutions of diffusion-convection equations would be a reality. There is some progress towards this goal, but much work needs to be done, and outstanding questions in the field need to be tackled. In the near term, detailed comparison between Monte Carlo simulations and analytic techniques on spectral index determinations is a viable possibility. So also are head-to-head distribution comparisons between PIC codes and Monte Carlo methods in the suprathermal regime, to elucidate the diffusive properties of shock layer turbulence.

One open question is *what is the real character of relativistic shock turbulence*? There is in all probability no single answer to this, either in one shock, let alone one cosmic source. PIC simulations are, at first sight the best tool to probe this issue, due to the richness of their turbulence information. Yet, this asset imposes significant computational constraints that at present preclude sampling large momenta or large spatial scales at a distance from the shock, particularly for electron–proton shocks where the disparate mass scales burden the computational speed. They can address Weibel instability physics, but do not cast light upon what turbulence is operating on large scales. This provides a niche for exploring injection issues, but restricts prospects in the near future for directly modeling energetic charges that emit in radio, optical, X-ray and gamma-ray bands of blazars and quasars. On the other hand, Monte Carlo techniques can consider such high energies, and therefore neatly interface with photon data, but must prescribe the turbulence or characterize it parametrically, and so have some difficulty diagnosing differences in the nature of turbulence on disparate spatial and momentum scales. From a theoretical and observational standpoint, it is highly desirable to discern whether Alfvén type turbulence is dominant on large scales in relativistic shocks, or if the Weibel instability can influence or spawn turbulence in these more remote shock locales. The answer may well require the contribution of both techniques.

Next, researchers on radio sources and blazars would like to know *how much does baryon loading reduce (or increase) shock efficiency*? For example, blazar modeling is currently biased toward a lepton jet paradigm, so that the acceleration results presented in Section 9.4 provide a relatively clear picture of what should be imported into emission models. However, the accretion material feeding black holes is predominantly hydrogenic in nature, so one anticipates that significant contamination of the jet by ions is possible. This mass loading incurs kinematic "drag" on the jet ballistics, with implications for the bulk motion: hydrogenic jets may well be slower than their pair counterparts. Yet, this ushers in the possibility of a reduced radiative energy budget and luminosity in relativistic jets, but does not directly curtail the efficiency of the acceleration process. In fact, as discussed at the end of Section 9.4.3, electron–ion equilibration due to shock layer electrostatic potentials offers the hope of redressing the inequity of the kinetic energy budget in $e-p$ shocks, so that electrons are no longer highly disfavored. Without the influence of cross-shock potentials, hydrogenic shocks should exhibit an enhancement of protons over electrons by a factor of $(m_p/m_e)^\sigma$ at a given energy above 1 GeV. With them, this enhancement may be reduced to a factor of merely $10-10^3$. Adding in the comparative radiative efficiency factor (typically $\sim [m_p/m_e]^3$) in favor of electrons, it becomes clear that this question may profoundly impact the hadronic vs. leptonic jet discussion. Hence, isolating the role of charge separation potentials in relativistic shocks is of paramount importance.

There are other theoretical avenues for exploration that are germane to extragalactic jets, such as determination of acceleration time scales in oblique shocks and comparing them with those known for parallel ones. Also, to aid spectral interpretation of blazar X-rays and gamma-rays, it is important to comprehend when nonlinear acceleration modifications, mentioned in Section 9.4.6, will alter the predictions of linear theory and impact the modeling of sources. These provide goals for theoretical investigations in the short term future. One can also ask *what observational diagnostics will precipitate advances on the theoretical front*? The answer is largely more of what has transpired over the last decade or so. The field has seen multiwavelength campaigns become prominent, almost ubiquitous, so that we have temporal snapshots of the broadband emission spectrum, and therefore less confused inferences concerning the underlying particle populations. This has propelled our understanding, and advances such as the recent contribution from the Fermi-LAT experiment in gamma-rays have helped fill in pieces of the blazar puzzle. Perhaps the most groundbreaking prospect on the short term horizon is that of X-ray polarimetry with Gravity and Extreme Magnetism SMEX (GEMS; see [112]), which will provide new insights into jet geometry. Such a contribution should afford diagnostics that reduce the degeneracies in source models, enabling more direct probes of the underlying physics of the jet plasma. This will set the scene for successor polarimetric initiatives in high-energy astrophysics, and continue the honing of our understanding of AGN jets that has improved dramatically over the last five decades.

References

1 Cohen, M.H. *et al.* (1977) *Nature*, **268**, 405.
2 Begelman, M.C., Blandford, R.D., and Rees, M.J. (1984) *Rev. Mod. Phys.*, **56**, 255.
3 Dondi, L. and Ghisellini, G. (1995) *Mon. Not. R. Astron. Soc.*, **273**, 583.
4 Ichimaru, S. (1973) *Basic Principles of Plasma Physics: a Statistical Approach*, W.A. Benjamin, Reading, Mass.
5 Stix, T.H. (1992) *Waves in Plasmas*, AIP, New York.
6 Alfvén, H. (1942) *Nature*, **150**, 405.
7 Weibel, E.S. (1959) *Phys. Rev. Lett.*, **2**, 83.
8 Fried, B.D. (1959) *Phys. Fluids*, **2**, 337.
9 Medvedev, M.V. and Loeb, A. (1999) *Astrophys. J.*, **526**, 697.
10 Silva, L.O., Fonseca, R.A., Tonge, J.W. *et al.* (2003) *Astrophys. J.*, **596**, L121.
11 Nishikawa, K.-I., Hardee, P., Richardson, G. *et al.* (2005) *Astrophys. J.*, **622**, 927.
12 Medvedev, M.V., Fiore, M., Fonseca, R.A. *et al.* (2005) *Astrophys. J.*, **618**, L75.
13 Spitkovsky, A. (2008) *Astrophys. J.*, **682**, L5.
14 Boyd, T.J.M. and Sanderson, J.J. (1969) *Plasma Dynamics*, Nelson, London.
15 Landau, L.D. and Lifshitz, E.M. (1959) *Fluid Mechanics*, Pergamon Press, London.
16 Gradshteyn, I.S. and Ryzhik, I.M. (1980) *Table of Integrals, Series and Products*, Academic Press, New York.
17 Jüttner, F. (1911) *Ann. Phys.*, **34**, 856.
18 Jüttner, F. (1911) *Ann. Phys.*, **35**, 145.
19 Synge, J.L. (1957) *The Relativistic Gas*, North Holland, Amsterdam.
20 de Groot, S.R., van Leeuwen, W.A., and van Weert, C.G. (1980) *Relativistic Kinetic Theory*, North Holland, Amsterdam.
21 Blandford, R.D. and McKee, C.F. (1976) *Phys. Fluids*, **19**(8), 1130.
22 Summerlin, E.J. (2009) Diffusive Acceleration of Particles at Collisionless Magnetohydrodynamic Shocks, PhD Thesis, Rice University.
23 Taub, A.H. (1948) *Phys. Rev.*, **74**, 328.
24 Weinberg, S. (1972) *Gravitation and Cosmology: Principles and Applications of the General Theory of Relativity*, John Wiley & Sons, New York.
25 Königl, A. (1980) *Phys. Fluids*, **23**, 1083.
26 de Hoffmann, F. and Teller, E. (1950) *Phys. Rev. D*, **80**, 692.
27 Ballard, K.R. and Heavens, A.F. (1991) *Mon. Not. R. Astron. Soc.*, **251**, 438.
28 Tolman, R.C. (1934) *Relativity, Thermodynamics and Cosmology*, University Press, Oxford.
Dover reprint, 1987.
29 Jackson, J.D. (1975) *Classical Electrodynamics*, John Wiley & Sons, New York.
30 Double, G.P., Baring, M.G., Ellison, D.C., and Jones, F.C. (2004) *Astrophys. J.*, **600**, 485.
31 Webb, G.M., Zank, G.P., and McKenzie, J.F. (1987) *J. Plasma Phys.*, **37**, 117.
32 Appl, S. and Camenzind, M. 1988, *Astron. Astrophys.*, **206**, 258.
33 Heavens, A.F. and Drury, L.O'C. (1988) *Mon. Not. R. Astron. Soc.*, **235**, 997.
34 Fermi, E. (1949) *Phys. Rev. Lett.*, **75**, 1169.
35 Krymsky, G.F. (1977) *Dokl. Akad. Naut. SSSR*, **234**, 1306.

36 Axford, W.I., Leer, E., and Skadron, G. (1977) Proc. 15th ICRC (Plovdiv), 13–26 August 1977, B'lgarska Akademiia na Naukite, Sofia, XI, 132.

37 Bell, A.R. (1978) *Mon. Not. R. Astron. Soc.*, **182**, 147.

38 Blandford, R.D. and Ostriker, J.P. 1978, *Astrophys. J.*, **221**, L29.

39 Drury, L.O'C. (1983) *Rep. Prog. Phys.*, **46**, 973.

40 Blandford, R.D. and Eichler, D. (1987) *Phys. Rep.*, **154**, 1.

41 Jones, F.C. and Ellison, D.C. (1991) *Space Sci. Rev.*, **58**, 259.

42 Baring, M.G. and Summerlin, E.J. (2009) in *Shock Waves in Space and Astrophysical Environments* (eds X. Ao, R. Burrows, and G.P. Zank), AIP Conf. Proc. 1183, 1–7 May 2009, New York, American Institute of Physics (AIP), p. 74, [astro-ph/0910.1072].

43 Jones, F.C. (1994) *Astrophys. J. Suppl. Ser.*, **90**, 561.

44 Pryadkho, J.M. and Petrosian, V. 1997, *Astrophys. J.*, **482**, 774.

45 Peacock, J.A. (1981) *Mon. Not. R. Astron. Soc.*, **196**, 135.

46 Kirk, J.G. and Schneider, P. (1987) *Astrophys. J.*, **315**, 425.

47 Ellison, D.C., Jones, F.C., and Reynolds, S.P. (1990) *Astrophys. J.*, **360**, 702. (EJR90).

48 Ostrowski, M. (1991) *Mon. Not. R. Astron. Soc.*, **249**, 551.

49 Bednarz, J. and Ostrowski, M. (1998) *Phys. Rev. Lett.*, **80**, 3911.

50 Kirk, J.G., Guthmann, A.W., Gallant, Y.A., and Achterberg, A. (2000) *Astrophys. J.*, **542**, 235.

51 Baring, M.G. (1999) Proc. 26th ICRC (eds D. Kieda, M. Salamon, and B. Dingus), 17–25 August 1999, Salt Lake City, International Union of Pure and Applied Physics (IUPAP), IV, 5.

52 Ellison, D.C. and Double, G.P. (2002) *Astrophys. J.*, **18**, 213.

53 Baring, M.G. (2004) *Nucl. Phys. B*, **136C**, 198.

54 Blasi, P. and Vietri, M. (2005) *Astrophys. J.*, **626**, 877.

55 Morlini, G., Blasi, P., and Vietri, M. 2007, *Astrophys. J.*, **658**, 1069.

56 Forman, M.A., Jokipii, J.R., and Owens, A.J. (1974) *Astrophys. J.*, **192**, 535.

57 Ellison, D.C., Baring, M.G., and Jones, F.C. (1995) *Astrophys. J.*, **453**, 873.

58 Ellison, D.C. and Double, G.P. (2004) *Astrophys. J.*, **22**, 323.

59 Stecker, F.W., Baring, M.G., and Summerlin, E.J. (2007) *Astrophys. J.*, **667**, L29.

60 Hoshino, M., Arons, J., Gallant, Y.A., and Langdon, A.B. (1992) *Astrophys. J.*, **390**, 454.

61 Ostrowski, M. (1990) *Astron. Astrophys.*, **238**, 435.

62 Gallant, Y.A. and Achterberg, A. (1999) *Mon. Not. R. Astron. Soc.*, **305**, L6.

63 Niemiec, J. and Ostrowski, M. (2004) *Astrophys. J.*, **610**, 851.

64 Martins, S.F., Fonseca, R.A., Silva, L.O., and Mori, W.B. (2009) *Astrophys. J.*, **695**, L189.

65 Sironi, L. and Spitkovsky, A. (2009) *Astrophys. J.*, **707**, L92.

66 Begelman, M.C. and Kirk, J.G. (1990) *Astrophys. J.*, **353**, 66.

67 Jokipii, J.R. (1987) *Astrophys. J.*, **313**, 842.

68 Kirk, J.G. and Heavens, A.F. (1989) *Mon. Not. R. Astron. Soc.*, **239**, 995.

69 Hartman, R.C., Bertsch, D.L., Fichtel, C.L. *et al.* (1992) *Astrophys. J.*, **385**, L1.

70 Punch, M., Akerlof, C.W., Cawley, M.F. *et al.* (1992) *Nature*, **358**, 477.

71 Krennrich, F., Bond, I.H., Bradbury, S.M. *et al.* (2002) *Astrophys. J.*, **575**, L9.

72 Aharonian, F., Akhperjanian, A.G., Beilicke, M. *et al.* (2003) *Astron. Astrophys.*, **410**, 813.

73 Aharonian, F., Akhperjanian, A.G., Anton, G. *et al.* (2009) *Astrophys. J.*, **696**, L150.

74 Abdo, A.A. *et al.* (2009) *Astrophys. J.*, **707**, 1310.

75 Baring, M.G. and Summerlin, E.J. (2007) *Astrophys. Space Sci.*, **307**, 165.

76 Sarris, E.T. and Van Allen, J.A. (1974) *J. Geophys. Res.*, **79**, 4,157.

77 Jokipii, J.R. (1982) *Astrophys. J.*, **255**, 716.

78 Pesses, M.E., Decker, R.B., and Armstrong, T.P. (1982) *Space Sci. Rev.*, **32**, 185.

79 Decker, R.B. and Vlahos, L. (1986) *Astrophys. J.*, **306**, 710.

80 Hall, D.E. and Sturrock, P.A. (1967) *Phys. Fluids*, **10**, 2620.
81 Jokipii, J.R. and Coleman, P.J. (1968) *J. Geophys. Res.*, **73**(5), 495.
82 Kulsrud, R. and Pearce, W.P. (1969) *Astrophys. J.*, **156**, 445.
83 Jones, F.C., Birmingham, T.J., and Kaiser, T.B. (1978) *Phys. Fluids*, **21**, 347.
84 Schlickeiser, R. (1989) *Astrophys. J.*, **336**, 243.
85 Baring, M.G., Ellison, D.C., Reynolds, S.R. *et al.* (1999) *Astrophys. J.*, **513**, 31.
86 Lucek, S.G. and Bell, A.R. (2000) *Mon. Not. R. Astron. Soc.*, **314**, 65.
87 Bednarz, J. and Ostrowski, M. (1996) *Mon. Not. R. Astron. Soc.*, **283**, 447.
88 Bednarz, J. (2000) *Mon. Not. R. Astron. Soc.*, **315**, L37.
89 Achterberg, A., Gallant, Y.A., Kirk, J.G. *et al.* (2001) *Mon. Not. R. Astron. Soc.*, **328**, 393.
90 Baring, M.G. (2002) *Publ. Astron. Soc. Aust.*, **19**, 60.
91 Meli, A. and Quenby, J.J. (2003) *Astrophys. J.*, **19**, 649.
92 De Jager, O.C. *et al.* (1996) *Astrophys. J.*, **457**, 253.
93 Inoue, S. and Takahara, F. (1996) *Astrophys. J.*, **463**, 555.
94 Garson, A.B., Baring, M.G., and Krawczynski, H. (2010) *Astrophys. J.*, **722**, 358.
95 Drury, L.O'C. and Völk, H.J. (1981) *Astrophys. J.*, **248**, 344.
96 Eichler, D. (1984) *Astrophys. J.*, **277**, 429.
97 Ellison, D.C. and Eichler, D. (1984) *Astrophys. J.*, **286**, 691.
98 Ellison, D.C., Baring, M.G., and Jones, F.C. (1996) *Astrophys. J.*, **473**, 1029.
99 Malkov, M.A. (1997) *Astrophys. J.*, **485**, 638.
100 Blasi, P. (2002) *Astrophys. J.*, **16**, 429.
101 Amato, E. and Blasi, P. (2005) *Mon. Not. R. Astron. Soc.*, **364**, L76.
102 Berezhko, E.G. and Ellison, D.C. (1999) *Astrophys. J.*, **526**, 385.
103 Zweibel, E.G. and Yamada, M. (2009) *Ann. Rev. Astron. Astr.*, **47**, 291.
104 Aschwanden M.J. (2002) *Space Sci. Rev.*, **101**, 1.
105 Lyubarsky, Y.E. and Kirk, J.G. (2001) *Astrophys. J.*, **547**, 437.
106 Pétri, J. and Lyubarsky, Y. (2007) *Astron. Astrophys.*, **473**, 683.
107 Nalewajko, K., Dimitrios, G., Begelman, M.C. *et al.* (2011) *Mon. Not. R. Astron. Soc.*, **413**, 333.
108 Abdo, A.A. *et al.* (2010) Nature, **463**, 919.
109 Larabee, D.A., Lovelace, R.V.E., and Romanova, M.M. (2003) *Astrophys. J.*, **586**, 72.
110 Jaroschek, C.H., Lesch, H., and Treumann, R.A. (2004) *Astrophys. J.*, **605**, L9.
111 Zenitani, S. and Hoshino, M. (2007) *Astrophys. J.*, **670**, 702.
112 Swank, J. *et al.* (2011) GEMS: Gravity and Extreme Magnetism SMEX, http://heasarc.gsfc.nasa.gov/docs/gems/ (accessed 29 April 2011).

10
Simulations of Jets from Active Galactic Nuclei and Gamma-Ray Bursts

Miguel A. Aloy and Petar Mimica

Together with theory and observations, numerical simulations have become a basic tool to deepen our understanding of astrophysical jets. The key processes taking place in jets are extraordinarily dynamic and complex, which limits purely analytic approaches to obtain order of magnitude estimates. Thus, we shall address with computer models highly nonlinear processes (e.g., jet generation, acceleration, collimation and propagation, jet/ambient interaction). An additional complication comes from the fact that the physics of jets involves a huge range of length and time scales. In practice, this fact prevents the utilization of first-principles simulations, based on a kinetic description of the jet plasma and of the ambient in which it is embedded, except for a handful of problems (e.g., particle acceleration at shocks, generation of magnetic fields, etc.). The plasma constituents are glued together by the action of magnetic fields of arbitrary topology, which set the Larmor gyration radius many orders of magnitude smaller than the typical size of the system (see [1]). Plasma electroneutrality at macroscopic scales is guaranteed because the Debye length of the constituent ions and electrons are also tiny compared with the jet radius. Under such conditions, a hydrodynamic viewpoint is accurate enough to model jets and their environments. Furthermore, if the magnetic field becomes dynamically important, that is, of energy content comparable to the kinetic or internal energy of the plasma, a magnetohydrodynamic description is adequate.

There are two basic types of numerical simulations in the context of astrophysical jets. On the one hand, we have simulations which look for the asymptotic (steady state) jet structure that arises from selected boundary conditions (see, e.g., Chapter 11). On the other hand, one can set up "time-dependent" simulations, which try to assess evolutionary/dynamic problems. Here we will focus on the results of the (magneto)hydrodynamic numerical modeling of relativistic jets, as well as on the future perspectives a detailed numerical work can yield. Among them, we elaborate in this chapter on some general aspects of the propagation of superluminal components in parsec-scale jets, the morphology, dynamics and stability properties of relativistic jets. We also consider in this chapter the most successful numerical algorithms, employed with the purpose of obtaining a refined understanding of

Relativistic Jets from Active Galactic Nuclei, First Edition. Edited by M. Böttcher, D.E. Harris, H. Krawczynski.

the physics of relativistic jets. Some selected recent results along with a number of semianalytic models, including general relativistic aspects, will be covered in Chapter 11.

Also, we will pay special attention to obtain the observational signature of relativistic flows, which is typically nonthermal, and thus, it needs to be carefully coupled to the thermal (i.e., hydrodynamical) evolution of the system. We finally note that simulations are also used to study the process of jet formation (see Chapter 4). This is a vibrant subfield with a lot of progress in recent years, which involves including a general relativistic treatment in the models, since the formation of relativistic jets is tightly linked to the existence of compact stars and black holes. Adequate coverage of the general relativistic notions needed to deal with jet formation is provided in Chapter 2.

10.1 Governing Equations

Most of the numerical work done in the field of relativistic jets assumes that they can be modeled as fluids. Thereby, the relativistic generalization of the Navier–Stokes equations governs the evolution of jets. These equations are, on the one hand, quite involved and, there is still some ongoing debate about whether different formalisms may lead to results that violate causality and/or generate unstable (unphysical) modes [2, 3]. On the other hand, the exact amount of viscosity in astrophysical jets is difficult to evaluate. Therefore, most of the work done in numerical modeling of astrophysical jets resorts to the relativistic extension of the Euler equations (or inviscid relativistic Navier–Stokes equations), which we will refer to as relativistic hydrodynamic (RHD) equations. If magnetic fields are dynamically relevant, and the plasma conductivity is large enough (ideal limit) these equations are supplemented with the corresponding (induction) Maxwell equations, and form the system of relativistic (ideal) magnetohydrodynamic (RMHD) equations. Furthermore, in the vicinity of extreme gravitational fields, general relativistic effects need to be accounted for. The equations of general relativistic (ideal) magnetohydrodynamics (GRMHD) form a system of coupled nonlinear, multidimensional, partial differential equations. These equations represent the physical fact that mass, momentum, energy and magnetic flux must be conserved. For numerical purposes, it is convenient to write the system of GRMHD equations as a first-order, flux-conservative system (but note that nonconservative formulations are also in use in the literature for example [4, 5]),[1] where the time variations of a vector of unknowns $\boldsymbol{U}(t, \boldsymbol{x})$ in every point of the space $\boldsymbol{x}$ are produced by the fluxes $\boldsymbol{F}^i(\boldsymbol{U}(t, \boldsymbol{x}))$ of such quantities across the surface of a sufficiently small control volume around $\boldsymbol{x}$, as well as, by the presence of source or sink terms $\boldsymbol{S}$ at that point. Mathematically, we

1) The fact that an algorithm is (flux-)conservative is relevant, not only because the underlying equations form a system of conservation laws, but also because by virtue of the Lax–Wendroff theorem [6], only conservative schemes can be expected to get the correct jump conditions and propagation speed for a discontinuous solution (e.g., a shock wave).

have (see, for example [7] or [8]):

$$\frac{1}{\sqrt{-g}}\left(\frac{\partial\sqrt{\gamma}\boldsymbol{U}(t,\boldsymbol{x})}{\partial t}+\frac{\partial\sqrt{-g}\boldsymbol{F}^i(\boldsymbol{U}(t,\boldsymbol{x}))}{\partial x^i}\right)=\boldsymbol{S}(t,\boldsymbol{x})\,, \tag{10.1}$$

where $i = 1, 2, 3$,

$$\boldsymbol{U}=\left[D, S_j, \tau, B^k\right]^T\,, \tag{10.2}$$

$$\boldsymbol{F}^i=\left[D\tilde{v}^i, S_j\tilde{v}^i + p^*\delta^i_j - b_j\frac{B^i}{W}, \tau\tilde{v}^i + p^*v^i - \alpha b^0\frac{B^i}{W}, \tilde{v}^i B^k - \tilde{v}^k B^i\right]^T\,, \tag{10.3}$$

$$\boldsymbol{S}=\left[0, T^{\mu\nu}\left(\frac{\partial g_{\nu j}}{\partial x^\mu}-\Gamma^\delta_{\nu\mu}g_{\delta j}\right), \alpha\left(T^{\mu 0}\frac{\partial\ln\alpha}{\partial x^\mu}-T^{\mu\nu}\Gamma^0_{\nu\mu}\right), 0, 0, 0\right]^T\,, \tag{10.4}$$

$g_{\mu\nu}$ ($\mu, \nu = 0, \ldots, 3$) being the metric of the space-time, whose line element, written according to a $3 + 1$ foliation reads as

$$g_{\mu\nu}dx^\mu dx^\nu = -(\alpha^2 - \beta_i\beta^i)dt^2c^2 + 2\beta_i c\,dt\,dx^i + \gamma_{ij}dx^i dx^j\,, \tag{10.5}$$

where α, β^i and γ_{ij} are the lapse function, the shift vector and the three-metric induced on each space-like slice of the foliation of the space-time. g and γ are the determinants of $g_{\mu\nu}$ and of $\gamma_{\mu\nu}$, respectively. $\Gamma^\delta_{\mu\nu}$ are the connection coefficients (see Section 11.1.1). The speed of light in vacuum is annotated by c. $T^{\mu\nu}$ refer to the components of the stress-energy tensor corresponding to a perfect, magnetized fluid (see Sections 11.1.2 and 9.3.3),

$$T^{\mu\nu} = \rho h^* u^\mu u^\nu + p^* g^{\mu\nu} - b^\mu b^\nu\,, \tag{10.6}$$

with $p^* = p + b_\mu b^\mu/2$ and $h^* = h + b_\mu b^\mu/(\rho c^2)$ being the total pressure and enthalpy per unit mass, respectively. p, ρ, v^i, u^μ, b^μ and h are the thermal pressure, the rest-mass density, the vector components of the three-velocity measured by the Eulerian observer,[2)] the components of the fluid four-velocity (with the normalization $g_{\mu\nu}u^\mu u^\nu = -c^2$), the components of the magnetic field four-vector, and the thermal specific enthalpy ($h = 1 + \epsilon/c^2 + p/(\rho c^2)$, ϵ being the internal energy per unit mass). The fluid Lorentz factor is defined as $W = 1/\sqrt{1 - v_i v^i/c^2}$, and, for the sake of brevity, we have introduced the variable $\tilde{v}^i = v^i - c\beta^i/\alpha$. Furthermore, D, S_j, τ, and B^k are the rest-mass density, the covariant components of the momentum density, the energy density (not including rest-mass density) and the vector components of the magnetic field measured by the Eulerian observer.

2) This observer is such that his four-velocity is perpendicular to the hypersurfaces of constant t at each event in the space-time. In the Kerr metric the Eulerian observer corresponds with the observer with zero azimuthal angular momentum (ZAMO) as measured from infinity.

Among the previous variables the following relations hold:

$$D = \rho W \,, \tag{10.7}$$

$$S_j = \rho h^* W^2 v_j - \alpha b^0 b_j \,, \tag{10.8}$$

$$\tau = \rho h^* W^2 c^2 - p^* - \alpha^2 (b^0)^2 - D c^2 \,, \tag{10.9}$$

$$b^0 = \frac{W B^i v_i}{c\alpha} \,, \tag{10.10}$$

$$b^i = \frac{B^i + \alpha b^0 u^i / c}{W} \,. \tag{10.11}$$

The special relativistic particularization of the system (10.1) (the RMHD system of equations), is readily obtained by taking (i) $g_{\mu\nu} = \eta_{\mu\nu}$, where $\eta_{\mu\nu}$ is the Minkowski metric of the (flat) space-time (or, equivalently $\alpha = 1$, $\beta^i = 0$; see also (9.27)–(9.31)). Making (ii) $b^\mu = 0$, one obtains the governing equations of general relativistic hydrodynamics (GRHD). Finally, if we take conditions (i) and (ii) simultaneously, the equations of RHD result.

10.2
Numerical Algorithms

For the numerical treatment of the system (10.1), there is a large diversity of numerical algorithms, a complete review of which is beyond the scope of this chapter (interested readers are referred to the reviews of [8, 9]). The aforementioned system together with the appropriate initial and boundary conditions is solved as an initial value boundary problem. This means that one must specify data and boundary conditions at a certain time t and update the value of the vector of unknowns up to $t' = t + dt$, dt being a *small* time interval.

To perform the numerical update of the solution, it is impossible to do it *everywhere* in the spatial domain of interest. Instead, one chooses a finite number of spatial points, where suitable approximations $\boldsymbol{U}^n_{i,j,k}$ to the actual solution of the equations $\boldsymbol{U}(t^n, x^1_i, x^2_j, x^3_k)$ are specified. Depending on the recipe chosen to perform the spatial discretization, the numerical algorithm can be cataloged as a finite differences, finite volumes, or finite elements method. While finite elements are quite popular in engineering problems, for astrophysical applications finite differences (FD) or finite volumes (FV) techniques are the most widely used (but see, e.g., [10–12]). The main difference between FD and FV techniques lies on whether the discrete solution approximates the point-values $\boldsymbol{U}(t^n, x^1_i, x^2_j, x^3_k)$ or averages of the exact solution $\boldsymbol{U}$ on control volumes (numerical cells) that tessellate the spatial region of interest. Regardless of the discretization method adopted, an accurate representation of the exact solution depends on the optimal choice of the numerical mesh, as well as on the number of points used to discretize the solution (the finer the grid, the better representation).

Though the uniform discretization of space and time is simple, it is not usually the choice that maximizes the efficiency and reliability of numerical simulations. Typically, we need finer grids in some regions, (e.g., where the fluid is nonsmooth), while at the same time, coarser grids are enough in smooth regions. A broadly used gridding scheme, which self-adapts to the physical conditions at each location of the computational domain, varying the size of the numerical cells according to the dynamics of the flow, is the Adaptive Mesh Refinement (AMR; [13]). There are many practical implementations of AMR for hydrodynamical codes developed for astrophysical or cosmological problems, both open source (e.g., Chombo, PARAMESH) or proprietary (e.g., AMRVAC [14, 15], AMRA [16], MASCLET [17]).

To advance forward in time the discrete solution $\boldsymbol{U}^n_{i,j,k}$ one may use implicit or explicit methods. In any case, the time advance algorithm has to pay special attention to the fact that the solutions of the system (10.1), may develop discontinuities (shock waves, contact or rotational discontinuities), even if the initial data is smooth, after a finite time (e.g., [18]). Thus, suitable dissipation terms must be included in the algorithm to control the numerical properties of nonsmooth flows. The most rudimentary form of dissipation adds, in a linear approximation, the same amount of *artificial viscosity* to all grid points. More elaborated methods (e.g., flux-corrected transport), include high-order (less dissipative) terms in smooth parts of the flow, while they resort to low-order (more dissipative) terms near discontinuities. The most robust formalism for treating shocks is the Godunov method, where the numerical solution is approximated as a piecewise uniform (discrete) function. Each cell takes a uniform value and the flux across the interface separating two adjacent numerical zones is computed by solving the Riemann problem posed for such an initial discontinuous distribution. High-order extensions of the Godunov method are very popular in astrophysics. They are built replacing the piecewise uniform approximation of the solution by higher-order polynomial interpolants, among which the parabolic piecewise method (PPM; [19]) is the basis of a number of hydrodynamic codes for astrophysical applications.

The smoothed particle hydrodynamics (SPH) algorithms are an alternative to grid-based methods. In SPH the fluid is represented by a Monte Carlo sampling of its mass elements (which can be thought of as particles). The motion and thermodynamics of these mass elements is then followed as they move under the influence of the hydrodynamic equations. Monaghan [20] devised the first algorithm to deal with special relativistic flows in SPH, and some recent application in the context of pulsar–wind-interaction and accretion-jet models has been performed by Romero *et al.* [21].

10.2.1 Specific Numerical Methods for MHD

While the governing equations of GRHD or RHD form a hyperbolic system of conservation laws, the GRMHD, RMHD or even the classical MHD system of equations may present singular points where the flow becomes parabolic. The latter happens only for certain degenerate configurations of the magnetic field (for a systematic review of these situations in the RMHD case see, for example, Antón *et al.* [22]).

Including the magnetic field both in the classical and relativistic hydrodynamic systems of partial differential equations involves adding, on the one hand, the Maxwell induction equations and, on the other hand, the constraint equation $\nabla \cdot \boldsymbol{B} = 0$. While the former equations are hyperbolic, the latter is elliptic. Analytically, a solution to the complete MHD system which starts from divergence-free magnetic data will remain divergence-free forever. Numerically, because of the finite accuracy of the integration methods, this is not guaranteed, and special care must be taken to keep the $\nabla \cdot \boldsymbol{B} = 0$ condition under control. Otherwise, severe stability problems pop up. Accumulating errors can lead to unphysical situations and may even produce the failure of a simulation.

The importance of properly dealing with the solenoidality constraint is paramount and can be used as a generic criterion to classify the existing numerical schemes for MHD.[3] Employing Balsara's [23] convention we find *divergence-free* and *divergence-cleaning* schemes. Divergence-free schemes preserve the divergenceless constraint to machine precision employing particular finite differencing representations of the induction equations, or they use an unconstrained vector potential $\boldsymbol{A}$ instead of the magnetic field $\boldsymbol{B} = \nabla \times \boldsymbol{A}$ (see [24] for some limitations of this method). The *constrained transport* [24, 25] (CT) type schemes are very popular among MHD practitioners using divergence-free algorithms. Divergence-cleaning methods rely upon a base scheme for the integration of the MHD equations, which do not care about the preservation of the divergenceless constraint beyond the truncation error. Violations of the $\nabla \cdot \boldsymbol{B} = 0$ condition are removed by the following means: (i) adding extra source terms (proportional to $\nabla \cdot \boldsymbol{B}$; *eight-wave formulation* [26, 27]); (i) projecting the magnetic field $\boldsymbol{B}^*$ provided by the base scheme onto the subspace of divergenceless solutions, such that $\boldsymbol{B} = \boldsymbol{B}^* - \nabla\phi$, by solving a Poisson (i.e., elliptic) equation for ϕ (*projection scheme* [28]); or (iii) transporting and/or dissipating the divergence errors making use of generalized Lagrange multipliers (*hyperbolic/parabolic divergence cleaning* [29]).

Except for (i), all the previous methods can be implemented in flux-conservative schemes. Numerical experience shows that the hyperbolic divergence cleaning decreases the divergence errors by up to two orders of magnitude compared with the eight-wave formulation. A comparison of the performance of several of the above mentioned methods can be found in Toth [30] (see also [31] for a more recent review). A very fruitful line of work has been to use high-order shock-capturing Godunov methods for the base scheme, together with the CT method for the solenoidal constraint. Initially, second-order schemes have been used both in the contexts of FV [27, 32–42] and FD [23, 43–50] schemes. Beyond second-order, the FV algorithms [23, 51–59] are far more involved to implement that FD ones [57, 60–63].

3) In RMHD the methods to deal with the divergence constraint are completely analogous to those of classical MHD, which, we recall, is the nonrelativistic limit of the (relativistic) Maxwell's equations combined with dynamical equations for a plasma in the single-fluid description.

Enforcing a divergence-free magnetic field in AMR is also possible, but the complexity of the resulting algorithms depends sensitively on the choice of the method to preserve the divergenceless constraint during the evolution. While AMR implementations using divergence-cleaning methods benefit from a relatively simple extension of AMR algorithms that work for nonmagnetized flows (e.g., [29, 64–66]), using AMR with divergence-free methods is typically much more complex [45, 67–69].

10.3 Basic Numerical Modeling

Even assuming that jets can be modeled as fluids, the complete study of their evolution (namely, from their birth/generation to their death/energy exhaustion) has to be broken down into evolutions at different scales. As we will show in the following sections, applied to AGN jets, one addresses with different types of simulations the generation and collimation of jets (happening at subparsec scales), the parsec scale phenomenology (VLBI jets, superluminal components, etc.) and the interaction of the jet and the surrounding environment at kiloparsec scales, probably shaping the observed jet and radio lobe morphologies. Due to the strong gravitational field of the central engine general relativity must be included to study the process of formation and collimation of AGN jets. Numerical simulations have been fundamental to test the different theoretical models of jet formation and collimation, and the interested reader is referred to Chapter 4, for a detailed review. In this chapter we will focus on the role played by numerical modeling of relativistic jets in other aspects of this involved astrophysical problem.

In the case of GRB jets, we would have a somewhat downscaled version of the previous regimes. At scales $\lesssim 10^8$ cm a GRMHD description of the process of jet formation is typically used. The propagation of the generated outflows in the optically thick regime ($10^8\,\text{cm} \lesssim r \lesssim 10^{11}\,\text{cm}$) is specifically modeled with RHD or RMHD simulations (if the magnetic field is important). The aim of such models is to understand the jet structure as it interacts with the progenitor and the circumstellar medium, until it reaches the coasting phase. Another set of simulations start from the coasting phase and cover the propagation of the ejecta for a few decades in radius in order to address the dynamics and observational signature in the prompt GRB and afterglow phases.

Adding to the differences in the relevant physical ingredients at different scales, there are questions, which are important at any scale, where numerical simulations are in common use. Among such questions, the stability of collimated relativistic pencils of plasma, is one of the most important. Thus, we consider this question first (Section 10.3.1), and later we give an overview of the results obtained with numerical modeling of relativistic jets at different scales (Sections 10.3.2 and 10.3.3).

10.3.1 Jet Stability

Lab experiments with fluids show that jets tend to expand sideways at the speed of sound and suffer various types of instabilities, developing vortices and internal shocks. After a few scale lengths (on the order of the cross-sectional radius), collimation is lost, velocities become subsonic, and flows are disrupted [70]. Contrarily, AGN jets propagate intact up to (in some cases) megaparsec scales without being greatly disrupted. This morphological difference initially prompted the idea that extragalactic jets were highly supersonic, freely expanding flows in underdense atmospheres with a negative density gradient [71].

Depending on whether a relativistic jet is magnetically (Poynting flux) dominated or kinetically dominated, it is prone to current- or pressure-driven or Kelvin–Helmholtz (KH) instabilities, respectively. The former instabilities may arise even in the absence of bulk motions and the driving factors are the equilibrium parallel current (current-driven) or the gas pressure versus field-line curvature (pressure-driven). Current-driven (CD) instabilities are considered in more detail in Section 10.3.1.1. Pressure-driven instabilities are left out of this chapter, but the interested reader can check the recent review by Longaretti [72]. KH instability can arise in both magnetized and unmagnetized jets, whenever a velocity gradient exists at the jet/external medium interface. This instability shall be relevant for jets, since the largest source of free energy in a jet is its bulk motion. Theoretically, the instability of relativistic jets is predicted from the linearized magnetofluid equations, and they also occur in the force-free approximation (see [73, 74] and the references therein for a review). Such instabilities result in fluid structures which have been observed in lab experiments, in numerical simulations and, very likely, in relativistic jets.

The key point in the theoretical analysis is to establish which are the most unstable modes, that is, the modes which grow faster or have the smaller growth times, and what are their corresponding wavelengths. It turns out that the instability time scales are quite short compared to the propagation time scale, and hence these modes can affect AGN jets very close to the central engine. Modes with the shorter wavelengths are a very likely source of turbulence in the jet flow, which eventually leads to thermal dissipation and nonthermal particle acceleration. Long-wavelength modes modulate the beam morphology and may be related with the observed structure of relativistic jets: quasi-periodic wiggles and knots, filaments, limb brightening or jet disruption [75–77].

The role played by numerical simulations to assess how unstable structures develop according to the theory, and saturate beyond the linear regime, modifying the jet dynamics is crucial for astrophysical jets, because the experimental conditions in laboratory jets (e.g., [78]) substantially depart from those expected in, for example, AGN jets. Indeed, one of the most important side effects of the KH or CD instability is the development of irregularities in the jet/ambient interface, which result in the entrainment of mass of the external medium into the jet, yielding a destabilization and deceleration of the beam flow. Only by means of sufficiently high-resolution numerical simulations, do we dare give a quantitative answer to the question of which are the most favorable physical conditions that prevent (enhance) mass entrainment (e.g., [79–84]). Furthermore, used in conjunction with the linear theory, numerical models of the jet instability can be confronted with actual observations to constrain jet properties (e.g., [85–88]).

Although we do not have a direct imaging of the flows that power GRBs, it is very likely that ultrarelativistic jets transport a fraction of the energy of the central engine to scales of ~ 0.1 pc. Thereby, the theoretical knowledge on AGN jet stability piled up over the years, can be transferred to GRB outflows. However, specific numerical simulations are necessary to address how nonlinear saturation acts in ultrarelativistic flows (e.g., [89, 90]).

10.3.1.1 Kelvin–Helmholtz Instabilities

The theoretical analysis of the stability of slab [91] and cylindrical jets [92], can be done in terms of small-amplitude Fourier components (normal modes) of the form $f(\boldsymbol{x}, t) = g(x) \exp[i(\boldsymbol{k} \cdot \boldsymbol{x} - \omega t)]$, with matched boundary conditions (e.g., the pressure) across the interface between the jet and the ambient medium. Plugging these perturbations into the linearized magnetofluid equations one obtains a dispersion relation $D(\omega, \boldsymbol{k}) = 0$, the solutions of which having $\operatorname{Im} \omega > 0$ or $\operatorname{Im} k < 0$ are locally or spatially growing modes, respectively. Among the growing modes we differentiate between *ordinary surface* modes, with amplitude steeply decreasing away from the interface, and *reflected body* modes, which affect the whole volume of the jet and grow because they become resonantly unstable as they reflect in a two-shear configuration.

A salient conclusion of the theoretical work developed in the late 1970s and the beginning of the 1980s was that the growth rate of the fastest growing modes at large beam Lorentz factor, and sound speeds is reduced w.r.t. Newtonian jets (i.e., with bulk velocities substantially smaller than c). Relativistic jet simulations have confirmed the growth rates and jet structures found theoretically (e.g., [93–97]). Thanks to the numerical analysis (e.g., [96, 97]), it has been established that the growth of the instabilities in relativistic (slab) jets goes through three stages (Figure 10.1). First, a linear phase where the KH modes grow as theoretically expected. Second, a saturation phase of the linear growth that finishes when the velocity of the perturbation reaches the speed of light in the reference frame of the jet. Third, a final nonlinear stage where the amplitude of the perturbations is such that shocks form, and small-scale (turbulent) motions develop, contributing to the conversion of the jet kinetic energy into thermal energy.

An important quantitative and qualitative difference in the analysis of KH-modes shows up when instead of a step density and velocity jump at the interface between the jet and the external medium (vortex-sheet approximation), a sheared structure exists. The dispersion relation is not algebraic, but instead it is an ordinary differential equation [98]. Ferrari *et al.* [99], Birkinshaw [98] and, more recently, Urpin [89] (in the ultrarelativistic limit) have studied analytically the growth of instabilities in sheared jets, finding that in this case short-wavelength instabilities may grow faster than the KH instabilities in the vortex-sheet approximation. Numerical simulations

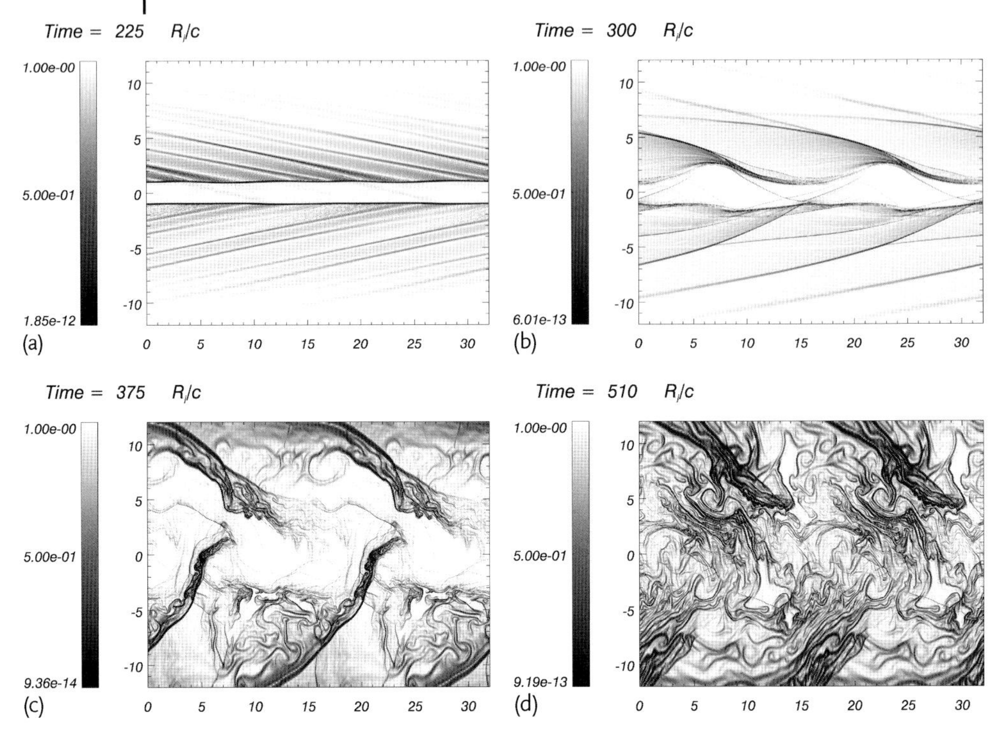

Figure 10.1 Plots of the magnitude of the rest-mass density gradient showing waves in the linear regime (a), saturation phase (b), transition to nonlinear regime, and in two instants of the nonlinear regime for a slab relativistic jet (c, d). The plots correspond to model B05 of the Ph.D. thesis of M. Perucho.

have shown that such shear-driven instabilities are triggered in the sheared beam of collapsar jets [90]. Indeed, a special class of shear-driven instabilities may become resonant and dominant at high-order modes for large bulk Lorentz factors and relativistic Mach numbers. In this regard, Perucho *et al.* [100] find that such resonant modes form hot shear layers which shield the jet core against other instabilities. Thus, the stabilizing properties of resonant modes could be at the basis of the small disruption that FR II sources suffer up to kpc scales.

As in the case of jets with shear layers, "spine-sheath" jet configurations display KH modes with different growth rates, wavelengths and propagation speeds. A thorough description of such structures can be found in Section 11.4.1. Again, in this case, numerical simulations help to assess the theoretical predictions beyond the linear regime.

Magnetized jets The radial structure induced by each normal mode and the motion of normal mode structures depends on whether the instability happens in Poynting flux-dominated or kinematically dominated jets.

From the analytic developments on the KH instability of Newtonian jets [101], a large beam density w.r.t. the external medium and strong magnetic fields reduce the effect of instability. The presence of magnetic fields has a stabilizing effect, in particular on small-wavelength modes. The result of such a stabilization is that turbulence and small-scale dissipation are greatly suppressed in strongly magnetized jets. Long-wavelength modes possess smaller growth rates, but dominate the dynamics of the flow. For ratios $c_A/c_s \geq 2$, where c_A and c_s are the Alfvén and sound speeds, KH modes are completely stabilized. Many authors have considered the magnetized KH instability of single-layer and double-layer (jets) nonrelativistic flows (see [59, 102] and references therein), where the main focus of such simulations is understanding the nonlinear mode saturation and final fate of KH unstable flows. A remarkable consequence that can be extrapolated from single-layer KH unstable flows [59] is that magnetic fields with a tiny initial magnetic energy, compared to the kinetic energy of the flow, are rapidly amplified in two spatial dimensions (2D), until the magnetic field energy reaches locally equipartition with the kinetic energy. Subsequently, it saturates due to resistive instabilities that disrupt the KH unstable vortex and decelerate the shear flow on a secular time scale. In three spatial dimensions (3D), the hydromagnetic mechanism seen in 2D may be dominated by purely hydrodynamic instabilities leading to less field amplification, thus, the magnetic energy is, at most, equal to the kinetic energy of the decaying shear flow. In the saturated state of most models, the magnetic field is mainly oriented parallel to the shear flow for rather strong initial fields, while weaker initial fields tend to lead to a more balanced distribution of the field energy among the components. In the relativistic regime, the same qualitative results remain valid, but additionally, it is found theoretically a stabilizing effect of high beam Lorentz factors (e.g., [73, 103]). Such trends have been confirmed in the nonlinear regime by numerical simulations [102, 103].

In current-carrying jets, in addition to the standard KH instabilities, CD and resistive instabilities may develop. Particularly disruptive among the CD modes is the so-called *kink* instability. In a jet bend the magnetic pressure associated to the azimuthal magnetic field component grows at its concave side, while it decreases at the convex side. The magnetic pressure gradient across the bend forces the bend to grow, creating a runaway kink that can only be compensated by the tension associated to poloidal magnetic field components. Since twisted structures are observed in many AGN jets from subparsec to kiloparsec scales (e.g., [85, 104, 105]), the CD kink instability is a likely explanation for such structures, but several other possibilities could also explain a similar phenomenology (e.g., resistive relaxation effects, KH helical modes, or the precession of the jet axis). Furthermore, helical structures have been found in nonrelativistic, as well as in relativistic simulations of magnetized jets (e.g., [106–112]). Identifying what kind of unstable mode prevails in the intricate dynamics found in such simulations is often a formidable task, since there are regimes in which jets can be both CD and KH unstable.

In relativistic force-free configurations, the linear instability criteria for the kink instability has been assessed in many papers [113–118]. Mizuno *et al.* [119] have studied the growth of the CD kink instabilities in cylindrical jets by means of 3D RMHD numerical simulations. Remarkably, their results show how the kink modes under consideration grow initially according to the linear theory. The nonlinear phase dynamics strongly depend on the magnetic pitch profile, being the case of an increased magnetic pitch with radius the one displaying a smaller growth rate. However, the growth of the instability is rather slow. Based on the numerical results, Mizuno *et al.* make a simple estimate, according to which, the jet disruption would happen beyond distances z, from the place where the jet flow forms, such that $z > 100a(z)\gamma(z)$, where a and γ are the jet radius and the jet Lorentz factor at a distance z. Thus, depending on the way a and γ vary with distance, it many happen that the instability does not develop at all. Such results are qualitatively consistent with the models of Moll *et al.* [110], whose simulations cover about three orders of magnitude in distance to capture the centrifugal acceleration, as well as the evolution past the Alfvén surface (see Figure 11.9 in Section 11.5).

10.3.2
Nonlinear Jet Dynamics

The nonlinear dynamical jet evolution is a field of the relativistic jet physics where numerical simulations have championed our knowledge. Depending on the scale to be modeled, two basic types of simulations have been performed. On the one hand, there are simulations in which an existing jet occupies a fraction of the computational grid, with the rest of the grid set up with a steady external medium. These simulations are more appropriate to model pc-scale jets (Section 10.3.2.1), in which the medium surrounding the jet is not disrupted by the perturbations triggered by the bow shock generated ahead of the jet, or by the turbulent cocoon developed as the jet drills its way through the ambient medium. On the other hand, to explore the dynamics of the jet propagation, the jet's head evolution and the large-scale morphology (cocoon, lobes, etc.), a slab or cylindrical flow is injected through a nozzle, typically located at one of the extremes of the computational grid. In Section 10.3.2.2, we sum up the foremost contributions using the latter numerical models.

10.3.2.1 **Parsec-Scale Jet Simulations**

The basis for the currently accepted interpretation of the phenomenology of relativistic jets was set by [120] and [121]. A number of analytic works have settled the basic understanding that accounts for the nonthermal synchrotron and inverse-Compton emission of extragalactic jets (e.g., [122]), as well as the spectral evolution of superluminal components in parsec-scale jets (e.g., [123–126]). At such scales the jet is modeled as a conical or cylindrical beam flowing in an unperturbed external medium. Superluminal components and other observed morphologies are understood as perturbations (e.g., shocks, blobs, bends, etc.) of the flow, which may travel along the jet. The advent of multidimensional relativistic (magneto)hydrodynamic numerical codes has allowed to perform time-dependent hydrodynamic models of the beam and of the perturbations that may travel downstream. Some practitioners [93, 127–130], compute the synchrotron emission of relativistic hydrodynamic jet models with suitable algorithms (see Section 10.4.2.1). These authors assumed that the magnetic field was dynamically negligible, that the emitted radiation had no back-reaction on the dynamics, and that synchrotron losses were negligible. All these assumptions are adequate for VLBI jets at radio observing frequencies if the jet magnetic field is sufficiently weak. A qualitative and quantitative improvement on the previous local approaches has been made in Mimica *et al.* [88], where the authors combine multidimensional relativistic models of compact jets with a new algorithm to compute the spectral evolution of suprathermal particles – namely, ultrarelativistic electrons and/or positrons – evolving in the jets, that is, including their radiative losses in addition to their adiabatic evolution. Furthermore, the spatial transport of the suprathermal particles is accounted for (see Section 10.4.2.2). Even in the simplest approximation to perform such a task, in which only the energy or the momentum of the suprathermal particles is fully evolved neglecting any angular evolution, an extra dimension (energy) is added to the evolutionary problem.

So far, all numerical models have been done in the reference frame of observers attached to the central engine (lab frame, hereafter). However, our observations are made in the reference frame of a distant observer, and thus, the relativistic time-dilation effect merges different evolutionary times in the lab frame, complicating substantially the calculation of the observational signature of the evolved hydrodynamic models. In fact, because of the relativistic time dilation, one adds an effective extra dimension to the two or three spatial dimensions of the numerical model. While in 2D models this can be achieved with the existing computational resources, in 3D simulations of superluminal sources (which are effectively 4D), the computational resources involved are much more demanding, and only a very reduced number of works have been able to tackle the study of the observational signature of 3D, pc-scale jets including all the relevant relativistic effects using a local approach [131]. It is obvious that 3D models that include nonthermal particle transport are, in fact, five-dimensional problems and, so far, they have not been tackled.

The predictive power of numerical simulations Numerical models of compact jets aim mainly to explain generic features observed in extragalactic jets. However, they have also an important predictive power. In the following, we identify several predictions that have arisen from simulations of pc-scale jets.

A salient result of the previous numerical models is that a single hydrodynamic perturbation injected through the jet inlet as, for example, variations of the flow velocity or of rest-mass density, may trigger multiple KH modes in the beam, which result in observable superluminal features. Remarkably, Agudo *et al.* [93] predicted the existence of *trailing* superluminal components, which develop in response to a pressure perturbation traveling downstream the beam of a pressure matched jet. Such components seem to have been later found in a number of sources (e.g., [132–135]). We point out that Mimica *et al.* [88] show that trailing components can be originated not only in pressure matched jets, but also in overpressured ones, where the existence of recollimation shocks does not allow for a direct identification of such features as KH modes. Aloy *et al.* [131] have also predicted the existence of superluminal components that may pop up from intermediate parts of the beam, instead of flowing from the core, as a result of the splitting of *large* hydrodynamic perturbations (or blobs) as they travel downstream. They are different from trailing components (which also appear away from the jet inlet) because they tend to move faster than a previously identified component, whose trajectory can be traced back to the jet core. Such effects are difficult to identify in current observations because their emissivity might be rather low and Doppler dimmed.

10.3.2.2 Kiloparsec-Scale Jet Simulations

Dynamics of the jet propagation The numerical modeling in the frame of classical fluid dynamics became a powerful tool in the study of extragalactic jets since the pioneering simulations by Rayburn [136] and Norman *et al.* [137] verified the jet model of Blandford and Rees [71] for classical double radio sources. In the 1990s, the numerical simulation of relativistic jets and GRBs was triggered by the development of new numerical techniques able to solve the equations of RHD under the extreme conditions found in these scenarios. Most of these new techniques (see [9, 138] for recent reviews) exploit the hyperbolic and conservative character of the equations of RHD and have allowed the numerical simulation of extragalactic relativistic jets in 2D – either planar or cylindrical symmetry – (e.g., [139–145]) or in 3D (e.g., [146–149]).

Martí *et al.* [144] performed a systematic parameter space coverage of the morphology and dynamics of relativistic cylindrical jets with relatively high Mach numbers (low Mach number jets have also been considered [141, 142]). Jets were injected with a pre-established velocity, rest-mass density contrast (w.r.t. the uniform ambient medium), sound speed (or specific internal energy), and adiabatic index (either considering $\gamma = 4/3$ or $\gamma = 5/3$). The most obvious results of these 2D RHD works is that relativistic jets display the same morphological elements as Newtonian jets. The interaction of the tip of the jet with the external medium drives two shocks separated by a contact discontinuity (the working surface [71]). A forward/bow shock sweeping the external medium is trailed by a reverse shock or Mach disc, which is identified with the hot-spot in powerful sources. The beam flow is efficiently decelerated at the Mach disc, where it is shocked (i.e., heated) and expands flowing sideways, inflating a cocoon where the wasted jet plasma is stored. Since the beam plasma has been thermalized at the Mach disc, and even deflected at the jet head, it may flow backwards (in the reference frame of the working surface). Such backflow is heterogeneous and composed of large-scale eddies, which result from the growth of Rayleigh–Taylor modes from the working surface that propagates at a time varying speed. A detailed modeling of the dynamics of such backflows has been recently presented by [150]. The unsteady backflow hits the beam stimulating pinching KH modes, which saturate in the form of biconical shocks.

In addition to the confirmation of the morphological elements present in the standard model for extragalactic jets, the most salient result of Martí *et al.* [144] is that increasing either the Lorentz factor or the specific enthalpy contributes to make more efficient the propagation of jets compared to *equivalent* Newtonian jet models. This is easy to understand if we realize that the momentum density (10.9) incorporates the contributions of W and h to the rest-mass density. The improved propagation efficiently results in practically featureless cocoons in the kinematic ($W \gg 1$) or thermodynamic ($h \gg 1$) ultrarelativistic limits. Contrarily, Lorentz factors $W \sim 2$–3 yield thick, turbulent cocoons and reduced advance speeds of the head of the jet. Another detailed comparison of relativistic and nonrelativistic axisymmetric jets was done by Rosen *et al.* [151]. It is important to notice that there is not a unique way of comparing relativistic and nonrelativistic jet models. Komissarov and Falle [130] pointed out that one should compare models with the same power, mass-flux or thrust at the inlet. In such a case, the morphodynamical differences between relativistic and nonrelativistic jet counterparts get substantially decreased (this is confirmed by [151]). However, Rosen *et al.* find that the velocity field of nonrelativistic numerical models cannot be scaled up to give the spatial distribution of Lorentz factors seen in relativistic simulations. Considering that the spatial distribution of Lorentz factor is determinant for shaping the total intensity for extragalactic jets, the authors conclude that nonrelativistic models are inadequate to account for the proper intensity distribution of a relativistic jet.

The parameters for setting up large-scale jet simulations have to be taken with special care. We have to choose the conditions that happened millions of years ago, in order to produce the currently observed morphologies. The way to set up the initial models is not unique, since, for example, the same kinetic power or thrust can be obtained by a number of combinations of W, ρ, c_s, and so on. Furthermore, some parameters (of the jet or of the external medium) are not directly accessible to observations (or they cannot be fixed with high accuracy). A detailed discussion of the constraints binding the initial jet parameters can be seen in Scheck *et al.* [152] (see also [153] for a nonrelativistic discussion).

A full 3D modeling of relativistic jets is necessary to account for the dynamics of for example, precessing jets (e.g., [146, 148]), deflection of beams by heterogeneous media (e.g., [148, 149]), or the growth of nonaxisymmetric KH modes (which are the most disruptive, see Section 10.3.2.1). Typically, 3D studies can be performed using lower resolution than those run in 2D, thereby, limiting their ability to fully capture the internal jet structure. However, such 3D studies can provide a reliable model of global morphology and dynamics. For instance, Aloy *et al.* [146] show that the coherent fast (mildly relativistic) backflows found in axisymmetric models are not present in 3D models (result was later confirmed using 3D RMHD models and improved resolution using AMR [148]). Due to small 3D effects in the relativistic beam, a lumpy distribution of apparent speeds results, which resembles, for example, that observed in M 87 [146]. Hughes *et al.* [148] argue that the convolution of rest-frame emissivity and Doppler boost in the case of a precessing jet leads to a core-jet-like structure. Nevertheless, the intensity fluctuations in the jet cannot be unanimously attributed to either the structural changes resulting from the helical modes or Doppler boost alone but, in general, they are a combination of both factors.

The beam of 3D models is typically surrounded by a hot boundary layer, where there is a mixture of jet and external medium plasma that results from the jet/ambient interaction (e.g., [147]). Although the origin of such a layer is purely numeric, the numerical diffusion inherent to any hydrodynamical code mimics, to some extent, that expected in actual jets. Indeed, many models of jet generation display a structure which consists of a fast spine surrounded by a thick, slowly moving sheath (see Chapters 4 and 11). Interestingly, the most extended view is that the interaction layer presents a monotonically decaying velocity from the jet core to the external medium. However, some numerical simulations of relativistic jets, both in the context of AGN jets [131], and in the context of progenitors of short GRBs [154], find that the shear layer between the jet and the ambient medium is nonmonotonic in some extended regions. Aloy and Rezzolla [155] interpreted such an anomalous behavior in terms of a novel relativistic hydrodynamic effect, by virtue of which, overpressured or hot jets are able to convert a part of its internal energy into kinetic energy. This is a purely RHD effect which gets amplified (even in cold, thermal pressure-matched jets) if strong toroidal fields are carried by the relativistic beam [156–158], and whose observational implications for kpc-scale jets have been investigated by Aloy and Mimica [156]. Unfortunately, at the observational level, the studies of the velocity structure of extragalactic jets and its relationship to their broadband emission are hampered by the difficulties in resolving the outflows transversely, especially at high energies (e.g., [159]). Thus, the existence of such anomalous layers remains untested so far.

In the last decade, a big endeavor has been initiated to incorporate dynamically relevant magnetic fields to the study of extragalactic jets by means of RMHD simulations (e.g., [43, 160–169]). In previous 2D RMHD simulations relativistic jets only served the purpose of demonstrating the capabilities of the RMHD code (e.g., [43], in slab geometry [170, 171], assuming axial symmetry and using cylindrical coordinates), or the simulations only covered a very limited range of jet parameters (e.g., in Komissarov [165], only two models were considered both involving only toroidal fields). An extensive parameter study of the morphology and dynamics of relativistic, magnetized, cylindrical jets has been performed by Leismann *et al.* [166], including both toroidal and poloidal field configurations of different strength, and beam plasmas of different adiabatic index. These authors find that the inclusion of the magnetic field leads to diverse effects, which contrary to Newtonian MHD models do not always scale linearly with the strength of the magnetic field. The relativistic models show, however, some clear trends. Axisymmetric jets with toroidal magnetic fields blow cavities with two differentiated parts: an inner one surrounding the beam which is compressed by magnetic forces, and an adjacent outer part which is inflated due to the action of the magnetic field. The outer border of the outer part of the cavity is given by the bow shock where its interaction with the external medium takes place. Toroidal magnetic fields well below equipartition (i.e., with a ratio of magnetic to thermal pressure $\beta = 0.05$) combined with a value of the adiabatic index $\gamma = 4/3$ yield extremely smooth jet cavities and stable beams. Prominent nose cones form when jets are confined by toroidal fields and carry a high Poynting flux (in this regard such results broaden previous findings [165]).[4]

4) In 3D, Mignone *et al.* [169] find that, for perturbed, current-carrying jets, such nose cones do not develop even for purely toroidal magnetic field topologies.

In contrast, models possessing a poloidal field do not develop such a nose cone. The size of the nose cone is correlated with the propagation speed of the Mach disc (the smaller the speed, the larger the size).

One of the goals of 3D (R)MHD simulations of astrophysical jets is to understand how they are able to survive the feared kink instability (see Section 10.3.1.1). Massaglia *et al.* [172] suggest that the formation of cocoons by the head of the jet shields the beam against such pathology. More recently, a quick jet expansion from the formation site has been suggested to explain the remarkable stability of observed jets [110, 111]. The high-resolution simulations of relativistic magnetized jets of Mignone *et al.* [169] agree well with the latter interpretation, namely that the kink instability shapes the overall jet morphology but it does not destroy it. However, the presented simulations are still too short to extract definitive conclusions.

From the knowledge obtained combining numerical models and theory, it seems that the development of harmful, nonaxisymmetric (e.g., helical) KH instabilities is counterbalanced by a number of processes, among which, Hardee [73] outlines the following: (i) strong or suitably ordered weak magnetic fields, (ii) steep gradients such as those associated with jet acceleration or expansion (this is a common request to hamper the destructive action of CD kink instabilities), (iii) a high density ratio between the jet and the external medium (because the amount of mixing is greatly suppressed), (iv) high Lorentz factor and/or high Mach number (because these elements reduce the growth rate), and (v) a sheath/shear velocity outflow around the jet, reducing velocity shear and increasing wave advection and growth lengths (which is something that naturally develops in many numerical simulations).

The FR I/FR II dichotomy Most of the numerical models of large-scale relativistic jets (see above), have been run for quite short periods of time, typically a fraction of 1 Myr. By such times, the jets have propagated only a few tens of kiloparsecs. Clearly, this is insufficient to extract firm conclusions of the large scale morphology and jet dynamics. Much longer simulations are needed to understand, for example, the deceleration of powerful sources, which is governed by its interaction with the external medium, or the feedback mechanism of jets on galaxy clusters or the intergalactic medium. Several theoretical models have considered the long-term evolution of kiloparsec-scale jets. Begelman and Cioffi ([173], hereafter BC89) devised a popular, simple model to describe the evolution of the cocoon. Komissarov and Falle [174, 175] explored the large-scale flow caused by classical and relativistic jets in a uniform external medium. Remarkably, they outline that jets with finite injection opening angles are recollimated by the high pressure in the cocoon, and that the flow becomes approximately self-similar at large times (in agreement with the self-similar expansion model of Falle [176]).

Two-dimensional Newtonian hydrodynamic simulations of axisymmetric light jets were performed by Cioffi and Blondin [177] in order to understand the evolution of the cocoon, and were compared with the simple analytic theory of BC89, but the simulations cover only a relatively short period of the cocoon's evolution. Only the Newtonian simulations of Hooda, Mangalam and Wiita [178] cover the evolution of axisymmetric extragalactic jets up to 100 Myr. Komissarov and Falle [174, 175] performed axisymmetric RHD simulations of powerful jets, focusing on the effects of the injection opening angle of the beam. Martí, Müller and Ibáñez [179] studied the long-term evolution of powerful (FR II) extragalactic jets on the basis of relativistic hydrodynamic simulations of cylindrical jets (up to an evolutionary time of 3 Myr with a relatively low numerical resolution) in a uniform atmosphere. Scheck *et al.* [152] extended the results of Martí *et al.* in two ways: (i) they consider longer evolutionary times (up to 6.6 Myr), and (ii) they payed attention to the influence of the composition of thermal matter in the morphology, dynamics and long term evolution of the jets (Figure 10.2).

Scheck *et al.* [152] find an evolution divided into two epochs. After a transient initial stage, in which the speed of propagation is $\simeq 0.2c$ (in agreement with the measured speeds of propagation of the hot spot in Compact Symmetric Objects (CSO) [180, 181]), the evolution of the jet is dominated by a strong deceleration. The jet advance speed becomes as small as $0.05c$ due to the degradation of the beam flow by means of internal shocks and the broadening of the beam cross-

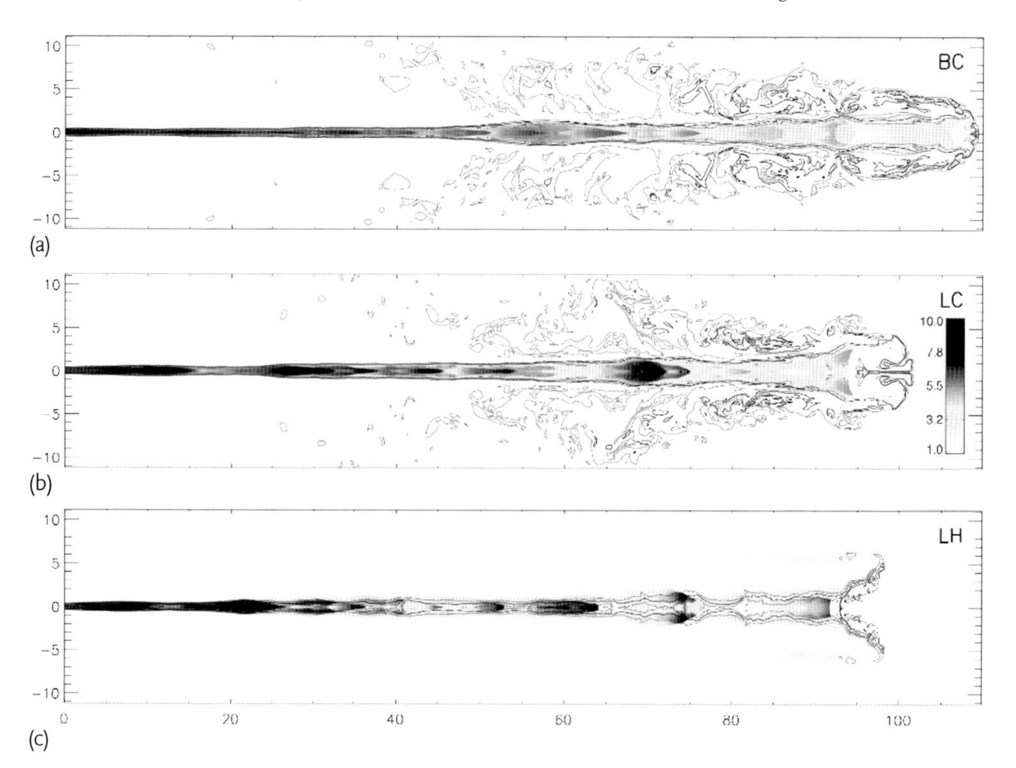

Figure 10.2 Snapshots of the Lorentz factor for the baryonic-cold (a), leptonic-cold (b), and leptonic-hot (c) at $t \simeq 6.4$ Myr. The black lines are isocontours of the beam mass fraction with values 0.1 (outermost) and 0.9 (innermost), which correspond to the limits of the cocoon and the beam, respectively. Reproduced from Scheck *et al.* [152] with permission from the Royal Astronomical Society.

section near the hot spot. This deceleration phase is consistent with the results of Komissarov and Falle. Later, the evolution becomes almost steady. The results of Scheck *et al.* nicely reproduce some of the gross morphodynamical features of FR II sources, namely, the lobe inflation, hot spot advance speeds and pressures, or the deceleration of the beam along the jet. Furthermore, since a realistic equation of state (EOS) is considered, these simulations have been used to restrict the values of basic parameters (e.g., the particle density and the kpc-scale flow speed) in the jets of real sources.

The results of Scheck *et al.* [152] are compared and interpreted with a simple generalization of the model of BC89, where the speed of propagation of the bow shock (v_{bs}) follows a power law with time ($v_{bs} \propto t^{\alpha}$), and thus, the cross-sectional radius of the cocoon (r_c), the average cavity pressure (P_c), and the aspect ratio (i.e., the ratio of jet length l_j to the cocoon radius) display the following scalings with time

$$r_c \propto t^{1/2-\alpha/4} \,, \quad P_c \propto t^{-1-\alpha/2} \,, \quad \frac{l_j}{r_c} \propto t^{1/2+5\alpha/4} \,. \tag{10.12}$$

From fits of the numerical models, $\alpha \sim -1/3$ and, accordingly, $r_c \propto t^{7/12}$, $P_c \propto t^{-5/6}$, $l_j/r_c \propto t^{1/12}$.

Unfortunately, large differences in the composition (electron–positron or leptonic jets vs. electron-proton or baryonic jets), which for the same kinetic luminosity and thrust yield also large variations in the injected specific internal energy and proper rest-mass density, do not bring noticeable differences between the morphology of jets. The only exception to such result comes from the cavity temperature. Whereas in the leptonic, cold model isothermal cocoons are formed, in the leptonic hot and baryonic cold models, the cocoon has a clumpy distribution of temperature. Likewise, Leismann [182] finds that the long-term evolution of magnetized jets (with an initial toroidal field, whose energy is in equipartition with the internal energy) does not differ much from the evolution of purely hydrodynamic models, although the beam and the cavity blown by the jet display some distinctive features. One of the reasons for such unexpected behavior being that after the first recollimation shock the magnetization of the beam becomes rather small, and remains so up to the Mach disc. Consequently, the magnetic pinching at the Mach disc is insufficient to force matter into a nose cone and flows backwards, as in the corresponding nonmagnetized jet models.

Our current understanding is that FR I jets are decelerating relativistic jets (e.g., [183]). Therefore, the evolution of FR I jets has been also studied using numerical simulations of relativistic jet models. Perucho and Martí [184] have tested the standard paradigm for jet deceleration employing the parameters derived by Laing and Bridle [183, 185] for 3C 31, including realistic density and pressure gradients for the atmosphere. A single simulation of a leptonic, cylindrical jet, spanning more than 7 Myr (i.e., $\sim$ 10% of the lifetime of 3C 31) and using the Synge's EOS was performed by the authors. The model agrees fairly well with the recent X-ray observations of the environment of FR I sources younger than 3C 31 (e.g., NGC 3801 [186], or Cen A [187]) since the bow shock evolves self-similarly, at almost constant speed. The inferred properties of the jets in 3C 31 are basically recovered by the numerical simulation. As for FR II sources (see above), the simulation is interpreted on the basis of a generalization of the BC89 model. From the power-law dependence of the bow shock speed with time displayed by the simulation, one gets an index $\alpha \propto -0.1$, which results in the following analytic estimates

$$r_c \propto t^{0.7} , \quad P_c \propto t^{-1.3} , \quad \frac{l_j}{r_c} \propto t^{0.2} , \tag{10.13}$$

that compare extremely well with fits to the simulation results yielding $r_c \propto t^{0.8}$, $P_c \propto t^{-1.3}$, and $l_s/R_s \propto t^{0.2}$.

Among the most likely possibilities for explaining the deceleration of extragalactic jets at kpc-scales, and thus explaining the observed FR I/FR II dichotomy, we find that either (i) the beam of the jet entrains mass from the external medium, or (ii) the jet encounters a much denser medium at some distance from the source. The first possibility, in its turn, may happen because of different physical processes. Perucho and Martí [184] showed that a recollimation shock in a light jet, in reaction to steep interstellar density and pressure gradients, may trigger nonlinear perturbations that lead to jet mixing and disruption with the external medium. On the other hand, Rossi *et al.* [84] with 3D RHD jet simulations in a uniform medium, and an approximation to the Synge's EOS with the code PLUTO, assume that velocity shear instabilities are likely the triggering mechanisms of entrainment [188, 189]. They find, consistently with previous studies, that the density contrast between the ambient medium and the jet (η) is a key parameter determining the (KH) instability evolution and entrainment properties. Indeed, they show how the interaction between the jet and the ambient leads to the development of a spine-sheath structure in agreement with the expectations (e.g., [147]). Rossi *et al.* note that lighter jets suffer stronger slowing down in the external layer than in the central part and conserve a central spine with a high Lorentz factor. However, it shall be pointed out that the simulations of Rossi *et al.*, which are scale-free, may not be properly scaled to obtain adequate values for the modeling of FR I jets. Using their Eq. (8), if the radius of the jet were $r_j \sim 100$ pc, the resulting power is excessive for a typical FR I source. On the other hand, if one would assume that the jet radius at kpc-scale be $r_j \simeq 10$ pc (to have a power in agreement with observed FR I jets), then the evolutionary times of the simulations, at most $760 r_j/c$ (in code units), become as small as 2.5×10^4 years, seemingly small to be reconciled with the age of typical radio sources.

The second possibility mentioned in the previous paragraph has been considered by Meliani *et al.* [190] (see Section 11.2.2), who have explored the FR I/FR II dichotomy using their RHD code AMRVAC, in which the Synge's EOS is also included. Because of the AMR capability of their code, they can follow the deceleration of jets for several hundred beam radii, but still, for evolutionary times a factor 6–40 times smaller than in [152], or [184]. Both cylindrical and conical jet models are analyzed, as well as the influence of different beam Lorentz factors (from 10 to 20). These jet simulations aim to model the Hybrid Morphology Radio Sources (HYMORS) as the results of the interaction with density discontinuities of one of the twin, powerful jets. The authors test several density jumps and realize that high enough density jumps may result into a sudden jet deceleration, whereby an originally set up FR II jet turns into a FR I type jet. It is, however, unlikely that a density jump, where the density of the outermost medium is larger than that of the medium closer to the jet engine, is a generic feature to explain the FR I/FR II dichotomy.

Finally, we point out that simulations have given insightful hints on what are the initial phases of the evolution of large-scale radio sources. For instance, Scheck *et al.* [152] showed a likely evolutionary track from CSOs, and Compact Steep Spectrum (CSS) sources to FR II sources (as suggested by, e.g., [191, 192]). On the other hand, Perucho and Martí [184] contributed to clarify that CSS sources are likely the young counterparts of FR I sources, as it was suggested by Drake *et al.* [193].

10.3.3 GRB Jets

In addition to extragalactic jets, galactic microquasars and pulsars, GRBs are astrophysical objects where, presumably, (ultra)relativistic jets are produced in nature. It turns out that we do not have a direct observation of collimated (i.e., jetted) outflows in GRBs. However, this can be attributed to several concurrent factors. Firstly, GRBs and their subsequent afterglows are located at cosmological distances, but happen at scales of 10^{12}–10^{18} cm from the energy source. Thus, we do not have the adequate observational resolution to spatially resolve them, even with the most advanced observational instruments (e.g., satellites Swift, HETE II or Fermi). Secondly, from theoretical reasons (compactness problem [194]) and observations, namely the detection of radio scintillation of the interstellar medium [195], and the measurement of superluminal proper motions in imaged afterglows [196], we know that GRBs are associated with ultrarelativistic flows, which hampers their detection, except if they are basically pointing directly towards the Earth. On the other hand, the lack of direct observation of collimated outflows is not exclusive of GRBs. Blazars are another example of sources where we believe a jetted flow is behind the observed phenomenology. Furthermore, the theoretical and, very specially, the numerical work carried out in the field of progenitors of GRBs shows overwhelming evidences that collimated channels of plasma are able to transport the energy produced in the central engine to the interstellar medium, where the dissipation of the kinetic energy of the flow brings GRBs and their associated afterglows.

The possibility that GRBs are not isotropic emissions was devised theoretically as a way to ameliorate the huge energetic budget implied by the standard fireball model for these powerful phenomena. However, the mechanism by which after the quasi-isotropic release of a few 10^{50} erg yields a collimated ejection of plasma could not be satisfactory explained analytically. The reason being that the collimation of an outflow by its progenitor system depends on very complex and nonlinear dynamics. That has made necessary the use of numerical simulations in order to shed some light on the viability of some likely progenitors of GRBs. We revise the most important aspects such simulations have revealed in Section 10.3.3.1, and leave for Sections 10.4.3.1 and 10.4.3.2 a deeper discussion on the observational signature that the generated ultrarelativistic jets produce.

10.3.3.1 Simulations of Progenitors of GRB Jets

The huge energy required to explain the observed fluence of many GRBs links them with the birth of stellar-mass black holes (BHs). However, even with the most advanced observational instruments we cannot reach the appropriate resolution to identify individual progenitors of GRBs. Our best hopes to unveil the central engines powering GRBs rely on the detection of gravitational waves produced by the formation of the central BH (e.g., in a merger of relativistic compact objects or in the course of the collapse of a rotating, massive stellar core) soon after the GRB radiation is produced.

In the field of GRBs, as they were originally associated with spherically symmetric relativistic fireballs, numerical relativistic hydrodynamic simulations of one-dimensional flows have been used to verify the standard model [197]. Relativistic outflows result in progenitor models of both long GRBs (e.g., [198, 199]) and short GRBs [200–203]. Only very recently has the fact that GRBs can be produced by *collimated* ultrarelativistic ejecta has been taken into account in multidimensional numerical simulations. These simulations have proven to be very useful to check the viability of the so-called collapsar model [198] as progenitors of long GRBs [90, 204–215] and of the merger of a binary system of compact objects for short GRBs [154, 216, 217]. The previous set of numerical models can be categorized into two basic groups: simulations dealing with the process of formation and collimation of ultrarelativistic jets, and a second group that focuses on the propagation of the jet through the progenitor or the interstellar medium.

Jet formation Regardless of whether they are long or short events, the central engine feeding GRBs is a stellar mass BH ($M_{\rm BH} \lesssim 10 M_{\odot}$) surrounded by a thick accretion disk. Several mechanisms have been considered to convert part of the central engine energy into an ultrarelativistic flow. The first one is a thermal mechanism. The viscous accretion of the disk matter onto the BH, which yields a strong heating that, in turn, produces a copious amount of thermal neutrinos and antineutrinos. These particles annihilate preferentially around the rotation axis producing a fireball of e^+e^-pairs and high-energy photons. Numerical simulations aiming to explore the process of (thermal) jet formation must include, at least, the effects of the strong (relativistic) gravitational field of the BH. Thus, multidimensional GRHD models are employed [154, 204]. In these simulations one finds that it is possible to form and accelerate a highly relativistic and well-collimated outflow under reasonable conditions for the energy release around the rotation axis of the star close to the event horizon of the BH. Alternatively, the accretion energy of the torus could be tapped by sufficiently strong magnetic fields (*hydromagnetic* generation) by means of a process similar to the Blandford–Payne-driven wind (see Section 4.3.1; and also [218]), or a nontrivial fraction of the rotational energy of the BH may also be converted into a Poynting flux (see Section 4.3.2; and [219]). Since for this kind of jet production mechanism, huge, dynamically important magnetic fields are necessary, multidimensional GRMHD models are necessary (e.g., [211–213, 215, 217]). These simulations also find that ultrarelativistic jets form, but their characteristics differ from the jets with thermal origin (see below). Some of the models of jet formation considered above encompass large enough physical domains to consider also the propagation of the formed jets to large distances, such that the jet reaches the surface of the progenitor star, with the goal of understanding the fiducial conditions in the jet when it begins its propagation on the external medium.

Jet propagation A different type of simulation focus on the propagation of a relativistic jet along the rotation axis of the progenitor star (i.e., the star whose center has collapsed to form the aforementioned central engine). In these simulations injection boundary conditions are imposed on a jet nozzle, typically located at $r_{\rm in} \sim 10^8$–10^9 cm from the center of the star. The aim of these models is to compute the properties of the ejected matter once transparency sets in, namely at distances $r_{\rm t} \sim 10^{12}$ cm, that is, 3–4 decades in radius away from the innermost boundary of the simulation. Since at distances $r > r_{\rm in}$ the gravitational field of the central object is rather small, and since the jet propagates to distances $r_{\rm t}$ in a few tens of seconds (a time which is much smaller than the sound-crossing time of the star) the self-gravity of the star can be neglected as well. Thus, RHD simulations suffice to model adequately the jet propagation.

What have we learned from the numerical modeling of progenitors of GRBs? In the following, some relevant aspects are discussed, focusing mostly on simulations of progenitors of long GRBs.

- *Collimation.* The generated outflows are inertially or magnetically confined. In the first case, the collimator is the low-density funnel, which is produced around the rotation axis due to the fast rotation of the star [204, 215], or the thick accretion disk in case of progenitors of short GRBs [154]. In the second case, the flow is self-collimated by its own magnetic field if it is strong enough [217]. Quite independently from the initial conditions and the inclusion of magnetic fields, the typical outflow half-opening angles, when the jet reaches the surface of the progenitor star, are $\theta_{\rm break} \lesssim 10°$. These small half-opening angles result from the recollimation of the outflow within the progenitor and they are independent of whether the boundary conditions are set to initiate the jet with much larger half-opening angles (e.g., $\theta_0 = 20°$; [205]), or on whether the jet is generated by an energy release into a volume spanning a half-opening angle $\theta_{\rm d} = 30°$ much larger than $\theta_{\rm break}$ [204], or an electromagnetic central engine is at work (e.g., [213]). In the course of their propagation through the progenitor thermally driven jets develop a nonhomogeneous structure, transverse to the direction of motion, whose main features are an internal ultrarelativistic spine (where the Lorentz factor may reach $\Gamma_{\rm core} \sim 30$–$50$ at jet breakout) within a half-opening angle of $< 2°$ laterally endowed by a moderately relativistic, hot shear layer ($\Gamma_{\rm shl} \sim 5$–10) extending up to $\theta_{\rm shl} < 20$–$30°$ [90]. The transverse structure of the jet is *nearly* Gaussian, regardless of the presence or not of dynamically important magnetic fields (e.g., [204, 217]), although a more accurate fit of the transverse structure of the jet, cannot be accommodated by a simple Gauss function [220].

- *Variability*. Our best hope is that the time variability found in the light curves of GRBs hides a key to unveil the physical properties of the inner engine. All produced outflows are highly variable due to the generation of KH [204, 221], shear-driven [90] or pinch magnetohydrodynamic (MHD) instabilities [211, 212, 215, 217]. Such *extrinsic* variability is independent of the (*intrinsic*) variability of the energy source and leads to the formation of irregularities in the flow which are the seeds of internal shocks in the outflow. Except in cases in which the source may produce quasi-periodic variability (perhaps induced by precession or nutation modes of the accretion disc), the extrinsic variability might be indistinguishable form the intrinsic one [222]. Very recently, Morsony *et al.* [209], using AMR simulations, have suggested that the pulses of several seconds duration often observed in GRBs could be produced by the interaction of the jet with the progenitor star, while the short time scale variability (with time scales of milliseconds) is most likely produced by the activity of the central engine. Numerical simulations of three-dimensional (3D) relativistic jets propagating through collapsar-like environments show that such jets are also stable [206], but it still remains unknown whether 3D relativistic, magnetohydrodynamic, collapsar jets will also be stable along its whole trajectory.
- *Jet breakout*. The jets generated are much lighter than their baryon reach environments. Hence, they propagate through the collapsar at moderate speeds (~ 0.3–$0.6c$) and fill up thick cigar-shaped cavities or cocoons of shocked matter that also propagate along with the beam of the jet and that, eventually may break the surface of the collapsar. Since the energy stored in the cocoon may amount to a few 10^{50} erg, it has been proposed that its eruption through the collapsar surface could yield a number of γ-ray/X-ray/UV-transients [223, 224]. Indeed, it has been suggested that GRBs are but one observable phenomenon accompanying BH birth and other possibilities may arise depending on the observer's viewing angle w.r.t. the propagation of the ultrarelativistic jet [225]. Thus, in a sort of unification theory for high-energy transients, one may see progressively softer events ranging from GRBs (when the jet emergence is seen almost head on) to UV flashes (if the jet eruption is seen at large polar angles) and accounting for X-ray-rich GRBs (XRR-GRBs) and X-ray flashes (XRF) at intermediate angles. Furthermore, ν-powered jets are very hot at breakout ($\sim$ 80% of the total energy is stored in the form of thermal energy), which implies that jets can still experience an additional acceleration by conversion of thermal into kinetic energy, even if the energy source has ceased its activity. Similarly, magnetically driven jets, may contain a substantial fraction of magnetic energy that can be converted to kinetic or thermal energy as the jet propagates.
- *Asymptotic Lorentz factors*. Purely hydrodynamic, ν-powered jets in collapsars seem to be able only to produce *moderate* terminal values of the bulk Lorentz factor ($\Gamma \sim 100$–400, e.g., [204, 205]), even if there is a further acceleration of the

forward shock as a result of an appropriate density gradient in the medium surrounding the progenitor [204]. From extrapolations of the numerical results of axisymmetric jets generated electromagnetically very large values of the asymptotic Lorentz factor of the outflow (e.g., $\Gamma_\infty \sim 1000$; [217] or $\Gamma_\infty \sim 5000$; [213]) can be attained. Thus, it seems rather plausible that MHD mechanisms have to be employed to generate jets with large Lorentz factors $\Gamma \sim 500$–1000, while ν-powered jets may still explain events with smaller Lorentz factors.

- *Angular energy distribution.* Lazzati and Begelman [226] theoretically estimated that the energy per unit solid angle ($dE/d\Omega$) in afterglow phase of typical GRBs displays a θ^{-2} dependence with the viewing angle after its eruption through the progenitor surface. Recently, Morsony *et al.* [208] have used hydrodynamic simulations to test the theoretical prediction of Lazzati and Begelman, and found that their numerical models do not follow the inferred theoretical angular energy distribution – a result that has been later confirmed with subsequent numerical models [209, 214]. Indeed, Mizuta and Aloy [214] find a strong correlation between the angular energy distribution of the jet, after its eruption from the progenitor surface, and the mass of the progenitors. The angular energy distribution of the jets from light progenitor stars is steeper than that of the jets injected in more massive progenitor models.

10.4
Numerics Confront Observations: Emission from Synthetic Jets

From Earth we cannot directly observe the jet fluid. Rather, the high-energy nonthermal particle population (Chapter 9) is directly responsible for the emitted radiation. Furthermore, relativistic effects (e.g., Doppler boosting, light aberration, time dilation; see Chapter 2) make the inference about the jet properties based on the observed emission a challenging task.

One of the ways to approach this problem is to develop numerical algorithms, which create synthetic observations based on the results of the numerical simulations. The first gain from coupling such algorithms to simulations is that it becomes easier to verify that the simulated dynamics reproduces observed features. However, even more important is the possibility to use simulations as *virtual laboratories* that enable us to gain a deeper understanding of the interplay of different physical assumptions and their influence on the observations.

10.4.1
Radiative Processes and Relativistic Effects

Our current understanding is that the observed radiation from relativistic jets is produced by nonthermal emission mechanisms (Chapter 3). As discussed in Sections 9.3 and 9.4, there exist a variety of processes in relativistic jets which are capable of accelerating particles to energies much higher than that of the background plasma. A common assumption in numerical simulations is that particles are accelerated at shocks or in the shear layers flanking relativistic jets (see Section 10.3.1.1). We can also assume that the energy distribution of accelerated particles is a nonthermal distribution, typically a power law or a combination of power laws (e.g., broken power law). On the other hand, we have strong indications that magnetic fields are present and/or can be generated in jets (see Chapter 3). Under this assumption, charged particles can gyroradiate around the magnetic field lines emitting synchrotron radiation. Finally, it is possible that the high-energy particles scatter the low-frequency photons to higher energies, i.e., the inverse-Compton is a radiation mechanism to be included in detailed models of jet emission.

As in Section 3.1.2, we introduce the following two quantities:

- *Emissivity* $j_\nu(\boldsymbol{x}, t; \boldsymbol{n})$. Defined as the amount of radiant energy emitted (or scattered in the case of the inverse-Compton process) by a volume element dV of material at position $\boldsymbol{x}$ into an elemental solid angle $d\omega$ around a direction $\boldsymbol{n}$, in a time interval dt, and in a frequency interval $d\nu$ around ν.
- *Absorption coefficient* $\alpha_\nu(\boldsymbol{x}, t; \boldsymbol{n})$. Defined as the fraction of the incident intensity absorbed per unit length ds of the volume element dV of material at position $\boldsymbol{x}$ from a beam of radiation propagating into a solid angle element $d\omega$ around a direction $\boldsymbol{n}$, in a time interval dt, and in a frequency interval $d\nu$ around ν.

Since most of our discussion about radiation processes takes place in the rest frame of the fluid, we will assume that these processes are isotropic in that frame and drop the explicit angular dependence (strictly speaking, we average j and α over all solid angles). We denote these averaged quantities as $j^{\rm em}_{\nu{\rm em}}(\boldsymbol{x}, t)$ and $\alpha^{\rm em}_{\nu{\rm em}}(\boldsymbol{x}, t)$.

Assuming that for each position $\boldsymbol{x}$ inside the jet we can compute $j^{\rm em}_{\nu{\rm em}}(\boldsymbol{x}, t)$ and $\alpha^{\rm em}_{\nu{\rm em}}(\boldsymbol{x}, t)$ in the local fluid rest frame, we still need to transform these quantities into the frame of a distant observer (e.g., an Earth observer). Analogously to the derivation of (2.54) the emissivity and absorption coefficient transform as

$$j^{\rm rec}_{\nu{\rm rec}}(\boldsymbol{x}, t) = j^{\rm em}_{\nu{\rm em}}(\boldsymbol{x}, t)\delta^2 \,, \tag{10.14}$$

$$\alpha^{\rm rec}_{\nu{\rm rec}}(\boldsymbol{x}, t) = \alpha^{\rm em}_{\nu{\rm em}}(\boldsymbol{x}, t)\delta^{-1} \,. \tag{10.15}$$

Using (10.14) and (10.15) with the appropriate expressions for the rest-frame emissivity and absorption coefficients we can, in principle, compute the time-dependent multiwavelength emission from a simulated jet. As we will see in the next section, this computation can be performed using different schemes of varying complexity, each with its own set of trade-offs between speed and accuracy.

10.4.2
Classification of Algorithms for Computing the Jet Emission

A method for computing the jet emission can be broadly characterized by the assumptions it makes about (i) the properties of the emitting fluid, and (ii) its coupling to the thermal fluid (i.e., the fluid actually modeled by (magne-

Table 10.1 Overview of the properties of emission computation algorithms needed to compute various astrophysical scenarios.

	Optically thin	Optically thick
Local	High-energy afterglow emission blazar X-ray emission	Steady jets
Transport	Optical/UV afterglow emission	Radio jets radio and afterglow emission

to)hydrodynamic simulations). In Table 10.1 we outline the properties needed to study different astrophysical scenarios.

With respect to the properties of the emitting fluid, we divide the algorithms into those which directly sum the emission from different parts of the fluid (*optically thin* in Table 10.1, see also (3.21)) and those which solve the radiative transfer equation (*optically thick* in Table 10.1, see also (3.22)). The emission summation is sufficient in those cases where the optical depth of the fluid can be neglected (usually for high-energy emission). In all other cases the equation of radiative transfer needs to be solved (see below and also Section 3.1).

The algorithms either use only the local fluid information to compute emissivity (*local* in Table 10.1) or they implement some form of the nonthermal particle transport (*transport* in Table 10.1).

In the following two sections we separately review the local models and those implementing the nonthermal particle transport.

10.4.2.1 Local Models

The first group of models in this class is that in which the underlying jet model is simple, usually stationary or based on a purely kinematical description of the beam. The radiative part, however, can be rather sophisticated [227–229]. These models take into account various relevant effects (Doppler effect, aberration of light, polarization of synchrotron emission, synchrotron self-absorption) and solve for the full radiative transfer equation. They usually do not need to take into account the time dependence of the jet dynamics, and are thus computationally less expensive.

There are, on the other hand, models which take into account the full hydrodynamical jet evolution, but use a simplified prescription for the radiative part [93, 127, 128, 131, 230–234]. Generally, nonthermal particle transport is not included, though in some of the models the cooling of high-energy electrons is followed [231, 233], or they solve the full radiative transfer equation (3.17) [128, 234]

$$\frac{d I_{\nu\mathrm{rec}}^{\mathrm{rec}}(\boldsymbol{x}, t)}{c d t} = j_{\nu\mathrm{rec}}^{\mathrm{rec}}(\boldsymbol{x}, t) - \alpha_{\nu rc}^{\mathrm{rec}}(\boldsymbol{x}, t) I_{\nu\mathrm{rec}}^{\mathrm{rec}}(\boldsymbol{x}, t) , \tag{10.16}$$

where $I_{\nu\mathrm{rec}}^{\mathrm{rec}}(\boldsymbol{x}, t)$ is the radiation intensity in the direction of the observer at a location $\boldsymbol{x}$. Note that ds in (3.17) has been replaced by $c\,dt$, which is the time measured by the observer along a particular line of sight. In this case $\boldsymbol{x}$ and t are not independent, since the radiation from the whole line of sight has to arrive simultaneously to the observer. Usually $j^{\mathrm{em}}_{\nu\mathrm{em}}$ and $\alpha^{\mathrm{em}}_{\nu\mathrm{em}}$ are computed in the fluid rest frame and then transformed into the observer frame using (10.14) and (10.15). See Section 3.1 for a more detailed discussion of the radiative transfer.

The integration of (10.16) has to be performed along the isochrones in the observer frame,

$$t_{\mathrm{obs}} = t - \frac{\boldsymbol{x}\boldsymbol{n}}{c} = t - \frac{r\mu}{c} , \tag{10.17}$$

where $\mu = \cos\theta$ is the angle between the line-of-sight towards the observer and the local radial direction. This means that $\boldsymbol{x}$ and t in (10.16) are not independent, and different lines-of-sight are selected by choosing the value of t_{obs}. Equation (10.17) is a good approximation when computing light curves of point sources. However, when computing images a slightly more complicated equation needs to be used (see, e.g., Appendix A1 of [88]).

Obviously, (10.16) has to be discretized in order for it to be integrated in a numerical simulation. In local emission models the discretization is usually done on the same grid that was used in the hydrodynamical simulation. This type of model is often used in the calculation of radiation from blast-waves, where most of the radiation is emitted close to strong shocks, and thus the effects of particle transport away from the site of their acceleration are not crucial [231].

10.4.2.2 **Nonthermal Particle Transport**

This group of algorithms attempts to model the transport of nonthermal particles in order to more accurately solve the radiation transfer (10.16). The basic difference with the previous model class is that the nonthermal particle population is treated explicitly.

On the one hand, the nonthermal particles spatial distribution can be discretized on the same grid that is used by the hydrodynamic simulation. In Mimica *et al.* [235] this was done by discretizing the nonthermal particle (electrons) energy distribution in each numerical cell in N bins and following the evolution of the number density n_i of electrons in each bin. By scaling the number density to the local fluid rest-mass density, tracer fractions $X_i := n_i m_{\mathrm{e}}/\rho;\ i = 1, \dots, N$ can be defined. Under the assumption that nonthermal electrons are advected with the thermal fluid, the equation of continuity can be used to track the spatial transport. In one dimension this equation reads as

$$\frac{\partial}{\partial t}(\rho \boldsymbol{X}\Gamma) + \frac{\partial}{\partial x}(\rho \boldsymbol{X}\Gamma\beta_\Gamma c) = 0 , \tag{10.18}$$

where $\boldsymbol{X} = (X_1, X_2, \dots, X_N)$. The changes in the energy distribution of particles due to radiative and adiabatic losses and gains are described by the kinetic equation [236]

$$\frac{\partial n(\gamma,t)}{\partial t} + \frac{\partial}{\partial \gamma}\left[\dot{\gamma}\, n(\gamma,t)\right] = Q(\gamma,t) , \tag{10.19}$$

where $n(\gamma, t)$ is the number density of particles with energy $\gamma m_{\mathrm{e}}c^2$ at the time t, $Q(\gamma, t)$ is the source term (e.g., a term that accounts for new, high-energy particles at shocks) and $\dot{\gamma} := d\gamma/dt$ are the energy losses and gains, that is, synchrotron (3.27) and Compton (3.63) cooling. Mimica *et al.* [235] discretized (10.18) and (10.19) using the same energy bins. Since for the synchrotron radiation the time scales of energy losses are usually much smaller than the hydrodynamic time scales, (10.19) can be solved separately from the magnetofluid equations. Finally, j_ν and α_ν can be computed in each numerical cell using the discretized nonthermal particle distribution and assuming a power-law distribution within each energy bin (i.e., the whole energy distribution is approximated as a piecewise power-law function).

More commonly, the nonthermal particle evolution is treated by attempting to solve the Boltzmann equation for the particle distribution function. In relativistic outflows an ensemble-averaged distribution function f follows the equation [88, 237, 238]

$$p^\beta \left(\frac{\partial f}{\partial x^\beta} - \Gamma^\alpha_{\beta\gamma} p^\gamma \frac{\partial f}{\partial p^\alpha} \right) = \left(\frac{df}{d\tau} \right)_{\text{coll}} , \tag{10.20}$$

where f is a function of the coordinates x^α and the four-momentum p^α of particles, and $\Gamma^\alpha_{\beta\gamma}$ are the Christoffel symbols. The collision terms are written on the right hand side, with τ being the particle proper time. This equation is almost never solved directly, since that involves a seven-dimensional integration. Usually different moments (integrals over the angular dependence of the particle momenta) of (10.20) are followed, and different assumptions about the microphysics are introduced (e.g., advection by a thermal fluid, diffusion-convection, etc.). In nonrelativistic hydrodynamics there has been a number of jet studies, which include the particle transport via solving for the distribution function f of nonthermal electrons [239–242]. In the relativistic case this has only been done by [88].

Finally, particle-in-cell (PIC) simulations have recently been used to compute synchrotron radiation [243, 244]. As opposed to a fluid description, PIC simulations use the kinetic description of the plasma (especially at shocks) to more accurately model its behavior (at the cost of only being able to simulate much smaller spatial and temporal time scales, namely a few hundred skin-depths of the constituent particles). To compute synchrotron radiation, they first perform a complete numerical simulation. Afterward instants of the evolution at which radiation is to be computed are chosen and the simulation restarted at those instants with much higher temporal resolution. This is necessary for the calculation, in the Fourier space, needed to compute the emission spectrum.

10.4.3 Applications

In this section we discuss some of the applications of the computation of synthetic emission, in the context of extragalactic and GRB jets.

10.4.3.1 Internal Shocks in Blazars and GRBs

The interaction of irregularities in the beam flow, traveling downstream of a jet is one of the candidate processes for the dissipation of the jet kinetic and magnetic energy into thermal energy and, ultimately, into radiation. We discuss here some recent applications of numerical models of the collision of shells with different speeds to the study of blazars, but the process is completely analogous in the prompt GRB case. The collision of such shells generates shocks inside the relativistic beam, so-called internal shocks (e.g., [245]).

As internal shocks propagate through the colliding shells, they are expected to accelerate electrons to high energies and to generate random magnetic fields. The numerical modeling of these two processes is often done by assuming that a fraction ϵ_e of the shocked fluid internal energy is passed on to the accelerated electrons, while another fraction ϵ_B goes into the random magnetic field component. Using this modeling of the shock microphysics, 2D RHD simulations of cylindrical shell collisions have been performed and computed the resulting light curves [235]. Their conclusion was that, contrary to the assumption made by semianalytic models that a single shell collision produces a single peak in the light curve, a multi-peaked light curve is possible. The main reason for this is that the shell interaction is influenced by the preheating of each shell resulting from the interaction with the surrounding medium before they collide. In follow-up work, Mimica *et al.* [246] performed a more systematic study, where the collisions of shells with different masses and energies was studied. It has been found that the light curves mostly depend on the ratio of shell Lorentz factors and shell geometry, while the exact distribution of the rest-mass density and the specific internal energy plays a secondary role, even though the hydrodynamic evolution might have some remarkable differences [247]. The magnetization σ (ratio of magnetic to kinetic energy density) and the ratio of magnetic to thermal energy density play an important role in the pre-collision phase, while in the shell collision (and hence for the light curves) only σ is important [248]. However, the latter results span only relatively small values of the shell magnetization. An interesting problem posed by collisions of very magnetized shells and, in general, for the internal shock model when magnetic fields are dynamically dominant (i.e., in the high-σ regime), is that the formation of strong shocks is extremely difficult. Indeed, it is known that depending on the magnetization of the colliding shells, instead of the standard double-shock (forward and reverse shocks) structure, a rarefaction-shock structure may result [249]. Such a situation brings a notorious reduction of the kinetic efficiency, that is, the efficiency of converting the initial kinetic and/or magnetic energy of the shells into thermal energy and, finally, into radiation. We shall point out that the radiative efficiency of the process is expected to be a fraction $f_r < 1$ of the estimated dynamic one. The exact value of f_r depends on the particularities of the emission processes, which radiate away the thermal or magnetic energy (e.g., due to reconnection events) of the shocked states. However, Mimica and Aloy [249] find that for moderately magnetized shocks ($\sigma \simeq 0.1$), the total dynamic efficiency[5] in a single two-shell interaction is optimal, arriving to be as large as 40%. Thus, the total dynamic efficiency of moderately magnetized shocks is larger than in the corresponding unmagnetized two-shell interaction. Furthermore, if the slower shell propagates with a sufficiently large velocity, the efficiency is only weakly dependent on its Lorentz factor [249]. Consequently, the dynamic efficiency of shell interactions in the magnetized flow of blazars and GRBs is effectively the same.

10.4.3.2 **External Shocks of GRBs**

It is generally assumed that the afterglow phase of GRBs is produced when the GRB ejecta starts interacting with the circumburst (external) medium surrounding the progenitor. The shocks propagating from the ejecta-medium interface dissipate the kinetic energy of the outflow. A forward shock (FS) propagates into the external medium, heating and accelerating it. In some cases there is a reverse shock (RS) propagating into the ejecta, heating and decelerating it. The ejecta deceleration thus occurs through a transfer of energy from the ejecta into the shocked external medium. As in the case of the internal shocks (Section 10.4.3.1), each shock can accelerate electrons and emit synchrotron radiation.

Downes *et al.* [231] have used one-dimensional RHD simulations coupled with a simple transport equation to study the afterglow emission as predicted by the standard fireball model (this code can be classified as the “local/thin” in Table 10.1). They found that, while the general results are similar to the analytic predictions, the details can differ, especially at large frequencies. This is because their numerical scheme can properly take into account the correct blast-wave geometry and consistently treat the adiabatic cooling of the shock-accelerated electrons.

The possible signature at optical wavelengths of a discontinuity in the external medium (such as a wind termination shock), has been considered by van Eerten *et al.* [250]. They have performed hydrodynamic simulations using the ARMVAC code [14, 15, 251] and then used the radiation code of van Eerten *et al.* [233] to compute the synthetic light curves of the corresponding hydrodynamic models. Their conclusion is that the signature of the ejecta encountering a wind termination shock is not significant enough to stand out. In a more recent study [234], the same numerical methodology has been applied to study the late phase of the afterglow by simulating in 2D the propagation of ejecta into a homogeneous medium. Remarkably, it has been found that the radiation flux may vary by an order of magnitude depending on the equation of state used in the simulations. Furthermore, they specifically considered the case of GRB 0303029 by using the physical parameters available in previous references, and realized significant discrepancies between the simulation output and the observations, mainly attributed to the different assumption about the blast wave geometry (hard-edged jet instead of a spherical explosion). Finally, they produced spatially resolved radio images of an afterglow, which show a ring-like structure at late times.

5) This quantity measures the conversion ratio of kinetic to thermal plus magnetic energy. Note that as a result of the collision, magnetic field amplification may happen. This magnetic field energy can be further reprocessed and converted in thermal energy by non-ideal mechanisms.

One of the most intriguing, and long-standing questions faced in the study of the GRB phenomenon is whether magnetic fields are dynamically relevant or not, and to which extent they shape the dynamics and the observed emission properties. Since we cannot directly perform measurements on the GRB central engine, we have to look for evidence of the role of the magnetic field either in the prompt GRB phase or in the afterglow. According to the ultra-high-resolution, one-dimensional, RMHD simulations of Mimica *et al.* [252], magnetic fields can leave their signature on the initial phases of the afterglow by substantially changing the back-reaction of the flow as a consequence of its interaction with the external medium. Ultrahigh numerical resolution is needed because of the extreme jump conditions in the region of interaction between the ejecta and the circumburst medium. Mimica *et al.* aim also to explain the observational signatures of GRB ejecta magnetization on the onset of the afterglow. In this regard, they point out that the reverse shock is typically weak or absent for ejecta with initial magnetization $\sigma \gtrsim 1$. Moreover, the onset of the forward shock emission is strongly dependent on the magnetization. On the other hand, the magnetic energy of the shell is transferred into the external medium on scales of several times the duration of the burst. The asymptotic evolution of strongly magnetized shells, after experiencing significant deceleration, resembles that of hydrodynamic shells, that is, they enter fully into the Blandford–McKee self-similar regime.[6] As a consequence of such asymptotic behavior, the later forward shock emission contains no information about the initial magnetization of the flow, which is unfortunate, since most of the available observational data does not refer to the early afterglow, but to the relatively late afterglow phases.

Very recently, refining the previous developments [252], Mimica *et al.* [253] have studied the early afterglow phase using the code MRGENESIS [248, 252] combined with the code SPEV [88]. They focused on the difference in observational signature between nonmagnetized and magnetized ejecta, whereby the flow magnetization parameter σ was varied between 0 (nonmagnetized ejecta) and 1 (corresponding to equipartition between the kinetic and the magnetic energies). They found that the RS optical signature (the so-called optical flash) is most prominent not for the nonmagnetized ejecta, as one would initially expect, but for mildly magnetized ejecta ($\sigma \simeq 0.01-0.1$). Furthermore, by simulating early optical afterglows of GRB 990123 and 090102 they find that the famous ninth magnitude optical flash of GRB 990123 is best explained by a mildly magnetized ejecta ($\sigma = 0.01$) while the 10% polarization of the optical emission of GRB 090102 [254] is best explained by $\sigma = 0.1$ ejecta. For $\sigma > 0.1$ they find that the RS becomes too weak to be observed in the optical band, that is, the FS emission dominates the afterglow emission.

6) The release of a large amount of thermal energy in a given medium triggers an explosion, where a blast wave enclosing the expanding ejecta develops. The asymptotic behavior of the ejecta was computed analytically by Blandford and McKee [123] generalizing to special relativity the (Newtonian) Sedov–Taylor solution.

10.4.3.3 **Superluminal Components**

Blobs of radiating plasma moving at apparent superluminal velocities are commonly observed in the radio jets of blazars and microquasars [255, 256]. The origin of the apparent superluminal motion is discussed in detail in Section 2.3.2.

As pointed out in Section 10.3.2.1 the observation of superluminal components is attributed to the downstream travel of blobs of plasma or shocks, or to the growth of (magneto)hydrodynamic instabilities. Following the former idea, Komissarov and Falle [230] simulated the interaction of a moving shock with a stationary reconfinement shock within a jet. By assuming that the fluid is optically thin they computed the observed emission, finding that a bright moving knot, associated with the moving shock, is much brighter and wider than the underlying jet, and also brighter than the stationary knot associated with the reconfinement shock. Furthermore, the moving knot's brightness increases until it reaches the stationary knot, after which it becomes more extended and gradually fades.

With a different model for the perturbation injected at the jet inlet, Gómez *et al.* [128] performed hydrodynamic simulations of axially symmetric parsec-scale jets. In this case, the quiescent beam is perturbed by temporarily increasing the velocity of the fluid at the boundary through which a jet is being injected. Because of the increased fluid velocity in the perturbation, a shocked state is created in its forward part, while a rarefaction develops at the rear end of it. Neglecting the synchrotron losses they computed the time-dependent radio images for a fixed viewing angle using the method of Gómez *et al.* [127]. The perturbation appears as a region of increased emission propagating down the jet at apparent superluminal velocity. Since it interacts with the underlying jet, its brightness fluctuates around the long-term trend of the decaying radio emission. A very recent improvement of this model has been considered in Mimica *et al.* [88] who self-consistently included the radiative losses in the calculations employing the SPEV code. The synchrotron radiative losses (Section 3.2.1) are found to be important for values of the magnetic field larger than those previously assumed [128]. As advanced in Section 10.3.2.1, a prediction made by Mimica *et al.* [88] is that the trailing components will have a frequency-dependent observational signature. This is still untested observationally, but it might give quite useful insights on the magnetic field energy associated to such radiating features. In Figure 10.3 a sequence of snapshots of the emission resulting from the passage of a perturbation through an overpressured jet is shown. The component is injected from the left side and travels towards the right. In the first snapshot it is seen as a dark region between 3 and 5 R_b. In the second snapshot it appears to split into two parts. This is because the component interacts with the standing (cross) shocks in the jet. In the third and fourth snapshots it can be seen that the component fades as it exits the computational domain. The last two snapshots show the temporary decrease of the emission in the wake of the component's passage and the gradual reestablishment of the quiescent jet.

10.4.3.4 **Emission from Large-Scale Jets**

A number of groups have computed the emission from large-scale jets in a variety of scenarios. On the one hand, there are nonrelativistic simulations of MHD jets

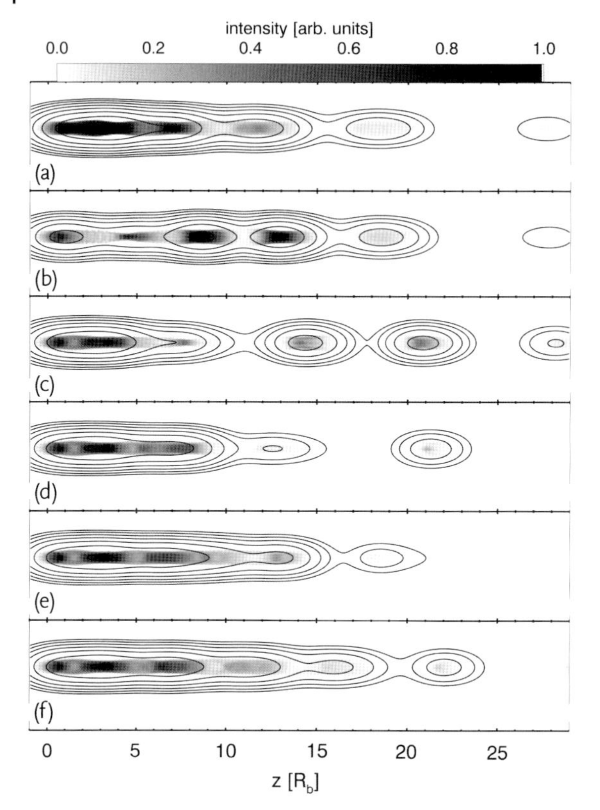

Figure 10.3 Snapshots of the emission at 43 GHz of the perturbed overpressured jet model computed by [88]. From (a) to (f), the observed emission 0.02, 0.39, 0.75, 1.12, 1.94 and 4.58 years after the component appears, respectively. The superimposed contours have been calculated by the convolution of the underlying image with a circular Gaussian beam whose radius at FWHM is $2.25R_b$. The contour levels are 0.005, 0.01, 0.02, 0.04, 0.08, 0.16, 0.32, 0.64 and 0.9 of the maximum of the convolved emission. The horizontal length scale is normalized so that the beam radius is $R_b = 0.1$ pc, while the vertical length scale has been compressed to show an interval of $10R_b$.

of radio galaxies performed by Jones *et al.* [239] and Tregillis *et al.* [241]. Jones *et al.* [239] performed axisymmetric simulations of Mach 20 light jets and focused on the synchrotron emission from the jet terminal shocks. They show that even for the steady inflowing jet the effects of shock acceleration and synchrotron cooling play an important role and that the properties of the emission are strongly influenced by the jet backflow. There is a wide range of complex synchrotron signatures which depend on the interplay between the adiabatic and synchrotron cooling with the underlying jet dynamics. Tregillis *et al.* [241] improved on the previous models of Jones *et al.* by performing 3D simulations of MHD jets in radio galaxies. They confirm that the jet head is a complex environment, with the backflowing material playing an important role. Most importantly, they find that once the effects of adiabatic expansion and synchrotron losses are taken into account, understanding the nature of jet emission requires a substantial knowledge of the details of the flow.

The first radio emission simulations from 3D relativistic hydrodynamic jets were computed in Aloy *et al.* [147]. These authors used a local approach [127, 128] to produce synthetic radio maps of large-scale jets, with the aim of studying the observational properties of the jet/ambient interaction layer (see Section 10.3.2.2). The authors find that if the magnetic field in the shear layer becomes helical, the emission shows a cross-section asymmetry, because of which either the top or the bottom of the jet dominates the emission. This, as well as limb or spine brightening, is a function of the viewing angle and flow velocity, and the top/bottom jet emission predominance can be reversed if the jet changes direction with respect to the observer or if it presents a change in velocity. The polarized flux displays a more prominent asymmetry than the total intensity, resulting from the field cancellation (or amplification) along the line-of-sight. Indeed, some observations of jet cross-section emission asymmetries in the blazar 1055+018 can be explained by assuming the existence of a shear layer with a helical magnetic field [147].

10.5 Summary and Outlook

In this chapter, we have shown how numerical simulations have allowed us to model the complex dynamics of relativistic jets. Together with theory and observations, numerical simulations are the third leg on which our knowledge of relativistic sources is built up. They are key to understanding why relativistic jets are not destroyed by the myriad of potentially harmful instabilities happening in supersonic beams of plasma (Section 10.3.1). Numerical models have been used to verify that a simple (magneto)hydrodynamic description may explain the typical structures observed both at pc- and kpc-scale (Sections 10.3.2.1 and 10.3.2.2). In particular, the study of superluminal components by means of numerical simulations is shedding some light on the nature of such components (i.e., whether they are simple traveling shocks, blobs of plasma or saturated flow instabilities). Also, substantial progress has been made on tackling the FR I/FR II morphological dichotomy with progressively refined numerical models. Furthermore, a notorious step forward has been accomplished by computing the nonthermal emission signature of (magneto)hydrodynamic models, which has allowed us to directly compare numerics and observations (Section 10.4). Not less important is the role played by simulations of relativistic jets to unveil the properties of outflows associated to GRBs. In spite of the lack of direct observations of jets in GRBs, the broad analogy between AGN or microquasar jets with GRB jets is, to a large extent, supported by numerical models of the latter (Section 10.3.3).

Regardless of the advances brought by simulations of relativistic jets, the numerical models of these sources are still relatively simple. The microphysics of relativistic jets is likely more complex than a fluid approximation can reproduce. As pointed out [101], kinetic effects define the transport coefficients and the development of perturbations and nonlinear structures inside the flow. As a result, the interaction layer between a jet and its environment is governed by these coefficients, and so mixing, entrainment, turbulence, etc., are critically dependent on them. A fully kinetic treatment is nowadays not possible, though RPIC simulations addressing the physics of shocks and their radiative properties are progressing fast [257]. Thus, our efforts shall probably aid to bridge the gap between kinetic (microscopic) and fluid descriptions. Several possibilities in such a direction (with different degrees of complexity) are:

- Incorporate subgrid resolution models in current simulations. First steps towards that direction have been taken by Casse and Marcowith [258] who applied a multiscale MHD kinetic scheme, coupling MHD simulations to kinetic computations using stochastic differential equations to include the unresolved numerical scales at the macroscopic level. Generalizing their approach to RMHD, would allow us to consistently include, for example, the yield and transport of high-energy particle within hot spots, or to better approximate the particle acceleration in relativistic jets.
- A possible step may be to implement a two-fluid model in numerical codes that have been adapted to the problem. Such fluids represent two separate plasma constituents with different charges (e.g., electrons and ions or electrons and positrons). Codes incorporating such an approach have recently been developed [259] and may soon be adapted to the problem at hand.

Furthermore, we shall include nonideal physics (e.g., resistivity, viscosity) in our models. A possible way is using two-fluid models (as mentioned above). Alternatively, one-fluid resistive RMHD (e.g., [260]) could be used to tackle the problem. In any case, resistive MHD problems in the limit of small resistivity are extremely stiff from the computational point of view, and thus, if we deal with them with explicit algorithms, we may face severe restrictions arising from the ultrahigh resolution needed to properly model the system. New, improved algorithms (e.g., [12, 261]) including massive parallelism and the ability of using hybrid computers (i.e., computers with architectures not only based on CPUs, but also on, e.g., GPUs) will probably determine our near future progress in this field.

References

1 Begelman, M.C., Blandford, R.D., and Rees, M.J. (1984) *Rev. Mod. Phys.*, **56**, 255.

2 Hiscock, W.A. and Lindblom, L. (1985) *Phys. Rev. D*, **31**, 725.

3 Sandoval-Villalbazo, A. and Garcia-Perciante, A.L. (2010) New Trends in Statistical Physics, World Scientific, pp. 293–301. Invited review for Festschrift in honor of Prof. Leopold Garcia-Colin Scherer. Festschrift in Honor of Leopold Garcia-Colin's 80th birthday. eprint arXiv:1001.4832.

4 De Villiers, J.P. and Hawley, J.F. (2003) *Astrophys. J.*, **589**, 458.

5 Anninos, P., Fragile, P.C., and Salmonson, J.D. (2005) *Astrophys. J.*, **635**, 723.

6 Lax, P.D. and Wendroff, B. (1960) *Commun. Pure Appl. Math.*, **13**, 217.

7 Anton, L., Zanotti, O., Miralles, J.A., Martí, J.M., Ibáñez, J.M., Font, J.A., and Pons, J.A. (2006) *Astrophys. J.*, **637**, 296.

8 Font, J.A. (2008) *Living Rev. Relat.*, **11**, 7.

9 Martí, J.M. and Müller, E. (2003) *Living Rev. Relat.*, **6**, 7.

10 Meier, D.L. (1999) *Astrophys. J.*, **518**, 788.

11 Richardson, G.A. and Chung, T.J. (2002) *Astrophys. J. Suppl. Ser.*, **139**, 539.

12 Dumbser, M. and Zanotti, O. (2009) *J. Comput. Phys.*, **228**, 6991.

13 Berger, M.J. and Colella, P. (1989) *J. Comput. Phys.*, **82**, 64.

14 Keppens, R., Baty, H., Bergmans, J., and Casse, F. (2004) *Astrophys. Space Sci.*, **293**, 217.

15 Keppens, R., Meliani, Z., van Marle, P., Delmont, A., Vlasis A., and van der Holst, B. (2011) *J. Comput. Phys.*, in press. doi:10.1016/j.jpc2011.01.020 http://www.sciencedirect.com/science/article/pii/S0021999111000386 (last accessed 14 September 2011), online published: 15 January 2011.

16 Plewa, T. and Müller, E. (2001) *Comput. Phys. Commun.*, **138**, 101.

17 Quilis, V. (2004) *Mon. Not. R. Astron. Soc.*, **352**, 1426.

18 LeVeque, R.J. (1990) *Numerical Methods for Conservation Laws*, Lectures in Mathematics, ETH-Zurich. Birkhauser-Verlag, Basel.

19 Colella, P. and Woodward, P.R. (1984) *J. Comput. Phys.*, **54**, 174.

20 Monaghan, J.J. (1985) *Comput. Phys. Rep.*, **3**, 71.

21 Romero, G.E., Okazaki, A.T., Orellana, M., and Owocki, S.P. (2007) *Astron. Astrophys.*, **474**, 15.

22 Anton, L., Miralles, J.A., Martí, J.M., Ibáñez, J.M., Aloy, M.A., and Mimica, P. (2010) *Astrophys. J. Suppl. Ser.*, **188**, 1.

23 Balsara, D.S. (2004) *Astrophys. J. Suppl. Ser.*, **151**, 149.

24 Evans, C.R. and Hawley, J.F. (1988) *Astrophys. J.*, **332**, 695.

25 Stone, J.M. and Norman, M.L. (1992) *Astrophys. J. Suppl. Ser.*, **80**, 791.

26 Powell, K.G. (1994) An approximate Riemann solver for magnetohydrodynamics (that works in more than one dimension), ICASE Technical Report 94-24, NASA Langley Research Center, VA.

27 Falle, S.A.E.G., Komissarov, S.S., and Joarder, P. (1998) *Mon. Not. R. Astron. Soc.*, **297** 265.

28 Brackbill, J.U. and Barnes, D.C. (1980) *J. Comput. Phys.*, **35**, 426.

29 Dedner, A., Kemm, F., Kröner, D., Munz, C.-D., Schnitzer, T., and Wesenberg, M. (2002) *J. Comput. Phys.*, **175**, 645.

30 Toth, G. (2000) *J. Comput. Phys.*, **161**, 605.

31 Mignone, A. and Bodo, G. (2008) Lecture Notes in Physics, vol. 754, Springer, Berlin, p. 71.

32 Zachary, A.L., Malagoli, A., and Colella, P. (1994) *SIAM J. Sci. Comput.*, **15**, 263.

33 Ryu, D., Jones, T.W., and Frank, A. (1995) *Astrophys. J.*, **452**, 785.

34 Clarke, D.A. (1996) *Astrophys. J.*, **457**, 291.

35 Balsara, D.S. (1998) *Astrophys. J. Suppl. Ser.*, **116**, 133.

36 Balsara, D.S. and Spicer, D.S. (1999) *J. Comput. Phys.*, **151**, 149.

37 Dai, W., and Woodward, P.R. (1998) *Astrophys. J.*, **494**, 317.

38 Ryu, D., Miniati, F., Jones, T.W., and Frank, A. (1998) *Astrophys. J.*, **509**, 244.

39 Powell, K.G., Roe, P.L., Linde, T.J., Gombosi, T.I., and De Zeeuw, D.L. (1999) *J. Comput. Phys.*, **154**, 284.

40 Mignone, A. and Bodo, G. (2006) *Mon. Not. R. Astron. Soc.*, **368**, 1040.

41 Obergaulinger, M., Aloy, M.A., and Müller, E. (2006) *Astron. Astrophys.*, **450**, 1107.

42 Beckwith, K. and Stone, J.M. (2011) eprint arXiv:1101.3573, accepted for publication on ApJS.

43 Komissarov, S.S. (1999) *Mon. Not. R. Astron. Soc.*, **303**, 343.

44 Crockett, R.K., Colella, P., Fisher, R.T., Klein, R.I., and McKee, C.F. (2005) *J. Comput. Phys.*, **203**, 422.

45 Fromang, S., Hennebelle, P., and Teyssier, R. (2006) *Astron. Astrophys.*, **457**, 371.

46 Lee, D. and Deane, A.E. (2009) *J. Comput. Phys.*, **228**, 952.

47 Londrillo, P. and Del Zanna, L. (2000) *Astrophys. J.*, **530**, 508.

48 Londrillo, P. and Del Zanna, L. (2004) *J. Comput. Phys.*, **195**, 17.

49 Torrilhon, M. (2005) *SIAM J. Sci. Comput.*, **26**, 1166.

50 Arterant, R. and Torrilhon, M. (2008) *J. Comput. Phys.*, **227**, 3405.

51 Dai, W. and Woodward, P.R. (1994) *J. Comput. Phys.*, **115**, 485.

52 Dai, W. and Woodward, P.R. (1997) *SIAM J. Sci. Comput.*, **18**, 957.

53 Torrilhon, M. and Balsara, D.S. (2004) *J. Comput. Phys.*, **201**, 586.

54 Gardiner, T.A. and Stone, J.M. (2005) *J. Comput. Phys.*, **205**, 509.

55 Gardiner, T.A. and Stone, J.M. (2008) *J. Comput. Phys.*, **227**, 4123.

56 Dumbser, M., Balsara, D.S., Toro, E.F., and Munz, C.-D. (2008) *J. Comput. Phys.*, **227**, 8209.

57 Balsara, D.S., Rumpf, T., Dumbser, M., and Munz, C.-S. (2009) *J. Comput. Phys.*, **228**, 2480.

58 Obergaulinger, M., Cerdá-Durán, P., Müller, E., and Aloy, M.A. (2009) *Astron. Astrophys.*, **498**, 241.

59 Obergaulinger, M., Aloy, M.A., and Müller, E. (2006) *Astron. Astrophys.*, **515**, 30.

60 Jiang, G.S., Wu, C.-C. (1999) *J. Comput. Phys.*, **150**, 561.

61 Del Zanna, L., Zanotti, O., Bucciantini, N., and Londrillo, P. (2007) *Astron. Astrophys.*, **473**, 11.

62 Balsara, D.S. (2009) *J. Comput. Phys.*, **22**, 552.

63 Mignone, A., Tzeferacos, P., and Bodo, G. (2010) *J. Comput. Phys.*, **229**, 5896.

64 Anderson, M., Hirschmann, E.W., Liebling, S.L., and Neilsen, D. (2006) *Class. Quantum Gravity*, **23**, 6503.

65 van der Holst, B., Keppens, R., and Meliani, Z. (2008) *Comput. Phys. Commun.*, **179**, 617.

66 Etienne, Z.B., Liu, Y.T., and Shapiro, S.L. (2010) *Phys. Rev. D*, **82**, 084031, eprint arXiv:1007.2848.

67 Balsara, D.S. (2001) *J. Comput. Phys.*, **174**, 614.

68 Collins, D.C., Xu, H., Norman, M.L., Li, H., and Li, S. (2010) *Astrophys. J. Suppl. Ser.*, **186**, 308.

69 Cunningham, A.J., Frank, A., Varnière, P., Mitran, S., and Jones, T.W. (2009) *Astrophys. J. Suppl. Ser.*, **182**, 519.

70 van Dyke, M. (1982) *An Album of Fluid Motion*, Parabolic Press, Stanford.

71 Blandford, R.D. and Rees, M.J. (1974) *Mon. Not. R. Astron. Soc.*, **169**, 395.

72 Longaretti, P.-Y. (2008) Lecture Notes in Physics, vol. 754, Springer, p. 131.

73 Hardee, P.E. (2006) *Relativistic Jets: The Common Physics of AGN, Microquasars, and Gamma-Ray Bursts*, AIP Conference Proceedings 856, 14–17 December 2005, Ann Arbor, Michigan, American Institute of Physics, p. 57, doi:10.1063/1.2356384.

74 Bonnano, A. and Urpin, V. (2011) *Astron. Astrophys.*, **525**, A100.

75 Hardee, P.E. (1981) *Astrophys. J.*, **250**, 9.

76 Hardee, P.E. and Norman, M.L. (1989) *Astrophys. J.*, **342**, 680.

77 Zhao, J.-H. *et al.* (1992) *Astrophys. J.*, **387**, 69.

78 Hsu, S.C. and Bellan, P.M. (2002) *Mon. Not. R. Astron. Soc.*, **334**, 257.

79 Norman, M.L., Burns, J.O., and Sulkanen, M.E. (1988) *Nature*, **335**, 146.

80 Bodo, G., Massaglia, S., Ferrari, A., and Trussoni, E. (1994) *Astron. Astrophys.*, **283**, 655.

81 Bodo, G., Rossi, P., Massaglia, S., Ferrari, A., Malagoli, A., and Rosner, R. (1998) *Astron. Astrophys.*, **333**, 1117.

82 Micono, M., Massaglia, S., Bodo, G., Rossi, P., and Ferrari A. (1998) *Astron. Astrophys.*, **333**, 989.

83 Rosen, A., Hardee, P.E., Clarke, D.A., and Johnson, A. (1999) *Astrophys. J.*, **510**, 136.

84 Rossi, P., Mignone, A., Bodo, G., Massaglia, S., and Ferrari, A. (2008) *Astron. Astrophys.*, **488**, 795.

85 Lobanov, A. and Zensus, J.A. (2001) *Science*, **294**, 128.

86 Lobanov, A., Hardee, P., and Eilek, J. (2003) *New Astron. Rev.*, **47**, 629.

87 Perucho, M. and Lobanov, A.P. (2007) *Astron. Astrophys.*, **469**, L23.

88 Mimica, P. *et al.* (2009) *Astrophys. J.*, **696**, 1142.
89 Urpin, V. (2002) *Astron. Astrophys.*, **385**, 14.
90 Aloy, M.A., Ibáñez, J.M., Miralles, J.A., and Urpin, V. (2002) *Astron. Astrophys.*, **396**, 693.
91 Birkinshaw, M. (1984) *Mon. Not. R. Astron. Soc.*, **208**, 887.
92 Hardee, P.E. (1979) *Astrophys. J.*, **234**, 47.
93 Agudo, I., Gómez, J.L., Martí, J.M., Ibáñez, J.M., Marscher, A.P., Alberdi, A., Aloy, M.A., and Hardee, P.E. (2001) *Astrophys. J. Lett.*, **549**, L183.
94 Hardee, P.E., Hughes, P.A., Rosen, A., and Gómez, E.A. (2001) *Astrophys. J.*, **555**, 744.
95 Hardee, P.E. and Hughes, P.A. (2003) *Astrophys. J.*, **583**, 116.
96 Perucho, M., Hanasz, M., Martí, J.M., and Sol, H. (2004) *Astron. Astrophys.*, **427**, 415.
97 Perucho, M., Martí, J.M., and Hanasz, M. (2004) *Astron. Astrophys.*, **427**, 431.
98 Birkinshaw, M. (1991) *Mon. Not. R. Astron. Soc.*, **252**, 73.
99 Ferrari, A., Massaglia, S., and Trussoni, E. (1982) *Mon. Not. R. Astron. Soc.*, **198**, 106.
100 Perucho, M., Hanasz, M., Martí, J.M., and Miralles, J.A. (2007) *Phys. Rev. E.*, **75**, 056312.
101 Ferrari, A. (1998) *Annu. Rev. Astron. Astrophys.*, **36**, 539.
102 Keppens, R., Meliani, Z., Baty, H., and van der Holst, B. (2009) Lecture Notes in Physics, vol. 791, Springer, p. 179.
103 Mizuno, Y., Hardee, P.E., and Nishikawa, K.-I. (2007) *Astrophys. J.*, **662**, 835.
104 Gomez, J.L. *et al.* (2000) *Science*, **289**, 2317.
105 Agudo, I. *et al.* (2007) *Astron. Astrophys.*, **476**, L17.
106 Lery, T., Baty, H., and Appl, S. (2000) *Astron. Astrophys.*, **355**, 1201.
107 Ouyed, R., Clarke, D.A., and Pudritz, R.E. (2003) *Astrophys. J.*, **582**, 292.
108 Nakamura, M. and Meier, D.L. (2004) *Astrophys. J.*, **617**, 123.
109 Nakamura, M., Li, H., and Li, S. (2007) *Astrophys. J.*, **656**, 721.
110 Moll, R., Spruit, H.C., and Obergaulinger, M. (2008) *Astron. Astrophys.*, **492**, 621.
111 McKinney, J.C. and Blandford, R.D. (2009) *Mon. Not. R. Astron. Soc.*, **394**, L126.
112 Carey, C.S. and Sovinec, C.R. (2009) *Astrophys. J.*, **699**, 362.
113 Istomin, Y.N. and Pariev, V.I. (1994) *Mon. Not. R. Astron. Soc.*, **267**, 629.
114 Istomin, Y.N. and Pariev, V.I. (1996) *Mon. Not. R. Astron. Soc.*, **281**, 1.
115 Begelman, M.C. (1998) *Astrophys. J.*, **493**, 291.
116 Lyubarsky, Y.E. (1999) *Mon. Not. R. Astron. Soc.*, **308**, 1006.
117 Tomimatsu, A., Matsuoka, T., and Takahashi, M. (2001) *Phys. Rev. D*, **64**, 123003.
118 Narayan, R., Li, J., and Tchekhovskoy, A. (2009) *Astrophys. J.*, **697**, 1681.
119 Mizuno, Y., Lyubarsky, Y., Nishikawa, K.-I., and Hardee, P.E. (2009) *Astrophys. J.*, **700**, 684.
120 Blandford, R.D. and Königl, A.F. (1979) *Astrophys. J.*, **232**, 34.
121 Königl, A. (1981) *Astrophys. J.*, **243**, 700.
122 Marscher, A.P. (1980) *Astrophys. J.*, **235**, 386.
123 Blandford, R.D. and McKee, C.F. (1976) *Phys. Fluids*, **19**, 1130.
124 Hughes, P.A., Aller, H.D., and Aller, M.F. (1985) *Astrophys. J.*, **298**, 301.
125 Marscher, A.P., Gear, W.K., and Travis, J.P. (1992) *Variability of Blazars* (eds E. Valtaoja and M. Valtonen), Cambridge University Press, Cambridge, p. 85.
126 Marscher, A.P. and Gear, W.K. (1985) *Astrophys. J.*, **298**, 114.
127 Gomez, J.L. *et al.* (1995) *Astrophys. J.*, **449**, L19.
128 Gomez, J.L. *et al.* (1997) *Astrophys. J.*, **482**, L33.
129 Duncan, G.C., Hughes, P.A., and Opperman, J. (1996) *Energy Transport in Radio Galaxies and Quasars* (eds P.E. Hardee, A.H. Bridle, and J.A. Zensus), ASP Conference Series, vol. 100, ASP, San Francisco, p. 143.
130 Komissarov, S.S. and Falle, S.A.E.G. (1996b) *Energy Transport in Radio Galaxies and Quasars* (eds P.E. Hardee,

A.H. Bridle, and J.A. Zensus), ASP Conference Series 100, San Francisco, p. 173.
131 Aloy, M.A. *et al.* (2003) *Astrophys. J.*, **585**, L109.
132 Gomez, J.L., Marscher, A.P., Alberdi, A., Jorstad, S.G., and Agudo, I. (2001) *Astrophys. J.*, **561**, L161.
133 Kadler, M., Ros, E., Perucho, M., Kovalev, Y.Y., Homan, D.C., Agudo, I., Kellermann, K.I., Aller, M.F., Aller, H.D., Lister, M.L., and Zensus, J.A. (2008) *Astron. Astrophys.*, **680**, 867.
134 Britzen, S., Vermeulen, R.C., Campbell, R.M., Taylor, G.B., Pearson, T.J., Readhead, A.C. S., Xu, W., Browne, I.W., Henstock, D.R., and Wilkinson, P. (2008) *Astron. Astrophys.*, **484**, 119.
135 Britzen, S., Kudryavtseva, N.A., Witzel, A., Campbell, R.M., Ros, E., Karouzos, M., Mehta, A., Aller, M.F., Aller, H.D., Beckert, T., and Zensus, J.A. (2010) *Astron. Astrophys.*, **511**, 57.
136 Rayburn, D.R. (1977) *Mon. Not. R. Astron. Soc.*, **179**, 603.
137 Norman, M.L., Smarr, L., Winkler, K.-H.A., and Smith, M.D. (1982) *Astron. Astrophys.*, **113**, 285.
138 Aloy, M.A. and Martí, J.M. (2002) Lecture Notes in Physics, vol. 589, Springer, p. 197.
139 van Putten, M.H.P.M. (1993) *Astrophys. J.*, **408**, L21.
140 Eulderink, F. and Mellema, G. (1994) *Astron. Astrophys.*, **284**, 654.
141 Duncan, G.C. and Hughes, P.A. (1994) *Astrophys. J.*, **436**, L119.
142 Martí, J.M., Müller, E., Font, J.A., and Ibáñez, J.M. (1994) *Astron. Astrophys.*, **281**, L9.
143 Martí, J.M., Müller, E., Font, J.A., and Ibáñez, J.M. (1995) *Astrophys. J.*, **448**, L105.
144 Martí, J.M., Müller, E., Font, J.A., Ibáñez, J.M., and Marquina, A. (1997) *Astrophys. J.*, **479**, 151.
145 Mizuta, A., Yamada, S., and Takabe, H. (2004) *Astrophys. J.*, **606**, 804.
146 Aloy, M.A., Ibáñez, J.M., Martí, J.M., Gómez, J.L., and Müller, E. (1999) *Astrophys. J.*, **523**, L125.
147 Aloy, M.A., Gómez, J.L., Ibáñez, J.M., Martí, J.M., and Müller, E. (2000a) *Astrophys. J.*, **528**, L85.
148 Hughes, P.A., Miller, M.A., and Duncan, G.C. (2002) *Astrophys. J.*, **572**, 713.
149 Choi, E., Wiita, P.J., and Ryu, D. (2005) *Bull. Am. Astron. Soc.*, **37**, 1296.
150 Mizuta, A., Kino, M., and Nagakura, H. (2010) *Astrophys. J.*, **709**, L83.
151 Rosen, A., Hughes, P.A., Duncan, G.C., and Hardee, P.E. (1999) *Astrophys. J.*, **516**, 729.
152 Scheck, L., Aloy, M.A., Martí, J.M., Gómez, J.L., and Müller, E. (2002) *Mon. Not. R. Astron. Soc.*, **331**, 615.
153 Krause, M. and Camenzind, M. (2003) *New Astron.*, **47**, 573.
154 Aloy, M.A., Janka, H.-Th., and Müller, E. (2005) *Astrophys. J.*, **436**, 273.
155 Aloy, M.A. and Rezzolla, L. (2006) *Astrophys. J.*, **640**, L115.
156 Aloy, M.A. and Mimica, P. (2008) *Astrophys. J.*, **681**, 84.
157 Mizuno, Y., Hardee, P., Hartmann, D.H., Nishikawa, K.-I., and Zhang, B. (2008) *Astrophys. J.*, **672**, 72.
158 Zenitani, S., Hesse, M., and Klimas, A. (2010) *Astrophys. J.*, **712**, 951.
159 Kataoka, J., Stawarz, L., Harris, D.E., Siemiginowska, A., Ostrowski, M., Swain, M.R., Hardcastle, M.J., Goodger, J.L., Iwasawa, K., and Edwards, P.G. (2008) *Astrophys. J.*, **685**, 839.
160 van Putten, M.H.P.M. (1996) *Astrophys. J.*, **467**, L57.
161 Koide, S., Nishikawa, K., and Mutel, R.L. (1996) *Astrophys. J.*, **463**, L71.
162 Koide, S. (1997) *Astrophys. J.*, **478**, 66.
163 Nishikawa, K., Koide, S., Sakai, J. *et al.* (1997) *Astrophys. J.*, **483**, L45.
164 Nishikawa, K., Koide, S., Sakai, J. *et al.* (1998) *Astrophys. J.*, **498**, 166.
165 Komissarov, S.S. (1999) *Mon. Not. R. Astron. Soc.*, **308**, 1069.
166 Leismann, T., Antón, L., Aloy, M.A., Müller, E., Martí, J.M., Miralles, J.A., and Ibáñez, J.M. (2005) *Astron. Astrophys.*, **436**, 503.
167 Mignone, A., Massaglia, S., and Bodo, G. (2005) *Space Sci. Rev.*, **121**, 21.
168 Keppens, R., Meliani, Z., van der Holst, B., and Casse, F. (2008) *Astron. Astrophys.*, **486**, 663.

169 Mignone, A., Rossi, P., Bodo, G., Ferrari, A., and Massaglia, S. (2010) *Mon. Not. R. Astron. Soc.*, **402**, 7.

170 Del Zanna, L., Bucciantini, N., and Londrillo, P. (2003) *Astron. Astrophys.*, **400**, 397.

171 Leismann, T., Aloy, M.A., and Müller, E. (2004) *Astrophys. Space Sci.*, **293**, 157.

172 Massaglia S., Bodo G., and Ferrari A. (1996) *Astron. Astrophys.*, **307**, 997.

173 Begelman, M.C. and Cioffi D. F. (1989) *Astrophys. J.*, **123**, 550.

174 Komissarov, S.S. and Falle, S.A.E.G. (1996) *Energy Transport in Radio Galaxies and Quasars* (eds P.E. Hardee, A.H. Bridle, and J.A. Zensus), ASP Conference Series, vol. 100, ASP, San Francisco, p. 165.

175 Komissarov, S.S. and Falle, S.A.E.G. (1998) *Mon. Not. R. Astron. Soc.*, **297**, 1087.

176 Falle, S.A.E.G. (1991) *Mon. Not. R. Astron. Soc.*, **250**, 581.

177 Cioffi, D.F. and Blondin, J.M. (1992) *Astrophys. J.*, **392**, 458.

178 Hooda J. S., Mangalam A. V., and Wiita P. J. (1994) *Astrophys. J.*, **423**, 116.

179 Martí, J.M., Müller, E., and Ibáñez, J.M. (1998) *Astrophysical Jets. Open Problems* (eds S. Massaglia and G. Bodo), Gordon and Breach Science Publishers, Amsterdam (MMI98), p. 149.

180 Owsianik, I., and Conway, J.E. (1998) *Astron. Astrophys.*, **337**, 69.

181 Taylor, G.B., Marr, J.M., Pearson, T.J., and Readhead, A.C. S. (2000) *Astrophys. J.*, **541**, 112.

182 Leismann T. (2004) Relativistic Magnetohydrodynamic Simulations of Extragalactic Jets, PhD Thesis, Technische Universität München.

183 Laing, R.A. and Bridle, A.H. (2002) *Mon. Not. R. Astron. Soc.*, **336**, 328.

184 Perucho, M. and Martí, J.M. (2007) *Mon. Not. R. Astron. Soc.*, **382**, 526.

185 Laing, R.A. and Bridle, A.H. (2002) *Mon. Not. R. Astron. Soc.*, **336**, 1161.

186 Croston, J.H., Kraft, R.P., and Hardcastle, M.J. (2007) *Astrophys. J.*, **660**, 191.

187 Kraft, R.P., Vázquez, S.E., Forman, W.R., Jones, C., Murray, S.S., Hardcastle, M.J., Worrall, D.M., and Churazov, E. (2003) *Astrophys. J.*, **592**, 129.

188 De Young, D.S. (1996) *Energy Transport in Radio Galaxies and Quasars* (eds P.E. Hardee, A.H. Bridle, and J.A. Zensus), ASP Conference Series, vol. 100, 19–23 September 1995, Tuscaloosa, Alabama, ASP, San Francisco, p. 261.

189 De Young, D.S. (2005) *X-Ray and Radio Connections* (eds L.O. Sjouwerman, K.K. Dyer), 3–6 February 2004, Santa Fe, New Mexico, Published electronically by NRAO. http://www.aoc.nrao.edu/events/xraydio (accessed 14 September 2011), presented 3–6 February 2004 in Santa Fe, New Mexico, USA, (E7.01).

190 Meliani, Z., Keppens, R., and Giacomazzo, B. (2008) *Astron. Astrophys.*, **491**, 321.

191 Fanti C., Fanti R., Dallacasa D., Schilizzi R.T., Spencer R.E., and Stanghellini C. (1995) *Astron. Astrophys.*, **302**, 317.

192 Perucho, M., and Martí, J.M. (2002) *Astrophys. J.*, **568**, 639.

193 Drake C.L., Bicknell G.V., McGregor P.J., and Dopita M.A. (2004) *Astron. J.*, **128**, 969.

194 Cavallo G. and Rees, M.J. (1978) *Mon. Not. R. Astron. Soc.*, **183**, 359.

195 Frail, D.A., Kulkarni, S.R., Nicastro, S.R., Feroci, M., and Taylor, G.B. (1997) *Nature*, **389**, 261.

196 Taylor, G.B., Frail, D.A., Berger, E., and Kulkarni, S.R. (2004) *Astrophys. J. Lett.*, **609**, L1.

197 Piran, T. (1999) *Phys. Rep.*, **314**, 575.

198 Woosley, S.E. (1993) *Astrophys. J.*, **405**, 273.

199 MacFadyen, A.I. and Woosley, S.E. (1999) *Astrophys. J.*, **524**, 262.

200 Paczynski, B. (1986) *Astrophys. J. Lett.*, **308**, L43.

201 Goodman, J. (1986) *Astrophys. J. Lett.*, **308**, L47.

202 Eichler, D., Livio, M., Piran, T., and Schramm, D.N. (1989) *Nature*, **340**, 126.

203 Mochkovitch, R., Hernanz, M., Isern, J., and Martin, X. (1993) *Nature*, **361**, 236.

204 Aloy, M.A., Müller, E., Ibáñez, J.M., Martí, J.M., and MacFadyen, A. (2000) *Astrophys. J.*, **531**, L119.

205 Zhang, W., Woosley, S.E., and MacFadyen, A.I. (2003) *Astrophys. J.*, **586**, 356.

206 Zhang, W., Woosley, S.E., and Heger, A. (2004) *Astrophys. J.*, **608**, 365.
207 Mizuta, A., Yamasaki, T., Nagataki, S., Mineshige, S. (2006) *Astrophys. J.*, **651**, 960.
208 Morsony, B.J., Lazzati, D., and Begelman, M.C. (2007) *Astrophys. J.*, **665**, 569.
209 Morsoni, B.J., Lazzati, D., and Begelman, M.C. (2010) *Astrophys. J.*, **723**, 267, eprint arXiv:1002.0361.
210 Tominaga, N., Maeda, K., Umeda, H., Nomoto, K., Tanaka, M., Iwamoto, N., Suzuki, T., Mazzali, P.A. (2007) *Astrophys. J.*, **657**, L77.
211 Barkov, M. and Komissarov, S.S. (2008) *Mon. Not. R. Astron. Soc.*, **385**, 28.
212 Barkov, M. and Komissarov, S.S. (2010) *Mon. Not. R. Astron. Soc.*, **401**, 1644.
213 Tchekhovskoy, A., McKinney, J.C., and Narayan, R. (2008) *Mon. Not. R. Astron. Soc.*, **388**, 551.
214 Mizuta, A. and Aloy, M.A. (2009) *Astrophys. J.*, **699**, 1261.
215 Nagataki, S. (2009) *Astrophys. J.*, **704**, 937.
216 Janka, H.-Th., Aloy, M.A., Mazzali, P.A., and Pian, E. (2006) *Astrophys. J.*, **645**, 1305.
217 McKinney, J.C. (2006) *Mon. Not. R. Astron. Soc.*, **368**, 1561.
218 Blandford, R.D. and Payne, D.G. (1982) *Mon. Not. R. Astron. Soc.*, **199**, 883.
219 Blandford, R.D. and Znajek, R.L. (1977) *Mon. Not. R. Astron. Soc.*, **179**, 433.
220 Aloy, M.A. (2001) in *Highlights of Spanish Astrophysics II* (eds J. Zamorano, J. Gorgas, and J. Gallego), Kluwer Academic, p. 33.
221 Gómez, E.A. and Hardee, P.E. (2004) *Gamma-Ray Bursts: 30 Years of Discovery* (eds E. Fenimore and M. Galassi), AIP Conference Proceedings 727, American Institute of Physics, p. 278, doi:10.1063/1.1810847.
222 Aloy, M.A. and Obergaulinger, M. (2007) *Rev. Mex. Astron. Astrophys.*, **30**, 96.
223 Ramirez-Ruiz, E., Celotti, A., and Rees, M.J. (2002) *Mon. Not. R. Astron. Soc.*, **337**, 1349.
224 Thöne, C.C. *et al.* (2011) eprint arXiv:1105.3015, accepted for publication in Nature..
225 Woosley, S.E. (2000) *Gamma-ray Bursts*, 5th Huntsville Symposium (eds R.M. Kippen, R.S. Mallozzi, and G.J. Fishman), AIP Conference Proceedings 526, American Institute of Physics, p. 555, doi:10.1063/1.1361599.
226 Lazzati, D. and Begelman, M.C. (2005) *Astrophys. J.*, **629**, 903.
227 Hughes, P.A. *et al.* (1989) *Astrophys. J.*, **391**, 74.
228 Gomez, J.L. *et al.* (1993) *Astron. Astrophys.*, **274**, 55.
229 Gomez, J.L. *et al.* (1994) *Astron. Astrophys.*, **292**, 33.
230 Komissarov, S.S. and Falle, S.A.E.G. (1997) *Mon. Not. R. Astron. Soc.*, **288**, 833.
231 Downes, T.P., Duffy, P., and Komissarov, S.S. (2002) *Mon. Not. R. Astron. Soc.*, **332**, 144.
232 Zhang, W. and MacFadyen, A. (2009) *Astrophys. J.*, **698**, 1261.
233 van Eerten, H.J. and Wijers, R.A.M. (2009) *Mon. Not. R. Astron. Soc.*, **394**, 2164.
234 van Eerten, H.J. *et al.* (2010) *Mon. Not. R. Astron. Soc.*, **403**, 300.
235 Mimica, P. *et al.* (2004) *Astron. Astrophys.*, **418**, 947.
236 Kardashev, N.S. (1962) *Sov. Astr.*, **6**, 317.
237 Webb, G.M. (1985) *Astrophys. J.*, **296**, 319.
238 Miralles, J.A., van Riper, K.A., and Lattimer, J.M. (1993) *Astrophys. J.*, **407**, 687.
239 Jones, T.W., Ryu, D., and Engel, A. (1999) *Astrophys. J.*, **512**, 105.
240 Micono, M. *et al.* (1999) *Astron. Astrophys.*, **349**, 323.
241 Tregillis, I.L., and Jones, T.W., and Ryu, D. (2001) *Astrophys. J.*, **557**, 475.
242 Casse, F. and Markowith, A. (2003) *Astron. Astrophys.*, **404**, 405.
243 Hededal, C. (2005) Gamma-Ray Bursts, Collisionless Shocks and Synthetic Spectra, PhD Thesis, Niels Bohr Institute.
244 Nishikawa, K.-I. *et al.* (2010) *Int. J. Mod. Phys. D*, **19**, 715.
245 Meszaros, P. and Rees, M.J. (1997) *Astrophys. J. Lett.*, **482**, L29.
246 Mimica, P. *et al.* (2005) *Astron. Astrophys.*, **441**, 103.

247 Kino, M., Mizuta, A., and Yamada, S. (2004) *Astrophys. J.*, **611**, 1021.

248 Mimica, P. *et al.* (2007) *Astron. Astrophys.*, **466**, 93.

249 Mimica, P. and Aloy, M.A. (2010) *Mon. Not. R. Astron. Soc.*, **401**, 525.

250 van Eerten, H.J. *et al.* (2009) *Mon. Not. R. Astron. Soc.*, **398**, L63.

251 Meliani, Z. *et al.* (2007) *Mon. Not. R. Astron. Soc.*, **376**, 1189.

252 Mimica, P. *et al.* (2009) *Astron. Astrophys.*, **494**, 879.

253 Mimica, P. *et al.* (2010) *Mon. Not. R. Astron. Soc.*, **407**, 2501.

254 Steele I.A. *et al.* (2009) *Nature*, **462**, 767.

255 Pearson, T.J. *et al.* (1981) *Nature*, **260**, 365.

256 Miralebl, I.F. and Rodriguez, L.F. (1994) *Nature*, **371**, 46.

257 Nishikawa, K.-I., Niemiec, J., Hardee, P.E., Medvedev, M., Sol, H., Mizuno, Y., Zhang, B., Pohl, M., Oka, M., and Hartmann, D.H. (2009) *Astrophys. J.*, **698**, L10.

258 Casse, F. and Markowith, A. (2005) *Astropart. Phys.*, **23**, 31.

259 Zenitani, S., Hesse, M., and Klimas, A. (2009) *Astrophys. J.*, **696**, 1385.

260 Komissarov, S.S. (2007) *Mon. Not. R. Astron. Soc.*, **382**, 995.

261 Palenzuela, C., Lehner, L., Reula, O., and Rezzolla, L. (2009) *Mon. Not. R. Astron. Soc.*, **394**, 1727.

11
Jet Structure, Collimation and Stability: Recent Results from Analytical Models and Simulations

Rony Keppens and Zakaria Meliani

This chapter will focus on fluid-dynamical aspects of the jet phenomenon, going from HD to MHD, classical to (general) relativistic. It will make links with, and infer interpretations for, observed jet phenomena. Phenomenological models for accretion disk-jet systems and plasma dynamical simulations have played crucial roles in studying AGN jet flows, as documented for example in the review by Ferrari [1]. These aspects have also been discussed in the previous chapter, along with novel insights already collected from them. In this chapter, we highlight selected recent results, and point out how advanced analytic models, covering general relativistic aspects, provide vital ingredients for computational modeling efforts.

In many such modeling efforts, one adopts a continuum description for jet flows, using a basic conservation law viewpoint. We start with formulating these laws as exploited in analytic jet models, and pay particular attention to what these imply for jet collimation scenarios. As analytic models make simplifying assumptions, most notably assuming stationary configurations, numerical modeling efforts are needed to gain insight in jet propagation and deceleration aspects. Modern high-resolution simulations allow to resolve the detailed growth and interaction of fluid instabilities in large-scale jet simulations. Their occurrence and evolution can be analyzed using instability criteria for magnetized jet flows from linear theory. Guided by observational and theoretical insights, modern jet studies encompass radially stratified jet flows, with "spine-sheath" or "two-component" morphologies. These models are discussed and compared in detail, with recent insights on how such two-component jet models, combined with their intrinsic liability to fluid instabilities, may explain the differing Faranoff–Riley (FR) I/II jet morphologies.

11.1
Exact Models for Collimated Jets

Exact, (semi)analytic models for astrophysical jet flows can provide quantitative insight in jet acceleration and collimation issues. For AGN jets which involve relativistic flows originating from near (supermassive) black hole surroundings, the

Relativistic Jets from Active Galactic Nuclei, First Edition. Edited by M. Böttcher, D.E. Harris, H. Krawczynski.

most advanced efforts have by now succeeded to connect general relativistic dynamics at the launching region with properties of the jet flows at distances from their source where special relativity suffices. Some exemplary efforts will be discussed next, and they all start from the governing equations for general relativistic magnetohydrodynamics. They express basic conservation laws in four-dimensional space-time, and in essence consist of Maxwell's equations with particle number as well as energy-momentum conservation. A brief excursion on the mathematics of four-dimensional, curved space-time (as opposed to the special relativistic concepts introduced in Chapter 2 which deal with flat space-time alone) is given first.

11.1.1
Concepts for Curved Space-Time

In four-dimensional space-time, we identify each point with a set of four coordinates, where $\underline{X}(ct, \boldsymbol{x}) \equiv \underline{X}(x_0, x_1, x_2, x_3)$ or $\underline{X}(x_\alpha)$ in short. Note how the greek index α now denotes a space-time index $\alpha = 0, 1, 2, 3$. A particle traveling through space-time follows what is referred to as its worldline, being the succession of coordinates $\underline{X}(x_\alpha)$ which it traverses. The concept of time is rather subtle in general relativity, and although we wrote $x_0 = ct$, one must realize that each particle worldline sees time progressing according to an idealized clock "carried" along. This is the proper time τ, and the rate of change in the space-time coordinates with respect to proper time is the particle four-velocity $\underline{U} = d\underline{X}/d\tau$. This is a four-vector quantity. Once a set of coordinates x_α are chosen to label space-time, vector quantities can be denoted by their four components in either of two dual basis vector sets $\underline{e}^\alpha = \underline{\nabla} x_\alpha$ (where $\underline{\nabla}$ denotes the gradient in four-space) and $\underline{e}_\beta \equiv \partial \underline{X}/\partial x_\beta$. Their dual nature implies that they indicate orthogonal directions in four-dimensional space-time, that is, $\underline{e}^\alpha \cdot \underline{e}_\beta = \delta^\beta_\alpha$, with δ^β_α the Kronecker delta (1 for $\alpha = \beta$, zero otherwise). In essence, these basis vectors define the metric through $\underline{e}_\alpha \cdot \underline{e}_\beta = g_{\alpha\beta}$ and $\underline{e}^\alpha \cdot \underline{e}^\beta = g^{\alpha\beta}$. This metric appears in the basic invariant quantity $ds^2 = d\underline{X} \cdot d\underline{X} = dX^\alpha \underline{e}_\alpha \cdot dX^\beta \underline{e}_\beta = g_{\alpha\beta} dX^\alpha dX^\beta$, which is the squared length of the infinitesimal displacement $d\underline{X}$. We here introduced the conventional Einstein summation, where repeated indices α and β infer that we sum up four occurrences through 0, 1, 2, 3. In writing the four-velocity as $\underline{U} = U^\alpha \underline{e}_\alpha$, we use contravariant components U^α, which can be converted to covariant components where $\underline{U} = U_\alpha \underline{e}^\alpha$ through $U_\alpha = g_{\alpha\beta} U^\beta$. An invariant quantity is found from $U^\alpha U_\alpha = -c^2$.

In general, the basis vectors $\underline{e}_\beta$ vary in direction and magnitude with respect to the coordinates, and by writing each derivative $\partial_\alpha \underline{e}_\beta$ as a linear combination of all basis vectors, we introduce the Christoffel symbols through $\partial_\alpha \underline{e}_\beta = \Gamma^\gamma_{\alpha\beta} \underline{e}_\gamma$. These Christoffel symbols are written in terms of the metric as $\Gamma^\gamma_{\alpha\beta} = \underline{e}^\gamma \cdot \partial_\alpha \underline{e}_\beta = 1/2 g^{\gamma\delta} \left(\partial_\alpha g_{\delta\beta} + \partial_\beta g_{\alpha\delta} - \partial_\delta g_{\alpha\beta}\right)$, with again summation implied over δ in the last expression. In flat space-time (special relativity), all these notions simplify drastically, since the squared line element becomes $d\underline{X} \cdot d\underline{X} = -c^2 dt^2 + dx_1^2 + dx_2^2 + dx_3^2$, making the metric $g_{\alpha\beta} = g^{\alpha\beta} = \mathrm{diag}(-1, +1, +1, +1)$ diagonal and of Minkowski type, with all Christoffel symbols vanishing exactly when x_1, x_2, x_3 denote orthogonal Cartesian coordinates.

In space-times, we encounter scalar (tensor rank 0), four-vector (tensor rank 1) and higher-order tensor quantities, with those of second rank indictable as $T^{\alpha\beta}\underline{e}_\alpha\underline{e}_\beta$. Due to the curvature of space-time, introducing derivatives in a manner which all observers can agree on, is not trivial. For a four-vector, we can write $\partial_\beta(U^\alpha\underline{e}_\alpha) = (\partial_\beta U^\alpha)\underline{e}_\alpha + U^\alpha\Gamma^\gamma_{\beta\alpha}\underline{e}_\gamma$, in turn giving the formula for a covariant derivative as $U^\alpha{}_{;\beta} = \partial_\beta U^\alpha + U^\mu\Gamma^\alpha_{\beta\mu}$. For a second rank tensor, the same reasoning reads as $T^{\alpha\beta}{}_{;\mu} = \partial_\mu T^{\alpha\beta} + T^{\gamma\beta}\Gamma^\alpha_{\mu\gamma} + T^{\alpha\gamma}\Gamma^\beta_{\mu\gamma}$. With these concepts, we can now state the governing equations for (general) relativistic magnetohydrodynamics.

11.1.2 General Relativistic Magnetohydrodynamics

Assuming that a jet description for a perfectly conducting, relativistic plasma flow is adequate, the conservation law equations can be written as follows. If ρ indicates the rest-mass density in the proper (comoving) frame of the jet matter, the vanishing divergence of the four-vector $\rho\underline{U}$ expresses particle conservation as

$$(\rho U^\alpha)_{;\alpha} = 0\,. \tag{11.1}$$

The second rank tensor for energy-momentum conservation has both fluid and electromagnetic field contributions. With proper thermodynamic quantities for pressure p and specific internal energy density ϵ, the fluid part reads as

$$T_{\mathrm{f}}^{\alpha\beta} = (\rho c^2 + \rho\epsilon + p)\frac{U^\alpha U^\beta}{c^2} + p g^{\alpha\beta}\,. \tag{11.2}$$

This form already assumes an isotropic pressure in the fluid frame, and governs an ideal fluid (neglecting viscosity and heat transfer) where an equation of state simply links $p(\epsilon,\rho)$.

The electromagnetic part of energy-momentum can, in the restrictive assumption of ideal MHD, be written by introducing the magnetic field four-vector $\underline{b} = b^\alpha\underline{e}_\alpha$ in the fluid rest frame. This is a vector orthogonal to the four-velocity, since $U^\alpha b_\alpha = 0$. For perfectly conducting media, the rest frame electric field four-vector vanishes, and it becomes possible to write the needed Maxwell equations as

$$(U^\alpha b^\beta - U^\beta b^\alpha)_{;\beta} = 0\,. \tag{11.3}$$

The scalar magnetic pressure is obtained from the invariant quantity $b^\alpha b_\alpha = 2p_{\mathrm{mag}}$ and the electromagnetic part to the stress-energy becomes

$$T_{\mathrm{em}}^{\alpha\beta} = 2p_{\mathrm{mag}}\frac{U^\alpha U^\beta}{c^2} + p_{\mathrm{mag}}g^{\alpha\beta} - b^\alpha b^\beta\,. \tag{11.4}$$

Conservation of energy-momentum follows from

$$\left(T_{\mathrm{f}}^{\alpha\beta} + T_{\mathrm{em}}^{\alpha\beta}\right)_{;\beta} = 0\,. \tag{11.5}$$

11.1.3
3 + 1 for Schwarzschild Black Hole Surroundings

The conservation laws from (11.1)–(11.5) and the electromagnetic evolution governed by (11.3) govern the evolution of the perfectly conducting plasma in a given metric $g^{\alpha\beta}$. In full general relativistic settings, one would need to solve the Einstein field equations, in which the stress-energy tensor appears as a source term for the Einstein tensor: a quantity completely determined by the space-time metric (and its derivatives with respect to the employed coordinates, such as those appearing in the Christoffel symbols). While meanwhile various numerical relativity groups have progressed to solve the full general relativistic MHD equations in dynamic space-times (see the review by Font [2], and in particular the `WHISKY-MHD` code by [3]), analytic work, as well as many contemporary simulation efforts, almost necessarily assume the test-fluid approximation. This boils down to fixing the metric to known form, for which the simplest one is that of a stationary, spherically symmetric solution for vacuum. This Schwarzschild metric applies for the (vacuum) surroundings of nonrotating, uncharged black holes, and the squared line element quantifying the metric is in Schwarzschild coordinates where $dX^\alpha = (cdt, dr, d\theta, d\varphi)$ given by

$$ds^2 = d\underline{X}\cdot d\underline{X} = -\alpha_l^2 c^2 (dt)^2 + \frac{1}{\alpha_l^2}(dr)^2 + r^2(d\theta)^2 + r^2 \sin^2\theta\,(d\varphi)^2 \,. \quad (11.6)$$

The metric is read off to be the diagonal $g_{\alpha\beta} = \mathrm{diag}\left(-\alpha_l^2, 1/\alpha_l^2, r^2, r^2\sin^2\theta\right)$. The g_{00} element involves the (square of the) lapse function $\alpha_l^2(r) = 1 - r_{\mathrm{H}}/r$ with event horizon $r_{\mathrm{H}} = 2GM/c^2$ for black hole mass M (with G the gravitational constant). This lapse function quantifies the variation between the proper time τ and the global Schwarzschild time coordinate t along the worldline of a "fiducial observer" (FIDO) whose four-velocity $\underline{U}_{\mathrm{FO}}$ is at all times orthogonal to a hypersurface of constant t in four-dimensional space-time. In formulae, these statements write as $\underline{U}_{\mathrm{FO}} = -\alpha_l c^2 \nabla t = -\alpha_l c\underline{e}^0 = (c/\alpha_l)\underline{e}_0$, and we note that $g^{00} = -1/\alpha_l^2 = \underline{e}^0\cdot\underline{e}^0 = c\nabla t\cdot c\nabla t$. All nonvanishing Christoffel symbols can be verified to be

$$\begin{aligned}
&\Gamma^t_{rt} = \Gamma^t_{tr} = \frac{r_{\mathrm{H}}}{2r(r-r_{\mathrm{H}})}\,, \quad \Gamma^r_{tt} = \frac{r_{\mathrm{H}}(r-r_{\mathrm{H}})}{2r^3}\,, \quad \Gamma^r_{rr} = \frac{-r_{\mathrm{H}}}{2r(r-r_{\mathrm{H}})}\,,\\
&\Gamma^r_{\theta\theta} = r_{\mathrm{H}} - r\,, \quad \Gamma^r_{\varphi\varphi} = (r_{\mathrm{H}} - r)\sin^2\theta\,,\\
&\Gamma^\theta_{\varphi\varphi} = -\sin\theta\cos\theta\,, \quad \Gamma^\varphi_{\theta\varphi} = \Gamma^\varphi_{\varphi\theta} = \frac{\cos\theta}{\sin\theta}\,,\\
&\Gamma^\theta_{r\theta} = \Gamma^\theta_{\theta r} = \Gamma^\varphi_{r\varphi} = \Gamma^\varphi_{\varphi r} = \frac{1}{r}\,.
\end{aligned} \quad (11.7)$$

It has become customary to exploit the so-called 3 + 1 formulation [4, 5] in both analytic and numerical approaches, which in this case means to exploit the FIDO viewpoint, so that perhaps more familiar three-vector quantities are always those found projected in space-like hypersurfaces orthogonal to the global time parameter t. These FIDO for Schwarzschild surroundings can be verified to experience no expansion of their worldline, meaning $U^{\alpha}_{\text{FO};\alpha} = 0$, but they do experience local gravitational acceleration since $U^{\alpha}_{\text{FO};\beta} U^{\beta}_{\text{FO}} = (0, c^2 r_{\text{H}}/2r^2, 0, 0) \equiv (0, -\boldsymbol{g})$. This gravitational acceleration is $\boldsymbol{g} = -c^2 \nabla \ln \alpha_l = -(GM/r^2)\underline{e}_1 = -(GM/r^2)\hat{e}_r/\alpha_l$ where $\hat{e}_r = \nabla r/\|\nabla r\| = \underline{e}^1/\sqrt{g^{11}} = \alpha_l \underline{e}_1$ denotes the familiar unit radial three-vector.

In practice, every four-vector quantity writes as

$$\underline{A} = \frac{[\underline{e}^0 \cdot \underline{A}]\underline{e}^0}{(\underline{e}^0 \cdot \underline{e}^0)} + \boldsymbol{a} , \tag{11.8}$$

where the three-vectors can be written in standard notation as $\boldsymbol{a} = A_r \hat{e}_r + A_\theta \hat{e}_\theta + A_\varphi \hat{e}_\varphi = A_r \alpha_l \underline{e}_1 + (A_\theta/r)\underline{e}_2 + (A_\varphi/(r \sin\theta))\underline{e}_3$. For four-velocity, this becomes $\underline{U} = U_0 \underline{e}^0 + \boldsymbol{u}$ which yields the Lorentz factor via $\Gamma = \sqrt{1 + \boldsymbol{u} \cdot \boldsymbol{u}/c^2} = 1/\sqrt{1 - \boldsymbol{v} \cdot \boldsymbol{v}/c^2}$ when the spatial part of the four-velocity $\boldsymbol{u} = \Gamma \boldsymbol{v}$ uses three-velocity $\boldsymbol{v}$. From the normalization $U^\alpha U_\alpha = -c^2$, one finds $\underline{U} = (c\Gamma/\alpha_l)\underline{e}_0 + \Gamma \boldsymbol{v}$. For the magnetic field four-vector, it becomes possible to write

$$\underline{b} = b^0 \underline{e}_0 + \boldsymbol{b} = \frac{\Gamma(\boldsymbol{v} \cdot \boldsymbol{B})}{\alpha_l c} \underline{e}_0 + \frac{\boldsymbol{B}}{\Gamma} + \Gamma(\boldsymbol{v} \cdot \boldsymbol{B}) \frac{\boldsymbol{v}}{c^2} . \tag{11.9}$$

This four-vector denotes the magnetic field measured in the local rest frame: where the local three-velocity vanishes, $\underline{b}$ reduces to $(0, \boldsymbol{B})$ in its temporal-spatial split. Once this realization in a split spatial-temporal $(3 + 1)$ form is achieved, one can work with the global time t and the three-vector quantities $\boldsymbol{v}$, $\boldsymbol{B}$ in relations that resemble Newtonian formulations. In particular, particle conservation from (11.1) can for example be rewritten in the form

$$\frac{\partial}{\partial t}(\Gamma \rho) + \nabla \cdot (\alpha_l \rho \Gamma \boldsymbol{v}) = 0 . \tag{11.10}$$

Note that the special relativistic case is incorporated by $\alpha_l \to 1$, and the Newtonian limit simply recognizes that $\Gamma \to 1$. Note that $\nabla = \underline{e}^i \partial_i$ denotes the gradient operator restricted to the spatial dimensions only. The temporal ($\alpha = 0$) component of Maxwell's relation (11.3) becomes $\nabla \cdot \boldsymbol{B} = 0$.

Mobarry and Lovelace [6] performed this 3 + 1 reformulation for the ideal MHD equations in a Schwarzschild metric, and analyzed the further restriction to steady MHD flows. In fact, for stationary conditions $\partial/\partial t = 0$, one can write force balance in terms of three-vector quantities as

$$\begin{aligned} \Gamma^2 \rho h \left(\frac{\boldsymbol{v}}{c} \cdot \nabla\right) \frac{\boldsymbol{v}}{c} &= -\rho h \Gamma^2 \nabla \ln \alpha_l + \Gamma^2 \rho h \left(\frac{\boldsymbol{v}}{c} \cdot \nabla \ln \alpha_l\right) \frac{\boldsymbol{v}}{c} \\ &\quad - \nabla p + \rho_c \boldsymbol{E} + \frac{1}{c} \boldsymbol{j} \times \boldsymbol{B} - \frac{1}{c} (\boldsymbol{E} \cdot \boldsymbol{j}) \frac{\boldsymbol{v}}{c} . \end{aligned} \tag{11.11}$$

In this form, we introduced specific enthalpy h defined as $\rho h = \rho c^2 + \rho \epsilon + p$, and the electric charge density ρ_c, electric field three-vector $\boldsymbol{E}$ as well as the charge current density $\boldsymbol{j}$ are found from $\boldsymbol{B}$ and $\boldsymbol{v}$ through the (stationary) Maxwell equations (and the infinite conductivity assumption), which in 3 + 1 for nonrotating black hole environments give

$$\nabla \cdot \boldsymbol{E} = 4\pi\rho_c \,, \quad \nabla \times (\alpha_l \boldsymbol{E}) = 0 \,, \quad \boldsymbol{E} = -\frac{\boldsymbol{v}}{c} \times \boldsymbol{B} \,,$$
$$\frac{4\pi\alpha_l}{c}\boldsymbol{j} = \nabla \times (\alpha_l \boldsymbol{B}) \,. \tag{11.12}$$

Equation (11.11) is fundamental to a quantitative analysis of possible MHD equilibrium configurations applicable to AGN jets, and can provide detailed insight into how conditions prevailing in the jet launch region impact the downstream jet structure. Mobarry and Lovelace [6] reworked this governing Euler equation under the extra assumption of axisymmetry into a generalized Grad–Shafranov type equation. The latter reexpresses this basic force balance into a single, second-order partial differential equation for the poloidal flux function Ψ, as axisymmetry $\partial/\partial\varphi = 0$ allows to decompose the magnetic field vector as

$$\boldsymbol{B} = \nabla\Psi \times \frac{\hat{e}_\varphi}{r\sin\theta} + \frac{4\pi I}{\alpha_l c}\frac{\hat{e}_\varphi}{r\sin\theta} \,. \tag{11.13}$$

The toroidal field is quantified by the axial current I flowing up along the symmetry axis. The precise dependence of $\Psi(r,\theta)$ (and from it $I(r,\theta)$) obtained from either analytic or numerical solution of the generalized Grad–Shafranov equation then completely determines an axisymmetric, stationary MHD flow pattern in Schwarzschild geometry. This realizes the immediate generalization of the standard (elliptic) Grad–Shafranov equation used in a laboratory plasma context, where a tokamak plasma equilibrium directly follows from $\nabla p = (1/(4\pi))(\nabla \times \boldsymbol{B}) \times \boldsymbol{B}$. Several limits of direct interest to astrophysical applications can be studied, such as the "force-free" limit where the electromagnetic energy dominates the matter kinetic energy, only leaving $\rho_c \boldsymbol{E} + (1/c)\boldsymbol{j} \times \boldsymbol{B} = 0$. This latter limit, where a black hole magnetosphere is reduced to solving an electrodynamic problem, is the one pioneered by Blandford and Znajek [7], in the more general case of a rotating black hole environment (Kerr metric). In the case of full MHD flow solutions, the governing second-order PDE can be locally elliptic or hyperbolic, depending on the local flow speed. Mathematically, the character changes at locations where the discriminant (composed from the coefficients of the highest, second-order derivatives) vanishes. To obtain well-behaved solutions, care must be taken at (a priori unknown) locations where the flow equals (poloidal) Alfvén, slow and fast magnetosonic speeds, just as in the classical limit of stationary nonrelativistic axisymmetric transmagnetosonic flows [8, 9]. Mostly as a result of these complications, analytic progress to gain insight from the governing Euler equation (11.11) for full MHD flow solutions has only recently been achieved, by making self-similarity assumptions to reduce the partial differential equations to ordinary differential equations, that are more easily handled. We discuss those approaches next.

11.1.4
Self-Similar Models: Classical to General Relativistic MHD

Steady-state transmagnetosonic flows described by the ideal MHD equations have been studied extensively in nonrelativistic classical settings (see, e.g., the review by Tsinganos [10]). To obtain exact analytic solutions, which can serve to benchmark numerical approaches, one realizes that the stationary axisymmetric limit leads to a natural split in poloidal versus toroidal vector components, as in (11.13), and allows to deduce various quantities that are invariant along a poloidal field line. It is straightforward to show that particle conservation from (11.10) reduces to $\nabla \cdot (\alpha_l \Gamma \rho \boldsymbol{v}_\mathrm{p}) = 0$, and that relations (11.12) necessitate that poloidal velocity $\boldsymbol{v}_\mathrm{p}$ is parallel to $\boldsymbol{B}_\mathrm{p}$, since the azimuthal electric field component must vanish $E_\varphi = 0$. In fact, the latter leads to isorotation expressed as $\boldsymbol{v} = k\boldsymbol{B} + r \sin\theta\, \Omega(\Psi)\hat{e}_\varphi/\alpha_l$. Since $\Omega(\Psi)$ is constant on a field line, it quantifies the angular speed from the field line viewpoint. More detailed analysis of all governing equations can be done under the additional assumption that the two-dimensional profiles of interest (in essence encoded in the flux function $\Psi(r, \theta)$) can be reexpressed in functions of a single variable only. In classical as well as in relativistic MHD, this additional assumption then distinguishes between either radial or meridional self-similarity. The self-similarity assumption implies that in each such case, knowing the variation of all physical quantities along a single poloidal field line, allows to reconstruct the full two-dimensional solution. Mathematically, the reason for adopting such assumptions is to transform the governing PDEs into separable equations, ending up with ordinary differential equations, which are more easily solvable by standard techniques.

For radial self-similarity, a rescaling based on conical (constant θ) surfaces is involved, while meridional self-similarity rescales through spheres (constant r). The prototype solutions are the Blandford and Payne [11] cold ($p = 0$) magnetocentrifugally accelerated wind in the radial category, while meridional self-similarity generalizes the hydro, transonic Parker wind [12] to axisymmetric transmagnetosonic MHD winds. Since the radial self-similarity class has divergent behavior on axis, it cannot describe (self-)collimated jet-like outflows that remain near the axis at far distances. The meridional self-similar models can give insight into how the asymptotic jet-like flows derive their asymptotic properties from the acceleration zones self-consistently. Already in nonrelativistic settings [13], these models can describe both cylindrically collimated flows or conically expanding flows far away, with transitions from one to another governed by energetic arguments prevailing in the launch region (at the base of the field/flow lines).

As a representative general relativistic MHD model for steady AGN jet flow, we revisit here the findings of Meliani *et al.* [14, 15], where meridional self-similarity was used to obtain semianalytical solutions for jets launched in the hot corona of a Schwarzschild black hole. To reduce the problem to governing ODEs with separable variations, some extra (on top of stationary axisymmetric flow) assumptions were introduced. The authors assumed that (i) the poloidal Alfvén number M found from $M^2 = 4\pi\alpha_l^2 \rho h \Gamma^2 v_\mathrm{p}^2 / B_\mathrm{p}^2 c^2$ varies with spherical radius only, and (ii) that the rotational speeds remain subrelativistic. This in turn allows to use approximate expressions for electric and magnetic contributions, for example, exploiting expansions which are consistent with describing only those streamlines that stay within the light "cylinder". This location is defined as where the rotation speed, as seen from a field line, equals c, hence where $r \sin\theta\, \Omega / \alpha_l = c$. In combination, these assumptions are adequate for describing the spine of observed jets, and the model applies to thermally driven jets only. The procedure to obtain full solutions ensures regularity at the (assumed spherical) Alfvén point where poloidal speeds go from sub- to super-Alfvénic, and then integrates numerically the governing (decoupled) ODEs. The relative ease with which solutions can be obtained allows for parametric analysis, and one can quantitatively analyze which forces aid to collimate the jet, how the flow accelerates from subslow to superfast magnetosonic speeds, and which launch region parameters play the most important role.

Figure 11.1 presents the streamline/field line morphology of two physically distinct solutions, namely an inefficient magnetic rotator (IMR) in Figure 11.1a, versus an efficient magnetic rotator (EMR) in Figure 11.1b. This distinction relates to a global constant for the solution, which quantifies whether off-polar streamlines have a deficit or excess magnetic energy compared to the polar one. An IMR must derive its flow collimation from thermal means (external medium pressure), while an EMR can self-collimate cylindrically with dominant magnetic pinching. The solutions shown in Figure 11.1 both go to cylindrical flows for the spine region asymptotically, but the IMR does so only by realizing a (mainly gas) pressure gradient that counteracts the outward centrifugal force, while the EMR has magnetic pinching playing that role. The different linestyles used merely serve to emphasize that by construction of these solutions, near-axis behavior is represented most accurately. The oscillatory expansion–recollimation behavior present for the IMR solution is a delicate interplay between magnetic, pressure and centrifugal forces. Both of these solutions have rather negligible Poynting flux. The presence of a hot corona surrounding the black hole is in these models assumed, and it causes the thermal driving. Parametric surveys of the solutions allowed to conclude that relativistic jets that resemble EMRs are less collimated than similar nonrelativistic models, as the electric force $\rho_c \boldsymbol{E}$ acts to decollimate and turns out on the same order as magnetic pinching and centrifugal force. In contrast, for IMRs, conclusions based on classical MHD jet flows also hold up for general relativistic cases.

These spine jet models only realize asymptotic Lorentz factors on the order of a few (the EMR up to $\Gamma \approx 2.8$, while the IMR up to $\Gamma \approx 2.4$). A correlation exists between (i) the asymptotic flow behavior and maximal flow speed obtained and with (ii) the departure of the field line rotation profile at the jet base from a Keplerian reference. In fact, this rotation is invariably sub-Keplerian in these solutions, but the more sub-Keplerian, slowly rotating cases end up in the IMR category, while less sub-Keplerian profiles relate to EMR solutions. This realization, combined with other properties distinguishing EMR from IMR jets, allowed to propose [15] how Fanaroff–Riley I (FR I) versus II (FR II) radio-loud galaxies with differing jet prop-

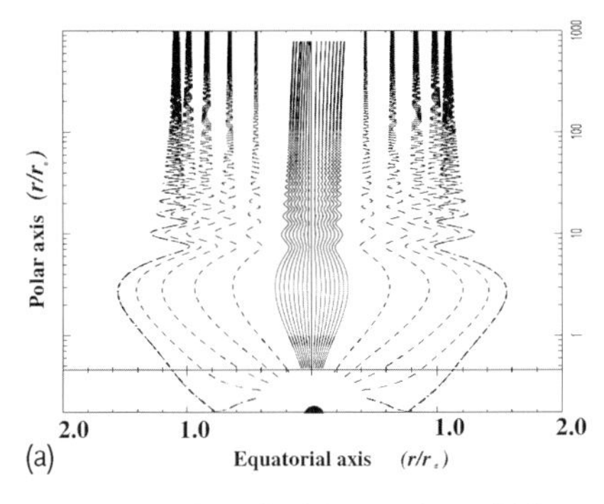

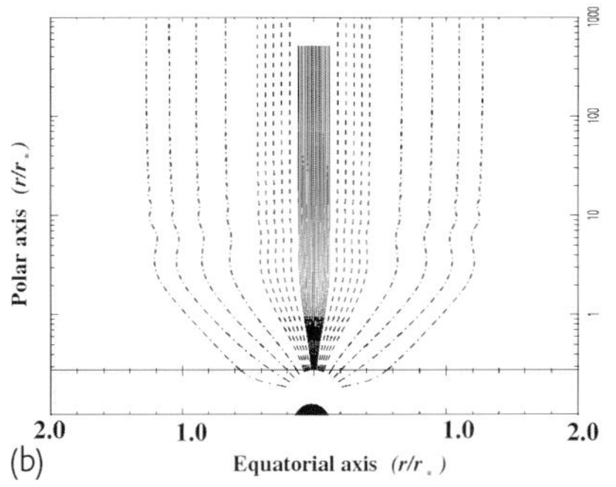

Figure 11.1 The poloidal streamlines/field lines for two representative analytic solutions for general relativistic MHD jet flows, from Meliani *et al.* [14]. (a) The case of an inefficient magnetic rotator, which asymptotically collimates to a cylindrical jet by a confining pressure gradient, while the (b) efficient magnetic rotator does so by magnetic pinching. Both solutions accelerate from subslow flow conditions mainly through thermal driving in the base region, representing a hot corona in the Schwarzschild metric. Distances are in units of the Alfvén radius r_*, denoting the distance along the polar axis where the speed equals the Alfvén speed. Reproduced from Meliani *et al.* [14] with permission from Astronomy and Astrophysics.

erties would relate to these spine jet models. FR I jets with low Lorentz factor flows share features with IMR solutions, such as indications for recollimation, and having the jet collimation likely aided by external pressure means. FR II jets (which are known to have higher speed flows than those found in these models) do magnetically self-collimate like EMR solutions. Since EMR solutions correlate with only slight sub-Keplerian flow in the launch region, these models suggest such launch flow conditions to prevail for FR II jets.

11.1.5
Models for Jets from Rotating Black Holes

In the more general case of a rotating black hole environment, the test fluid approximation solves the GRMHD equations in the Kerr metric (a stationary, axisymmetric solution of the Einstein equations for vacuum). The $3 + 1$ formulation for describing space-time around rotating black holes involves in addition to the black hole mass M, the parameter $a \equiv J/Mc$ which has a length dimension and quantifies the black hole angular momentum J. One frequently uses Boyer–Lindquist coordinates, where the metric is diagonal in the purely spatial coordinates (r, θ, φ). In space-time, off-diagonal elements exist due to the black hole rotation, making $g_{03} = g_{30} \neq 0$, hence the Boyer–Lindquist expression for the squared line element is of the form

$$ds^2 = g_{00}c^2(dt)^2 + 2g_{03}c\,dt\,d\varphi + g_{11}(dr)^2 + g_{22}(d\theta)^2 + g_{33}(d\varphi)^2 \,. \quad (11.14)$$

One can again introduce fiducial observers, in this case called "Zero Angular Momentum Observers" or ZAMOs, which allow to define three-vector quantities in local flat Minkowski sense. While for a Schwarzschild black hole, fiducial observers are at rest in absolute space, ZAMOs for Kerr move at angular velocity ω, where $\omega(r,\theta) = -g_{03}c/g_{33}$. The difference between global time t and proper time τ for ZAMOs is again quantified by a lapse function α_l, whereby $\alpha_l^2 = -g_{00} + g_{33}\omega^2/c^2$. This lapse function generalizes the one given for Schwarzschild, and physically arises from the gravitational redshift of fiducial observer clocks, so they still experience gravitational acceleration $\boldsymbol{g} = -c^2\nabla \ln \alpha_l$. The hole event horizon where $g_{11} \to \infty$ corresponds to $r_+ = GM/c^2 + \sqrt{(G^2M^2)/c^4 - a^2}$. As $0 \leq a \leq GM/c^2$, for $a = 0$ one retrieves the Schwarzschild radius value, hence $r_+(a = 0) = r_{\mathrm{H}}$. The surface where $g_{00} = 0$ given by $r_{\mathrm{E}} = GM/c^2 + \sqrt{(G^2M^2)/c^4 - a^2\cos^2\theta_{\mathrm{E}}}$ defines the border of the ergosphere region between r_+ and this surface. There, corotation is enforced (frame-dragging effect), and the black hole rotation rate relates to $\omega(r_{\mathrm{E}}(\theta_{\mathrm{E}} = 0°) = r_+, \theta_{\mathrm{E}} = 0) = ac^3/2GMr_+ \equiv \Omega_{\mathrm{BH}}$.

The analysis of full GRMHD flows about Kerr black holes is nowadays mostly performed by direct numerical simulations [16, 17]. Analytic work has been restricted to the case of a force-free magnetosphere, hence retaining only $\rho_c\boldsymbol{E} + (1/c)\boldsymbol{j}\times\boldsymbol{B} = 0$ from the full force balance, and invoking stationarity and axisymmetry. This is adequate for Poynting flux-dominated scenarios, as it ignores all dynamical effects of the flow. The pioneering solution in this regime is the Blandford–Znajek [7] analysis, which demonstrated that rotational energy of the black hole can be tapped to power a jet. A noteworthy extension was performed by Fendt [18], where the governing partial differential equation for the flux function Ψ was solved numerically. The force-free limit of the generalized Grad–Shafranov equation expressing cross-field force balance is again a second-order PDE for Ψ, where the coefficient in the highest derivative now introduces two light surfaces where the equation becomes singular (and reduces to a PDE of first order). These should be distinguished from the (asymptotic) light cylinder $r\sin\theta = R_{\mathrm{L}} = c/\Omega = 1$, which gives the location where the angular velocity of field lines Ω equals c. Fendt [18] assumed that the rotational velocity of all field lines is identical, and set equal to a fraction of the black hole spin Ω_{BH}. Under this assumption, the location of the two singular light surfaces for the governing PDE is known a priori, and it became possible to obtain smooth solutions, passing through both singular surfaces, by suitable iteration on the current distribution. Using finite element techniques (as frequently employed in computations for the standard Grad–Shafranov equilibrium in tokamak geometries), global jet solutions could be parametrically explored. Figure 11.2 gives two representative solutions of force-free Kerr magnetospheres obtained with this technique, where all field lines rotate at $0.4\Omega_{\mathrm{BH}}$, and the parameter $a = 0.8R_{\mathrm{L}}$. The jets obtained behave as cylindrical jets, with asymptotic widths prescribed to $3R_{\mathrm{L}}$. This shape is assumed to be maintained by an external pressure from surrounding material. Indeed, these jets are not self-collimated, but within the domain shown, they realize full internal force-balance. These solutions clearly go beyond self-similarity assumptions, as they incorporate field topologies allowing both in-fall to the hole, as well as outflow to the jet.

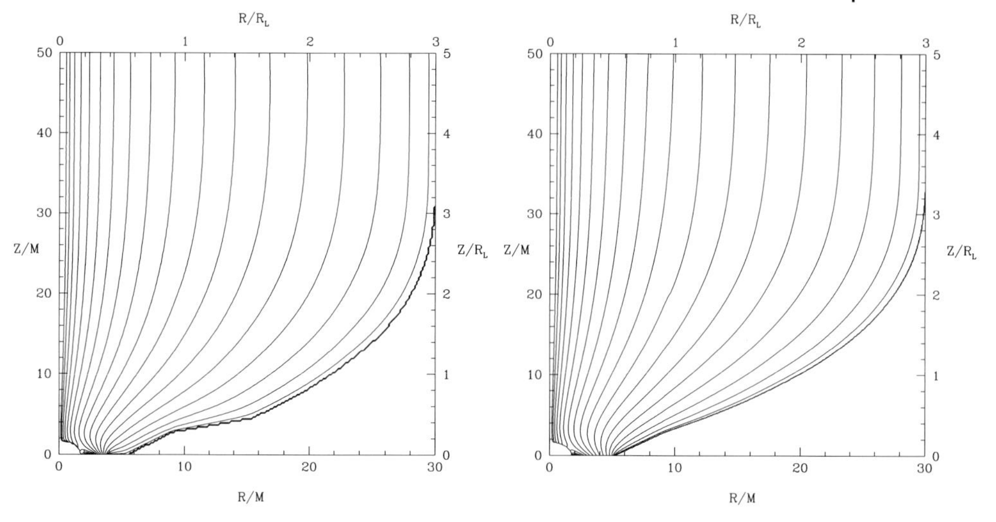

Figure 11.2 (a, b) The poloidal field lines for two representative solutions of a force-free magnetosphere in a Kerr metric, from Fendt [18]. The two cases differ in their internal current distribution realizing internal force balance, and lead to cylindrical jets (purely Poynting flux) of fixed cylindrical asymptotic shape. Reproduced from Fendt [18] with permission from Astronomy and Astrophysics.

11.2
Numerical Findings on Propagation, Deceleration, Collimation

The AGN jet models discussed in the previous section provided detailed insight into how the near black hole conditions conspire to realize an asymptotically collimated, steady configuration. The concrete examples we discussed for a Schwarzschild and Kerr metric also serve to demonstrate the shortcomings of such models. The MHD model from Meliani *et al.* [14] realizes a smooth acceleration of plasma flow from subslow speeds to superfast flows that acquire modest Lorentz factors far downstream self-consistently, with the potential to self-collimate. However, it assumes a hot corona near the black hole/inner accretion disk, and gives by construction accurate solutions for the spine of the jet only. All force-free models, for which the one by Fendt [18] represents a fully multidimensional, self-consistent example, ignore all plasma flow altogether, and hence need to infer Lorentz factors from charged test-particle considerations since they only include Poynting flux. How this Poynting flux is converted to kinetic/internal energy of high Lorentz factor flows is less understood, and only full general relativistic MHD simulations, going beyond pure ideal MHD to invoke Ohmic dissipation processes, can ultimately answer these issues. On the other hand, much has been done to understand how already accelerated, collimated flows (with kinetic or Poynting flux-dominant contributions) propagate, maintain their collimating properties and decelerate by interaction with surrounding interstellar medium (ISM). These kinds of studies are mentioned next, for which special relativistic treatments are frequently adopted.

11.2.1
Entrainment and Deceleration

A representative study targeting the physics of relativistic jet propagation and deceleration is the one by Rossi *et al.* [19]. Their motivation comes from the observational fact that FR I type jets (associated radio sources with luminosity $L_{178\,\mathrm{Mhz}} \leq 2 \times 10^{25}\ \mathrm{W\,Hz^{-1}\,sr^{-1}}$, according to the original Fanaroff–Riley [20] categorization) can be decelerated from typical bulk Lorentz factor 3–10 flows at parsec scale, to subrelativistic speeds at kiloparsec distances from the source. This calls for multidimensional studies of relativistic flows, which can quantify how jet-ISM interaction can lead to rapid energy transfer.

Rossi *et al.* [19] performed three-dimensional, relativistic hydro studies of jet propagation in idealized settings, where the jet flow is assumed to be uniform, cylindrical, and at prescribed inlet beam Lorentz factor of 10. This kinetic energy-dominated flow then enters a uniform, static ISM under pressure-matched conditions. The main parameter determining the overall morphology of the jet as it propagates through the ISM is then its density contrast between jet beam and ambient medium $\eta = \rho_{\mathrm{ISM}}/\rho_{\mathrm{beam}}$, and in their study, the authors assumed underdense jets with $\eta = 10^2$–10^4. The classical Mach number $M = v_{\mathrm{beam}}/c_{\mathrm{s}}$ of the beam was varied from 3 to 30, restricting the models to supersonic flows, and the jet dynamics was followed up to a maximal distance of 200 jet radii. Due to the interaction with the ambient medium, mitigated by the establishment of a shear flow layer, leading to Kelvin–Helmholtz instabilities which favor mixing, a kind of "spine-layer" structure developed self-consistently further downstream. In fact, the inner core (spine) of the jet can maintain highly relativistic speeds especially in the more underdense jets injected, while surrounding layers represent more slowly moving material. Specifically, at distances of about 20 or more jet radii from the injection point, the temporal evolution of the mass fraction in the jet versus four-velocity $\Gamma v/c$ clearly shows a double-peaked structure developing. Corresponding to a central spine, one such peak is near-axial matter continuing to travel at $\Gamma v/c \approx 5$ far downstream, while a growing fraction attains lower velocities that are also found throughout the expanding cocoon of the jet matter. This latter cocoon bounds the external ISM material shocked by the jet injection, while the instabilities responsible for mixing occur at the highly dynamic contact interface between jet and ISM matter and in the region of backflows surrounding the jet beam. Figure 11.3 shows a 3D rendering of the jet flow for a highly underdense $\eta = 10^4$ reference case, visualized here using a passive tracer allowing to distinguish jet beam from external ISM material. The mixing is clearly operating from about 20 jet radii in this case, and a synthetic observation of this jet at inclination 60° to the line-of-sight (LOS) is given as well. The latter represents a line-integrated synchrotron emissivity measure, which from these pure hydrodynamic studies is taken to be $\rho/\left[\Gamma(1 - v\cos\theta/c)\right]^{2.5}$, with θ the inclination between jet axis and LOS. For this high inclination angle, the dominant emission then stems from the slow surrounding material, and some evidence for limb brightening can be argued from these maps, since the relativistic core (spine) is deboosted in this view.

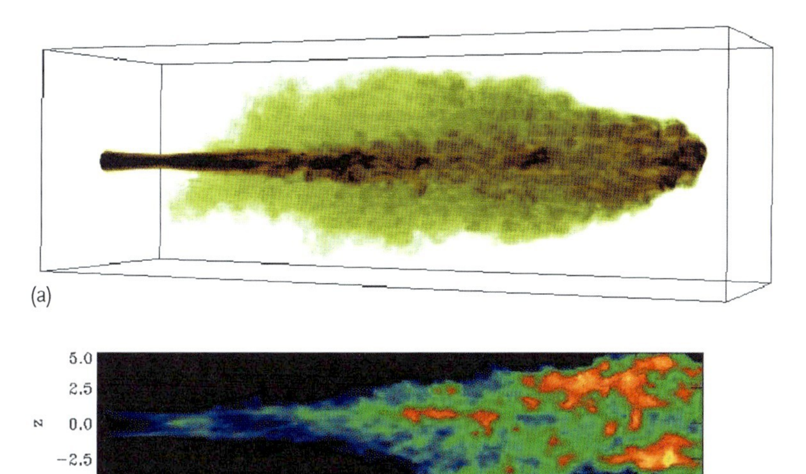

Figure 11.3 (a) A relativistic hydro jet study [19] demonstrated how underdense jets mix in their uniform ISM surroundings (colors relate to material opacity when rendering), and (b) how a synthetic observation at a fixed 60° angle between jet axis and line-of-sight would appear. Mixed material then shows up clearly in green to red color, for increased emission, with indications for a (deboosted) relativistic spine. Reproduced from Rossi *et al.* [19] with permission from Astronomy and Astrophysics.

The role of entrainment on jet deceleration, in relation to the FR I/II dichotomy, returns in many simulations done in the last decade, and a more complete overview is given in Section 10.3.2.

11.2.2
Fanaroff–Riley I/II and HYMORS: ISM Influences

While the previous study investigated jet deceleration by interaction with uniform ISM surroundings, a related study by Meliani *et al.* [21] augmented our insights into how ISM density variations influence relativistic jet flows. The existence of Hybrid Morphology Radio Sources (HYMORS [22]) inspired the authors to perform a parametric survey of axisymmetric hydro jets that start with either FR I or FR II energy contents. The HYMORS form a peculiar subclass of radio sources where an FR I type lobe exists on one side of the nucleus, with an FR II one at the other side. Since both jets are launched under presumably identical launch conditions (in particular, leading to powerful FR II type jets), their simultaneous presence strongly hints at ISM influences leading to rapid deceleration. In Meliani *et al.* [21], both cylindrical and conical jets were followed up to 400 jet beam radii into ISM surroundings that have either uniform or decreasing density profiles, in which additionally a contact discontinuity is encountered along the jet path. As in the Rossi *et al.* [19] study, an equation of state which accounts for effectively locally varying polytropic index (directly impacting the achievable compression ratio at shocks) was incorporated.

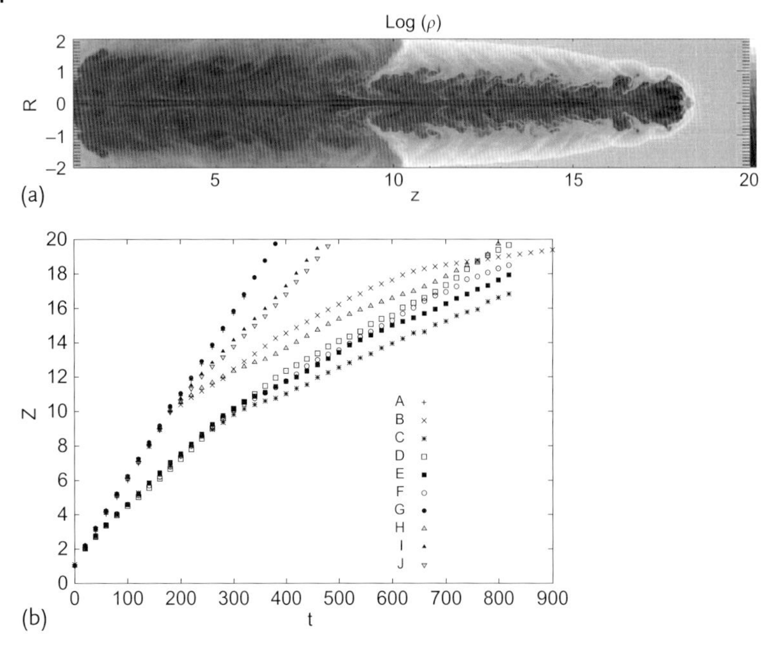

Figure 11.4 (a) The proper density in a logarithmic plot for a relativistic hydro jet traversing varying ISM conditions. (b) For all ten relativistic hydro models from Meliani *et al.* [21], the deceleration as quantified by jet head propagation as function of time. Distances R and Z are expressed in parsec, while time t is expressed in jet beam light crossing time, with assumed beam radius 0.05 parsec. Reproduced from Meliani *et al.* [21] with permission from Astronomy and Astrophysics.

Ten model runs were performed, and a plot collecting their main propagation characteristics is shown in Figure 11.4. The figure gives the position of the jet head as a function of time, and one notes how the models separate into six FR II type jets (with integrated kinetic energy flux over the jet beam up to 10^{46} erg/s) as well as four FR I jets (corresponding to 10^{43} erg/s). They all encounter a contact discontinuity, after times indicated as $t \approx 200$, and it is seen how sudden decelerations can result from this interaction. The precise details on overall jet deceleration are again significantly determined by jet/ISM density ratios, with rapid deceleration inevitably resulting from sudden density increases in the ISM encountered. It is worthwhile noting that the survey included transitions from overdense to underdense jet conditions, and that reflection-transmission effects were quantified at density contrasts up to a factor of 1000. This kind of locally discontinuous ISM stratification was found to be particularly dramatic, allowing to even realize FR II to FR I type jet changeover. A large part of the jet-directed energy is then reflected back into the first medium, while the underdense density situation in the second medium leads to heavy mixing and deceleration by entrainment. Related insights on the FR I/II dichotomy, also obtained through numerical modeling, have been discussed in detail in Section 10.3.2.2.

11.2.3
Jet Composition and EOS

The simulations from Meliani *et al.* [21] adopted pressure matched jet conditions (like the 3D study from Rossi *et al.* [19]), and varied the density of the ISM medium in continuously decreasing manners, as well as in discontinuous contact interfaces. When jets with initial opening angles up to 1° were injected in the constant pressure ISM, the initial expansion of the injected jet led to a corresponding decrease in its internal density and pressure. The surrounding cocoon pressure then ultimately induces an oblique shock in the near-inlet region. The jet is decelerated there, and becomes cylindrically collimated beyond it. Those cases which simulated the jet as an overdense structure are motivated by jet launch models where accretion disk material gets lifted out of the disk magnetocentrifugally [11, 23, 24], leading to a dense (hollow) jet structure as compared to the surrounding medium. The equation of state (EOS) in both works discussed was of Synge type, adopting a convenient and numerically inexpensive approximation due to Mathews [25]. In this approximation, one avoids the evaluation of Bessel functions that appear in the Synge EOS, governing a relativistic gas. Both the exact and the approximate Mathews EOS have the local effective polytropic index varying from 4/3 (relativistically hot matter) to 5/3 for cold, classical internal temperature regimes. The knowledge of the jet composition only enters numerical studies through this EOS, and baryonic jet flows will adopt $\rho = m_\mathrm{p} n$ with proton mass m_p and proper number density n.

A noteworthy extension of the EOS prescription used in relativistic jet simulations is the one employed by Perucho and Martí [26] (and references therein). In that study, the specific case of the FR I jet associated with radio galaxy 3C 31 was investigated with one high-resolution axisymmetric simulation. Taking into account all observational knowledge of the jet and ISM surroundings, the simulation injected a cylindrical jet at modest Lorentz factor $\Gamma \approx 2$ in a radially structured atmosphere, with both external density and pressure decreasing continuously. To keep this atmosphere in equilibrium, an effective gravity field is assumed for the ISM. The jet was injected at an extreme density contrast of $\eta = 10^5$ (this ultralight nature of the jet has no clear relation to theorized jet launch scenarios), and initially overpressured with respect to the ambient medium. This overpressured, cylindrical jet behaves similar to pressure-matched, conical jets when expanding in the surroundings, in so far that recollimation shocks appear, acting to slow down the jet beam. Their EOS prescription uses the actual Synge variation, allowing to distinguish leptonic (electrons and positrons) from baryonic (proton) matter contributions. This comes at the expense of solving for another proper rest-mass density evolution like (11.10), namely one for leptonic, and one for total proper rest-mass densities. The jet was injected as a pure leptonic flow, and baryons appear in the jet only from mixing with ambient matter (no nuclear physics is accounted for in this simulation). The simulation followed the FR I jet dynamics until the bow shock location was at 15 kpc from the source, and confirmed scenarios for the temporal evolution of FR I sources.

11.2.4 Magnetic Field Topologies

The numerical studies for jet propagation from the previous sections all ignore the dynamical influence of magnetic fields, and model kinetic energy-dominated flows. The synchrotron emission through which we observationally infer jet properties derives from a relativistic electron population spiraling about magnetic fields, and polarization maps have given direct clues on the prevailing magnetic field topologies. Many multidimensional MHD simulations have targeted flow dynamics where the dynamical role of $\boldsymbol{B}$ can be quantified. Some relevant insights are collected in the following, derived from selected recent studies.

A nonrelativistic study of magnetized jet propagation through both uniform and stratified environments is the work by O'Neill *et al.* [27]. Fully three-dimensional simulations followed helically magnetized jets with a uniform light $\eta = 100$ core (surrounded by a thin transition annulus), pressure-matched with the ambient medium. The inlet jet power was invariably dominated by kinetic and thermal components. In the stratified cases, the ISM model has an isothermal density/pressure profile decreasing quadratically (King type atmosphere). Light supersonic jets advanced more rapidly through such stratified atmospheres, with hardly any deceleration of the jet head velocity. The study quantified the efficiency of energy transfer from the jet to the ambient medium, and concluded that reheating of this medium through shock dissipation is a recurring feature of such jets. This is an important topic of active research, which aims to quantify the role of turbulent, shock-dominated cocoons in reenergizing the intracluster medium. In Figure 11.5, a volume rendering of the compression rate in the magnetized jet path is shown, clearly revealing the internal jet beam shocks, as well as the reheating in the surrounding cocoon. A shock-web structure appears in the tip of the jet, which is more pronounced along the entire jet path when uniform surroundings are modeled. Jet-cocoon interactions in the backflow regions surrounding the jet beam are directly responsible for this flow complexity. The magnetic field is undergoing local strengthening correlated with the shocks, with flux stretching regions that follow.

The extension to relativistic MHD jet studies started in earnest in the late 1990s, but many works have assumed restricted field topologies, with either purely toroidal [28], or purely poloidal field components [29]. A study by Keppens *et al.* [30] targeted the morphology of AGN jets pervaded by helical field and flow topologies, using grid-adaptive, special relativistic MHD models. Assuming axisymmetric conditions, kinetic energy-dominant jet flows with near-equipartition magnetic fields were followed for 140 light crossing times of the jet radius. Jet rotation is included, and the combined pressure, magnetic field and rotation profiles throughout the jet radius enforce radial force balance in the inlet. The rotation at the inlet is entirely within the light cylinder, and the full profiles relate to the self-similar spine jet models discussed earlier. Modest density contrasts $\eta = 10$ are adopted, while the axial jet velocity has a smoothly varying Lorentz factor variation from a central spine up to $\Gamma \simeq 22$ down to $\Gamma \simeq 5$ at the jet edge. This variation already incorporates the insights gained from the relativistic hydro models, where evidence for

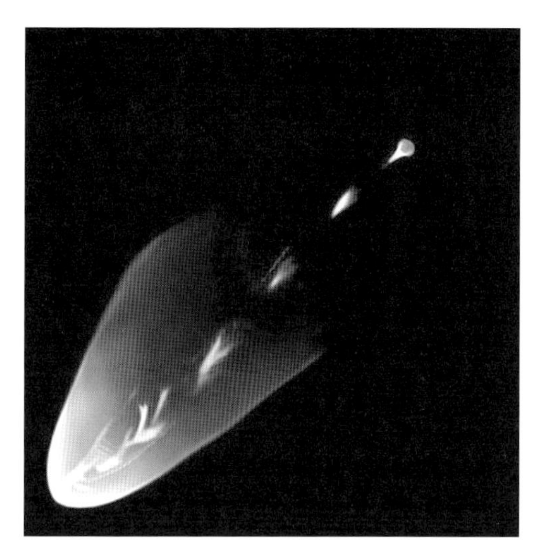

Figure 11.5 The compression rate (bright indicating higher compression rates) in a helically magnetized jet penetrating a stratified atmosphere, when the jet has propagated up to 100 jet radii. This 3D MHD study from [27] points out the role of jet cocoons for reheating ambient ISM. Reproduced from O'Neill *et al.* [27] with permission from the AAS.

a spine-layer structure was found. The magnetic field topology of a representative case is shown in Figure 11.6a. The study in essence confirmed the classical MHD results from O'Neill *et al.* [27] as the internal shocks in the jet beam lead to locally enhanced helicity, pinching the flow. This pinching acts to locally reaccelerate the jet beam. The helically magnetized jets were found to decelerate only beyond 100 jet radii, mitigated by this reacceleration. Figure 11.6b gives a synchrotron emissivity measure, directly derived from the local flow plus field properties, which in this helical field case highlights the edges of the jet beam and its many internal cross-shocks. By varying the field structure from toroidal, to helical, to poloidal topologies, the sites with enhanced synchrotron emissivity measure were shown to shift from the edge (toroidal) to the jet core (poloidal), with different distribution between the termination shock (hot spot) versus jet beam. At the same time, more advanced treatments are needed to translate the emissivities shown here to actual emission maps, a topic discussed in more detail in Section 10.4.

Relativistic MHD modeling has meanwhile progressed to demonstrate jet formation in three-dimensional, general relativistic settings about rotating Kerr black holes [16]. Truly parametric surveys are still to be performed in earnest, but McKinney and Blandford [16] showed that depending on the disk field topology assumed, disk turbulence would not prevent jets reaching up to Lorentz factor 10. Their 3D results also indicated that the mainly toroidal field topology would survive distortions due to current-driven kink modes out to 1000 gravitational radii. Clearly, the ultimate stability of jets from AGN cores is a complex interplay between (linear and nonlinear) modes where the detailed shear flow and magnetic field (hence current) distribution plays a crucial role, with as yet only rough estimates for stability limits derived for simple top-hat type profiles. At the same time, insights gained

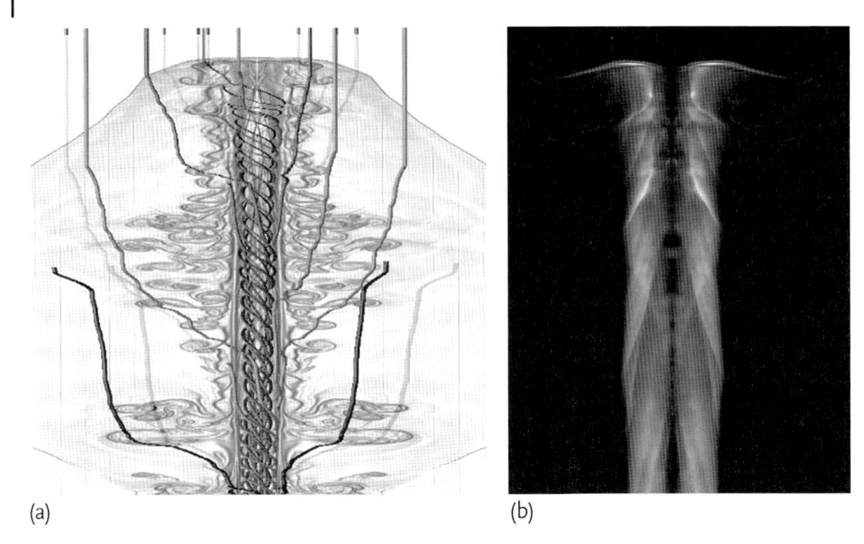

Figure 11.6 The helical field topology for a relativistic MHD jet [30]. (a) Selected field lines are visualized, together with a translucent Schlieren plot of density. A deduced synchrotron emissivity map of a similar case is shown in (b). Reproduced from Keppens *et al.* [30] with permission from Astronomy and Astrophysics.

from simplified analytic models as well as from modern high-resolution simulations combine in jet studies incorporating multiple jet components, to which we turn our attention in the final sections of this chapter.

11.3 Two-Component Jets: a Recurring Paradigm

11.3.1 Observational and Theoretical Arguments

The advent of detailed multiwavelength observations of radio-loud sources provides unprecedented views on extragalactic jets. X-ray observations augment broadband radio interferometry, and parameters inferred from X-ray jet studies have been reviewed by Harris and Krawczynski [31]. For the specific case of the radio-loud quasar PKS 1127-145, a study [32] of the 300 kpc jet detectable in X-ray by Chandra, combined with radio VLA data, put forth convincing arguments for a two-component model. Significant differences in the spectral properties of knots and features within the inner 14″ exist as compared to the outer portions (up to 30″). Both parts have projected sizes up to 100 kpc, and hence are clearly extended jet flows. The authors concluded that a "proper jet" consisting of a fast inner spine, together with its sheared, mixed layer, is responsible for the X-ray components, while the "jet-sheath" from the extended jet delivers the radio photons. This outer "sheath" is more slowly moving, and different cooling processes can explain why inner jet knots have a weak and steep-spectrum characteristic (radiative cooling acting in the shear layer surrounding the inner fast spine), while adiabatic cooling occurs in the outer extended slow jet, leading to a bright and flat-spectrum in radio. Another object studied at high angular resolution in radio data is the BL Lac object Markarian 501, and clear evidence for limb brightening was presented by Giroletti *et al.* [33]. Radial velocity structure, again split into an inner fast spine ($\Gamma \approx 15$) and outer jet layer ($\Gamma \approx 3$) at parsec distance from the core could explain this effect, when the jet is viewed at an inclination angle of about $\theta \approx 20°$. The spine would then be deboosted, similar to the effect seen in the synthetic image shown in Figure 11.3. With data for spine-layer speed estimates from regions at projected distances spanning 0.0001–50 parsecs, the case of MKN501 suggests that radial velocity structure already exists at close (parsec) distances from the source. However, the inferred inclination angle has been questioned, and an alternative explanation for the MKN501 limb brightening was given by Sahayanathan [34]. This invokes the diffusion of electrons that are accelerated in a single component, jet boundary shear layer, with a sharp velocity gradient.

Radial structure, implying two-component flow topologies, is actually expected on theoretical grounds. The outer accretion disk regions, in combination with a favorable global magnetic field topology threading the disk, can give rise to an extended outflow which in essence lifts disk matter magnetocentrifugally into a wide jet-wind flow [11, 23, 24]. A dense disk wind, which can be self-collimated into a jet flow, then corresponds to the outer jet region, and its rotation and density structure must relate to the underlying near-Keplerian disk profile. This disk jet/wind can also be aided by thermal driving, when the disk establishes a hot corona above it. For the near black hole disk regions, where ultimately general relativistic effects come into play, the faster rotation and the magnetic field advection and amplification will rather give rise to a relativistically hot, inner flow. These could start as a Poynting flux-dominated jet relying on frame-dragging [7] for which two-dimensional solutions were shown in Figure 11.2. These can continue as electromagnetically dominated spines, but likely dissipate to provide hot coronal conditions from which spine jets can be launched (Figure 11.1). In any case, the matter state of this inner jet is relativistic, and faster flow speeds would result.

There is a clear analogy between such two-component jets for AGNs, and those diagnosed extensively in young stellar objects (YSOs), where the forming star has a stellar wind, surrounded by a wider accretion-ejection structure. Radially resolved rotation profiles through YSO jets already provided quantitative evidence for their magnetocentrifugal driving. An overview of model as well as observational aspects by Ferreira *et al.* [35] concluded that for jets associated with accreting T Tauri stars, up to three dynamical components were needed: a steady self-collimated disk wind, a pressure-driven coronal stellar wind, and variable hot "blobs" associated with reconnection events. The latter form YSO analogs of solar coronal mass ejections, and can originate whenever antiparallel magnetic fields make contact (e.g., when the stellar dipole is reversed with respect to the outer disk magnetic field [36]). The outer disk wind would in fact act to collimate the inner stellar outflow, an aspect which returns in AGN context as well.

11.3.2
Aspects Deduced from Modern Simulations

From high-resolution relativistic jet simulations, further evidence for the existence of significant radial velocity structure can be collected as well. As indicated earlier, even jets injected without radial structure will self-consistently establish a spine-layer morphology further down the jet path. However, this corresponds to the layer of heavy mixing and entrainment of ambient, shocked matter, and is thereby not necessarily arising from a differently launched jet component.

Such a difference in the underlying launch mechanism is invoked in the study by Bogovalov and Tsinganos [37], where the authors investigated the self-collimating properties of relativistic, superfast magnetized flows numerically. In fact, starting from a monopolar magnetic field topology, in which a radially expanding, two-component flow is inserted, the authors followed the evolution to a truly axisymmetric steady configuration. The inner flow started off as a $\Gamma = 5$ jet flow, smoothly decreasing into an extended nonrelativistic "disk wind". The final solution demonstrates a well-collimated relativistic jet flow at far distances, up to $10^5 r_{\text{fast}}$ where r_{fast} denotes the radius where speeds equal the local fast speed. Without the surrounding nonrelativistic wind zone, no such well-collimated jets were established. In the externally collimated solutions, steady shock fronts in the inner relativistic wind zone form, due to the wind-wind collision dynamics. Hence, in direct analogy with YSO systems, surrounding disk winds may be a crucial ingredient to explain the collimation of relativistic AGN jets. A more parametric study of relativistic MHD winds has been performed by Porth and Fendt [38], where a disk wind was taken as a physical boundary condition, and both split monopole and an hourglass-shaped potential field was adopted initially. Similar to the work by Bogovalov and Tsinganos, steady solutions were obtained with jets having half-opening angles of $3-7°$, while the split monopole initial fields led to less-collimated solutions. Outflow solutions were characterized by a narrow, hollow relativistic jet (Lorentz factors up to 6), surrounded by subrelativistic flow from the outer disk regions. Parameter studies then quantified the role of rotation and magnetization at the disk boundary in accelerating the jets.

11.4
Stability Studies for Radially Structured Jets

11.4.1
Spine-Sheath Models

Motivated by the observational arguments in favor of significant radial velocity structure in relativistic jets, analytic as well as numerical studies have concentrated on the dynamical properties of *spine-sheath* jet flows. These could be the individual components of a two-component outflow structure, or merely refer to the self-emerging split into a fast spine, and surrounding shear layer (possibly coinciding with a backflow). A study by Hardee [39] adopted such doubly uniform, magnetized spine-sheath jet structure, and rigorously manipulated the linearized relativistic MHD equations to a dispersion relation governing the normal modes. In the equilibrium model, flow and magnetic field are purely aligned with the jet axis, and the "sheath" is handled as a uniform magnetized flow layer extending to infinity. Those approximations neglect many potentially important instabilities, for example all those related to helical fields (current-driven modes), as well as ignoring all rotation-driven effects, and the presence of a second bounding interface when a two-component structure applies. Despite these shortcomings, the model gives accurate information on dominant shear-flow-driven Kelvin–Helmholtz modes. Since the governing dispersion relation is a transcendental equation in the eigenfrequency ω for given axial and azimuthal wavenumbers k_Z and m in a Fourier mode variation $\exp[i(k_Z Z + m\varphi - \omega t)]$, approximate expressions for limiting cases were given. Parametric studies of the dispersion relation solutions were performed numerically. The knowledge on instability thresholds, and ultimately the structure of the eigenmode perturbations, can be used to identify observed jet distortions in specific radio sources as arising from dominantly unstable modes. In this way, standard "magnetoseismological" techniques are introduced in astrophysical jet contexts.

The interpreting power of normal mode studies was subsequently employed in fully 3D, relativistic MHD simulations of such spine-sheath structures. Mizuno *et al.* [40] performed 10 studies of preexisting cylindrical jet flows in 3D Cartesian domains, and actively perturbed the flow at one boundary. The equilibrium jets assume a fixed density contrast of $\eta = 2$, with a spine jet flow where $\Gamma = 2.5$, surrounded in either static or nonrelativistic wind surroundings (where the flow is $0.5c$). The thermodynamic conditions represent a cold jet in a hot surrounding medium. Figure 11.7 shows selected field lines and an isosurface relating to the

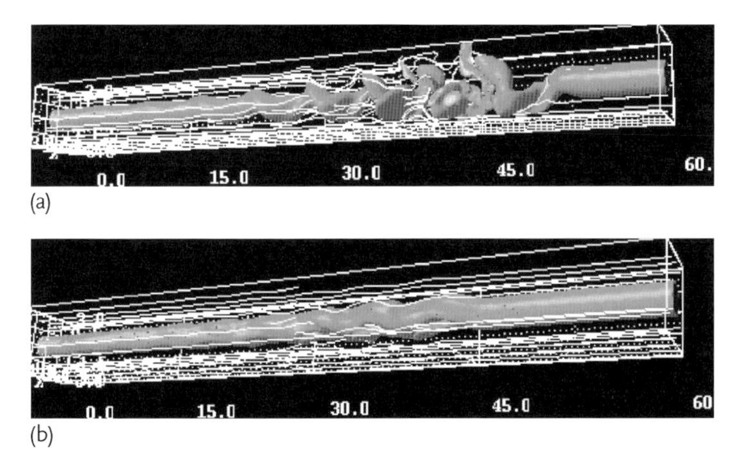

Figure 11.7 Simulations of 3D spine-sheath relativistic MHD jets [40] revealed pronounced differences between weakly magnetized, wind embedded flows (a) vs. strongly magnetized flows in static surroundings (b). Reproduced from Mizuno *et al.* [40] with permission from the AAS.

spine-sheath transition, for a weakly magnetized jet in wind surroundings, versus a strongly magnetized jet in a static environment. The figure shows that spine-sheath structures can be distorted significantly for weakly magnetized conditions, while strong fields, as well as the presence of a wind environment, act to stabilize the flow. These 3D nonlinear evolutions were verified against normal mode predictions, and wavelengths as well as unstable mode numbers were found to agree with the linear predictions (see also Section 10.3.1.1).

11.4.2
Two-Component Jets and FR I/II Classification

As indicated, the spine-sheath investigations discussed in the previous section made simplifying assumptions on the actual jet structure. In an attempt to bring into account analytic knowledge to asymptotic relativistic MHD jet behavior, a realistic two-component jet should obey the force balance equation (11.11). In the spine-sheath structures from Mizuno *et al.* [40], this reduces to total pressure balance between inner spine and outer sheath, without a dynamical role by the flow. This force balance equation, at far distances from the black hole such that the effective gravitational term containing $\nabla \ln \alpha_l$ can be neglected, can be reworked to

$$\Gamma \rho \left(\frac{\boldsymbol{v}}{c} \cdot \nabla\right) \left(\frac{\Gamma h \boldsymbol{v}}{c}\right) = -\nabla p + \rho_c \boldsymbol{E} + \frac{1}{c} \boldsymbol{j} \times \boldsymbol{B} \,. \tag{11.15}$$

All symbols have their usual meaning, with thermodynamic quantities ρ, p, h indicating proper density, pressure, and specific enthalpy. This used the stationary energy and Maxwell equations, to rewrite the $\boldsymbol{E} \cdot \boldsymbol{j}$ contribution appearing in (11.11). Starting from this form, exact analytic solutions could be obtained [41] for axisymmetric, purely cylindrical configurations where $\boldsymbol{B} = B_\varphi(R)\hat{e}_\varphi + B_Z(R)\hat{e}_Z$ as well as $\boldsymbol{v} = v_\varphi(R)\hat{e}_\varphi + v_Z(R)\hat{e}_Z$. These then represent the asymptotically cylindrical jets, self-consistently launched and accelerated from near black hole surroundings, such as those shown in Figure 11.1. In Meliani and Keppens [41], a two-component jet structure was then initialized with these radially varying, force-balanced solutions applying in both an inner, fast, hot jet, as well as in an outer, cold, slow jet. Across the inner-outer jet interface, effective total pressure-matching completed the full force balanced two-component jet equilibrium.

The stability of these two-component jets was subsequently studied in both 2.5D and 3D relativistic simulations. In [41], the simulations assumed invariance along the jet axis (Z-direction) and thus could reduce the computational effort to a 2.5D study of the long-term evolution of jets in a transverse cross-section. The inclusion of rotation, combined with the split in a fast, hot inner jet, surrounded by a cold, dense outer jet, turned out to be a decisive factor into the overall dynamical behavior of jets. While two-component jets with a low inner kinetic energy flux contribution remained stable and highly relativistic for long distances, jets with the inner jet contributing strongly to the total jet kinetic energy flux are subject to a relativistic Rayleigh–Taylor type instability. Figure 11.8 shows the proper density in a cross-section of the jet, at a time corresponding to 163 years of jet propagation.

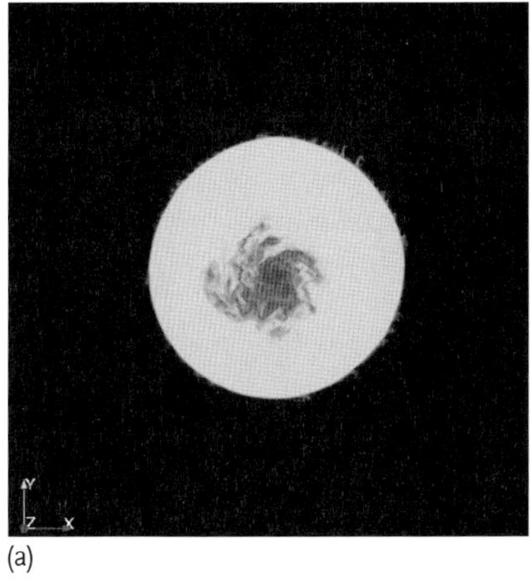

(a)

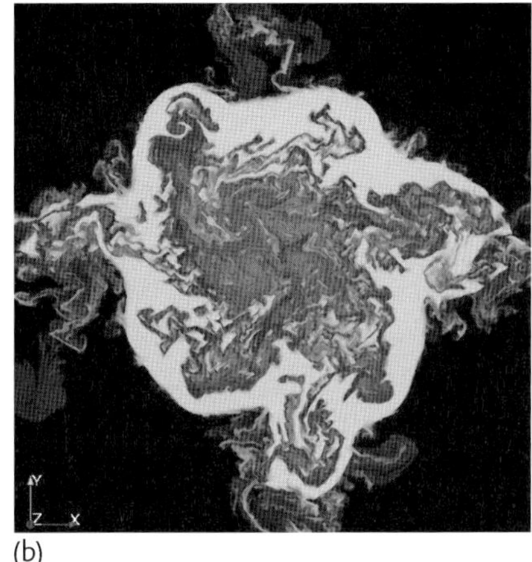

(b)

Figure 11.8 Two-component jets seen in cross-section, after 163 years of propagation (up to about 30 parsec). Case (a) remains a clearly discernible two-component flow, while (b) has been subject to a relativistically enhanced Rayleigh–Taylor mode, discussed and analyzed in detail in Meliani and Keppens [41]. Reproduced from Meliani and Keppens [41] with permission from the AAS.

It compares a case where the jet remained stable (Figure 11.8a), versus a situation where the instability has set in, and has led to complete mixing of both jet components. The instability criterion reads simply as

$$\left(\Gamma^2 \rho h + B_Z^2\right)_{\text{inner}} > \left(\Gamma^2 \rho h + B_Z^2\right)_{\text{outer}} , \tag{11.16}$$

and is a relativistically enhanced, rotation-induced linear mode. It should be noted that the rotational speeds needed to trigger the mode can be strongly subrelativistic, with all models simulated having a maximal local $v_\varphi = 0.01c$ (subsonic rotation). The criterion (11.16) was derived from appropriately linearizing the relativistic MHD equations, neglecting toroidal fields but including rotation. The instability exists in pure hydrojets as well, but is an intrinsically relativistic effect due to the $\Gamma^2 h$ factor appearing. Indeed, the low density, relativistically hot, inner jets studied had Lorentz factors of $\Gamma \approx 30$, while the outer jet started at $\Gamma \approx 3$. Cases liable to the instability would experience a deceleration of the mean Lorentz factor down to Lorentz factor 8 within 160 years (or after propagating 30 pc), while others would remain at high ($\Gamma \geq 20$) flow speeds throughout.

These new insights suggest that the intrinsic properties of the inner engine (black hole and accretion disk) are the key element which control jet stability and its kinematics at the parsec scale. According to the contribution of various regions of the accretion disk-black hole to the total energy flux of the jet, the jet may remain stable and relativistic or become unstable and decelerate. This then gives a new interpretation to the radio source FR I/II dichotomy. Since jets in FR II cases stay relativistic and narrow on all scales, they could correspond to two-component jet configurations with low-energy flux contribution from the inner jet. In contrast, jets in FR I are brighter near the center and fade out towards the edge, and decelerate from pc to kpc scales. This matches with two-component jet configurations with high-energy flux contribution from the inner jet. In this case, the relativistic Rayleigh–Taylor type instability develops at the interface between the two components. This induces rapid deceleration of the inner component and the formation of a wide and turbulent shear region.

11.5
Further Challenges for Modern Simulations

The results obtained with advanced semianalytic modeling of AGN jets, combined with modern high-resolution special or even general relativistic hydro or MHD simulations, have been important complementary tools to study jet structure, collimation, and propagation aspects. By highlighting selected studies from the literature, this chapter made it clear that various aspects still deserve further study. In this section, we give some guidelines for future investigations.

A particularly challenging aspect concerns the role of the magnetic field and its (varying) topology from the launch region near the black hole-accretion disk, to the asymptotically organized flows far-field. If jets (or in a two-component framework, the inner jets) start as Poynting flux-dominated structures, how do they match up asymptotically with organized fields in kinetic energy-dominated flows? The current generation of GRMHD studies already provides hints that such Poynting flux-dominated flows can start off successfully, despite dominant toroidal magnetic field concentrations. From laboratory plasma experience, such Z-pinch like topologies are violently unstable to current-driven modes. Important clues to resolve this discrepancy have already been obtained in nonrelativistic MHD simulations. The nature of kink instabilities is inherently 3D, so these insights are once more driven by the significant advance in computing power within the last decades.

Studies of current-driven kink instabilities in 3D classical MHD simulations can be done in complimentary ways. Local simulations concentrate on a periodic segment of an already organized, helically magnetized jet flow. These can make direct use of knowledge from linear MHD computations, for example exploiting the exact eigenfunctions and/or to verify agreement on linear growth rates. Baty and Keppens [42] adopted this approach to investigate the nonlinear dynamics in helically magnetized jets, subject to both Kelvin–Helmholtz and current-driven kink deformations. In the weak magnetic field regime studied, it was shown that both modes can interact nonlinearly, in such a way that the helical deformation of the central jet portion due to the kink mode, acts to stabilize the Kelvin–Helmholtz roll-up and mixing occurring at the jet edge. How this mode-mode interplay may manifest itself in relativistic (two-component) jet scenarios is yet to be studied. One may even suspect that the detailed jet composition can significantly affect the liability of jets to such kink deformations, and also that aspect would require further parametric exploration in (special) relativistic MHD studies. A first step has recently been taken by Mizuno *et al.* [43], where the current-driven kink was studied in a static (no flow) force-free magnetic configuration, in special relativity. As expected, the kink

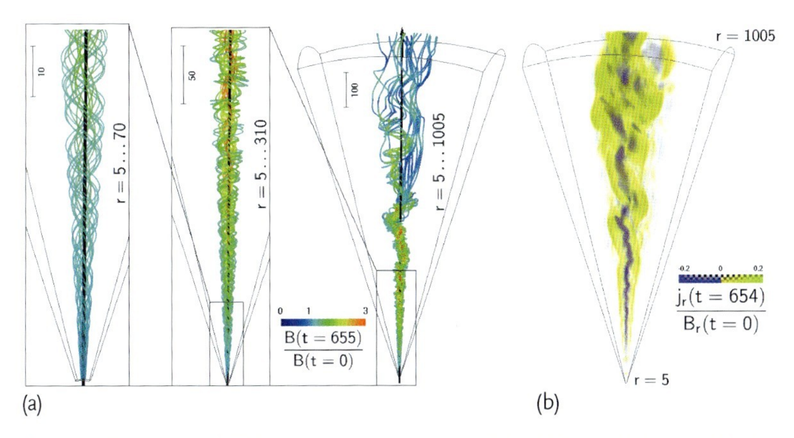

Figure 11.9 Magnetic field lines in progressively larger radial sections (a) and radial current density view (b) from a 3D nonrelativistic MHD study by Moll [44], where the decay of toroidal field due to kink mode development and dissipation is demonstrated. Reproduced from Moll [44] with permission from Astronomy and Astrophysics.

evolution is strongly influenced by the precise magnetic pitch profile, and thus it is imperative to establish how realistically launched jets are organized in terms of magnetic field helicity variation, as well as in their detailed flow topology.

Clues on how the magnetic field topology may self-organize from more toroidal, to more poloidal topologies at far distances, and how kink instabilities may help dissipate the toroidal field to then drive and accelerate the flow, can be deduced from classical 3D MHD studies by Moll [44]. Continuing on an earlier study [45], the author compares axially symmetric, with full 3D simulations of magnetically driven jets propagating in stratified atmospheres. In the axisymmetric case, kink instability is prohibited, and about 20% of the Poynting flux remains in the flow, which accelerates through sonic, Alfvén and fast speeds. In 3D, this is no longer the case, since current-driven helical deformations arise, ultimately dissipating magnetic energy to internal energy. A figure demonstrating the global jet magnetic field structure on increasing length scales is given in Figure 11.9. The length unit is here the diameter of a rigidly rotating disk imposed as bottom boundary condition. This twists up the initially poloidal field, so that jet propagation and kink development can be studied. Although the simulations clearly demonstrate how potentially vital roles are played by 3D kink modes, it is to be stressed that in both these nonrelativistic [42, 44], as well as in many general relativistic MHD studies [16], ideal MHD assumptions are used, letting the dissipation occur through numerical truncation/discretization errors. The details of these dissipative effects will require fully 3D, resistive MHD studies, for which solvers have only recently been adopted [46].

In conclusion, our knowledge on how instabilities known from nonrelativistic settings return in relativistic disguise, driven by combinations of shear flows, pressure/density variations, rotation and currents, is still being charted in modern investigations. Moreover, ultrarelativistic flows with Lorentz factors of 100 or more are known to exist and be jet-like in the extreme astrophysical events accompanying gamma-ray bursts. Such high Lorentz factor flows may well be subject to as yet unknown linear as well as nonlinear instabilities or mode-mode interactions. Insights gained from laboratory jet studies, over YSO systems, to relativistic AGN jets and ultimately GRB scenarios, compliment one another in many ways. Numerical simulations, and synthetic observations derived from these virtual jets, will thereby continue to give direct feedback to observational campaigns.

References

1 Ferrari, A. (1998) Modeling extragalactic jets. *Annu. Rev. Astron. Astrophys.*, **36**, 539–598.

2 Font, J.A. (2008), Numerical hydrodynamics and magnetohydrodynamics in general relativity. *Liv. Rev. Relat.*, **11**, 7. http://www.livingreviews.org/lrr-2008-7 (accessed 6 April 2011).

3 Giacomazzo, B. and Rezzolla, L. (2007) `WhiskyMHD`: a new numerical code for general relativistic magnetohydrodynamics. *Class. Quantum Gravity*, **24**, S235–S258.

4 Thorne, K.S. and Macdonald, D. (1982) Electrodynamics in curved spacetime: $3 + 1$ formulation. *Mon. Not. R. Astron. Soc.*, **198**, 339–343 (and Microfiche MN 198/1).

5 Macdonald, D. and Thorne, K.S. (1982) Black-hole electrodynamics: an absolute-space/universal-time formulation. *Mon. Not. R. Astron. Soc.*, **198**, 345–382.

6 Mobarry, C.M. and Lovelace, R.V.E. (1986) Magnetohydrodynamic flows in Schwarzschild geometry. *Astrophys. J.*, **309**, 455–466.

7 Blandford, R.D. and Znajek, R.L. (1977) Electromagnetic extraction of energy from Kerr black holes. *Mon. Not. R. Astron. Soc.*, **179**, 433–456.

8 Heinemann, M. and Olbert, S. (1978) Axisymmetric ideal MHD stellar wind flow. *J. Geophys. Res.*, **83**, 2457–2460.

9 Goedbloed, J.P., Keppens, R., and Poedts, S. (2010) *Advanced Magnetohydrodynamics. With Applications to Laboratory and Astrophysical Plasmas*, Cambridge University Press, Cambridge.

10 Tsinganos, K. (2007) Theory of MHD jets and outflows, in *Jets from Young Stars*, Lecture Notes in Physics, vol. 723, Springer, pp. 117–160.

11 Blandford, R.D. and Payne, D.G. (1982) Hydromagnetic flows from accretion discs and the production of radio jets. *Mon. Not. R. Astron. Soc.*, **199**, 883–903.

12 Parker, E.N. (1958) Dynamics of the interplanetary gas and magnetic fields. *Astrophys. J.*, **128**, 664–676.

13 Sauty, C. and Tsinganos, K. (1994) Nonradial and nonpolytropic astrophysical outflows III. A criterion for the transition from jets to winds. *Astron. Astrophys.*, **287**, 893–926.

14 Meliani, Z., Sauty, C., Vlahakis, N., Tsinganos, K., and Trussoni, E. (2006) Nonradial and nonpolytropic astrophysical outflows. VIII. A GRMHD generalization for relativistic jets. *Astron. Astrophys.*, **447**, 797–812.

15 Meliani, Z., Sauty, C., Tsinganos, K, Trussoni, E., and Cayatte, V. (2010) Relativistic spine jets from Schwarzschild black holes: Application to AGN radioloud sources. *Astron. Astrophys.*, **521**, A67.

16 McKinney, J.C. and Blandford, R.D. (2009) Stability of relativistic jets from rotating, accreting black holes via fully three-dimensional magnetohydrodynamic simulations. *Mon. Not. R. Astron. Soc.*, **394**, L126–L130.

17 Hawley, J.F. and Krolik, J.H. (2006) Magnetically driven jets in the Kerr metric. *Astrophys. J.*, **641**, 103–116.

18 Fendt, C. (1997) Collimated jet magnetospheres around rotating black holes. General relativistic force-free 2D equilibrium. *Astron. Astrophys.*, **319**, 1025–1035.

19 Rossi, P., Mignone, A., Bodo, G., Massaglia, S. and Ferrari, A. (2008) Formation of dynamical structures in relativistic jets: the FR I case. *Astron. Astrophys.*, **488**, 795–806.

20 Fanaroff, B.L. and Riley, J.M. (1974) The morphology of extragalactic radio sources of high and low luminosity. *Mon. Not. R. Astron. Soc.*, **167**, 31P–35P.

21 Meliani, Z., Keppens, R., and Giacomazzo, B. (2008) Fanaroff–Riley type I jet deceleration at density discontinuities. Relativistic hydrodynamics with a realistic equation of state. *Astron. Astrophys.*, **491**, 321–337.

22 Gopal-Krishna and Wiita P.J. (2002) Hybrid morphology radio sources and the Fanaroff–Riley dichotomy. *New Astron. Rev.*, **46**, 357–360.

23 Casse, F. and Keppens, R. (2002) Magnetized accretion-ejection structures: 2.5-dimensional magnetohydrodynamic simulations of continuous ideal jet launching from resistive accretion disks. *Astrophys. J.*, **581**, 988–1001.

24 Casse, F. and Keppens, R. (2004) Radiatively inefficient magnetohydrodynamic accretion-ejection structures. *Astrophys. J.*, **601**, 90–103.

25 Mathews, W.G. (1971) The hydromagnetic free expansion of a relativistic gas. *Astrophys. J.*, **165**, 147–164.

26 Perucho, M. and Martí, J.M. (2007) A numerical simulation of the evolution and fate of a Fanaroff–Riley type I jet. The case of 3C 31. *Mon. Not. R. Astron. Soc.*, **382**, 526–542.

27 O'Neill, S.M., Tregillis, I.L., Jones, T.W. and Ryu, D. (2005) Three-dimensional simulations of MHD jet propagation through uniform and stratified external environments. *Astrophys. J.*, **633**, 717–732.

28 Komissarov, S.S. (1999) Numerical simulations of relativistic magnetized jets. *Mon. Not. R. Astron. Soc.*, **308**, 1069–1076.

29 Leismann, T., Antón, L., Aloy, M.A., Müller, E., Martí, J.M., Miralles, J.A., and Ibánez, J.M. (2005) Relativistic MHD simulations of extragalactic jets. *Astron. Astrophys.*, **436**, 503–526.

30 Keppens, R., Meliani, Z., van der Holst, B., and Casse, F. (2008) Extragalactic jets with helical magnetic fields: relativistic MHD simulations. *Astron. Astrophys.*, **486**, 663–678.

31 Harris, D.E. and Krawczynski, H. (2006) X-ray emission from extragalactic jets. *Annu. Rev. Astron. Astrophys.*, **44**, 463–506.

32 Siemiginowska, A., Stawarz, L., Cheung, C.C., Haris, D.E., Sikora, M., Aldcroft, T.L., and Bechtold, J. (2007) The 300 kpc long X-ray jet in PKS 1127-145, $z = 1.18$ quasar: constraining X-ray emission models. *Astrophys. J.*, **657**, 145–158.

33 Giroletti, M., Giovannini, G., Feretti, L., Cotton, W.D., Edwards, P.G., Lara, L., Marscher, A.P., Mattox, J.R., Piner, B.G., and Venturi, T. (2004) Parsec-scale properties of Markarian 501. *Astrophys. J.*, **600**, 127–140.

34 Sahayanathan, S. (2009) Boundary shear acceleration in the jet of MKN501. *Mon. Not. R. Astron. Soc.*, **398**, L49-L53.

35 Ferreira, J., Dougados, C., and Cabrit, S. (2006) Which jet launching mechanism(s) in T Tauri stars? *Astron. Astrophys.*, **453**, 758–796.

36 Fendt, C. (2009) Formation of protostellar jets as two-component outflows from star-disk magnetospheres. *Astrophys. J.*, **692**, 346–363.

37 Bogovalov, S. and Tsinganos, K. (2005) Shock formation at the magnetic collimation of relativistic jets. *Mon. Not. R. Astron. Soc.*, **357**, 47, 918–928.

38 Porth, O. and Fendt, C. (2010) Acceleration and collimation of relativistic magnetohydrodynamic disk winds. *Astrophys. J.*, **709**, 1100–1118.

39 Hardee, P.E. (2007) Stability properties of strongly magnetized spine-sheath relativistic jets. *Astrophys. J.*, **664**, 26–46.

40 Mizuno, Y., Hardee, P., and Nishikawa, K.-I. (2007) Three-dimensional relativistic magnetohydrodynamic simulations of magnetized spine-sheath relativistic jets. *Astrophys. J.*, **662**, 835–850.

41 Meliani, Z. and Keppens, R. (2009) Decelerating relativistic two-component jets. *Astrophys. J.*, **705**, 1594–1606.

42 Baty, H. and Keppens, R. (2002) Interplay between Kelvin–Helmholtz and current-driven instabilities in jets. *Astrophys. J.*, **580**, 800–814.

43 Mizuno, Y., Lyubarsky, Y., Nishikawa, K.-I., and Hardee, P.E. (2009) Three-dimensional relativistic magnetohydrodynamic simulations of current-driven instability. I. Instability of a static column. *Astrophys. J.*, **700**, 684–693.

44 Moll, R. (2009) Decay of the toroidal field in magnetically driven jets. *Astron. Astrophys.*, **507**, 1203–1210.

45 Moll, R., Spruit, H.C., and Obergaulinger, M. (2008) Kink instabilities in jets from rotating magnetic fields. *Astron. Astrophys.*, **492**, 621–630.

46 Komissarov, S.S. (2007) Multidimensional numerical scheme for resistive relativistic magnetohydrodynamics. *Mon. Not. R. Astron. Soc.*, **382**, 995–1004.

12
Jets and AGN Feedback

Christopher S. Reynolds

12.1
Introduction

SMBHs and, in particular, the jets that they produce may have a profound effect on the way that galaxies have evolved over cosmic time. In a nutshell, the paradigm that has emerged over the past decade is that energy released by processes associated with a central SMBH act as a "thermostat", stifling or even completely stopping the growth of massive galaxies. The generic term given to this phenomenon is *AGN feedback*.

Observations clearly show that the growth of a SMBH and the growth of its host galaxy are related. For the nearby galaxies that can be studied in detail, there are striking correlations between properties of a galaxy and the measured mass of the SMBH that it hosts. It is found that the SMBH mass, $M_{\rm BH}$, is approximately 0.1% of the stellar mass in the galactic bulge, M_* [1]. Thus SMBHs are not randomly distributed between galaxies – low-mass SMBHs are in small galaxies and massive SMBHs are in massive galaxies. In fact, the best correlation appears to be between $M_{\rm BH}$ and the *velocity dispersion* of stars in the body of the galactic bulge, σ – as shown in Figure 12.1, it has been found that $M_{\rm BH} \propto \sigma^\beta$ where $\beta = 4$–5 [2–4]. While the observations are clear, the meaning of these correlations is not. Today's massive galaxies are thought to have been built by the successive merger of smaller galaxies, each of which may have contained a SMBH and, during such a merger, the SMBHs probably also merge together. Thus, even in the absence of a physical connection between the SMBH and the host galaxy, the averaging process inherent in these many merger events would lead to larger SMBHs in larger galaxies. However, many astrophysicists believe that a full explanation for the SMBH-galaxy correlations requires a direct physical link between the growth of the SMBH and the galaxy. Indeed, as we discuss in this chapter, there are multiple reasons to think that SMBHs and, in particular, the jets that they create can actively regulate galaxy growth. Given that the formation of jets is governed by processes occurring in the immediate vicinity of the black hole (i.e., within 0.01 pc) whereas a galaxy has char-

Relativistic Jets from Active Galactic Nuclei, First Edition. Edited by M. Böttcher, D.E. Harris, H. Krawczynski.

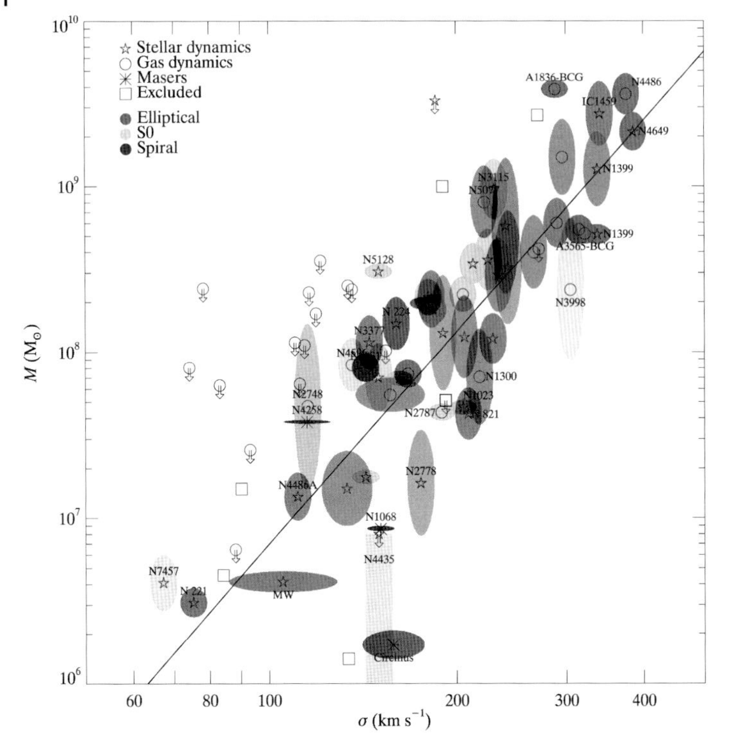

Figure 12.1 Mass of the central supermassive black hole, M, as a function of the stellar velocity dispersion of the galactic bulge, σ. The ellipses indicate the uncertainty in M and σ. The line represents the best fit given by $M = 10^{8.12}(\sigma/200\ \mathrm{km\,s^{-1}})^{4.24}\ M_\odot$. Reproduced from Gultekin *et al.* [2] with permission from the AAS.

acteristic size scales of 10 kpc or more, AGN feedback is a truly remarkable example of multiscale physics.

Putting aside the multiscale nature of the problem for now, it is easy to see from simple energy arguments that SMBHs have the *potential* to significantly affect the growth of a galaxy. Consider a galaxy that has a total stellar mass M_*. Since our calculation is purely for illustrative purposes, let us restrict attention to a bulge-dominated galaxy. Thus, using the correlation between the mass of the SMBH and the stellar mass noted above, we can write $M_{\mathrm{BH}} = f_{\mathrm{BH}} M_*$, where $f_{\mathrm{BH}} \approx 10^{-3}$. If the black hole was formed by the accretion of gas which released its gravitational potential energy to its surroundings with efficiency η, then the total energy released building the SMBH is

$$E_{\mathrm{BH}} = \frac{\eta M_{\mathrm{BH}} c^2}{1-\eta} = \frac{f_{\mathrm{BH}}\eta}{1-\eta} M_* c^2 \approx 10^{-4} M_* c^2 \,, \tag{12.1}$$

where we have taken a typical value for the efficiency of black hole accretion as $\eta \approx 0.1$.

In a somewhat extreme thought experiment, suppose that this energy was used to uniformly raise the temperature of *all* of the gas that was destined to form the stars we see in the galaxy. Treating the gas as a monoatomic ideal gas, the energy per particle at a temperature T is $3/2 k_B T$. Ignoring the starting temperature of the gas (i.e., assuming that it starts from absolute zero rather than stellar temperatures), the temperature resulting from the SMBH heating can be derived by equating E_{BH} with the thermal energy of the final gas mass M_*

$$E_{BH} = \frac{f_{BH}\eta}{1-\eta} M_* c^2 = \frac{3}{2}\left(\frac{M_*}{\mu m_p}\right) k_B T \,, \tag{12.2}$$

where μ is the mean mass of a particle in the gas (summing over free electrons and ions) in units of the proton mass m_p; we find that $\mu \approx 0.7$ for ionized gas with standard cosmic composition. Solving this equation for temperature, we find $T \approx 5 \times 10^8$ K independent of the mass M_*. This is *easily* sufficient to expel the gas from the gravitational potential well of any known galaxy, as we shall see in Section 12.2, thereby completely preventing any star formation!

This chapter presents the current (circa 2010) viewpoint of AGN feedback, focusing on the jet-driven feedback that we believe is important in the more massive of galaxies. We start with a brief review of galaxy formation theory, including a discussion of two classical problems that (within the current paradigm) maybe solved by invoking AGN feedback. We shall then discuss interactions between AGN jets and surrounding gas both from an observational and theoretical viewpoint. Finally, as both a pointer to the future and as a word of caution, we will explore very recent developments in the basic physics of hot gaseous atmospheres, of which the consequences for the AGN feedback paradigm are hard to predict.

12.2 Galaxy Formation and Two Classic Problems

Our understanding of galaxy formation and evolution is still very much a work in progress. Here we give a basic review of the aspects of galaxy formation that are needed to motivate and understand the current paradigm for AGN feedback. For a detailed review of galaxy formation theory, see the review by Benson [5].

12.2.1 Cosmological Background

The Big Bang resulted in a distribution of matter that was spatially homogeneous to a very high degree as well as expanding/diluting with cosmic time.[1] Approxi-

1) Given our discussion of special relativity in Chapter 2, one might be surprised that it is possible to define a unique "cosmic time". In fact, by considering relativistic simultaneity, it can be seen that there can be at most one inertial reference frame in which an expanding Universe appears spatially homogenous. This is the frame with which we define cosmic time. In terms of the formal construction of cosmological models, it is the *assumption* of homogeneous spatial sections of space-time that breaks the equivalence of reference frames in an otherwise fully relativistic theory.

mately 20% of matter is comprised of particles found in the *standard model of particle physics*; from a second after the big bang onwards, this is mostly in the form of protons, neutrons and electrons. While strictly a misuse of particle physics terminology, it is customary to use the term *baryonic matter* to describe this entire component. However, we know that the remaining 80% of the matter in the Universe is in some form not found within the standard model; this is the *nonbaryonic dark matter*. Dark matter particles are massive yet interact only via the gravitational and weak forces. Indeed, for the epoch of galaxy formation, it is likely that only gravitational forces are relevant for the behavior of dark matter particles.

Dark matter is a central player in the story of galaxy formation. While the matter distribution in the early Universe was close to being homogeneous, it was not precisely so; the distribution was imprinted by perturbations resulting from quantum fluctuations during inflation (at approximately $t \sim 10^{-34}$ s after the Big Bang). Galaxy formation theory attempts to unravel the incredible story of how random quantum fluctuations become the rich and complex Universe that we see around us today.

From the earliest times, the fluctuations in the dark matter start to grow in amplitude. The basic physics of this process is straightforward; slightly overdense regions create local gravitational fields that attract more matter thereby becoming denser, leading to a runaway process. At early times, the baryonic matter behaves somewhat differently. For the first 380 000 years after the big bang, the baryonic matter is tightly coupled to the radiation field and, hence, fluctuations in the baryonic matter density resist gravitational collapse due to the action of radiation pressure. This era ends when the Universe cools to the point where protons and electrons combine to form neutral hydrogen atoms; this is known as the *epoch of recombination*. As the baryonic matter becomes predominantly neutral, the opacity of the matter plummets and the radiation field decouples from the matter (these free-streaming photons continue to redshift as the Universe expands and are observed today as the cosmic microwave background). With the sudden loss of radiation pressure, fluctuations in the baryonic matter density gravitationally collapse. After about 50 Myr (corresponding to a redshift of $z \sim 50$), the fluctuations in both the dark matter and the baryons represent major perturbations on the background density field. Before this time, one can follow the evolution of structure using precise analytical techniques (linear perturbation theory); after this time, the evolution is nonlinear and numerical simulations are required.

Simulations show that, on large scales ($> 10\,\mathrm{Mpc}$), the nonlinear evolution of the dark matter forms a system of interconnected filaments and sheets; see Figure 12.2. This has become known as the *cosmic web*. On smaller scales, however, dark matter forms gravitationally bound systems, *dark matter halos*. Dark matter halos are believed to be in virial equilibrium meaning that $E_K + 2E_G \approx 0$ where E_K is the total kinetic energy of the dark matter particles in the halo and E_G is the total gravitational potential energy of the halo. Both direct numerical simulations and analytic arguments (based on the spherical top-hat collapse model [6]) suggest that dark matter halos attain a characteristic overdensity factor $\Delta \approx 200$ compared with the average density of the Universe evaluated at the time/redshift that the halo

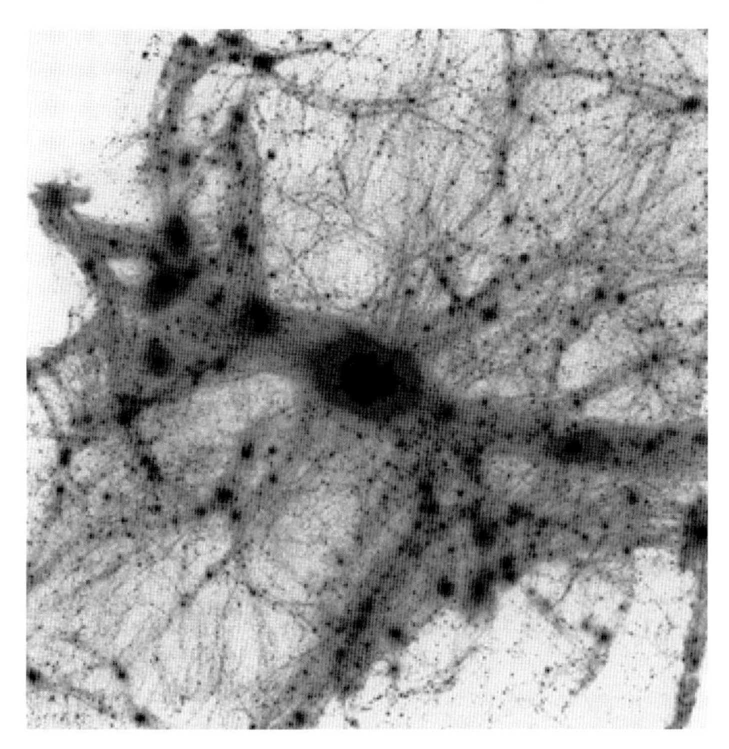

Figure 12.2 Computer simulation of dark matter structures in and around a Milky Way like halo. The simulated region shown is 8 Mpc on a side. On scales comparable to this region and larger, dark matter forms filamentary and sheet-like structures. On the smaller scales shown here, the dark matter forms well defined, gravitationally bound dark halos. Figure kindly supplied by M. Ricotti and E. Polisenky.

forms, $\rho_0(z)$. Thus, the characteristic radius of a halo with mass M is

$$r_{\text{halo}} = \left(\frac{3M}{4\pi\rho_0(z)\Delta}\right)^{1/3} . \tag{12.3}$$

Dark matter halos are the sites for galaxy formation. Baryons within the sphere of influence of the dark matter halo accrete into the halo, radiate away their thermal and gravitational potential energy, and form cold/molecular gas which in turn forms stars (for a review of the complex topic of star formation, see [7]). Through this process, the stellar mass of a galaxy is built up.

There is overwhelming evidence that this basic framework is correct (e.g., see [5]). However, there are important aspects of galaxy formation that appear not to fit with the simplest implementations of these ideas. In the rest of this section, we focus on two long-standing problems for which AGN feedback appears to provide a solution.

12.2.2
The Overcooling Problem

Galaxy formation theorists are very fortunate that the matter content of the Universe is dominated by dark matter. The fact that, for all practical purposes, dark matter particles only interact gravitationally allows for a much simplified calculation of the formation of structure in a dark matter-dominated Universe – one does not have to account for the action of thermal pressure, magnetic or radiation forces when modeling the gross evolution of structure. An important quantity is the *mass function* of the dark matter halos, $n(M, z)$, defined such that the number of halos per unit volume in the mass range $M \to M + \delta M$ at a redshift of z is given by $(\partial n/\partial M)\delta M$. The simplicity induced by the dark matter dominance of the Universe allows the halo mass function to be predicted as both a function of mass and redshift with a high degree of confidence.

To predict the stellar mass of a galaxy given a dark matter halo of some mass and size, we need to consider how baryonic matter cools and flows into the central regions of the dark matter halo. This problem was examined in a seminal paper by Rees and Ostriker [8] and has been subject to intense work since that time. The general picture that emerges is as follows. In sufficiently massive dark matter halos ($M > 10^{12}\ M_\odot$), a converging supersonic inflow of baryonic matter encounters an *accretion shock* which raises the temperature of the gas to the virial temperature

$$T_\mathrm{v} = \frac{2GM\mu m_\mathrm{p}}{3r_\mathrm{halo}k_\mathrm{B}} . \tag{12.4}$$

For illustrative purposes, we can evaluate T_v for a halo that forms at the present time such that ρ_0 is the current average density of the Universe; the result is $T \approx 3.5 \times 10^5 (M/10^{12}\ M_\odot)^{2/3}$ K. In the central regions of the halo, the shocked baryons form a hot atmosphere. This atmosphere is in a state of *hydrostatic equilibrium* where the pressure and gravitational forces are in balance,

$$\nabla p = -\rho \nabla \Phi , \tag{12.5}$$

where p is the pressure of the baryons, ρ is the mass density of the baryons, and Φ is the gravitational potential (which can be dominated by the dark matter). The central regions of the hydrostatic atmosphere then cool relatively gradually (principally via emission lines of carbon, oxygen, neon and iron) and eventually forms molecular gas clouds which become the site of star formation.

In less massive halos ($M < 10^{12}\ M_\odot$), the situation can be rather different. Here, the cooling of the postshock gas is so fast that it can never form a hydrostatic equilibrium atmosphere. Without such an atmosphere, the standing accretion shock is unstable [9], and it seems likely that a large fraction of the baryons can accrete into the central regions of the cluster without shocking, that is, they accrete "cold" [10, 11] and can rapidly contribute to star formation within the galaxy.

The basic outcome of these considerations is that, in all but the most massive halos in the Universe (corresponding to clusters of galaxies), the majority of baryons

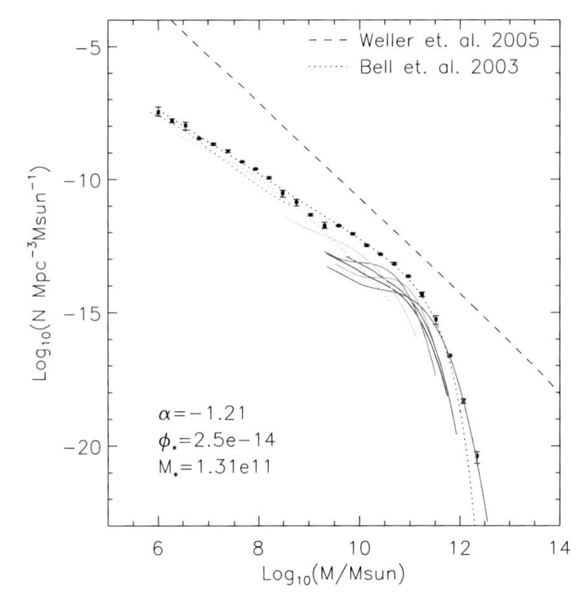

Figure 12.3 Baryonic mass functions for field galaxies (data points) compared with the dark matter halo mass function (dashed line). Reproduced from Read and Trentham [12] with permission from the Royal Astronomical Society.

within the sphere of influence of dark matter halos have had the chance to cool to low temperatures and form stars. Thus, if the dominant physics at work is that of gravitational collapse/infall and cooling, the stellar mass function (defined such that the number of galaxies per unit volume with stellar mass in the range $M_* \rightarrow M_* + \delta M_*$ is given by $[\partial n_*/\partial M_*]\delta M_*$) should simply mirror the dark matter halo mass function, $n_*(M) \propto n(M)$.

This simple expectation fails; compared to the population of dark matter halos, we observe too few low-mass galaxies and too few high-mass galaxies (see Figure 12.3). We conclude that there are agents at work that are preventing baryons within the halo from cooling or, alternatively, ejecting the baryons from the halo altogether. This mis-match between the expected efficiency of baryon cooling and the observed inefficiency of cooling is the *overcooling problem*.

An obvious and important agent is energy injection into the baryonic matter from star formation itself, specifically energy injection from radiation pressure, stellar winds, and supernova explosions. Calculations (e.g., see [5]) suggest that of order 10^{42} J of energy is injected into the interstellar medium for each solar mass of stars formed, that is, a fraction $\eta_* = 5 \times 10^{-6}$ of the rest-mass energy of the star-forming gas is fed back into the interstellar medium. In the optimal case where half of the baryonic matter turns into stars and heats the other half, we can follow arguments similar to those presented in Section 12.1 to show that *stellar feedback* would heat the interstellar medium to at most $T_* \sim 3 \times 10^6$ K. Thus, we can see that stellar feedback can have a dramatic impact on low- to moderate-mass galaxies (where the virial temperature of the dark matter halo is $T_\text{v} \ll T_*$) but is insufficient to heat baryons in the most massive galaxies. An energy source which is more efficient than star formation (in the sense of producing more heating per baryon than possible via star formation) is required to explain the deficit of high-mass galaxies. As we saw in Section 12.1, AGN feedback satisfies this requirement.

12.2.3
The Cooling Flow Problem

A related but distinct problem is the *cooling flow problem* which manifests itself on a variety of scales, from (large) isolated elliptical galaxies to large clusters of galaxies [35]. In all of these systems, a substantial fraction of the baryonic mass is observed to be in the form of hot gas close to the virial temperature of the system. In isolated elliptical galaxies ($T_{\rm v} \sim 1\text{–}5 \times 10^6$ K), much of this hot *interstellar medium* may result from stellar winds. However, in clusters of galaxies ($T_{\rm v} > 10^7$ K), this *intracluster medium* (ICM) is precisely the hydrostatic atmosphere of postshock accreted intergalactic baryons that was discussed in Section 12.2.2.

The hot ISM/ICM can be studied using X-ray observations. Under the conditions found in these ISM/ICM atmospheres, X-rays are produced by thermal bremsstrahlung and collisionally excited line emission. Both of these are two-body processes and, hence, the rate of emission of radiation with frequency ν per unit volume of gas is given by $\mathcal{L}(\nu)d\nu = n_{\rm e}^2 \Lambda(T, Z; \nu)d\nu$, where $n_{\rm e}$ is the number density of electrons in the gas and Λ is the so-called cooling function which depends only on the gas temperature T, metallicity Z, and X-ray frequency ν. Furthermore, the atmosphere as a whole is optically thin, that is, essentially all emitted photons can leave the cluster and be observed. Modern X-ray observatories such as the Chandra X-ray Observatory and the *XMM-Newton* Observatory have the capability to simultaneously image and perform spectroscopy of these ISM/ICM atmospheres. At each point in the image, the observed X-ray surface brightness SB is just the sum of the emission along the line-of-sight, that is,

$$\text{SB} \propto \iint n_{\rm e}^2 \Lambda(T, Z; \nu) A(\nu) d\nu dl\,, \tag{12.6}$$

where $A(\nu)$ is the effective collecting area of the observatory to X-rays of frequency ν, and the l-integral is along the line-of-sight through the galaxy/cluster. Given knowledge of the temperature T and metallicity Z, and an assumed geometry (e.g., ellipsoidal) the observed X-ray image can be "deprojected" in order to determine the electron density as a function of radius from the galaxy/cluster center, $n_{\rm e}(r)$. The temperature $T(r)$ and metallicity $Z(r)$ can be determined using the X-ray spectrum, again requiring an analysis that explicitly accounts for projection effects.

Once we have determined $n_{\rm e}(r)$, $T(r)$, and $Z(r)$ from X-ray observations, we can compute some interesting derived quantities. Of particular interest is the thermal enthalpy content per unit volume of the gas (i.e., the energy extracted per unit volume of gas if it cools to $T = 0$ at constant pressure) $\epsilon = (5/2)nk_{\rm B}T$, and the cooling rate $\mathcal{L} = n_{\rm e}^2 \Lambda(T, Z)$, where $\Lambda(T, Z) = \int \Lambda(T, Z; \nu)d\nu$. The ratio of these

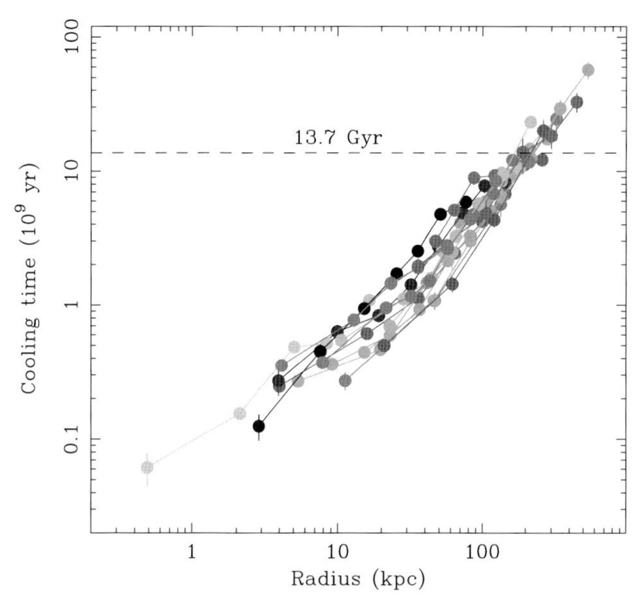

Figure 12.4 Cooling times as a function of radius for a sample of cooling core clusters as determined by the Chandra X-ray Observatory. The dashed line represents a cooling time equal to the age of the Universe according to the WMAP standard cosmology, 13.7 Gyr. Reproduced from Voigt and Fabian [13] with permission from the Royal Astronomical Society.

quantities gives the cooling time of the gas

$$t_{\text{cool}} = \frac{5k_{\text{B}}T}{2n\Lambda(T,Z)} . \tag{12.7}$$

Figure 12.4 shows the measured cooling times of the ICM as a function of radius for a sample of relaxed galaxy clusters, that is, clusters that appear *not* to have recently undergone a major disturbance due to a cluster-cluster merger event. Based on models of cosmological structure formation, it is reasonable to assume that most of these clusters have been in a relaxed state (suffering at most minor mergers) for the past 5 Gyr or so – the cooling time falls below 5 Gyr within approximately 100 kpc. This is traditionally taken to define the *cooling radius*; the region within the cooling radius is the *cooling core*. Within the cooling core, the cooling times can be even shorter, with times as short as 0.1 Gyr in the central-most regions. Left unchecked, this cooling would lead to large quantities of cold gas and, presumably, associated star formation.

A naive calculation can be used to estimate the rate at which mass should be cooling and "dropping out" of the hot ISM/ICM phase in the cooling core. If we assume that the observed luminosity of the cooling core L_{C} is dominated by gas cooling from the virial temperature T_{v} down to zero, then we have

$$L_{\text{C}} = \frac{5\dot{M}k_{\text{B}}T_{\text{v}}}{2\mu m_{\text{p}}} , \tag{12.8}$$

where $\dot{M}$ is the rate at which mass is cooling. This simple formula predicts mass cooling rates of $\sim 10\,M_\odot\,\mathrm{yr}^{-1}$ in isolated elliptical galaxies and $\sim$ 100–1000 $M_\odot\,\mathrm{yr}^{-1}$ in large clusters of galaxies. Integrated over 1 Gyr, this amounts to $10^{10}\,M_\odot$ of cold gas in elliptical galaxies and 10^{11}–$10^{12}\,M_\odot$ of cold gas in the cores of large clusters.

However, observations show no indications for large deposits of cold gas that are this large. While some star formation is sometimes seen in these systems, the star formation rate is at least an order of magnitude below that suggested by the X-ray cooling (e.g., see [14]). Similarly, while there are significant quantities of cold gas seen in the form of extensive systems of H_α-emitting filaments, the amount of cold gas is an order of magnitude smaller than suggested by the X-ray cooling. This discrepancy between the observed cooling and the lack of cold/star-forming gas is the *mass-sink cooling flow problem* [15].

A more direct form of the cooling flow problem results from high-resolution X-ray spectroscopy of cooling core clusters. For this discussion, it is useful to define the emission measure of a volume V of gas, $\mathrm{EM} = \int n_\mathrm{e}^2 dV$; the luminosity from a region with a given temperature and metallicity is then given by $L = \mathrm{EM}\Lambda(T, Z)$. If the cooling core of the galaxy/cluster is in steady state, with a continuous and steady supply of gas cooling radiatively from the virial temperature down to zero, the emission measure of gas at each temperature will be given by

$$\frac{d\mathrm{EM}}{dT} = \frac{5\dot{M}k_\mathrm{B}}{2\mu m_\mathrm{p}\Lambda}\,, \tag{12.9}$$

where the integral defining $d\mathrm{EM}(T)$ is only taken over regions with temperature $T \rightarrow T + dT$, and $\dot{M}$ is the rate at which mass is being supplied to this process. High-resolution spectroscopy of galaxy/cluster cores, such as is currently possible using the Reflection Grating Spectrometer (RGS) on board the *XMM-Newton* Observatory, allow us to identify specific X-ray emission lines resulting from gas cooling through a specific range of temperatures. This, in turn, allows us to study the distribution of emission measures with temperature, $d\mathrm{EM}/dT$. The results are surprising. Gas appears to cool unchecked from the virial temperature T_v down to $T \sim 0.1$–$0.3\,T_\mathrm{v}$; but there is no evidence for gas cooling below this temperature floor. This form of the cooling problem has been called the *soft X-ray cooling flow problem* [15].

Again, we are led to the conclusion that some agent is preventing hot baryons in these galaxies/clusters from cooling. The agent must be efficient, allowing the baryons to maintain temperatures of 10^7 K or more, again implicating AGN feedback. Unlike the overcooling problem discussed in Section 12.2.2 which largely relates to the epoch of major galaxy formation, the cooling flow problem is relevant here and now to nearby systems that can be studied in great detail. As we shall see in Sections 12.3.2 and 12.3.3, detailed observations of nearby galaxy clusters provide the most direct evidence to date that AGN jets have significant impact on their galactic scale environment.

12.3 Jet–ICM Interactions in Galaxy Clusters

In the proceeding section, we have described how the overcooling and cooling flow problems lead us to the notion that, in the more massive dark matter halos, some fraction of the accreted baryons are prevented from cooling. We found that the energy source must be efficient (in terms of energy released per baryon) and noted that AGN feedback fits the bill. However, the efficiency argument merely shows that AGN feedback is plausible. We must seek more direct evidence that AGN feedback is actually at play in real galaxies.

In this section, we discuss direct evidence of powerful interactions between jetted AGN and the surrounding ICM in nearby galaxy clusters. These observations are made possible by the modern generation of high-resolution imaging X-ray observatories, currently exemplified by the Chandra X-ray Observatory, that allow us to map out the ICM of nearby galaxy clusters in exquisite detail. Through these kinds of data, we can gain insights into whether AGN are truly the agents regulating baryon cooling and, if so, how the feedback loop really works.

12.3.1 Theoretical Expectations

We begin by discussing the theory of how relativistic jets from an AGN interact with a surrounding ISM/ICM atmosphere. For now, we treat the ICM using the framework of ideal hydrodynamics. We shall show later that the assumption of ideal hydrodynamics misses some important aspects of the system (especially the role of conduction and magnetic fields), but it is a useful starting point.

Consider a galaxy containing a central SMBH surrounded by an atmosphere of hot ISM/ICM which is initially quiescent, that is, in a state of hydrostatic equilibrium. We suppose that at time $t = 0$, some event triggers accretion onto the central SMBH and the formation of back-to-back powerful relativistic jets. For powerful jets (or in the early stages of a less-powerful jet), the jet propagates at relativistic speeds in an almost lossless manner before terminating in a series of shocks. At the terminal shocks, the bulk flow of the jet plasma is slowed from relativistic to nonrelativistic speeds. The shocked jet plasma then produces a backflow which inflates a cocoon of relativistic plasma[2)] (see Figure 12.5). At least at early times, the cocoon can have a significantly higher pressure than the surrounding ISM/ICM atmosphere. Consequently, the cocoon undergoes a pressure-driven expansion into the surrounding ISM/ICM which is supersonic with respect to the sound speed in the ISM/ICM atmosphere. This expansion drives a shock wave into the atmosphere, and the expanding cocoon sweeps up a shell of shocked ISM/ICM. The cocoon of relativistic plasma is separated from the shocked ISM/ICM by a *contact*

2) This plasma is relativistic in the sense that the random velocities of the particles is relativistic. Indeed, in many observed radio galaxies, the electrons in the freshly shocked jet plasma produces X-ray synchrotron radiation, which suggests that the electrons are ultrarelativistic with Lorentz factors $\gamma \sim 10^6$ or more.

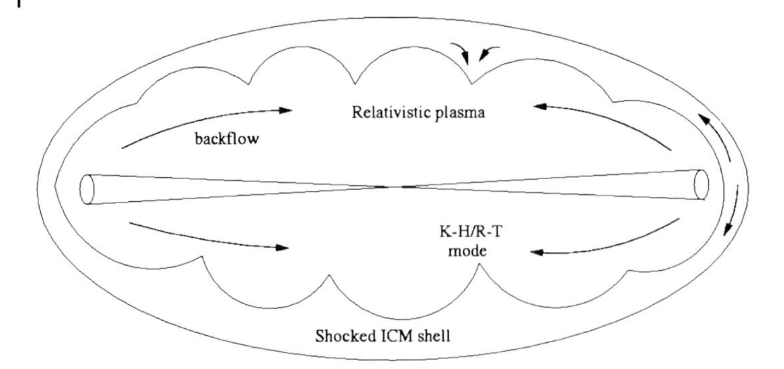

Figure 12.5 Schematic representation of the cocoon/shell structure associated with a powerful AGN jet. Back-to-back jets are created by a central AGN and propagate in a lossless manner until they terminate in a strong shock. The shocked jet plasma forms a backflow that inflates a cocoon. At early times, this cocoon is overpressured with respect to the surrounding ICM and, hence, expands supersonically into the ICM. The result is a shocked ICM shell, separated from the cocoon by a contact discontinuity. Over time, Kelvin–Helmholtz instabilities act to destroy the contact discontinuity, mixing the shocked ICM into the cocoon.

discontinuity; this interface is unstable to the *Kelvin–Helmholtz (K–H) instability* which results from the strong backflow of relativistic plasma parallel to the interface. This is fundamentally the same instability that produces surface waves on the ocean when the wind blows horizontally across the water. In the cocoon/ICM case, the K–H instability produces rolls in the contact discontinuity which act to mix the shocked ICM and the relativistic cocoon plasma.

If the jet power is constant or declining with time, the cocoon expansion decelerates unless the density gradient in the surrounding atmosphere is very steep. This can be seen from elementary considerations. Suppose the jet power is $L_{\rm j}$ which is constant in time, and the cocoon that it drives expands into an ISM/ICM atmosphere with density $\rho_{\rm icm}$. Suppose that the atmosphere has a density that declines with radius in a manner described by $\rho_{\rm icm,0}(r/r_0)^{-\beta}$, where $\rho_{\rm icm,0}$ is the density at some reference point $r = r_0$ and the parameter $\beta > 0$ describes the steepness of the density gradient. Finally, suppose the cocoon that the jet inflates has a fixed shape (i.e., a spheroid with fixed ratios between the major and minor axes) with length $x_{\rm c}(t)$ which grows with time; thus the volume of the cocoon as a function of time is $V_{\rm c} \propto x_{\rm c}^3$. Roughly, the energy required to inflate the cocoon to a volume $V_{\rm c}$ is given by $E_{\rm c} \sim P_{\rm c} V_{\rm c} \propto P_{\rm c} x_{\rm c}^3$, where $P_{\rm c}$ is the pressure in the cocoon. If the cocoon is highly overpressured relative to the background atmosphere, momentum balance across the shock front allows us to say that $P_{\rm c} = \rho_{\rm icm} v_{\rm c}^2$, where $v_{\rm c} = dx_{\rm c}/dt$ is the expansion velocity of the cocoon. Putting together these pieces, we can eliminate $E_{\rm c}$ and $P_{\rm c}$ and find that

$$L_{\rm j} t = A\rho_{\rm icm,0} \left(\frac{dx_{\rm c}}{dt} \right)^2 x_{\rm c}^{3-\beta} , \tag{12.10}$$

where A is a constant of proportionality. Solving this differential equation (and noting that $x_c = 0$ when $t = 0$) gives

$$x_c \propto \left(\frac{L_j}{\rho_{icm,0}}\right)^{1/(5-\beta)} t^{3/(5-\beta)} , \qquad (12.11)$$

or, differentiating with respect to time, $v \propto t^{(\beta-2)/(5-\beta)}$. Thus, provided that $\beta < 2$ (almost always the case in real astrophysical systems), we see that the cocoon slows at it expands. Of course, the deceleration of the cocoon is even stronger if the jet power declines with time.

It is therefore inevitable that the cocoon expansion ceases to be supersonic, at which point the bulk of the cocoon will come into pressure balance with the surrounding atmosphere. There are several consequences of this transition. Firstly, without a strong driver, the strong shock that is being driven into the ISM/ICM will weaken as it propagates outwards, eventually becoming better described as a strong sound wave. Secondly, the contact discontinuity now becomes subject to the *Rayleigh–Taylor (R–T) instability*. The relativistic plasma in the cocoon has very low density and has a much denser shell of shocked ISM/ICM resting on top of it. The R–T instability is fundamentally driven by the tendency for the dense fluid to fall through the less dense fluid. The combined action of the R–T and K–H can largely destroy the contact discontinuity, leading to a mixing of the relativistic plasma with the shocked ISM/ICM as well as back-to-back buoyantly rising "plumes" or "bubbles" of low-density mixed jet-plasma/ICM material (see Figure 12.6).

Within this theoretical framework, how does AGN feedback occur? There are two sides of AGN feedback: (i) how does the energy of the AGN jet get converted into thermal energy of the ISM/ICM?, and (ii) how is the power of the AGN jets regulated so as to, on average, offset the large scale radiative cooling? In general terms, it is relatively easy to thermalize the jet energy in the ISM/ICM through a combination of strong shock heating, dissipation of weak shocks/sound waves, and cocoon-ISM/ICM mixing. The hard part is understanding how the jet power (which is determined by conditions within 0.01 pc of the SMBH) is regulated so as to match the radiative cooling of the ISM/ICM core (on scales of 1–100 kpc). The most common viewpoint is that the system maintains a *regulated cooling flow*; the jet heating allows just enough net cooling to occur in the ISM/ICM to maintain a small trickle of mass to flow from the atmosphere into the sphere of influence of the black hole and the subsequent accretion into the black hole powers the jets. If the jet power is temporarily increased, the ISM/ICM atmosphere is overheated, the accretion of mass into the SMBH is reduced and the jet power reduced. Conversely, if the jet power is temporally reduced, the ISM/ICM undergoes excess cooling, the accretion of mass into the SMBH is increased, and the jet power increases.

Having laid out the theoretical framework used to describe the interaction of a jetted AGN with the ICM, we turn to observations of real systems.

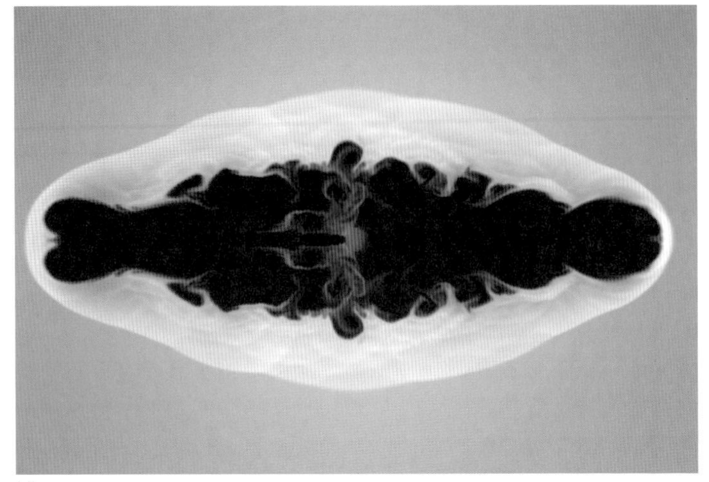

(a)

(b)

Figure 12.6 Two snapshots from a computer simulation showing the evolution of a jet-blown cocoon. Shown here is a rendering of the gas density on a slice that bisects the cocoon along the jet axis. (a) The cocoon during its overpressured expansion phase. The jets themselves have a very low density and are not readily apparent in this figure. The shocked shell of ICM is very apparent. (b) The evolution of the system long after the jet activity has ceased. The cocoon has broken into two buoyantly rising plumes. The (axisymmetric) simulations are described in [16].

12.3.2
Jet-Blown Cavities

A jet-blown cocoon and resulting buoyant plumes displace the X-ray-emitting ISM/ICM and, hence, show up as distinct depressions in maps of X-ray surface brightness. In fact, these *X-ray cavities* are one of the clearest indications of a jet/ICM interaction.

The Perseus cluster (at a distance of 75 Mpc from Earth) is the galaxy cluster with the highest X-ray surface brightness on the sky and displays a spectacular set of jet-blown cavities associated with the radio-loud AGN hosted by the central galaxy of this cluster, NGC 1275 (see Figure 12.7). By looking at clusters such as Perseus, we can identify two types of cavities. *Active cavities* are coincident with the jet-fed radio-emitting cocoon/lobes. They are usually very distinct in surface brightness maps and extend almost down to the center of the cluster. In Perseus, the central "figure of eight" structure in Figure 12.7 defines the active cavities; the cavities are coincident with the bright radio lobes powered by the jets from the core of NGC 1275. It is tempting to identify the shells surrounding these active cavities as the shocked shell of ICM. However, X-ray spectroscopy reveals that these shells are actually cooler than the ambient ICM and, hence, they appear to be shells of ICM that have been uplifted without being strongly shocked – this demonstrates that the cocoon of this radio galaxy is expanding subsonically. Indeed, the vast majority of observed active cavities appear to be expanding subsonically and hence must be in approximate pressure equilibrium with the surrounding ISM/ICM. The second type of cavity are the *ghost cavities*. These are found further from the cluster and appear disconnected from the active jet-fed cocoon. In Perseus, several ghost cavities have been identified but the clearest is the crescent shaped depression visible towards the top-right of Figure 12.7. These structures are previous jet-blown cavities that have now become disconnected from the currently active cocoon and are buoyantly rising within the ICM atmosphere; direct evidence for this comes from the discovery of low-frequency radio emission coincident with ghost cavities corresponding to synchrotron radiation from an old population of electrons. It is currently unclear whether the existence of distinct ghost cavities is due to discrete bursts of activity by the AGN, or random/turbulent motions within the cluster ("ICM weather") that can cause a cocoon to separate from a continuously active jet.

ICM cavities have taken on a central importance in studies of AGN feedback because they provide an opportunity to estimate jet power. More precisely, we can use X-ray observations to estimate the power needed to inflate the cavities within the observed atmosphere. To compute a power, we need to determine the energy required and the time scale over which that energy was delivered. Firstly, let us discuss the determination of the energy required to inflate the cavities. Suppose that a cavity is inflated to a volume V_c in an atmosphere with pressure P_a; both of these quantities can be estimated from X-ray observations of the cavity and the surrounding ICM atmosphere. Furthermore, let us suppose that most of this inflation occurs in the subsonic regime, when the cavity is in approximate pressure equilibrium with the surrounding atmosphere. An estimate for the energy required to inflate the cavity (including the internal energy of the cavity and the work done to displace the ICM) is $E_c = \gamma_c P_a V_c/(\gamma_c - 1)$ where γ_c is the adiabatic index (or ratio of specific heats) for the material within the cocoon. If the cocoon material is highly relativistic, we can take $\gamma_c = 4/3$, giving to $E_c = 4P_a V_c$; if large amounts of ISM/ICM have been mixed into the cocoon hence making it nonrelativistic, it is more appropriate to take $\gamma_c = 5/3$ in which case $E_c = 5/2 P_a V_c$.

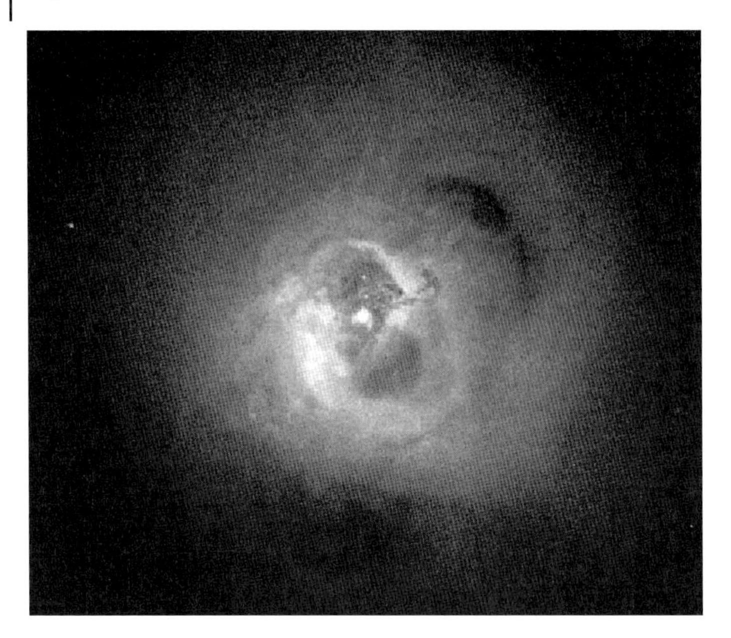

Figure 12.7 X-ray image of the central regions of the Perseus cluster of galaxies as seen by the Chandra X-ray Observatory. The central active cavities are clearly seen, as is the crescent shaped ghost cavity to the northwest (upper right) of the cluster center. Reproduced from Fabian *et al.* [17] with permission from the Royal Astronomical Society.

The time taken to inflate the cavities can be estimated in two different ways. We can calculate the time required for the outer edge of the cavity to travel from the center of the cluster to its present location assuming that it traveled at the sound speed of the ICM. The sound speed, c_s, is related to the temperature of the gas by $c_s = \sqrt{\gamma_{icm} k_B T_{icm}/\mu m_p}$ where $\gamma_{icm} = 5/3$ is the adiabatic index of the ICM and T_{icm} is the characteristic temperature of the inner parts of the ISM/ICM atmosphere (which can be determined observationally). If the leading edge of the cavity is a distance R from the cluster center, we can estimate the cavity age as $\tau_s = R/c_s$. This method is best suited for active cavities. Alternatively, we can assume that the cavity is buoyantly rising and has achieved its buoyant terminal velocity, ν_{buoy}. The theory of buoyantly rising bubbles in a dense fluid gives that $\nu_{buoy} = \sqrt{2gV_c/SC}$ where g is the local acceleration due to gravity, S is the cross-sectional area of the cavity, and C is the "drag coefficient" which computer simulations suggest has the value $C = 0.75$. The age of the cavity is then $\tau_{buoy} = R/\nu_{buoy}$. Since it relies upon the assumption that the bubble has achieved its buoyant terminal velocity, this technique is best suited to ghost cavities. For a given cavity, the age estimate τ_s is usually a factor of two smaller than the age estimate τ_{buoy}. Once an appropriate estimate for the lifetime of the bubble has been computed (τ_{buoy} or τ_s), this can be combined with the estimate of the energy needed to inflate the bubble in order to compute the power needed to inflate the bubble.

Armed with an estimate of the jet-power from an analysis of X-ray cavities, we can test a central tenet of the AGN feedback hypothesis: that the time-average jet power

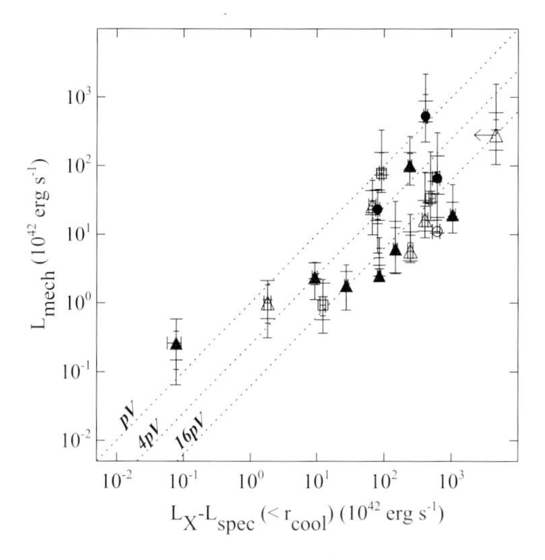

Figure 12.8 Comparison of the inferred mechanical luminosity from the AGN jets (L_{mech}, inferred from the properties of the X-ray cavities) and the "missing" cooling luminosity of the core for a sample of cooling core clusters. The missing cooling luminosity is expressed in terms of difference between the total observed X-ray luminosity (L_X) and the luminosity corresponding to the matter that is actually seen to cool to low temperatures (L_{spec}, determined spectroscopically). The lines show the loci of points for which $L_{mech} = L_X - L_{spec}$ under the assumption that the energy content of the cavities is given by pV, $4pV$ and $16pV$, respectively. Reproduced from Birzan *et al.* [18] with permission from the AAS.

is comparable to the radiative luminosity of the ISM/ICM within the cooling core. Figure 12.8 shows the results of such an analysis of a sample of 16 galaxy clusters by Birzan and collaborators [18]. Indeed, we see the expected correlation between the "mechanically" injected energy and the cooling luminosity although, in many clusters, the power inferred from the cavities is a factor of a few-to-10 too small to offset the cooling. This mismatch does not significantly stress the AGN feedback paradigm; however, it is easy to envisage that a significant fraction (50–90%) of the jet power is *not* used to inflate the cavities but, for example, drives compressional disturbances into the ICM (i.e., weak/strong shocks and sound waves). Indeed, the AGN feedback that we seek as an answer to the overcooling or cooling-flow problems must heat the bulk of the ICM in the cooling ISM/ICM core; the inflation of well-defined cavities is insufficient. Thus, it is interesting to discuss compressional disturbances in more detail.

12.3.3 Shocks and Sound Waves

Figure 12.9 shows the X-ray image of the Perseus cluster, using a technique known as *unsharp masking* to enhance the "edge-like" disturbances in the ICM. In addition to the cavities that were discussed above, these data reveal an an extensive set of

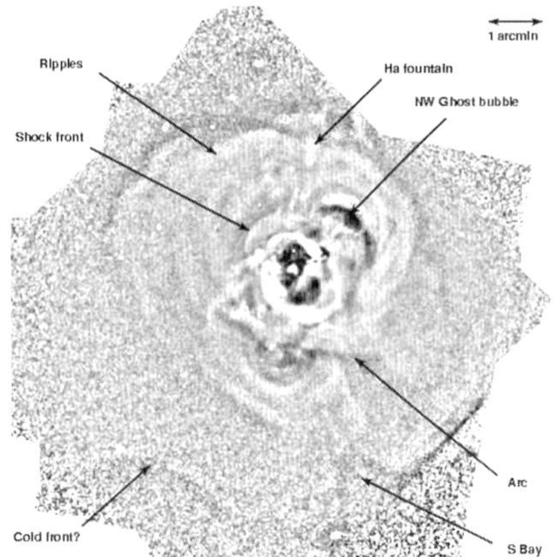

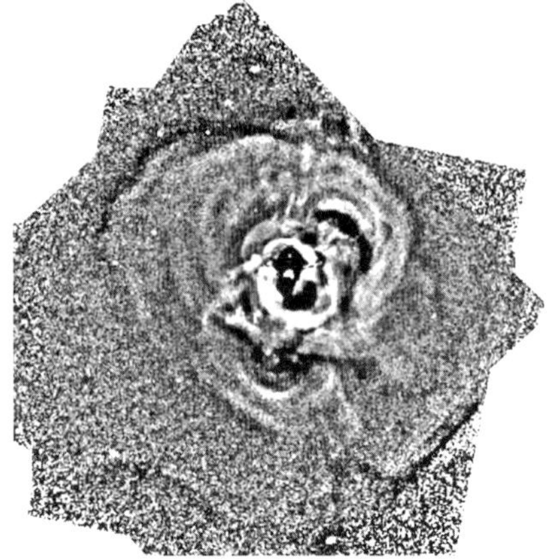

Figure 12.9 Unsharp mask image of the central regions of the Perseus cluster from the Chandra X-ray Observatory. To form this image, a heavily smoothed image (Gaussian smoothing with a dispersion of 10 arcsec) is subtracted from a lightly smoothed image (Gaussian smoothing with dispersion of 1 arcmin). This technique enhances edges and other small scale structure. Reproduced from Fabian *et al.* [17] with permission from the Royal Astronomical Society.

"ripples" that are clearly centered on the AGN. Here we discuss these structures and the role that they may play in mediating AGN heating of the ICM.

Firstly, we will discuss shock waves. As was already discussed in Section 12.3.1, a very powerful radio galaxy (or the early stages of a less-powerful radio galaxy) can produce a jet-blown cocoon which is strongly overpressured with respect to the surrounding ISM/ICM. The result is that the cocoon expands supersonically (with respect to the sound speed in the ambient ISM/ICM) and drives a strong shock wave into this atmosphere (see Figure 12.5).

It is useful to briefly review the physics of hydrodynamic shocks. The density, temperature, and pressure of the gas change discontinuously across the shock front. Suppose that the unshocked ISM/ICM has a sound speed c_s, and that a shock is sweeping across this material at a speed $v_{\rm shock}$; thus, the *Mach number* of the shock is $\mathcal{M} = v_{\rm shock}/c_s$. Suppose that the density and pressure of the unshocked (ambient) gas is ρ_1 and p_1, and the corresponding quantities for the shocked gas are ρ_2 and p_2. Across the shock front, there is a discontinuous change in density and pressure between these two sets of values. By transforming to a frame of reference in which the shock front is stationary and the analyzing the equations of mass, momentum, and energy conservation, we find

$$\frac{\rho_2}{\rho_1} = \frac{(\gamma+1)\mathcal{M}^2}{(\gamma+1)+(\gamma-1)(\mathcal{M}^2-1)} = \frac{4\mathcal{M}^2}{\mathcal{M}^2+3}\,, \tag{12.12}$$

$$\frac{p_2}{p_1} = \frac{(\gamma+1)+2\gamma(\mathcal{M}^2-1)}{(\gamma+1)} = \frac{1}{4}(5\mathcal{M}^2-1)\,, \tag{12.13}$$

where the second equality of each line assumes that the adiabatic constant of the gas is $\gamma = 5/3$. It is interesting to also look at the jump of entropy across the shock

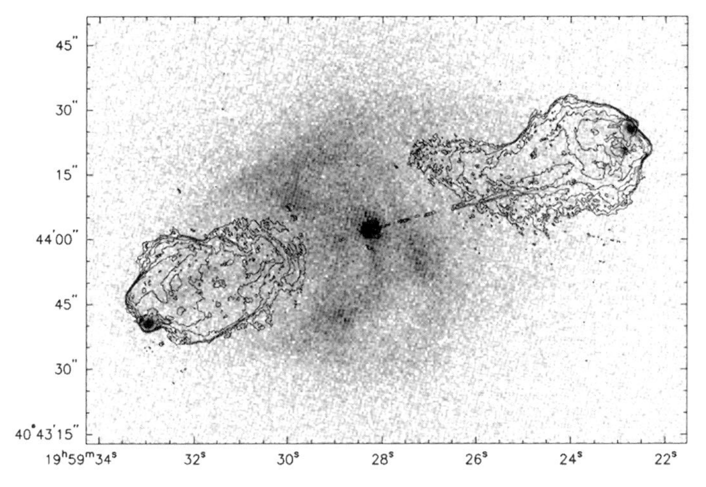

Figure 12.10 Contours of 6-cm radio intensity (from the Very Large Array) overlaid on the X-ray image (from the Chandra X-ray Observatory) of the powerful radio galaxy Cygnus A. Note the ellipsoidal enhancement in X-ray flux corresponding to the cocoon-bounding ICM shock. Reproduced from Wilson *et al.* [19] with permission from the AAS.

front since it is this quantity that corresponds to *irreversible* heating of the gas. For an ideal gas, the entropy per particle is $S = k_B \ln(p\rho^{-\gamma})$. Thus, let us examine the jump of the quantity $s = p\rho^{-\gamma}$ across the shock front. For strong shocks ($\mathcal{M} \gg 1$), we find that the fractional entropy jump is given by

$$\frac{s_2 - s_1}{s_1} \approx \frac{5}{4^{8/3}}\mathcal{M}^2 \gg 1 \,, \tag{12.14}$$

demonstrating that strong shocks are, in principle, an effective way of irreversibly heating the ISM/ICM.

Despite the theoretical appeal, however, it is unclear that heating by strong shocks plays an important role in real AGN feedback. Figure 12.10 shows the Chandra image of the famous radio galaxy Cygnus A and reveals an enhanced emission from a prolate-spheroidal region which [19] identify as the cocoon-driven ICM shock. However, even in Cygnus A, the most powerful radio galaxy in our local Universe, detailed analysis of the shock suggests a Mach number of only $\mathcal{M} \approx 1.3$ [19] – this is hardly a very strong shock! There are several other clusters that seems to display AGN-driven shocks of similar strengths but, in aggregate, heating by strong AGN-driven shocks appears not to be a common or important process.

Let us turn our attention to the weaker compressional disturbances, that is, weak shocks and sound waves. If we assume ideal hydrodynamics (i.e., neglecting viscosity and thermal conduction) sound waves are dissipationless and hence cannot heat the ICM. Even weak shocks are only very weakly dissipated. To see this, let us examine the shock-jump equations (12.12) in the case where the shock front is only barely supersonic, $\mathcal{M}^2 = 1 + \epsilon$, with $\epsilon \ll 1$. By substituting this form into (12.12) and Taylor expanding, we get

$$\frac{s_2 - s_1}{s_1} = \frac{5}{48}\epsilon^3 + O(\epsilon^4) , \quad (12.15)$$

that is, the entropy jump in weak shocks is third order in the small parameter ϵ and hence is very small.

However, ideal hydrodynamics is not an adequate description of the ISM/ICM. Of importance for the present discussion is the fact that both the viscosity and the thermal conductivity of the gas can be significant.[3] Both of these processes can dissipate even linear (small amplitude) sound waves. We define the dissipation length ℓ to be the distance over which the energy flux carried by the disturbance decreases by a factor of e. For a sound wave of frequency $f = 10^{-5} f_{-5}$ yr, the dissipation length is

$$\ell = 6.97 \frac{n_e T_7^{-1} f_{-5}^{-2}}{\left(\frac{\xi_\nu}{0.1}\right) + 11.8\left(\frac{\xi_\kappa}{0.1}\right)} \text{ kpc} , \quad (12.16)$$

where $T = 10^7 T_7$ K is the temperature of the gas, and ξ_ν and ξ_κ are the "suppression factors" for the viscosity and thermal conductivity, respectively. These suppression factors account, in a phenomenological way, for the suppression of viscosity and conduction by magnetic fields. The fact that the dissipation length is comparable to or shorter than a typical cooling core tells us that sound waves and weak shocks can be dissipated within the cooling core and hence may be a relevant mechanism for AGN feedback. Indeed, in contrast to the case for strong shocks, weak shocks and sound waves appear to be very common, being seen in essentially every ICM atmosphere for which sufficiently sensitive data have been obtained.

Let us explore this idea a little further, following the treatment of Fabian *et al.* [20]. Suppose that the central AGN injects an *acoustic luminosity* of L_{inj} at some radius $r = r_{\text{inj}}$ (which we may identify with the size of the jet-blown cocoon) and, furthermore, we assume that the injection of this acoustic energy is spherically symmetric and steady in time. The acoustic luminosity $L_s(r)$ at radius $r > r_{\text{inj}}$ decreases with increasing r due to the action of dissipation, and this dissipated energy heats the atmosphere. We now derive an expression for the heating rate per unit volume of the ISM/ICM. From the definition of dissipation length, we have

$$\frac{dL_s}{dr} = -\frac{L_s}{\ell} , \quad (12.17)$$

which has the solution

$$L_s(r) = L_{\text{inj}} \exp\left(-\int_{r_{\text{inj}}}^{r} \frac{1}{\ell(r')} dr'\right) . \quad (12.18)$$

3) In fact, due to the presence of a magnetic field in the ISM/ICM, both the viscosity and thermal conductivity become tensorial quantities with a high degree of anisotropy. The anisotropic conduction, in particular, renders the atmosphere susceptible to a surprising host of instabilities. We shall briefly discuss these in Section 12.4.

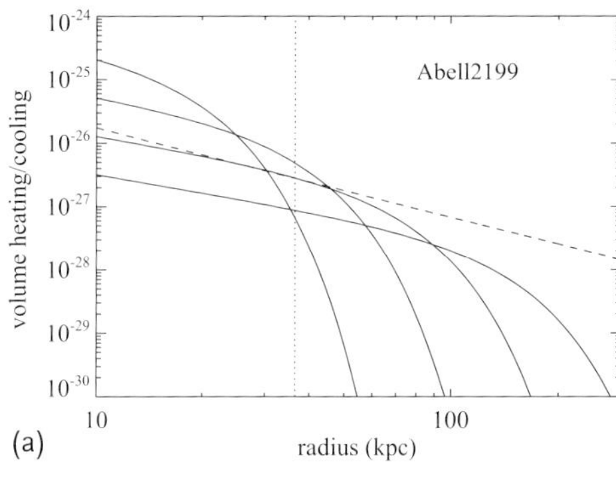

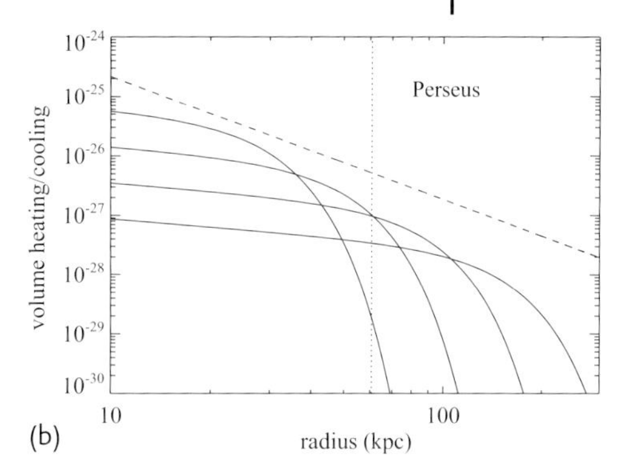

Figure 12.11 Dissipative volume heating rates (solid lines) compared with radiative cooling rates (dashed lines) in the steady-state spherical acoustic heating model described in the text and applied to two clusters, (a) Abell 2199 and (b) Perseus. The heavy solid line is for the following set of illustrative set of parameters for which the heating and cooling approximately balance: $L_{\rm inj} = 10^{44}$ erg s^{-1} for A 2199, $L_{\rm inj} = 5 \times 10^{44}$ erg s^{-1} for Perseus, $f_{-5} = 1$, $\xi_\nu = 0.1$, $\xi_\kappa = 0$. Other choices of f_{-5}, $\xi_\nu = 0.1$, and $\xi_\kappa = 0$ change the dissipation lengthscale – the family of additional (thin) solid lines shows the effect of successive doublings of frequency ($f_{-5} = 2$, $f_{-5} = 4$, and $f_{-5} = 8$). The vertical dashed lines indicate a fiducial cooling radius where the radiative cooling time of the gas is 3 Gyr. Reproduced from Fabian *et al.* [20] with permission from the Royal Astronomical Society.

By considering a shell of thickness δr, we see that the volume heating rate is given by

$$\epsilon_{\rm heat} = -\left(\frac{1}{4\pi r^2 \delta r}\right)\left(\frac{d L_{\rm s}}{d r}\right)\delta r = \frac{L_{\rm s}(r)}{4\pi r^2 \ell} \,. \tag{12.19}$$

Figure 12.11 shows this volume heating rate evaluated for the density and temperature structure of two well-studied clusters (Abell 2199 and the Perseus cluster). In both cases, we see that a reasonable choice of model parameters ($L_{\rm inj}$, f_{-5}, ϵ_ν and ϵ_κ) leads to a heating profile that approximately matches the radiative loses.

12.4
Thermal Conduction, MHD Instabilities, and an Alternative View of AGN Feedback

Above, we have presented the standard view of jet-mediated AGN feedback in elliptical galaxies and clusters: in essence the mechanical energy from the AGN jet is used to heat the ISM/ICM directly, thereby preventing cooling of the baryons. Underlying this discussion was the treatment of the ICM as a fluid that can be described by the equations of hydrodynamics. However, this is strictly not the correct framework for describing the ICM. The ICM is almost fully ionized and, at some level, magnetized. Thus one is led to consider the ICM as a magnetohydrodynamic (MHD) system.

Furthermore, as we have seen already in our discussion of sound wave dissipation, viscosity and thermal conduction can be significant within the ICM. Here, we shall be especially concerned with the physics of thermal conduction. Conduction is due to the streaming/diffusion of electrons from hot regions to colder regions. The gyroradius of an electron within the ICM moving in a magnetic field B is

$$r_e \approx 1 \times 10^8 \left(\frac{k_B T}{3\,\text{keV}}\right)^{1/2} \left(\frac{B}{1\,\mu\text{G}}\right)^{-1} \text{cm} , \tag{12.20}$$

where T is the temperature of the ICM. On the other hand, the mean free path of the electron to Coulomb collisions is

$$\lambda_e \approx 2 \left(\frac{k_B T}{3\,\text{keV}}\right)^{2} \left(\frac{n}{10^{-2}\,\text{cm}^{-3}}\right)^{-1} \text{kpc} , \tag{12.21}$$

where n_e is the electron number density of the ICM. We see that $r_e \ll \lambda_e$ for any plausible magnetic field strengths. Thus, electrons are strongly channeled along magnetic field lines and the thermal conduction is strongly anisotropic. An ionized gas in this regime is known as a *dilute plasma*.

Within the MHD framework, the dynamics of the ICM are described by the mass, momentum, and entropy equations which are, respectively,

$$\frac{\partial \rho}{\partial t} + \nabla \cdot (\rho \boldsymbol{v}) = 0 , \tag{12.22}$$

$$\rho \frac{D\boldsymbol{v}}{Dt} = \frac{(\nabla \times \boldsymbol{B}) \times \boldsymbol{B}}{4\pi} - \nabla P + \rho \boldsymbol{g} , \tag{12.23}$$

$$\frac{\partial \boldsymbol{B}}{\partial t} = \nabla \times (\boldsymbol{v} \times \boldsymbol{B}) , \tag{12.24}$$

$$\frac{D \ln P\rho^{-\gamma}}{Dt} = -\frac{\gamma - 1}{P} \left[\nabla \cdot \boldsymbol{Q} + \rho \mathcal{L}\right] , \tag{12.25}$$

where ρ is the mass density, $\boldsymbol{v}$ is the fluid velocity, $\boldsymbol{B}$ is the magnetic field, $\boldsymbol{g}$ is the local gravitational acceleration, P is the gas pressure, γ is the adiabatic index, $\boldsymbol{Q}$ is the conductive heat flux, and $\mathcal{L}$ is the radiative energy loss per unit mass of fluid. D/Dt is the Lagrangian derivative, $\partial/\partial t + \boldsymbol{v} \cdot \nabla$. Accounting for the anisotropic nature of the thermal conduction, the conductive heat flux is given by

$$\boldsymbol{Q} = -\chi \boldsymbol{b}(\boldsymbol{b} \cdot \nabla) T , \tag{12.26}$$

where $\boldsymbol{b}$ is a unit vector in the direction of the local magnetic field, T is the temperature and χ is the thermal conductivity:

$$\chi \simeq 6 \times 10^{-7} T^{5/2}\,\text{erg}\,\text{cm}^{-1}\,\text{s}^{-1}\,\text{K}^{-1} . \tag{12.27}$$

We shall see that the combination of MHD and anisotropic conduction has profound implications for the physics of the ICM, and may impact our view of AGN feedback.

12.4.1
The Near Impossibility of a Stable Hydrostatic Equilibrium

We can ask the following simple formal question. Suppose we set up a (weakly magnetized) ICM atmosphere in a state of hydrostatic and thermal equilibrium ($v = 0$, $B^2 \ll P$, $\nabla P = \rho \mathbf{g}$, $\nabla \cdot \mathbf{Q} = -\rho \mathcal{L}$). Then under what circumstances is this equilibrium a stable state?

To determine the stability of the equilibrium atmosphere, one supposes that the system is given a small perturbation away from the equilibrium state and then expands the equations describing the system to first order in the small perturbation. The resulting linear equations can then be analyzed to determine if the perturbation will grow monotonically with time (in which case the system is *unstable*), decay with time (in which case the system is *stable*), or undergo growing oscillations (in which case the system is *overstable*). In the case of pure hydrodynamics with no conduction and cooling ($\mathbf{B} = 0$, $\chi = 0$ and $\mathcal{L} = 0$), this procedure yields the result that a hydrostatic atmosphere is stable if and only if the entropy of the gas is increasing upwards. Violation of this criterion (known as the Schwarzschild criterion) leads to *convection*.

The stability of a MHD atmosphere with anisotropic conduction has been analyzed by a number of authors [21–25] and shown to be a rich and subtle subject. Provided one is in the regime where conduction along the field lines is effective, the Schwarzschild (entropy gradient) condition is shown to be irrelevant to the convective stability. Instead, convection is driven by radial temperature gradients. Remarkably, a temperature gradient of any sign can drive convective instability. For temperature decreasing upwards (hot on bottom, cold on top), the convective instability is known as the *magnetothermal instability* (MTI) and is essentially a conduction-modified form of the normal convection process [21, 22]. For temperature increasing upwards (cold on bottom, hot on top; this is the situation found in cluster cooling cores), the instability is known as the *heat flux-driven buoyancy instability* (HBI) and, as the name suggests, is actually driven by the conductive heat flux in the atmosphere [23]. Both of these instabilities lead to ICM turbulence. It is found that both of these instabilities can be shut off for special magnetic field geometries: the MTI no longer operates when the magnetic field lines are vertical, and the HBI shuts off for horizontal magnetic field lines. However, even these special configurations have recently been shown to be overstable (i.e., oscillate with exponentially growing amplitude) unless the temperature gradient is extremely shallow compared with the pressure gradient [25]. In essence, if one sets up a system in one of these special configurations, they possess g-modes (i.e., fluid waves in which the restoring force is due to gravity/buoyancy), which grow exponentially in amplitude. Again, it would be expected that turbulence would result.

Thus, we come to the remarkable conclusion that it is impossible for the ICM to form a hydrostatic atmosphere that is stable if there are significant temperature gradients. This conclusion is a formal one and does not imply that real atmospheres cannot be in a state that closely approximates hydrostatic equilibrium – they can! But those (almost) hydrostatic atmospheres will inevitably possess perturbations that lead to turbulence, even in the absence of any action by an AGN or subcluster mergers.

12.4.2 MHD Models of Cluster Cooling Cores and an Alternative Role for AGN

The potential importance of thermal conduction in cooling flow/core clusters has been recognized for a long time [26]. The reservoir of thermal energy in the ICM beyond the cooling radius is huge, leading one to naturally ask whether cluster cooling cores are entirely heated and stabilized against thermal collapse by the conduction of heat from the hotter outer regions of the ICM. Indeed, if one neglects any magnetic suppression, the conduction in all but the coolest clusters would be so powerful as to erase the temperature gradient entirely. The fact that cooling cores exist in the ICM of hot clusters at all shows that conduction must be suppressed below the unmagnetized-plasma value.

Early attempts to account for the magnetic suppression of conduction assumed that, on global scales, the fact that the conduction was forced to follow tangled magnetic field lines led to conduction that was effectively isotropic but scaled down by some global, fixed suppression factor $f \sim 0.01{-}0.3$ [27, 28]. However, to achieve thermal equilibrium, such models require fine-tuning of the suppression factor on a cluster-by-cluster basis [29]. Furthermore, assuming f is constant for a given atmosphere still leaves the atmospheres globally thermally unstable [30]. The nature of the thermal instability is easy to see. Suppose that the cooling core of an ICM atmosphere is in thermal equilibrium, with radiative loses being offset by the conductive heat flux. Consider a small isobaric perturbation which makes the core a little cooler and denser. The radiative luminosity of the core would increase, but the conductive heat flux (which is proportional to $T^{5/2}$) decreases. Thus the core will continue to cool, become denser in order to maintain hydrostatic equilibrium, and thus thermally collapse.

The realization that anisotropic conduction induces a host of MHD instabilities has reinvigorated the study of conduction within the ICM. The conductive heat flux associated with temperature gradients in the ICM now drive instabilities which produce turbulence, thereby tangling the magnetic field and altering the conductive flux. In principle, one could imagine that cooling cores are inherently self-regulating, with MHD turbulence (driven by the HBI and associated overstabilities) regulating the conductive heat flux into the core at a level required to stabilize the core against thermal collapse [24]. Global MHD simulations of cooling cores [31, 32] suggest that this inherent self-regulation is not achieved – the nonlinear evolution of the HBI acts to reorient the field to be orthogonal to the temperature gradient thereby shutting off the conductive heat flux and inducing thermal core collapse (Figure 12.12).

However, one must now look to the other agents that can introduce disturbances into the ICM, namely, subcluster mergers, the "wakes" induced by galaxies orbiting within the cluster, and the central AGN. These external disturbances act to scramble the magnetic field structure away from the "insulating" geometry in-

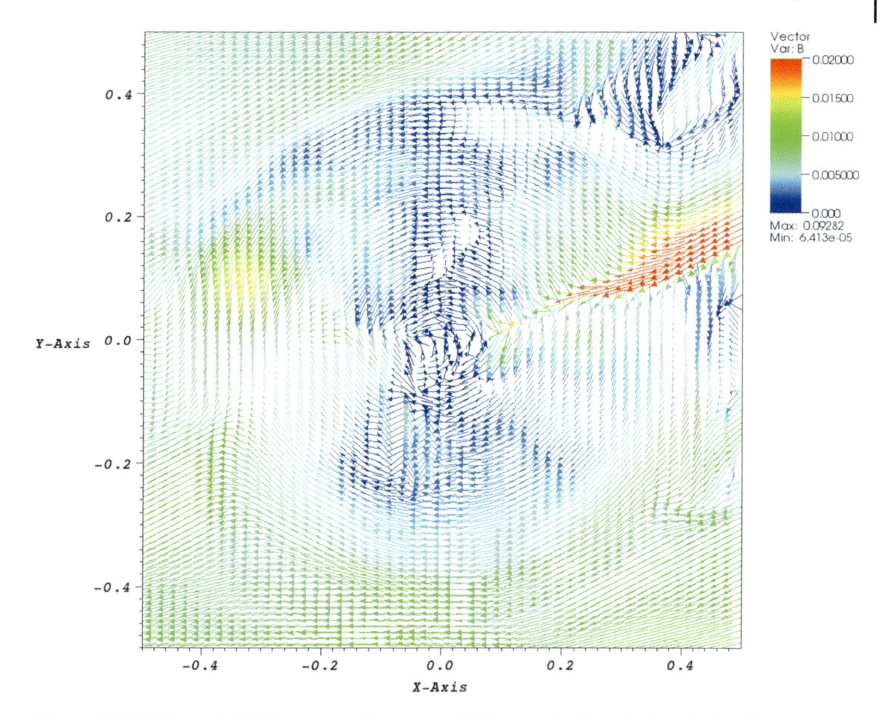

Figure 12.12 Magnetic field structure from a midplane slice through a three-dimensional computer simulation which follows the magnetohydrodynamics including the effects of anisotropic thermal conduction. The color represents the strength of the magnetic field. The action of the heat flux-driven buoyancy instability (HBI) has wrapped the magnetic field onto shells such that the local magnetic field is orthogonal to the temperature gradient (which is in the radial direction). In this particular simulation, this field line wrapping insulates the core of the cluster from the conductive heat flux causing it to undergo thermal collapse. Reproduced from Bogdanovic *et al.* [31] with permission from the AAS.

duced by the HBI, once again allowing conductive heat flux into the cluster cooling core [33, 34].

The role of conduction and these MHD instabilities is still very much a work in progress and it remains to be seen whether this physics is really of relevance to the cluster cooling core problem. However, it does raise the fascinating possibility that the AGN plays a crucial role, not as a *heater*, but as a "stirrer".

References

1 Magorrian, J. *et al.* (1998) *Astron. J.*, **115**, 2285.
2 Gultekin, K. *et al.* (2009) *Astrophys. J.*, **698**, 198.
3 Gebhardt, K. *et al.* (2000) *Astrophys. J.*, **539**, L13.
4 Ferrarese, L. and Merritt, D. (2000) *Astrophys. J.*, **539**, L9.
5 Benson, A.J. (2010) *Phys. Rep.*, **495**, 33.
6 Gunn, J.E. and Gott, J.R. (1972) *Astrophys. J.*, **176**, 1.

7 McKee, C.F. and Ostriker, E.C. (2007) *Annu. Rev. Astron. Astrophys.*, **45**, 565.

8 Rees, M.J. and Ostriker, J.P. (1977) *Mon. Not. R. Astron. Soc.*, **179**, 541.

9 Birnboim, Y. and Dekel, A. (2003) *Mon. Not. R. Astron. Soc.*, **345**, 349.

10 Fardal, M.A., Katz, N., Gardner, J.P., Hernquist, L., Weinberg, D.H., and Dave, R. (2001) *Astrophys. J.*, **562**, 605.

11 Keres, D., Katz, N., Weinberg, D.H. and Dave, R. (2005) *Mon. Not. R. Astron. Soc.*, **363**, 2.

12 Read, J.I. and Trentham, N. (2005) *Phil. Trans. R. Soc. A*, **363**, 2693.

13 Voigt, L.M. and Fabian, A.C. (2004) *Mon. Not. R. Astron. Soc.*, **347**, 1130.

14 Hicks, A.K. and Mushotzky, R.F. (2005) *Astrophys. J.*, **635**, L9.

15 Peterson, J.R. *et al.* (2003) *Astrophys. J.*, **590**, 207.

16 Reynolds, C.S., Heinz, S., and Begelman, M.C. (2002) *Mon. Not. R. Astron. Soc.*, **332**, 271.

17 Fabian, A.C., Sanders, J.S., Taylor, G.B., Allen, S.W., Crawford, C.S., Johnstone, R.M., and Iwasawa, K. (2006) *Mon. Not. R. Astron. Soc.*, **366**, 417.

18 Birzan, L., Rafferty, D.A., McNamara, B.R., Wise, M.W., and Nulsen, P.E.J. (2004) *Astrophys. J.*, **607**, 800.

19 Wilson, A.S., Smith, D.A., and Young, A.J. (2006) *Astrophys. J.*, **644**, L9.

20 Fabian, A.C., Reynolds, C.S., Taylor, G.B., and Dunn, R.H.J. (2005) *Mon. Not. R. Astron. Soc.*, **363**, 891.

21 Balbus, S.A. (2000) *Astrophys. J.*, **534**, 420.

22 Balbus, S.A. (2001) *Astrophys. J.*, **562**, 909.

23 Quataert, E. (2008) *Astrophys. J.*, **673**, 758.

24 Balbus, S.A. and Reynolds, C.S. (2008) *Astrophys. J.*, **681**, L65.

25 Balbus, S.A. and Reynolds, C.S. (2010) *Astrophys. J.*, **720**, L97.

26 Binney, J. and Cowie, L.L. (1981) *Mon. Not. R. Astron. Soc.*, **247**, 464.

27 Churazov, E., Bruggen, M., Kaiser, C.R., Bohringer, H., and Forman, W. (2001) *Astrophys. J.*, **554**, 261.

28 Narayan, R. and Medvedev, M.V. (2001) *Astrophys. J.*, **562**, L129.

29 Conroy, C. and Ostriker, J.P. (2008) *Astrophys. J.*, **681**, 151.

30 Kim, W.T. and Narayan, R. (2003) *Astrophys. J.*, **596**, 889.

31 Bogdanovic, T., Reynolds, C.S., Balbus, S.A., and Parrish, I.J. (2009) *Astrophys. J.*, **704**, 211.

32 Parrish, I.J., Quataert, E., and Sharma, P. (2009) *Astrophys. J.*, **703**, 108.

33 Ruszkowski, M. and Oh, S.P. (2010) *Astrophys. J.*, **713**, 1332.

34 Parrish, I.J., Quataert, E., and Sharma, P. (2010) *Astrophys. J.*, **712**, L194.

35 Peterson, J.R. and Fabian, A.C., (2006) *Phys. Rep.*, **427**, 1.

13
Summary and Outlook

Markus Böttcher, Daniel E. Harris, and Henric Krawczynski

In this book, we have given an overview of the current state of our understanding of relativistic, extragalactic jets. In this section, we briefly highlight open questions and particularly promising avenues for progress.

13.1
The Core: Insights into the Processes of Jet Formation, Acceleration, and Collimation

The last two decades have brought tremendous progress regarding our knowledge of mechanisms involved in AGN accretion and jet formation. Observations of the relativistically broadened Fe K_α line have allowed us to make first estimates of the spins of supermassive black holes. Combined radio imaging, optical polarimetry, and X-ray and γ-ray observations have established jet properties at the very base of the jet.

As discussed in Chapter 8, the particle acceleration and radiation mechanisms dominating in the immediate vicinity of the central engine of extragalactic jet sources can most directly be investigated through the study of blazars. The radio through optical/UV/X-ray emission is well explained as synchrotron emission from relativistic electrons. Concerning the high-energy emission, the literature is currently strongly biased toward a framework explaining high-energy emission as Compton emission from the same relativistic electrons, scattering either internal (synchrotron) or external photons. However, hadronic processes, in which the high-energy emission is initiated by ultrarelativistic protons, remain a viable alternative. The simplest models make strong predictions concerning temporal correlations between the emission in different wavebands due to them being produced by the same electrons. However, such correlations are often either not seen at all, or they do not follow the spectral trends predicted in simple one-zone leptonic models. Additional complications arise from extreme Doppler factor requirements to explain the rapid VHE γ-ray variability of some HBLs, which do not conform with Doppler factor estimates based on superluminal motion seen on larger scales. More complicated models, such as multizone models, spine-sheath models, or internal-shock

Relativistic Jets from Active Galactic Nuclei, First Edition. Edited by M. Böttcher, D.E. Harris, H. Krawczynski.

models are capable of solving some of these problems, but at the expense of introducing additional, usually poorly constrained parameters. The lack of consistent correlated (or not) variability patterns even among different flaring episodes of the same object has so far made it impossible to develop a single framework that would be able to explain all observed aspects of the blazar phenomenon.

The problem resides in good part in the lack of continuous monitoring of blazars throughout the electromagnetic spectrum. While Fermi provides such continuous monitoring now in the > 100 MeV–GeV energy regime, X-ray all-sky monitors are generally only detecting the X-ray brightest blazars. In the optical regime, the situation might very soon improve with the operation of PANSTARRS and the Large Synoptic Survey Telescope (LSST). However, in the radio, X-ray and VHE γ-ray regime, pointed observations by a small number of facilities will, for the foreseeable future, remain the only way to study blazar variability. An additional observational challenge is incomplete wavelength coverage. In particular, LBLs and FSRQs have their synchrotron peak in the infrared, which is notoriously difficult to access observationally. Observational resources with infrared satellites (Spitzer, Herschel) are very limited and are only very rarely devoted to blazar studies.

The past several years have revealed that not only high-frequency peaked blazars are emitters of VHE γ-rays, but the VHE blazar list now includes a few LBLs and FSRQs as well. The current generation of ACT facilities is only able to scratch the surface of the VHE γ-ray phenomenology, only detecting what are probably the brightest flares. There are only a handful of VHE blazars for which a quiescent-state VHE γ-ray flux can be detected. The synergy between Fermi and current ACT facilities is now providing continuous energy coverage from 100 MeV to > 1 TeV. However, a big problem lies in the still long (typically at least a few days, often weeks) integration time required to obtain useful spectral information in the Fermi energy range. This contrasts with the spectral information obtained during VHE γ-ray flares in integrations of often only a few hours or less. Any spectral features seen in the transition between the Fermi and ACT regimes therefore need to be taken with great caution.

Substantial progress is expected to come from a fleet of new experiments which will perform very sensitive observations of individual objects. The hard X-ray telescopes of the NuSTAR and ASTRO-H- missions will scrutinize the 5–80 keV emission from AGNs with unprecedented sensitivity. In the γ-ray band, the US-European Cherenkov Telescope Array (CTA) project will improve on the sensitivity of current ground-based γ-ray telescopes by one order of magnitude. The Atacama Large Millimeter Array (ALMA) in Chile will provide a valuable resource in the future to extend the radio coverage of many low-frequency peaked blazars toward their synchrotron peak. These telescopes will allow us to scrutinize the temporal evolution of the spectral energy distribution of blazars with unprecedented sensitivity. The sensitive observations will also allow us to constrain any quiescent jet emission component. The data will afford the possibility to unambiguously establish the emission mechanism, to identify the particle acceleration mechanism, and to further constrain the properties of jets at their bases. The higher sensitivity of CTA might allow for the detection of $\gamma\gamma$ absorption features in the γ-ray spectra of blazars that can be associated with radiation fields local to the source. This would provide valuable input to the discussion of the location of the blazar zone.

The X-ray polarimetry mission GEMS has the potential to make several breakthrough observations. GEMS observations of Seyfert galaxies will constrain the properties of the hot coronae of AGN accreting disks and will make substantial contributions to our knowledge of blazar jets. For high-frequency peaked BL Lac objects, GEMS observations probe smaller emission volumes than optical polarimetry observations, as the electrons emitting synchrotron radiation in the X-ray band cool much faster than the electrons responsible for the optical synchrotron emission. GEMS has the potential to reveal the helical structure of the magnetic field at the bases of jets by discovering that the X-ray emission from extreme synchrotron blazars exhibits ubiquitous polarization swings. Even if such swings are not observed, the degree of the polarization, and its correlation with the X-ray spectral index can be used to validate the hypothesis of the synchrotron origin of X-rays and to infer properties of the magnetic field topology in the X-ray-emitting volumes. In the case of Flat-Spectrum Radio Quasars, GEMS samples the high-energy emission component, and will be able to distinguish between a synchrotron self-Compton and an external Compton origin of the X-rays. A high polarization degree ($\gg 10\%$) would clearly favor a synchrotron self-Compton origin as external Compton models predict polarization degrees $\ll 10\%$.

13.2 Large-Scale Jets: Insights into Their Structure and Make-Up and Their Impact on Their Hosts

In Chapters 5–7 we have seen that large-scale jets are often knotty structures with offsets and progressions. The existence of optical synchrotron emission at kpc scales from the jets implies that *in-situ* acceleration must take place thousands of parsecs from the central engine. The nature of the acceleration process responsible for the energization of the relativistic electrons remains unknown. This ties into our general ignorance of the nature of the knots – simply regions of increased magnetic field, stationary shocks, density inhomogeneities propagating along the jet, or the signatures of interactions of the jet with inhomogeneities in the external medium.

While the synchrotron origin of the radio through optical emission seems fairly well established, the origin of the X-ray emission in large-scale jets (synchrotron or IC/CMB being the prime candidates) is still a hotly debated issue, and it seems that the answer is not the same for all jets – and possibly not even for all knots within the same jet. Some of the problems of the IC/CMB model can be ameliorated if the emission comes from relativistic minijets inside the relativistic jet. The double Doppler boosting leads to a very high effective Doppler factor even for relatively modestly relativistic propagation and large angles of the large-scale jet to the line-of-sight. Future studies of magnetic reconnection might make it possible to understand how minijets can form and how they can accelerate particles. If the X-ray emission is of synchrotron origin, the required ultra-high-energy electrons should emit observable inverse-Compton γ-rays. The discovery of the inverse-Compton emission by Fermi and/or CTA would vindicate this model. Observationally, the emission would manifest itself as a stationary component. At higher redshifts, the inverse-Compton component might even dominate over the core components because of the $(1 + z)^4$-increase of the CMB energy density.

If, on the other hand, the X-rays are produced via IC/CMB, then they would constitute a signature of relatively low-energy electrons, which radiate their synchrotron emission at low-frequency radio emission, which is notoriously difficult to measure. A high-sensitivity, high-angular-resolution low-frequency radio array like LOFAR may be instrumental in investigating the expected spatial correlation between low-frequency radio and X-ray emission in the IC/CMB scenario.

Continued X-ray observations of X-ray cavities with Chandra and XMM-Newton will allow us to study the jet heating of the intracluster medium. It might be possible to use the observations to constrain the jet duty cycle and the jet composition. The microcalorimeter on board the ASTRO-H satellite will afford for the first time high-throughput spectroscopic observations of galaxy clusters with a sufficiently high-energy resolution to constrain the turbulence of the intracluster gas and the ways in which the gas is energized.

13.3 Theory and Simulations

The rediscovery of the magnetorotational instability in numerical accretion disk simulations has provided us for the first time with a first-principle explanation of the source of viscosity in accretion disks. General relativistic magnetohydrodynamical simulations are now being used to verify previously developed analytical models of jet formation. Models of magnetic jet acceleration and confinement seem to be able to explain how accretion leads to jets with bulk Lorentz factors of ~ 50.

Our understanding of the collimation and stability of kpc jets (as observed in FR II radio galaxies), is still limited. A recurring paradigm appears to lie in two-component (spine-sheath) jets which can be stabilized against Rayleigh–Taylor type and current-driven instabilities by poloidal magnetic fields. A crucial realization was that much of the relevant physics is inherently 3D and cannot be captured in 2D simulations. The significant advance in computing power over the past decade has been instrumental for the progress in this field. Numerical simulations have shed light on the nature of superluminal components and the physical origin of the FR I/FR II dichotomy. Further advances in computer performance will allow us to make end-to-end 3D simulations of the accretion disk, the launching, acceleration, and collimation of jets. Hopefully, the simulations will shed light on how different processes in the surroundings of the accreting supermassive black hole conspire to establish a geometry capable of launching, collimating and accelerating jets.

Another important aspect to be studied further is how a magnetic field-dominated outflow transforms into a particle-dominated flux at larger spatial scales. If the magnetic field is anchored in the accretion disk and wound up into a helical structure, substantial toroidal magnetic fields are expected whose role is not yet understood. It has been shown in Chapter 10 that detailed 3D MHD simulations of the expected current-driven kink instabilities might be the key to answering these questions. MHD simulations of relativistic jets rely on macroscopic properties of the fluid resulting from microscopic processes which are not evaluated self-consistently in the framework of the MHD simulation. Thus, such properties have to be prespecified. They can, in principle, be investigated locally through PIC simulations, but those cannot be extended to the large scales covered in the MHD simulations.

In Chapter 9, we have seen that Monte Carlo simulations are an efficient tool to probe diffusive particle acceleration at relativistic shocks to ultrarelativistic energies. However, they are not able to provide a self-consistent description of relativistic shock turbulence needed for diffusive acceleration processes. PIC simulations are the tool of choice to investigate the microphysics of the development of Alfvén and/or Weibel turbulence in relativistic shock environments, but are not well suited to be extended to larger scales and ultrarelativistic particle energies. Substantial progress on this front will require a synergy between PIC and Monte Carlo simulations as well as analytical approaches. A self-consistent treatment of all aspects of particle acceleration will be essential to make robust predictions concerning the injection spectra and composition of relativistic particle populations at relativistic shocks, which can then be tested by means of their radiative signatures.

Bridging the gap between the kinetic description of the microphysical processes in PIC and Monte Carlo simulations and the macroscopic (fluid) description of MHD will be a major challenge for the foreseeable future.

Appendix A
Physical and Astrophysical Constants

Markus Böttcher, Daniel E. Harris, and Henric Krawczynski

Quantity	Symbol	Value
Speed of light in vacuum	c	$2.997\,924\,580\,0 \times 10^{10}$ cm s^{-1}
Planck constant	h	$6.626\,068\,96(33) \times 10^{-27}$ erg s
Planck constant, reduced	$\hbar = h/2\pi$	$1.054\,571\,628(53) \times 10^{-27}$ erg s
Electron charge	e	$4.803\,204\,27(12) \times 10^{-10}$ esu
Electron mass	m_e	$9.109\,382\,15(45) \times 10^{-28}$ g
Proton mass	m_p	$1.672\,621\,637(83) \times 10^{-24}$ g
Classical electron radius	r_e	$2.817\,940\,289\,4(58) \times 10^{-13}$ cm
Thomson cross-section	σ_T	$6.652\,458\,558(27) \times 10^{-25}$ cm^2
Gravitational constant	G_N	$6.674\,28(67) \times 10^{-8}$ cm^3 g^{-1} s^{-2}
Boltzmann constant	k_B	$1.380\,650\,4(24) \times 10^{-16}$ erg K^{-1}
Solar mass	$M_\odot$	$1.988\,4(2) \times 10^{33}$ g
Astronomical unit	AU	$1.495\,978\,707\,00(3) \times 10^{13}$ cm
Parsec	pc	$3.085\,677\,6 \times 10^{18}$ cm

(from: Review of Particle Physics 2008)

Conversion equations:
1 erg = 10^{-7} J
1 G = 10^{-4} T

Electromagnetism in CGS units:
Maxwell equations:

$$\nabla \cdot \boldsymbol{E} = 4\pi\rho$$

$$\nabla \times \boldsymbol{B} = \frac{4\pi}{c}\boldsymbol{j} + \frac{1}{c}\frac{\partial \boldsymbol{E}}{\partial t}$$

$$\nabla \cdot \boldsymbol{B} = 0$$

$$\nabla \times \boldsymbol{E} = -\frac{1}{c}\frac{\partial \boldsymbol{B}}{\partial t}$$

Relativistic Jets from Active Galactic Nuclei, First Edition. Edited by M. Böttcher, D.E. Harris, H. Krawczynski.

Lorentz force:

$$\boldsymbol{F} = q\left(\boldsymbol{E} + \frac{\boldsymbol{v}}{c} \times \boldsymbol{B}\right)$$

Energy density:

$$u_{\rm em} = \frac{1}{8\pi}\left(E^2 + B^2\right)$$

Index

Relativistic Jets from Active Galactic Nuclei, First Edition. Edited by M. Böttcher, D.E. Harris, H. Krawczynski.